Effects of Digitalization and Circular Economy on Sustainable Policy and Climate Change Prevention

Durmuş Çağrı Yıldırım
Tekirdag Namık Kemal University, Turkey

Seda Yıldırım
Tekirdağ Namık Kemal University, Turkey

Vinay Kandpal
Graphic Era University, India

IGI Global
Scientific Publishing
Publishing Tomorrow's Research Today

Published in the United States of America by
IGI Global Scientific Publishing
701 East Chocolate Avenue
Hershey, PA, 17033, USA
Tel: 717-533-8845
Fax: 717-533-7115
Website: https://www.igi-global.com E-mail: cust@igi-global.com

Library of Congress Cataloging-in-Publication Data

Names: Yildirim, Durmus Çagri, 1981- editor. | Yıldırım, Seda, editor. | Kandpal, Vinay, editor.
Title: Effects of digitalization and circular economy on sustainable policy and climate change prevention / edited by Durmus Çagri Yildirim, Seda Yıldırım, Vinay Kandpal.
Description: Hershey, PA : IGI Global Scientific Publishing, [2025] | Includes bibliographical references and index. | Summary: "This book aims to explore and collect original and review studies related to digitalization and sustainability. As it is proposed, this book will present how digitalization influences circular practices and 2030 Sustainable Development Goals"-- Provided by publisher.
Identifiers: LCCN 2024054400 (print) | LCCN 2024054401 (ebook) | ISBN 9798369399095 (hardcover) | ISBN 9798369399101 (paperback) | ISBN 9798369399118 (ebook)
Subjects: LCSH: Sustainable development--Technological innovations--Case studies. | Environmental economics--Technological innovations--Case studies. | Circular economy--Case studies.
Classification: LCC HC79.E5 .E3317 2025 (print) | LCC HC79.E5 (ebook) | DDC 338.9/27--dc23/eng/20250210
LC record available at https://lccn.loc.gov/2024054400
LC ebook record available at https://lccn.loc.gov/2024054401

British Cataloguing in Publication Data
A Cataloguing in Publication record for this book is available from the British Library.

Table of Contents

Detailed Table of Contents

innovations like digitalization and the circular economy. This study examines how digital technologies support sustainability and advance the circular economy and SDGs. Tools like AI, blockchain, and IoT transform industries by reducing energy use, waste, and enabling predictive maintenance. Solutions like product passports boost recycling, while green finance, such as bonds and carbon credits, supports climate action. Initiatives like smart cities, precision farming, and digital healthcare lower emissions and improve resource use. Challenges include data privacy, energy consumption, and e-waste. This study aids policymakers, researchers, and practitioners in harnessing digitalization for sustainable growth.

In the face of global challenges, achieving the United Nation's Sustainable Development Goals (SDGs) requires innovative and collaborative efforts across sectors. However, there is a paucity of literature to study the effectiveness of collaboration and initiatives that foster SDG achievement. This study explores the role of digital technologies in enhancing partnerships that are aimed at achieving SDGs. Using a case study approach, the research examines two initiatives (i.e., UNDP's Digital SDG Partnership and WEF's Digital Agenda) where digital tools and platforms have facilitated effective multi-stakeholder collaboration, improved resource mobilization and enhanced transparency and accountability. The findings highlight the transformative potential of digital tools in bridging gaps, fostering innovation, and scaling impact for SDGs. The paper concludes by mentioning significance, limitations, and recommendations for future directions of the study for leveraging digital technologies in future SDG partnerships to maximise their potential for sustainable development.

Digital technology with circular economy concepts is one of the game-changing strategies for sustainable development goals (SDGs). The digital circular economy presents possibilities and threats, which this section discusses. Concerns about data privacy and security, regulatory and economic hurdles, and the need for a shift in culture and behavior are among the most pressing problems. However, there are

Covid era has also seen a rise in the use of Artificial Intelligence and Internet of Things for better monitoring of environmental issues. The chapter also talks about how various governments across the globe have been implementing various policies and initiatives with better environmental implications. With rapid rise in waste management and associated sectors, multiple challenges including budgeting and research & development arise frequently; and overcoming these challenges is the call of the hour.

The digitalization wave has unlocked new opportunities for sustainability, particularly in urban waste management, where blockchain technology offers innovative solutions to circular economy challenges. This chapter explores blockchain's transformative potential in advancing sustainable waste management and mitigating climate change. Traditional systems face inefficiencies in tracking, recycling, and transparency, which blockchain addresses through decentralized, immutable records that enhance waste tracking, incentivize recycling, and ensure process transparency. Case studies from leading cities illustrate the environmental benefits, carbon footprint reduction, and alignment with Sustainable Development Goals (SDGs) 11 and 12. This chapter aims to highlight blockchain's role in fostering accountability among stakeholders and provides policy recommendations for integrating blockchain to establish a digitalized, circular economy framework.

This chapter focuses on the juncture at which climate change, digitalization, and sustainable urban development converge, with an emphasis on the role of cities in climate change mitigation. On the one hand, urban areas are a great contributor to global energy consumption and greenhouse gas emissions; on the other hand, they also pose opportunities. The chapter explores how digital technologies such as AI, IoT, blockchain, and digital twins are now changing the profile of cities to make them efficient, resilient, and eco-friendly. Cases from Indian cities—Bengaluru, Pune, and Delhi—demonstrate the impact of digital tools on optimizing energy management, resource allocation, and environmental stewardship. It also discusses

What measures should be taken to effectively address the negative effects of the digitalization of ride-hailing services? Let's ponder some thoughts!! Findings validate that the digitalization of ride-hailing services remains a threat to environmental sustainability. Despite this, ride-hailing services have successfully handled the issue of rising concern of unemployment in the urban setting of Kolkata. The study was conducted at the traffic signals of the Jadavpur and Rabindra Bharti areas, Kolkata. Using random sampling 150 OLA and Uber drivers of two-wheelers and four-wheelers and 150 riders, especially those belonging to the working age group (i.e. 15- 64 years). An analysis shows that the average Air Quality Index (AQI) was maximum during winter i.e. above 140 AQI and reduced during the rainy season in both study locations. Still, there is apprehension that the rise in ride-hailing services could adversely impact the Air Quality Index. Hence, smart use of digitalization and ride-hailing services (i.e. carpooling and electrification) can address air pollution.

This chapter examines how sustainable IoT technologies influence climate policy, emphasizing pre- and post-COVID changes. The pandemic served as a catalyst, speeding up the use of IoT for environmental monitoring, resource management, and emissions reduction. The IoT has significantly impacted climate governance, shifting from reactive to proactive tactics post-COVID due to real-time monitoring and data-driven decision-making. This integration has also improved sustainability in agriculture, energy, urban management, and disaster resilience. Notwithstanding these developments, there are still issues, such as inadequate infrastructure, exorbitant prices, and worries about data privacy, especially in developing countries. Future development, international cooperation, and cross-sector alliances will be necessary to fully realize IoT's potential in mitigating climate change. Integrating the IoT with new technologies to create more robust and sustainable systems will be crucial to the future of sustainable IoT in climate policy.

to all those manufacturers, policymakers etc. who have a great interest in moving towards the sustainable environment in the textile industry.

The purpose of this study is to spread knowledge regarding how Indonesia's good governance will be impacted by the rule of law in 2021. Following that, we have determined the effects of the Rule of Law on Indonesian residents. Citizenry and the alterations to their mental processes toward the rule of law. “Qualitative research” is the methodology employed by the researchers working in this field. The World Bank website, Dimension AI, and Google Scholar provided the research's data, which was then presented descriptively. This paper's research helps us understand how a robust legal system might promote effective governance in a system or society. Moreover, how the government can effectively and transparently carry out the rule of law, this paper will only cover the extent to which each aspect of the change is explained in detail; it will also examine potential solutions to problems raised by the public and how best to put them into practice; and lastly, it will discuss the beneficial effects that the change in the laws has had on good governance.

Higher education has a big influence on sustainability and supports international initiatives to advance sustainable development and the digital revolution. Higher education's digital revolution has a big impact on sustainability since it supports international initiatives to advance sustainable development. Despite the significant regional differences in the SDGs' implementation, it is evident that universities play a critical role in fostering an attitude that makes the concepts of the SDGs easier to

Preface

The 21st century is a digital age, and it is a century that increases digitalization and sustainability together. One of the main reasons why digitalization is also addressed within the scope of sustainability policies and strategies is the fight against climate change, the pandemic (COVID-19), and the increasing need for environmental value and social welfare. There is little time left for the 2030 Sustainable Development Goals to be realized, but improvements can continue in some basic areas with the help of digital solutions and technology.

In this edited volume, " Effects of Digitalization and Circular Economy on Sustainable Policy and Climate Change Prevention," we aim to explore and collect original, and review studies related to digitalization and sustainability. As it is proposed, this book will present how digitalization influences circular practices and 2030 Sustainable Development Goals. It is thought that the book will be useful primarily for academic and policy makers and will bring a new perspective to students in higher education and readers interested in the field of sustainability. As a result, the reflection of the relationship between digitalization and sustainability on circular applications will be explained because of the collection of research from different countries and with different themes.

We believe that this book will be an important reference source for academics, policy makers and professionals working on digitalization, digital transformation and sustainable development. The chapters in the book provide both theoretical and practical information. As editors, we are proud to bring together and present to the market various international publications that we believe will be useful for sustainable development. The benefits we will provide for sustainable development make us proud.

CHAPTER OVERVIEW

Chapter 1: Digital Pathways to Sustainability: Integrating Technology and Innovation for a Greener Future

Abstract: Digitalization has emerged as a revolutionary element across various sectors and communities, altering conventional practices and creating new avenues for sustainability. This chapter delves into how digital technologies intersect with circular economy principles and sustainable development, emphasizing their contribution to combating climate change and furthering the United Nations' 2030 Sustainable Development Goals (SDGs). By evaluating case studies, theoretical models, and empirical evidence, this chapter investigates the ways in which cutting-edge technologies like artificial intelligence (AI), blockchain, and the Internet of Things (IoT) promote sustainable practices and foster environmental resilience. The chapter concludes by providing practical insights for policymakers, industry executives, and researchers on leveraging digital innovation for a greener future.

Chapter 2: Digitalization, Sustainability, and Strategy - An Analytical Model of Green Investment Dynamics

Abstract: This chapter explores the dynamics of subsidies, digitalization, and stakeholder behavior in the green investment ecosystem through a game-theoretic framework. Governments optimize subsidies and penalties, firms decide on green technology investments, and consumers make adoption choices based on costs, benefits, and digital awareness. The model incorporates digitalization as a key variable influencing costs, efficiency, and decision-making. The numerical experiment reveals that digitalization amplifies the effects of subsidies, increasing consumer adoption and firm investments while enhancing government policy efficiency. Key findings include: (1) consumer adoption grows synergistically with subsidies and digitalization; (2) firm investments increase with financial and technological incentives but show diminishing returns at higher subsidies; and (3) government payoffs are optimized at moderate subsidy levels coupled with digital tools. Thus, this chapter highlights how policymakers can design strategies that integrate financial incentives with digital innovation.

Chapter 3: Digital Transformation and Sustainability Artificial Intelligence and Sustainable Development

Abstract: Digital transformation and sustainability are increasingly intertwined in today's rapidly evolving landscape, where technological advancements and environmental concerns converge to shape the future of industries, communities, and societies. Operations, strategies and This transformation can enhance efficiency, improve customer experiences, and foster innovation. However, it also carries the responsibility of addressing the pressing sustainability social inequities. The relationship between digital transformation and sustainability is complex, yet synergistic. On one hand, blockchain—provides powerful tools for enhancing sustainability efforts across various sectors. For example, can consumption be smart while improving efficiencies reducing waste? Additionally, big data analytics can help organizations assess their environmental impact, enabling informed decision-making that supports sustainable practices. This data-driven approach allows companies to track their carbon footprints, manage resources more effectively, and identify areas for improvement.

Chapter 4: The Intersection of Digitalization and Circular Economy, Advancing Sustainable Development and Climate Change Mitigation

Abstract: Digitalization and circular economy are slowly being adopted by industries and are starting to reshape sustainable development and climate change mitigation. Technologies such as Artificial Intelligence (AI), Internet of Things (IoT) and blockchain are presenting new methodologies in handling resources, waste, and sustainability systems. At the same time, Circular Economy principles of reduce, reuse, and recycle are guiding sustainable production and consumption. In combination, these forces serve to shape proper strategies in the face of environmental threats, as well as contribute toward the achievement of the Sustainable Development Goals set by the United Nations. In this book, the link between digitalization and circular economy is explored with regards to the ways in which the two paradigms can help the development of sustainable solutions to address resource management and climate change.

Chapter 5: Digital Transformation as a Catalyst for Sustainability for Advancing Circular Economy and Environmental Innovation: Harnessing Emerging Technologies for Sustainable Growth and Green Innovation

Abstract: As the world adapts to modern norms which recognises sustainability as key to the progress of the world, digital transformation has consequently become a vitality requisite for ruining sustainability in companies. This chapter examines the relationship between digital technologies including artificial intelligence, blockchain, internet of things and big data to sustainable development in relation to environmental problems including climate change and resource optimization as well as waste management. By extracting sustainable consumption and production patterns in the business environment, enterprise and policymakers are exploring new options on using digital approaches to advance the 2030 Sustainable Development Goals (SDGs). This chapter aims at exploring the potential the relationship between digital transformation and advancing circular economy. It shows where and how digital technology improves productivity, minimises waste and brings innovation to industries like energy, production, logistics and supply chains.

Chapter 6: Digitalization and Circular Economy: Pioneering Sustainable Policies for Climate Change Mitigation and 2030 Sustainable Development Goals

Abstract: The 21st century has driven rapid changes in technology, economy, and environmental policies. Addressing climate change, resource depletion, and inequality requires innovations like digitalization and the circular economy. This study examines how digital technologies support sustainability and advance the circular economy and SDGs. Tools like AI, blockchain, and IoT transform industries by reducing energy use, waste, and enabling predictive maintenance. Solutions like product passports boost recycling, while green finance, such as bonds and carbon credits, supports climate action. Initiatives like smart cities, precision farming, and digital healthcare lower emissions and improve resource use. Challenges include data privacy, energy consumption, and e-waste. This study aids policymakers, researchers, and practitioners in harnessing digitalization for sustainable growth.

Chapter 7: Leveraging Digital Technologies for Effective SDG Partnerships: A Case Study Approach

Abstract: In the face of global challenges, achieving the United Nations Sustainable Development Goals (SDGs) requires innovative and collaborative efforts across sectors. However, there is a paucity of literature to study the effectiveness of collaboration and initiatives that foster SDG achievement. This study explores the role of digital technologies in enhancing partnerships that are aimed at achieving SDGs. Using a case study approach, the research examines two initiatives (i.e., UNDP's Digital SDG Partnership and WEF's Digital Agenda) where digital tools and platforms have facilitated effective multi-stakeholder collaboration, improved resource mobilization and enhanced transparency and accountability. The findings highlight the transformative potential of digital tools in bridging gaps, fostering innovation, and scaling impact for SDGs. The paper concludes by mentioning significance, limitations, and recommendations for future directions of the study for leveraging digital technologies in future SDG partnerships to maximise their potential for sustainable development.

Chapter 8: Digital Circular Economy for Attaining Sustainable Development Goals: Technologies for Global Sustainability

Abstract: Digital technology with circular economy concepts is one of the game-changing strategies for sustainable development goals (SDGs). The digital circular economy presents possibilities and threats, which this section discusses. Concerns about data privacy and security, regulatory and economic hurdles, and the need for a shift in culture and behavior are among the most pressing problems. However, there are many chances to be innovative through working with others, forming partnerships, and expanding into new areas. Stakeholders may make strides towards a more sustainable and resilient economy by responding to these threats and capitalizing on these possibilities. To better understand and implement digital circular practices, future research should focus on creating standardized impact metrics, developing cutting-edge technologies, establishing supportive policy frameworks, gaining insights into consumer behavior, creating innovative business models, conducting sector-specific studies, considering social equity, collaborating globally, and ensuring long-term sustainability.

Chapter 9: The Role of Circular Economy in Advancing Sustainability

Abstract: This chapter investigates the role of the circular economy (CE) in advancing sustainability, focusing on how circular models can drive environmental, economic, and social benefits by minimizing waste, conserving resources, and promoting regenerative practices. The study reviews core CE strategies such as product lifecycle extension, resource recovery, and sustainable material sourcing and assesses their effectiveness in achieving sustainable development goals (SDGs). Through industry case studies and policy analyses, it identifies key challenges, including financial, regulatory, and cultural barriers that hinder CE adoption. This chapter highlights the need for collaborative frameworks, supportive government policies, and innovative business models to enable a transition from linear to circular systems, emphasizing that CE is not only essential for reducing environmental impact but also crucial for building resilient, sustainable economies. The findings underscore the importance of CE as a fundamental approach to sustainability and climate resilience.

Chapter 10: Circular Waste Management through Decentralized Recycling in the Digital Era

Abstract: With rapid advancements in the world of technology, a rise in convenience for the human species has also increased wastage. As a result of this waste increment, this chapter focuses on the concepts of Circular Economy and Waste Management across the globe, with focus on the era of Digital Platforms. Governments worldwide are using digital concepts in decision-making as resource allocation has been focussed upon maximising return while keeping the environmental capital in mind. Post Covid era has also seen a rise in the use of Artificial Intelligence and Internet of Things for better monitoring of environmental issues. The chapter also talks about how various governments across the globe have been implementing various policies and initiatives with better environmental implications. With rapid rise in waste management and associated sectors, multiple challenges including budgeting and research & development arise frequently; and overcoming these challenges is the call of the hour.

Chapter 11: Harnessing Blockchain Technology for Circular Economy Practices in Sustainable Urban Waste Management: A Digital Solution for Climate Change Mitigation

Abstract: The digitalization wave has unlocked new opportunities for sustainability, particularly in urban waste management, where blockchain technology offers innovative solutions to circular economy challenges. This chapter explores blockchain's transformative potential in advancing sustainable waste management and mitigating climate change. Traditional systems face inefficiencies in tracking, recycling, and transparency, which blockchain addresses through decentralized, immutable records that enhance waste tracking, incentivize recycling, and ensure process transparency. Case studies from leading cities illustrate the environmental benefits, carbon footprint reduction, and alignment with Sustainable Development Goals (SDGs) 11 and 12. This chapter aims to highlight blockchain's role in fostering accountability among stakeholders and provides policy recommendations for integrating blockchain to establish a digitalized, circular economy framework.

Chapter 12: Interaction Among Climate Change, Digitalisation, and Sustainable Urban Development

Abstract: This chapter focuses on the juncture at which climate change, digitalization, and sustainable urban development converge, with an emphasis on the role of cities in climate change mitigation. On the one hand, urban areas are a great contributor to global energy consumption and greenhouse gas emissions; on the other hand, they also pose opportunities. The chapter explores how digital technologies such as AI, IoT, blockchain, and digital twins are now changing the profile of cities to make them efficient, resilient, and eco-friendly. Cases from Indian cities—Bengaluru, Pune, and Delhi—demonstrate the impact of digital tools on optimizing energy management, resource allocation, and environmental stewardship. It also discusses challenges such as digital divides and privacy concerns, requiring strong policies to ensure that technology access is equitable. The chapter concludes by calling for integrating digitalization with sustainable urban development in order to create economically, socially, and environmentally sustainable cities.

Chapter 13: AI-Driven Solutions for Sustainable Urban Development: Smart Cities and Resource Management

Abstract: The unprecedented expansion in cities throughout the world has actually made the need for sustainable city growth solutions more essential. Artificial Intelligence (AI) has been established as a disruptive and gene change technology that proposes something new in solving urban and resource planning problems as well as plight to sustainability. This chapter discusses the relationship between AI and urban sustainability by considering smart cities as the conceptual context to embed AI applications. Areas of focus are energy and environment, water and waste, ITS and system management, and governance of cities. The role of ethics in the implementation of AI is pointed out as the breeze turns to biases in the data used – specifically, data privacy issues. The chapter features various and profound cases and proofs of implementations from around the world in support of the key points made and offers suggestions for practical recommendations for policymakers as well as urban planners.

Chapter 14: Green Transition In The Automotive Industry: Turkey Case

Abstract: Policy makers are turning to alternative and sustainable practices to slow down the effects of climate change and global warming. The transition from practices that increase carbon emissions, cause air pollution and rapidly consume natural resources, which are part of the traditional economy, to circular practices has begun. At this point, the automobile market has been selected as one of the leading sectors for green transition. The development of the automobile sector, where green transition and circular practices are seen, is observed in various publications and reports. In this study, the circular changes experienced in the automobile sector in Turkey will be examined. In the study, tables and graphics will be interpreted from the sources obtained from current databases.

Chapter 15: Digitalization of Ride-Hailing Services: Revisiting Economic and Environment Sustainability

Abstract: What measures should be taken to effectively address the negative effects of the digitalization of ride-hailing services? Let's ponder some thoughts!! Findings validate that the digitalization of ride-hailing services remains a threat to environmental sustainability. Despite this, ride-hailing services have successfully handled the issue of rising concern of unemployment in the urban setting of Kolkata. The study was conducted at the traffic signals of the Jadavpur and Rabindra Bharti areas, Kolkata. Using random sampling 150 OLA and Uber drivers of two-wheelers and four-wheelers and 150 riders, especially those belonging to the working age group (i.e. 15- 64 years). An analysis shows that the average Air Quality Index (AQI) was maximum during winter i.e. above 140 AQI and reduced during the rainy season in both study locations. Still, there is apprehension that the rise in ride-hailing services could adversely impact the Air Quality Index. Hence, smart use of digitalization and ride-hailing services (i.e. carpooling and electrification) can address air pollution.

Chapter 16: Comprehensive Overview of Sustainable IoT for Innovations and Their Impact on Climate Policy: A Pre and Post COVID Comparison

Abstract: This chapter examines how sustainable IoT technologies influence climate policy, emphasizing pre- and post-COVID changes. The pandemic served as a catalyst, speeding up the use of IoT for environmental monitoring, resource management, and emissions reduction. The IoT has significantly impacted climate governance, shifting from reactive to proactive tactics post-COVID due to real-time monitoring and data-driven decision-making. This integration has also improved sustainability in agriculture, energy, urban management, and disaster resilience. Notwithstanding these developments, there are still issues, such as inadequate infrastructure, exorbitant prices, and worries about data privacy, especially in developing countries. Future development, international cooperation, and cross-sector alliances will be necessary to fully realize IoT's potential in mitigating climate change. Integrating the IoT with new technologies to create more robust and sustainable systems will be crucial to the future of sustainable IoT in climate policy.

Chapter 17: Digitalisation, a Key Used to Fill the Void Amid COVID-19

Abstract: COVID-19 as infections continued to rise globally, uncertainty over economic growth deepened. Ways to tackle the pandemic were made with the correspondence of technological development. There was alarming need to implement technologies that meet the contactless demand to prioritise health and safety. Countries enhanced digital infrastructure, adopted online business models, and expanded digital payment systems. Remote working became standard in many sectors, while cloud computing and artificial intelligence gained traction to support operations. Rapid digital transformation has been pivotal in managing pandemic disruptions, fostering resilience, and reshaping how industries function in a more tech-integrated world. Study analysis the interplay of digitalisation during COVID-19 pandemic and its related effects. A qualitative approach based on secondary data has been applied in this study. Moreover, the findings used in the study correlate with the recent time. This study concludes that technological advancements have been evident, to mitigate and handle the impacts of the pandemic.

Chapter 18: Implementing Conceptual Framework into Textile Industry for Sustainable Climate Change Prevention

Abstract: The textile industry plays a very important and a crucial role in the factors that can affect the environment globally, and due to such factors, it faces immense pressure over themselves to adopt various sustainable methods. This paper examines the development of various conceptual frameworks that can be implemented to adopt sustainability in the textile industry. By the adoption of the framework, it would help us to tackle multiple problems such as carbon emission, extensive utilization of the resources and waste production. The study explores multiple supply chain strategies such as use of renewable energy, cutting edge technology such as biodegradable material and use of dyes that are not harming the environment. The research suggests that a business that is well structured improves the its stability and profitability while reducing the environmental damage. The research provides insightful information to all those manufacturers, policymakers etc. who have a great interest in moving towards the sustainable environment in the textile industry.

Chapter 19: Mapping the Nexus Rule of Law's Influence on Good Governance: A Bibliometric Exploration

Abstract: The purpose of this study is to spread knowledge regarding how Indonesia's good governance will be impacted by the rule of law in 2021. Following that, we have determined the effects of the Rule of Law on Indonesian residents. Citizenry and the alterations to their mental processes toward the rule of law. "Qualitative research" is the methodology employed by the researchers working in this field. The World Bank website, Dimension AI, and Google Scholar provided the research's data, which was then presented descriptively. This paper's research helps us understand how a robust legal system might promote effective governance in a system or society. Moreover, how the government can effectively and transparently carry out the rule of law, this paper will only cover the extent to which each aspect of the change is explained in detail; it will also examine potential solutions to problems raised by the public and how best to put them into practice; and lastly, it will discuss the beneficial effects that the change in the laws has had on good governance.

Chapter 20: Implementing Sustainability Into Higher Education Challenges & Prospects

Abstract: Higher education has a big influence on sustainability and supports international initiatives to advance sustainable development and the digital revolution. Higher education's digital revolution has a big impact on sustainability since it supports international initiatives to advance sustainable development. Despite the significant regional differences in the SDGs' implementation, it is evident that universities play a critical role in fostering an attitude that makes the concepts of the SDGs easier to spread. This study examines the challenges and benefits of integrating sustainability into higher education. Higher education is crucial to achieving the SDGs, particularly Goal 1 (end poverty in all of its manifestations), Goal 3 (guarantee healthy lifestyles and promote well-being for all at all ages), and Goal 5 (gender equality). Objectives 8 (economic expansion and skilled labor), 12 (conscientious production and consumption), 13 (climate change), and 16 (peace, justice, and robust institutions).

In this book, which contains 41 publications in total, the close relationships between digitalization and sustainability are examined theoretically and empirically. The road to sustainable development is a long and challenging process. This volume, titled "Effects of Digitalization and Circular Economy on Sustainable Policy and Climate Change Prevention", sheds light with useful information for the digital transformation required for the construction of a sustainable future.

The chapters in this book show the connection between various sectors and digital applications and sustainability. In terms of content, we can be a guiding source for both the academy and the market.

As editors, we have revealed the important role of applications such as digital technologies, digital transformation, innovative approaches and green innovation for a sustainable future. We believe that the book is also valuable in terms of sustainability, environmental economics and digital transformation literature.

As a result, we have shown the readers the relationship between digital technologies, sustainable development and green applications from a broad perspective. We would like to thank all the authors who contributed to the book and hope that our book will be useful to a wide range of audiences, such as literature and policy makers, young researchers, and students.

Durmus Çagri Yildirim
Tekirdag Namik Kemal University, Turkey

Seda Yildirim
Tekirdag Namik Kemal University, Turkey

Vinay Kandpal
Graphic Era University, Dehradun, India

Acknowledgement

We would like to thank the authors and referees who supported us in the hope that this book will contribute to the green economy and sustainable development literature and provide useful information to researchers and readers. Since the responsibility of the book chapters belongs to the corresponding author and co-authors of each chapter, you can contact the chapter authors for information.

Durmus Çagri Yildirim
Tekirdag Namik Kemal University, Turkey

Seda Yildirim
Tekirdag Namik Kemal University, Turkey

Vinay Kandpal
Graphic Era University, Dehradun, India

Chapter 1
Digital Pathways to Sustainability:
Integrating Technology and Innovation for a Greener Future

Marina Shaji
https://orcid.org/0009-0008-1007-3031
Christ University, India

Amit Kumar
Christ University, India

ABSTRACT

Digitalization has emerged as a revolutionary element across various sectors and communities, altering conventional practices and creating new avenues for sustainability. This chapter delves into how digital technologies intersect with circular economy principles and sustainable development, emphasizing their contribution to combating climate change and furthering the United Nations' 2030 Sustainable Development Goals (SDGs). By evaluating case studies, theoretical models, and empirical evidence, this chapter investigates the ways in which cutting-edge technologies like artificial intelligence (AI), blockchain, and the Internet of Things (IoT) promote sustainable practices and foster environmental resilience. The chapter concludes by providing practical insights for policymakers, industry executives, and researchers on leveraging digital innovation for a greener future.

DOI: 10.4018/979-8-3693-9909-5.ch001

INTRODUCTION

The 21st century has marked the beginning of a time where digital transformation and sustainability intersect, leading to pioneering solutions for some of the most critical global issues such as climate change, resource depletion, and social inequality. The rapid advancement of digitalization—driven by innovations in artificial intelligence (AI), the Internet of Things (IoT), blockchain, and big data—provides transformative opportunities to rethink sustainability practices. When effectively integrated, digital solutions can propel progress toward the United Nations' 2030 Sustainable Development Goals (SDGs) by improving efficiency, enabling circular economy strategies, and encouraging environmentally conscious behaviors.

The combination of digitalization and sustainability is particularly promising when viewed through the lens of circular economy principles. Circular economies prioritize reducing waste and optimizing resource use by reimagining production, consumption, and recycling systems. New digital tools play a central role in these initiatives by offering real-time monitoring, predictive analytics, and smart automation that improve resource management (Geissdoerfer et al., 2017). Furthermore, digitalization empowers stakeholders by enhancing transparency, promoting data-driven decision-making, and facilitating innovative collaborations across various industries (Parida et al., 2019).

This chapter, titled "Digital Pathways to Sustainability: Integrating Technology and Innovation for a Greener Future," explores the convergence of technology, innovation, and sustainability. It investigates how digital tools are assisting businesses, policymakers, and communities in tackling environmental and social issues through transformative practices.

The goals of this chapter are threefold:

- To examine the theoretical and practical implications of digital technologies in promoting sustainable practices.
- To showcase case studies and empirical research that illustrate successful implementation of digital solutions within circular economy initiatives.
- To pinpoint future prospects and obstacles in utilizing digitalization to achieve the 2030 SDGs.

By connecting technological advancements with sustainability objectives, this chapter seeks to offer meaningful insights for academics, practitioners, and policymakers looking for innovative routes to sustainability. Through a combination of literature review and case study evaluation, it emphasizes the necessity for a strategic integration of digitalization into sustainability frameworks to develop resilient and inclusive systems that serve both people and the planet.

Key Concepts

1. Digitalization: The incorporation of digital technologies into routine operations, improving efficiency and enabling innovative approaches that contribute to sustainability objectives.
2. Sustainability Transition Theory (STT): A framework that investigates how societies move away from unsustainable practices toward sustainable options, highlighting the impact of technology and social dynamics in this transition.
3. Circular Economy (CE): An economic framework that focuses on resource reclamation, recycling, and remanufacturing instead of the conventional linear model of "take, make, dispose," encouraging sustainability through closed-loop mechanisms.
4. Digital Innovation Theory: Examines how technological progress drives changes within organizations and society, especially in promoting sustainable practices through efficient and transparent processes.
5. Diffusion of Innovations Theory: A theory that explores how new ideas and technologies disseminate within societies, concentrating on factors that affect the uptake of digital solutions for sustainability.
6. Triple Bottom Line (TBL): A framework that stresses the importance for organizations to balance social, environmental, and economic results, promoting comprehensive approaches to sustainability.
7. Closed-Loop Systems: Systems intended to maintain products and materials in use as long as possible via recycling and reuse, minimizing waste and environmental harm.
8. Business Model Innovation: The creation of new business models that are in alignment with sustainability principles, such as product-as-a-service, bolstered by digital platforms for greater efficiency.
9. Technological Innovation Systems (TIS): A theory focusing on the development and dissemination of new technologies within a system, highlighting their role in facilitating transitions to sustainable practices.
10. Systems Thinking: An approach that views sustainability challenges as interconnected systems, requiring holistic solutions.
11. Digital Divide: The gap in access to digital technologies between different socio-economic and geographic groups.
12. Internet of Things (IoT): Networks of connected devices that gather and exchange data to optimize processes and enhance sustainability.
13. Big Data: The analysis of large data sets to derive insights for improving resource management and decision-making.
14. Supply Chain Transparency: Ensuring visibility into the origin, production, and distribution of products to promote ethical and sustainable practices.

15. Predictive Maintenance: The use of AI and IoT to anticipate equipment failures, reducing downtime and waste.

THEORETICAL AND CONCEPTUAL FOUNDATIONS

The intertwining of digitalization, circular economy initiatives, and sustainability objectives is supported by various theoretical frameworks that elucidate the connections among these ideas and their consequences for environmental and socio-economic changes. Below are significant theories and models pertinent to this chapter:

1. Sustainability Transition Theory

Overview:

Sustainability Transition Theory (STT) is a framework that investigates how societies evolve from unsustainable habits to more sustainable alternatives, often within the context of technological advancements, policy modifications, and social dynamics. This theory is particularly useful for understanding how digital technologies can expedite sustainable development and facilitate circular economy frameworks.

Key Concepts:

- Sociotechnical Systems: STT highlights the significance of sociotechnical systems, which combine technology with societal structures, practices, and policies. Digital instruments (e.g., IoT, AI, blockchain) have the potential to disrupt current systems and forge new paths toward sustainability.
- Path Dependency: The theory addresses how historical decisions influence present trajectories, which can be altered through innovations like digitalization that disrupt existing paths toward more sustainable methods.

Relevance to the Chapter:

- This theory illustrates how digital technologies, such as AI, blockchain, and IoT, function as niche innovations that disrupt traditional systems and advance sustainability initiatives.
- It offers a framework for comprehending how digitalization aids the transition to a circular economy, impacting both technological progress and policymaking.

2. Circular Economy Theory

Overview:

The Circular Economy (CE) model contests the conventional linear economy, which adheres to a "take, make, dispose" approach. CE promotes a regenerative system where resources are utilized for as long as feasible through recycling, remanufacturing, and sharing. Digitalization is crucial in facilitating these circular practices by enhancing resource tracking, waste management, and product lifecycle oversight.

Key Concepts:

- Closed-Loop Systems: The circular economy focuses on establishing closed-loop systems where products are crafted for durability, reuse, and recyclability. Digital technologies such as IoT, blockchain, and AI can facilitate these systems by monitoring resources and guaranteed supply chain transparency.
- Product Life Cycle Management: The circular economy advocates for effective management of products from inception to disposal, with digital tools improving the capability to monitor and refine product lifecycles.

Relevance to the Chapter:

- This model forms the basis for understanding how digital technologies contribute to the circular economy by promoting sustainable production, consumption, and waste management.
- It connects digital solutions to sustainable methods by demonstrating how technologies like AI, blockchain, and IoT can strengthen circular economy approaches.

3. Digital Innovation Theory

Overview:

Digital Innovation Theory examines how advancements in technology—particularly in artificial intelligence, IoT, blockchain, and big data—propel organizational and societal transformations. Regarding sustainability and the circular economy, digital innovation facilitates more efficient, transparent, and scalable approaches to addressing environmental issues.

Key Concepts:

- Disruptive Innovation: Digital advancements frequently disrupt established business models, providing more efficient, cost-effective, and sustainable alternatives to conventional methods.

- Technology Adoption and Diffusion: This theory also examines how innovations spread throughout society, impacting both the private and public sectors. The incorporation of digital tools within sustainability and circular economy practices exemplifies such adoption.
- Open Innovation: In the digital context, open innovation allows organizations to collaborate across various sectors, integrating solutions to sustainability challenges, including resource management, waste reduction, and environmental monitoring.

Relevance to the Chapter:

- Digital Innovation Theory elucidates how digital technologies are not only creating new business models but also facilitating the development of sustainable practices and solutions across industries.
- It underscores the significance of digital tools in forming and expanding circular economy models, particularly through smart technologies and data-driven decision-making.

4. Diffusion of Innovations Theory

Overview:

The Diffusion of Innovations Theory, formulated by Everett Rogers, aims to comprehend how, why, and at what pace new ideas and technologies permeate cultures and social systems. This theory is especially pertinent when analyzing the adoption of digital technologies for sustainability, as it delineates the factors influencing the distribution of digital solutions for climate change and circular economy initiatives.

Key Concepts:

- Innovation Characteristics: Rogers identifies several traits that impact the adoption of innovations, such as relative advantage, compatibility, complexity, trialability, and observability. These traits help clarify why certain digital tools (for instance, blockchain for supply chain transparency or AI for resource optimization) are more readily embraced in sustainability efforts.
- Adopter Categories: The theory classifies adopters into innovators, early adopters, early majority, late majority, and laggards, aiding in understanding how digital solutions propagate among various sectors and regions.

Relevance to the Chapter:

- This theory is pivotal for understanding how digital innovations concerning sustainability—such as AI for climate forecasting or blockchain for circular supply chains—are accepted by different stakeholders (like governments, businesses, and consumers).
- It also aids in pinpointing strategies to expedite the adoption of sustainable digital technologies, which is a primary focus of the chapter.

5. The Triple Bottom Line (TBL) Theory

Overview:

The Triple Bottom Line (TBL) theory, proposed by John Elkington, stresses the necessity for organizations to balance three essential pillars: People, Planet, and Profit. In terms of digitalization and sustainability, TBL suggests that businesses should prioritize not only financial outcomes but also social and environmental results, ensuring their activities contribute to overarching societal objectives.

Key Concepts:

- Environmental Sustainability: Businesses are urged to diminish their environmental impact by implementing practices that conserve natural resources, minimize waste, and alleviate climate change.
- Social Responsibility: TBL underscores the significance of societal welfare and community influence, ensuring that businesses have a positive effect on societal prosperity.
- Economic Viability: The theory emphasizes that sustainability practices must be economically feasible, ensuring alignment between environmental and social goals and long-term business success.

Relevance to the Chapter:

- TBL is vital for grasping the broader objectives of sustainability and how digitalization can facilitate achieving these goals. By leveraging digital tools, organizations and governments can enhance environmental and social outcomes while also sustaining economic viability.
- This framework aligns closely with the chapter's emphasis on how digitalization and the circular economy can concurrently tackle environmental, social, and economic sustainability.

KEY CONCEPTUAL FOUNDATIONS FOR DIGITALIZATION, CIRCULAR ECONOMY, AND SUSTAINABILITY

The combination of digitalization and circular economy principles with sustainability is based on several fundamental concepts. These concepts illustrate how digital technologies can facilitate more efficient and sustainable systems and practices, leading to long-lasting environmental, economic, and social advantages. Below is an in-depth discussion of the main conceptual pillars, along with pertinent literature that offers further exploration of each idea.

1. Digitalization and Sustainability: Fostering Resource Efficiency and Innovation

Conceptual Foundation:

Digitalization entails the incorporation of digital technologies—such as Artificial Intelligence (AI), the Internet of Things (IoT), blockchain, and big data—into regular operations and industries. When applied to sustainability, digitalization opens new opportunities for enhancing resource efficiency, streamlining production processes, and minimizing waste, all of which are vital for meeting sustainable development goals (SDGs).

- Resource Efficiency: Digital tools are capable of gathering and analyzing large volumes of data to optimize resource utilization, lessen waste, and boost energy efficiency. For instance, AI can be utilized in manufacturing to enhance energy consumption patterns or in agriculture to maximize water use.
- Innovation in Sustainability: Digitalization encourages innovation by allowing for new business models, including the sharing economy and circular economy. These frameworks can lower the need for fresh resources by advocating for reuse, remanufacturing, and recycling.

Literature:

- Kagermann et al. (2013) examine how the "Industrie 4.0" (Fourth Industrial Revolution) concept integrates digital technologies to enhance industrial systems, aligning them with sustainable practices. The transition toward intelligent manufacturing systems is seen as crucial for improving resource efficiency and reducing environmental harm.
- Müller et al. (2020) contend that digitalization is a significant driver of sustainability transitions, particularly through smart systems that assist companies in monitoring and minimizing their environmental footprints.

2. Circular Economy (CE): Redefining Value through Digitalization

Conceptual Foundation:

The Circular Economy (CE) opposes the conventional linear economy, characterized by the production, usage, and disposal of products, by advocating for a regenerative framework. CE focuses on the persistent use of resources through recycling, reuse, and remanufacturing, aiming to minimize waste and lessen environmental consequences. Closed-loop Systems: In CE, products and materials are designed for reuse or recycling, preventing waste from entering the environment. Digital tools such as IoT and blockchain can monitor materials across the supply chain, enhancing the effectiveness of recycling processes and aiding in the management of end-of-life products.

- Business Model Innovation: CE also encourages the development of new business models like product-as-a-service (e.g., leasing or sharing) as opposed to ownership. Digital platforms can support these models by linking service providers with consumers, enabling greater efficiency and resource sharing.

Literature:

- Geissdoerfer et al. (2017) offer a comprehensive analysis of the interaction between digital technologies and the circular economy, elucidating how innovations such as IoT, AI, and blockchain can bolster circular business models by enhancing the tracking and management of materials and products.
- The Ellen MacArthur Foundation (2013) emphasizes how the circular economy, aided by digital technologies, can generate sustainable value by reducing material waste and improving resource efficiency. 3. Technological Innovation Systems (TIS): Facilitating Sustainability Transitions Conceptual Foundation: The theory of Technological Innovation Systems (TIS) centers on how new technologies are developed and distributed within a defined system, and how they can propel transitions towards more sustainable practices. When considering digitalization and sustainability, TIS illustrates how the incorporation of digital technologies (such as AI, IoT, and blockchain) can trigger movements towards sustainability by altering industries and forming new markets. • Systemic Change: TIS theory highlights the significance of changes across the entire system, proposing that digital technologies can stimulate such transformations through their connections with other technological, institutional, and market factors.

Networks of Actors: The theory also underscores the critical role of networks comprising various actors (like businesses, governments, and research organizations) in fostering innovation and dissemination. In the realm of digital tools aimed at sustainability, these networks can promote the adoption of new practices and business models across different sectors.

Literature

- Bergek et al. (2008) examine the dynamics of innovation systems, stressing the roles that technology, institutions, and policies play in encouraging transitions toward sustainability. They depict how emerging digital technologies can reshape sectors like energy, waste management, and manufacturing.
- Schot and Steinmueller (2018) investigate how digital technologies aid in systemic innovation, suggesting that digitalization serves as a pivotal element in facilitating sustainability transitions by linking diverse actors within the innovation ecosystem. 4. Triple Bottom Line (TBL): Harmonizing People, Planet, and Profit Conceptual Foundation: The Triple Bottom Line (TBL) framework, put forth by John Elkington, highlights the necessity for businesses and organizations to concentrate on three performance dimensions: social (People), environmental (Planet), and economic (Profit). In the context of digitalization, TBL advocates for the implementation of technologies that not only yield profits but also have a beneficial impact on society and the environment.

3. Sustainability

- Social Sustainability: Digital solutions can tackle social issues like poverty, inequality, and resource access by enhancing communication, education, and service availability.
- Environmental Sustainability: Digitalization can promote environmental sustainability through innovations such as energy efficiency, waste minimization, and carbon footprint reduction.
- Economic Sustainability: By boosting efficiency and encouraging new business models (like digital services and shared economies), digitalization can create new economic opportunities.

Literature:

- Elkington (1999) argues that organizations must integrate TBL principles into their strategies to achieve sustainable results across all three dimensions.

- Hart and Milstein (2003) explore how the TBL framework can be utilized in business innovation, emphasizing that digital technologies play a vital role in advancing the social, environmental, and economic aspects of sustainability. 5. Systems Thinking: Comprehensive Approach to Sustainability and Circular Economy Conceptual Foundation: Systems Thinking is a conceptual model that regards sustainability challenges as part of intricate, interconnected systems. It is based on the principle that the actions of a system cannot be comprehended by examining its components separately; instead, it requires an understanding of the connections and feedback loops within the system. Digital technologies can assist in mapping these systems and simulating the effects of various decisions, making them essential for addressing sustainability issues. • Feedback Loops: Digital tools can oversee and manage feedback loops within systems (such as energy usage, resource consumption, and waste production) to enhance sustainability practices.

Interconnectedness: Systems thinking stresses that environmental, economic, and social issues are interrelated. Digitalization can provide a comprehensive perspective on these challenges and aid in developing integrated solutions.

Literature:

- Senge (2006) emphasizes the significance of systems thinking for sustainability, noting that digital technologies facilitate a better understanding of the dynamic interactions among different elements affecting sustainability.
- Berkhout et al. (2004) apply systems thinking to sustainability transitions, highlighting the necessity for digital tools to analyze and manage the complexities of ecological, social, and economic systems.

CONCEPTUAL DEVELOPMENT:

The convergence of digital technology and sustainability presents a groundbreaking approach to tackling global issues like climate change, resource exhaustion, and socio-economic disparities. This theoretical foundation relies on established frameworks, including Sustainability Transition Theory, the Circular Economy Model, the Digitalization for Sustainable Development Framework, and the Sociotechnical Systems Approach. By integrating perspectives from these theories, this section formulates a well-rounded argument and hypothesis about how digitalization can promote sustainability, emphasizing the principles of a circular economy and the 2030 Sustainable Development Goals (SDGs).

Argument: Digitalization as a Catalyst for Sustainability

Digitalization serves as both a catalyst and a facilitator for sustainable practices by improving efficiency, transparency, and creativity. The incorporation of digital technologies within sustainability frameworks enables significant transformations across various sectors, promoting circular economy initiatives and advancing progress toward the Sustainable Development Goals (SDGs).

1. Digitalization Promotes Efficiency and Creativity

Current research highlights the ability of digital technologies to enhance resource efficiency, minimize waste, and optimize energy consumption. For instance, IoT sensors in smart cities track energy usage, allowing for data-driven decisions that can decrease greenhouse gas emissions (Zhou et al., 2021). Likewise, blockchain technology offers complete traceability in supply chains, minimizing inefficiencies and boosting accountability (Kouhizadeh et al., 2020).

From a theoretical perspective, the Digitalization for Sustainable Development Framework (Hilty & Aebischer, 2015) emphasizes the supportive effects of technology. However, it also recognizes potential rebound effects, where improvements in efficiency might result in higher consumption. This highlights the crucial need to sync digital solutions with strong governance frameworks to avoid unintended outcomes.

2. Circular Economy Concepts Enhanced by Digitalization

The Circular Economy Model offers a foundational perspective for reimagining production and consumption systems. Digitalization bolsters circular practices by enabling real-time resource monitoring, predictive maintenance, and innovative business models such as product-as-a-service (Geissdoerfer et al., 2017). Examples include:

- Predictive Maintenance: AI-driven algorithms in manufacturing forecast equipment failures, lowering downtime and waste.
- Resource Optimization: Digital twins replicate product life cycles, allowing manufacturers to enhance resource utilization and prolong product lifespan.

Although the Circular Economy Model focuses on material reuse and waste reduction, its effectiveness relies on the widespread use of digital tools. Nonetheless, the significant energy consumption associated with technologies like blockchain and the digital divide in developing areas present substantial hurdles.

3. Enabling Stakeholders and Enhancing Transparency

Sustainability Transition Theory emphasizes the importance of niche innovations in challenging established systems and driving systemic change (Geels, 2002). Digital tools empower various stakeholders, including governments, businesses, and consumers, by fostering transparency and enabling informed decision-making. For example:

- Governments: Digital platforms such as open data portals enhance public access to environmental information, increasing accountability and participation.
- Businesses: AI-driven analytics offer insights into sustainable practices, assisting companies in adhering to environmental regulations.
- Consumers: Mobile applications allow individuals to monitor their carbon footprints and make eco-friendly choices.

However, to achieve systemic change, it is essential to confront resistance at the regime level, where established systems frequently obstruct the adoption of digital innovations.

HYPOTHESIS: DIGITALIZATION AS A CATALYST FOR CIRCULAR ECONOMY PRACTICES AND PROGRESS TOWARD THE SDGS

Drawing from the existing theories and literature, the following hypothesis is proposed:

Digitalization, enabled by advanced technologies like IoT, AI, and blockchain, facilitates the adoption of circular economy practices, thereby hastening the achievement of the 2030 Sustainable Development Goals. This process is influenced by factors including stakeholder collaboration, regulatory frameworks, and the prevention of rebound effects.

Key Components of the Hypothesis:

- Enabler Technologies: Digital solutions lay the groundwork for the execution of circular practices, boosting efficiency and encouraging transparency.
- Mediating Factors: Collaboration among stakeholders, favorable policies, and awareness initiatives are essential for the effective use of digital solutions.

- Rebound Mitigation: Tackling unintended consequences, such as heightened consumption or energy utilization, is crucial for harnessing the complete potential of digitalization.

Critical Discussion

Though the hypothesis suggests a promising journey, several obstacles need to be tackled to realize a successful merger of digitalization and sustainability:

- Digital Divide: Imbalanced access to digital technologies worsens inequalities between developed and developing areas, hindering the worldwide adoption of sustainable practices.
- Energy and Resource Demands: The creation and functioning of digital technologies can frequently leave a considerable environmental footprint. For example, the energy requirements of data centers and blockchain infrastructures can be at odds with sustainability objectives.
- Regulatory and Ethical Challenges: Issues surrounding data governance, cybersecurity, and individual privacy present significant hurdles to the broader use of digital tools. Policymakers must implement strong regulatory frameworks to address these challenges.

Overcoming these challenges necessitates a comprehensive approach:

- Investing in Green Technologies: Encouraging the development of energy-efficient data centers, environmentally-friendly blockchain systems, and principles of eco-design.
- Capacity Building: Improving digital literacy and access in marginalized regions to close the digital gap.
- Collaborative Governance: Building partnerships between governments, industries, and civil society to co-create solutions that support sustainability objectives.

Supporting Logical Explanations, Reasoning, and Examples

1. Digitalization Increases Efficiency and Minimizes Waste

Logical Explanation:

Digital technologies, including the Internet of Things (IoT), artificial intelligence (AI), and big data analytics, improve resource utilization, reduce waste, and lower environmental impact. By supplying real-time data and forecasting insights, these tools allow organizations to spot inefficiencies, cut down on material usage, and create sustainable processes.

Example:

- Smart Grids and Energy Management: IoT-enhanced smart grids enable utility companies to dynamically oversee and control electricity distribution. This leads to decreased energy losses and the incorporation of renewable energy sources such as wind and solar. For example, the implementation of smart grids in various European cities has led to a 10% decrease in energy consumption (European Commission, 2020).
- Predictive Maintenance in Manufacturing: AI-driven systems forecast equipment failures in manufacturing facilities, minimizing downtime and averting material waste. General Electric (GE) has implemented these systems in its aviation sector, resulting in annual savings of millions of dollars while significantly reducing production-related waste.

2. Advancing Circular Economy through Digitalization

Logical Explanation:

Circular economy strategies concentrate on keeping materials in their highest possible utility for extended periods. Digitalization supports these strategies by enabling resource tracking, recycling, and reuse through cutting-edge technologies such as blockchain, digital twins, and cloud-based services.

Example:

- Blockchain for Supply Chain Transparency: Blockchain technology facilitates real-time tracking of products and materials, ensuring resources are reused or recycled effectively. For instance, IBM's Food Trust blockchain provides visibility in the food supply chain, decreasing waste by as much as 30% by pinpointing inefficiencies and speeding up recalls.
- Digital Twins in Product Lifecycle Management: Digital twins serve as virtual representations of physical assets that simulate and forecast product lifecycles. Firms like Siemens utilize digital twins to optimize resource usage, prolong product lifespan, and minimize waste during the design and production phases.

3. Encouraging Behavioral Change through Digital Tools

Logical Explanation:

Digital platforms and applications enable consumers to make environmentally friendly choices by offering actionable insights into their ecological footprint. This is in line with Sustainability Transition Theory, which highlights the influence of innovation on transforming social norms and behaviors.

Example:

- Carbon Tracking Applications: Applications such as “Oroeco” and “JouleBug” assist users in monitoring their carbon emissions and propose methods for reduction, such as utilizing public transport or lowering energy usage at home.
- Gamification for Recycling: In South Korea, municipalities employ gamified applications to motivate recycling efforts, where residents can receive rewards for correctly sorting waste. These programs have resulted in a 44% rise in recycling rates in the areas involved (OECD, 2019).

4. Addressing Structural Barriers and Promoting Systemic Change

Logical Explanation:

Digitalization can tackle structural obstacles, like regulatory issues, by enhancing collaboration and enabling decisions based on evidence. AI- and big data-enabled platforms assist policymakers by offering simulations and scenario analyses for sustainable urban planning and resource management.

Example:

- Digital Urban Planning in Singapore: Singapore’s Urban Redevelopment Authority utilizes AI and 3D modeling to devise plans for sustainable urban growth, harmonizing housing, transit, and green spaces. This initiative has significantly diminished energy consumption and improved urban sustainability.
- Digital Solutions for Water Management: In India, the digital platform “AQUASAVE” aids in monitoring water usage and mitigating waste through IoT sensors. This initiative has played a role in achieving sustainable water management in areas prone to drought.

5. Addressing Rebound Effects and Mitigating Risks

Logical Explanation:

Although digitalization brings many advantages, it can also result in unintended outcomes, such as rebound effects where improvements in efficiency lead to a rise in overall consumption. Mitigating these risks involves combining digital solutions with governance structures and public awareness initiatives.

Example:

- Energy Use in Data Centers: The swift expansion of cloud computing has heightened the energy requirements of data centers. Firms like Google and Microsoft are addressing this challenge by investing in energy-efficient infrastructures and renewable energy sources. Google's "carbon-intelligent computing" initiative has cut down data center emissions by 30%.
- Regulatory Measures for Blockchain: The energy-heavy characteristics of blockchain activities, particularly in cryptocurrency mining, present sustainability hurdles. Countries like Sweden and Canada are implementing regulatory actions, such as providing incentives for utilizing renewable energy in blockchain operations.

6. Collaborative Governance and Stakeholder Involvement

Logical Explanation:

Cooperation among governments, industries, and civil society is vital for amplifying digital solutions and ensuring fair access. Digital platforms enable this collaboration by establishing shared environments for innovation, knowledge sharing, and resource distribution.

Example:

- Public-Private Partnerships (PPP): The EU's Horizon 2020 initiative promotes PPPs to develop sustainable digital solutions, including smart city projects and green energy initiatives. These joint efforts have resulted in significant advancements in urban sustainability across Europe.
- Digital Literacy Initiatives: Organizations like the United Nations Development Programme (UNDP) implement digital literacy initiatives in developing nations to close the digital gap and encourage inclusive involvement in sustainability efforts.

Potential Implications and Applications of the Conceptual Model

The integrated conceptual model that combines digitalization and sustainability outlines a strategy for using technology to expedite the shift towards a circular economy and attain the 2030 Sustainable Development Goals (SDGs Below, we delve deeper into its possible implications and uses.

1. Implications for Policy Development

Potential Implications:

Policymakers can utilize the findings from this model to create data-driven policies that promote the use of digital technologies for sustainability. By acknowledging the importance of digital solutions in optimizing resources, minimizing waste, and empowering stakeholders, governments can establish conducive frameworks that encourage innovation and fair access.

Applications:

- Smart Regulatory Mechanisms: Governments could establish policies that incentivize green technologies, such as tax incentives for businesses that implement IoT-based resource management or blockchain systems for transparency.
- National Digital Sustainability Strategies: Nations like Singapore and Estonia have rolled out comprehensive plans that merge digitalization with sustainability aims. This model provides a framework for adopting similar strategies worldwide.
- Carbon Pricing with Digital Platforms: Digital solutions can facilitate precise emissions monitoring, allowing for real-time adjustments to carbon pricing and trading mechanisms.

2. Advancing Corporate Sustainability Practices

Potential Implications:

Organizations can utilize the model to synchronize their operations with sustainability goals while preserving their competitive edge. Digital tools support the creation of circular business models, improve supply chain transparency, and drive innovation.

Applications:

- Circular Business Models: Companies may embrace product-as-a-service (PaaS) models, wherein ownership transitions from consumers back to manufacturers, thereby promoting material reuse and recycling.
    - Example: Philips' "Pay per Lux" lighting solution model maximizes resource efficiency and minimizes waste.
- Enhanced ESG Reporting: Digital platforms powered by AI and big data offer real-time insights into environmental, social, and governance (ESG) performance, allowing companies to fulfill investor and regulatory requirements.
- Blockchain for Ethical Sourcing: Large corporations can deploy blockchain technology to ensure the ethical procurement of raw materials, like cobalt for batteries, addressing both environmental and social issues.

3. Empowering Local Communities and Individuals

Potential Implications:

The model underlines how digital technologies can empower communities and individuals to engage actively in sustainability initiatives. By making sustainability attainable and actionable, digital tools encourage widespread behavioral change.

Applications:

- Digital Inclusion Programs: Targeted initiatives to bridge the digital gap ensure that marginalized communities gain access to the advantages of sustainability-focused digitalization.
    - Example: India's Digital India initiative has enabled rural populations with resources for digital literacy and sustainable agricultural practices.
- Citizen Science Initiatives: Mobile applications and IoT devices can involve citizens in environmental monitoring, such as tracking air quality or reporting incidents of illegal deforestation.
- Personalized Carbon Footprint Trackers: Apps that assess individual carbon footprints and offer practical suggestions promote sustainable lifestyle choices.

4. Transforming Urban Environments

Potential Implications:

The combination of digitalization and circular economy strategies can turn cities into centers of sustainability. Smart city technologies enhance resource efficiency, improve quality of life, and mitigate environmental impacts.

Applications:

- Smart Waste Management Systems: IoT-enabled waste bins with sensors can optimize collection routes, leading to reduced emissions and lower operational costs.
  - o Example: Barcelona has introduced smart waste bins, resulting in a 20% reduction in waste management expenses.
- Green Infrastructure Planning: Utilizing digital twins of urban areas allows planners to simulate the effects of green infrastructure, such as parks and green roofs, on urban heat islands and stormwater management.
- Renewable Energy Integration: Smart grids facilitate the smooth incorporation of renewable energy sources, dynamically balancing supply and demand.

5. Expanding Sustainable Development in Developing Countries

Potential Implications:

Developing countries can utilize the conceptual framework to bypass conventional development routes by directly incorporating digital and sustainable methodologies. By focusing on investments in digital infrastructure, these countries can tackle socio-economic inequalities and environmental issues concurrently.

Applications:

- AgriTech Innovations: Platforms powered by IoT and AI enhance agricultural methods, decreasing water and pesticide consumption while boosting yields.
  - o Example: The Indian digital initiative "e-Choupal" links farmers with markets and weather information, enhancing productivity and minimizing waste.
- Decentralized Renewable Energy Solutions: Systems for energy trading based on blockchain allow for decentralized renewable energy networks in areas lacking centralized grid access.
  - o Example: In Kenya, companies like Power Ledger enable peer-to-peer energy trading, broadening access to renewable energy.
- Mobile Banking Solutions: Digital payment platforms, such as Kenya's M-Pesa, empower unbanked individuals and promote investments in sustainable businesses.

6. Tackling Global Environmental Issues

Potential Implications:

The conceptual framework emphasizes the potential of scalable digital solutions to confront intricate global issues like climate change, loss of biodiversity, and water scarcity.

Applications:

- International Climate Monitoring Systems: AI and satellite technology deliver precise and timely information on climatic variables, aiding policymakers in proactive responses.
  - o Example: NASA's Earth Science Division employs AI to observe deforestation and forecast climate change repercussions.
- Biodiversity Preservation: AI-driven image recognition technologies can monitor wildlife numbers and alert to threats, supporting conservation initiatives.
  - o Example: Google AI's Wildlife Insights platform assists in global biodiversity surveillance.
- Water Resource Administration: IoT devices track water consumption and quality, optimizing resource distribution and curbing overuse.

7. Cooperation Among Stakeholders

Potential Implications:

The model highlights the importance of collaboration among various stakeholders to create systemic transformation. Digital platforms enable collaboration by granting shared access to data, insights, and resources.

Applications:

- Public-Private Collaborations (PPCs): Governments and enterprises can jointly develop digital solutions for sustainability, such as renewable energy endeavors or digital waste management systems.
- Global Knowledge Sharing Networks: Initiatives like the United Nations' "Partnerships for SDGs Platform" allow stakeholders to exchange best practices and amplify innovations on a global scale.
- Joint Research and Innovation: Academia, industry, and governmental entities can collaborate to create digital instruments suited to local contexts, increasing their effectiveness.

IMPLICATIONS AND FUTURE RESEARCH:

This chapter delves into the convergence of digitalization, sustainability, and circular economy practices, holding considerable promise for influencing both theoretical progress and practical implementations. Below, we highlight the primary implications for academia, industry, policy, and society.

1. Enhancing Academic Discussion

Potential Impact:

The conceptual amalgamation of digitalization and sustainability enhances academic literature by providing a comprehensive framework that merges cutting-edge technologies with circular economy concepts. This aids in the development of interdisciplinary research that connects technology, environmental sciences, and socio-economic studies.

Examples:

- Refinement of Theories: The integration of Sustainability Transition Theory with Digital Innovation Theory in this chapter creates a more sophisticated theoretical foundation for investigating sustainable digital transformations.
- Innovation in Research Methodologies: The focus on utilizing big data analytics, AI, and blockchain for sustainability encourages methodological advancements, promoting the adoption of sophisticated tools for the empirical examination of sustainability initiatives.

Future Research Paths:

- Scholars could broaden the model by investigating the long-term socio-economic ramifications of digital solutions in relation to achieving the Sustainable Development Goals (SDGs).
- Comparative analyses between developed and developing nations could illuminate the scalability and adaptability of digital sustainability initiatives.

2. Shaping Policy Development

Potential Impact:

Insights from this chapter can guide policymakers in formulating strategies that prioritize digital instruments as facilitators of sustainable development. By addressing regulatory obstacles and fostering digital equity, governments can enable comprehensive transformations across industries and communities.

Examples:

- Strategies for Digital Sustainability: National governments may create policies that embed smart technologies within urban planning, waste management, and energy systems.

- International Collaborations: The chapter's focus on blockchain and IoT for enhancing transparency can inform global agreements aimed at promoting sustainable supply chain operations.

Future Research Directions:

- Additional studies could explore the success of public-private partnerships (PPPs) in amplifying digital solutions for sustainability.
- Research on regulatory frameworks to oversee the energy consumption of digital infrastructures, such as data centers, could tackle significant challenges.

3. Promoting Industrial Innovation

Potential Impact:

Industries can gain considerably from employing digital tools in their circular economy practices. By aligning their corporate strategies with sustainable development goals, companies can improve operational efficiency, lessen waste, and secure competitive advantages.

Examples:

- Management of Product Lifecycles: Organizations can utilize digital twins and blockchain technology to monitor product lifecycles, ensuring effective recycling and reuse of materials.
- Reporting on ESG and Building Investor Trust: The emphasis on real-time tracking of sustainability metrics in the model corresponds with the increasing demand from investors for transparency regarding environmental, social, and governance (ESG) performance.

Future Research Paths:

- Additional investigation is required to understand how digitalization influences the supply chain dynamics of small and medium-sized enterprises (SMEs).
- Examining how digital literacy can facilitate the adoption of sustainability practices in industry may provide practical insights.

4. Empowering Communities and Individuals

Potential Impact:

Digital technologies can enable individuals and communities by widening access to sustainability solutions and promoting behavioral change. The chapter's focus on gamification, mobile applications, and digital inclusion illustrates ways to engage citizens in the transition toward sustainability.

Examples:

- Citizen Engagement: Applications that monitor carbon footprints or reward recycling efforts can inspire significant behavioral changes, lessening ecological impacts.
- Digital Inclusion: Closing the digital divide ensures that marginalized communities can contribute to and gain from the shift toward a sustainable economy.

Future Research Paths:

- Investigating the lasting effects of gamified sustainability programs on community behaviors is necessary.
- Research examining the overlap between digital inclusion and gender equality in sustainability could yield important insights.

Future Research Opportunities

1. Energy Efficiency of Digital Infrastructure

As digitalization expands, so does the demand for energy from the supporting infrastructure, such as data centers and blockchain systems. Research is needed to investigate renewable energy approaches and energy-efficient technologies to counteract these effects.

2. Measuring the Impact of Digital Tools on SDGs

Quantitative research assessing how digital tools contribute to particular SDGs, like poverty alleviation (SDG 1) or clean energy (SDG 7), would offer valuable data for stakeholders.

3. Addressing Rebound Effects

Future studies should aim to comprehend and minimize rebound effects, such as the rise in resource consumption stemming from efficiency improvements realized through digital tools.

4. Cross-Sectoral Collaboration Models

Exploring effective collaboration models between government, industry, and academia to broaden the reach of digital sustainability initiatives worldwide is an encouraging direction for research.

Suggested Areas for Further Research or Exploration

Based on the conceptual model that unites digitalization, circular economy practices, and sustainability objectives, the following domains provide opportunities for additional research. Each domain addresses existing gaps or insufficiently studied elements within the framework, supplying comprehensive explanations and potential research methods for future exploration.

1. Energy Efficiency and Environmental Effects of Digital Technologies

Justification:

Although digital technologies are vital for sustainability, they also pose environmental challenges such as excessive energy use and the generation of electronic waste. Investigating energy-efficient systems and environmentally friendly digital infrastructures is crucial for alleviating these unintended drawbacks.

Research Questions:

- What are the energy-related effects of new digital technologies like blockchain, IoT, and AI in sustainable applications?
- How can incorporating renewable energy with digital infrastructure lessen carbon emissions?

Potential Methodologies:

- Lifecycle Assessment (LCA): Assess the environmental effects of digital infrastructure throughout its life cycle, from creation to disposal.

- Simulation Models: Implement computational models to project energy savings from the adoption of eco-friendly technologies in data centers and IoT networks.

2. Socioeconomic Effects of Digital Sustainability Initiatives in Developing Nations

Justification:

Developing countries frequently encounter challenges such as digital divides, inadequate infrastructure, and social and economic disparities. Exploring how digital tools can help overcome these hurdles while advancing sustainability could yield transformative insights.

Research Questions:

- What effects do digital sustainability initiatives have on economic development and social equity in emerging markets?
- What influence does digital literacy have on the effectiveness of sustainability initiatives?

Potential Methodologies:

- Field Studies: Carry out community-based research to assess the socio-economic effects of digital tools like agricultural mobile applications or renewable energy technologies.
- Comparative Analyses: Investigate case studies from various countries to uncover best practices and scalable models.

3. Governance and Policy Frameworks for Digital Sustainability

Justification:

Governance structures and regulatory frameworks must adapt to promote the use of digital tools in sustainability efforts. This area examines how policies can align with technological progress and societal objectives.

Research Questions:

- What regulatory challenges impede the implementation of digital tools for sustainable practices?
- How can governments promote collaboration between the private sector and public institutions to expand digital sustainability solutions?

Potential Methodologies:

- Policy Analysis: Review existing policies in nations with strong digital sustainability practices to identify gaps and opportunities for improvement.
- Stakeholder Engagement: Employ participatory techniques to gather perspectives from policymakers, businesses, and civil society regarding effective governance models.

4. Behavioral Change and Public Engagement through Digital Tools

Justification:

Digital technologies such as gamification, social media, and applications can impact individual actions and community involvement in sustainability. Researching their effectiveness could improve engagement strategies.

Research Questions:

- What elements influence the success of gamification in encouraging sustainable habits?
- How do digital platforms affect public views and behaviors concerning environmental issues?

Potential Methodologies:

- Behavioral Experiments: Evaluate the effect of gamified applications on recycling rates or energy usage in controlled environments.
- Sentiment Analysis: Examine social media data to gain insights into public attitudes toward sustainability initiatives.

5. Circular Economy and Digitalization in Specific Sectors

Justification:

The use of digital technologies differs by industry. Investigating sector-specific applications, including manufacturing, agriculture, and healthcare, can yield customized approaches to sustainability.

Research Questions:

- In what ways can IoT and AI improve circular practices within the manufacturing sector?
- What function do digital platforms serve in minimizing food waste throughout the agricultural supply chain?

Potential Methodologies:

- Case Studies: Analyze organizations that have effectively integrated digital technologies into circular economy frameworks.
- Sectoral Analyses: Assess the influence of digital tools on key performance indicators such as waste reduction and resource efficiency across various sectors.

6. Ethical Considerations in Digital Sustainability

Justification:

The rise of digitalization introduces ethical dilemmas concerning data privacy, surveillance, and fair access to technology. Examining these challenges is vital to ensure that digital sustainability efforts respect human rights and promote inclusivity.

Research Questions:

- How can concerns regarding data privacy be addressed in sustainability initiatives that rely on IoT and big data?
- What approaches can guarantee equitable access to digital tools for sustainability among various socio-economic groups?

Potential Methodologies:

- Ethnographic Studies: Investigate the experiences of underrepresented communities in utilizing digital tools for sustainability.
- Legal Analysis: Review current legal frameworks that govern data privacy and recommend enhancements for sustainability contexts.

7. Measuring the Impact of Digital Tools on SDGs

Justification:

While digital tools play a role in advancing the SDGs, measuring their impact accurately poses challenges. This area emphasizes the development of metrics and methodologies to effectively track progress.

Research Questions:

- Which indicators are most effective at gauging the influence of digital technologies on specific SDGs?
- In what ways do digital tools expedite or impede progress toward sustainability goals?

Potential Methodologies:

- Indicator Development: Create and validate novel metrics to evaluate the effectiveness of digital solutions.
- Longitudinal Studies: Monitor the effects of digital sustainability initiatives over time to identify trends and obstacles.

8. Addressing Rebound Effects of Digitalization

Justification:

Gains in efficiency from digital technologies can lead to heightened resource consumption, a phenomenon known as rebound effects. Investigating methods to lessen these effects is essential for achieving sustainable outcomes.

Research Questions:

- What are the main factors driving rebound effects in digital sustainability applications?
- How can regulations and practices reduce these unintended consequences?

Potential Methodologies:

- Economic Modeling: Employ input-output models to examine resource flows and pinpoint rebound effects.
- Scenario Analysis: Simulate various policy interventions to evaluate their effectiveness in alleviating rebound effects.

9. Multi-Stakeholder Collaboration Models for Scaling Solutions

Justification:

Cooperation among governments, industries, academia, and civil society is crucial for expanding digital sustainability initiatives. Researching effective collaboration models can amplify their impact.

Research Questions:

- What elements contribute to the success of multi-stakeholder collaborations in digital sustainability endeavors?
- How can global platforms promote knowledge sharing and collective action?

Potential Methodologies:

- Network Analysis: Investigate current collaborations to identify key participants and their roles.
- Interviews and Surveys: Collect perspectives from stakeholders regarding the facilitators and challenges of collaboration.

These research areas offer a thorough framework for examining the opportunities and challenges posed by digitalization in fostering sustainability. By tackling these questions, researchers can enhance understanding, inform policies, and guide practices, ensuring that digital tools act as a catalyst for a fair and sustainable future.

CASE STUDIES: DIGITALIZATION AND SUSTAINABILITY

The subsequent case studies illustrate how digital tools and methodologies contribute to initiatives for a circular economy, sustainable development, and climate change mitigation. These examples provide practical insights into the real-world applications of the concepts explored in this chapter.

1. Blockchain for Supply Chain Clarity: Everledger

Overview:

Everledger operates a digital platform that utilizes blockchain technology to improve transparency and traceability in supply chains, especially within the diamond sector. Blockchain guarantees that every transaction within the supply chain is recorded in an immutable manner, enabling stakeholders to verify ethical sourcing and sustainability efforts.

Impact:

- Prevents "blood diamonds" from entering the market through verification of their origins.
- Facilitates circular economy initiatives by monitoring recycled materials in industries such as electronics.
- Enhances consumer trust by providing transparent data.

Key Takeaways:

This case showcases the potential of blockchain in fostering accountability and ethical practices across global supply chains.

2. Smart Cities and IoT: Barcelona's Urban Sustainability

Overview:

Barcelona has embraced Internet of Things (IoT) technologies to enhance urban sustainability. Projects include intelligent lighting, waste management, and traffic monitoring systems. IoT-enabled sensors gather data to improve city services and minimize resource use.

Impact:

- Achieved a 30% reduction in energy consumption through smart street lighting systems.
- Increased waste collection efficiency by 20% via sensor-equipped bins.
- Improved air quality monitoring, leading to better public health results.

Key Takeaways:

Barcelona exemplifies how IoT technologies can convert urban environments into more sustainable, efficient, and livable spaces.

3. Circular Economy in Manufacturing: Philips Lighting

Overview:

Philips Lighting (now known as Signify) has embraced a circular economy model by providing "lighting as a service" (LaaS). Rather than selling lightbulbs, the company installs and manages lighting systems while retaining ownership, ensuring recycling and reuse at the end of their life cycle.

Impact:

- Decreased raw material usage by 28% through effective product lifecycle management.
- Reduced waste by refurbishing and reusing components.
- Improved customer satisfaction with cost-effective lighting options.

Key Takeaways:

This approach demonstrates how digitalization can promote circular practices in manufacturing, benefiting both companies and the environment.

4. AI for Climate Risk Assessment: Microsoft AI for Earth

Overview:

Microsoft's AI for Earth program leverages artificial intelligence to tackle significant environmental issues worldwide. In partnership with various environmental organizations, Microsoft creates AI tools designed to forecast climate risks, including deforestation and water shortages.

Impact:

- Enabled precision agriculture initiatives, resulting in a 35% reduction in water consumption during pilot programs.
- Enhanced conservation efforts through advanced predictive modeling of deforestation trends.
- Facilitated climate adaptation strategies for at-risk areas.

Key Takeaways:

AI technologies can significantly contribute to anticipating and alleviating climate-related threats, thereby promoting resilience and sustainability.

5. Big Data for Water Resource Management: Cape Town's Day Zero Crisis

Overview:

In response to the water crisis in Cape Town from 2017 to 2018, big data analytics were utilized to track water consumption and forecast shortages. The city implemented digital tools to offer real-time information and educate citizens on water conservation practices.

Impact:

- Achieved a 55% decrease in water use in critical areas through focused intervention efforts.
- Enhanced resource management by pinpointing high-risk regions using predictive analytics.
- Prevented Day Zero by promoting community cooperation through transparency in data sharing.

Key Takeaways:

This example illustrates how data-driven strategies can effectively combat resource scarcity and bolster community resilience against environmental challenges.

KEY INSIGHTS AND CONTRIBUTIONS

This chapter delves into the convergence of digitalization, the circular economy (CE), and sustainability, particularly examining how digital technologies can promote sustainable business practices and aid in achieving the 2030 Sustainable Development Goals. A number of significant insights and contributions have surfaced throughout this discussion:

- Digitalization as a Driver of Sustainability: Digital technologies serve a crucial role in promoting sustainability by allowing for improved resource efficiency, increasing transparency, and encouraging innovation. The application of digital tools such as artificial intelligence (AI), blockchain, and the Internet of Things (IoT) optimizes processes across supply chains, energy management, waste minimization, and circular economy initiatives. By lowering resource use and enabling waste repurposing, digitalization facilitates the shift towards a circular economy, which aids in addressing climate change.
- The Importance of Circular Economy Frameworks: When circular economy principles are combined with digital technologies, they create an effective approach to separating economic growth from environmental harm. Digital solutions, including blockchain for traceability and data-centric circular business strategies, significantly enhance the efficiency and scalability of CE initiatives. Nonetheless, obstacles persist in evaluating and standardizing the social and environmental impacts of these activities. Therefore, merging digital tools with circular economy frameworks improves both environmental effectiveness and business profitability, aligning with the Triple Bottom Line (TBL) model of sustainable development.
- Multi-Level Frameworks for Sustainability Transitions: The chapter underscores the necessity of employing multi-level theories, such as Sustainability Transition Theory (STT), to comprehend the intricate processes involved in sustainability transitions. Digital technologies operate at the niche level, spurring innovation while broader societal systems (e.g., policies and cultural norms) adapt to these transformations.
- Critical Examination of Frameworks: A thorough analysis of existing frameworks, including STT, CE, and TBL, revealed several advantages, such as the comprehensive nature of TBL and its alignment with global sustainability objectives, but also pinpointed noteworthy drawbacks. These include difficulties associated with the absence of standardization in evaluating social and environmental impacts and the potential for greenwashing.
- Directions for Future Research: The chapter highlights several avenues for further investigation, especially in creating integrated models that merge the

benefits of digital innovation, circular economy, and sustainability transition theories. Future exploration should tackle the challenges related to scaling digital solutions for sustainability, weaving social and environmental considerations into business frameworks, and ensuring the effective use of digital tools to achieve both economic and social advantages.

CONCLUSION

This chapter adds to the expanding body of literature on digitalization and sustainability by providing a critical assessment of key frameworks and offering actionable insights into how digital technologies can enhance circular economy practices. By emphasizing the essential role of innovation and the necessity for an integrated, systemic approach, it lays the foundation for future research and practices that can support the realization of the 2030 Sustainable Development Goals and mitigate climate change. The combination of digital technologies with circular economy models represents not only an opportunity for businesses to boost efficiency but also a vital step towards transforming systems for a more sustainable and environmentally friendly future.

As the global community faces intricate and interconnected issues like climate change, resource depletion, and social inequality, the insights provided here contribute meaningfully to the ongoing discussions about how to align digital technologies with sustainability initiatives. Below are several key ways the ideas and findings presented in this chapter are pertinent to different disciplines:

1. Sustainability and Environmental Studies

The combination of digital technologies with sustainability initiatives holds substantial promise for tackling global environmental issues. As researchers and practitioners in environmental studies continue to seek solutions for minimizing carbon emissions, managing waste, and improving resource efficiency, the framework outlined in this chapter illustrates how digitalization—utilizing tools such as IoT, AI, and blockchain—can promote improved monitoring, data gathering, and optimization of environmental strategies.

Relevance: This chapter enhances the understanding of digital sustainability, establishing a practical basis for merging digital technologies with sustainability objectives. It invites deeper inquiry into the influence of digital innovation on ecosystem preservation, biodiversity, and climate resilience.

2. Business and Management Disciplines

Digital transformation serves as a vital catalyst for innovation in the business sector. As organizations increasingly embrace circular economy principles to boost efficiency and minimize waste, the insights offered in this chapter are particularly significant to the domain of business management. For businesses aiming to maintain their competitive edge while committing to sustainable development principles, this work presents a conceptual framework for embedding sustainability within their strategic and operational plans.

Relevance: The chapter connects business strategy with sustainability management, introducing digital solutions as essential facilitators of circular economy practices, making it particularly relevant to corporate sustainability and operations management research.

3. Technological Innovation and Digital Transformation

From the perspective of technological innovation, this chapter adds to ongoing research regarding digitalization and its applications in addressing real-world challenges, especially in the realm of sustainable development. The chapter offers evidence that digital technologies can considerably improve resource efficiency, enhance waste management, and simplify the execution of circular economy practices. This is especially relevant for engineering and information technology fields, where research into smart systems, artificial intelligence, and big data analytics can advance sustainability objectives.

Relevance: For academics focused on technology studies, this chapter emphasizes the importance of digital innovation as a key force for promoting sustainability, showcasing the potential of emerging technologies to deliver long-term environmental and social advantages. It also creates new opportunities for research into smart technologies and the management of sustainable innovation.

4. Public Policy and Governance

The insights outlined in this chapter carry significant implications for public policy and governance. As governments globally implement regulations and policies aimed at achieving sustainability targets—such as those outlined in the Paris Agreement and the 2030 SDGs—the relevance of digital tools in meeting these goals becomes increasingly pronounced. Policymakers and government agencies are progressively looking to digital solutions to assess environmental effects, enforce regulations, and foster transparent reporting across industries. The chapter advocates for further exploration into how policy frameworks can encourage the adoption of digital tools that bolster sustainable practices and how to weave these solutions into existing governance structures.

Relevance: This chapter is important for researchers in policy development who are focused on crafting future sustainability initiatives that leverage the potential of digital technologies. It lays a conceptual groundwork for examining regulatory approaches that can facilitate the shift towards a more circular and sustainable economy.

5. Interdisciplinary Collaboration and Systems Thinking

Ultimately, this chapter advocates for a collaborative approach to sustainability. It highlights the necessity of cooperation among scientists, engineers, business executives, policymakers, and social scientists to attain long-term sustainability objectives. The merging of digitalization with circular economy practices necessitates a systems-oriented strategy that takes into account social, technological, economic, and environmental factors. The intricate and interdependent nature of these challenges demands a comprehensive viewpoint, which is increasingly acknowledged in sustainability science as a vital component for effective change.

Relevance: This chapter presents a compelling case for systems thinking and collaborative research, which are increasingly regarded as essential for addressing complex sustainability issues. Researchers from various disciplines can utilize these insights to promote joint initiatives and cohesive solutions aimed at achieving sustainability.

The concepts explored in this chapter carry considerable significance for the progression of sustainability research and implementation. By investigating the convergence of digital technologies, the circular economy, and sustainability, this chapter offers valuable perspectives that span numerous fields, including environmental studies, business management, technology, and policy. Its emphasis on the function of digital innovation in fostering sustainability not only enriches academic discussions but also provides practical insights for industries, governments, and communities working to fulfill the 2030 Sustainable Development Goals.

REFERENCES

Bakici, T., Almirall, E., & Wareham, J. (2013). A smart city initiative: The case of Barcelona. *Journal of the Knowledge Economy*, *4*(2), 135–148. DOI: 10.1007/s13132-012-0084-9

Bijker, W. E., Hughes, T. P., & Pinch, T. (1987). *The social construction of technological systems: New directions in the sociology and history of technology*. MIT Press.

Davis, F. D. (1989). Perceived usefulness, perceived ease of use, and user acceptance of information technology. *Management Information Systems Quarterly*, *13*(3), 319–340. DOI: 10.2307/249008

Ellen MacArthur Foundation. (2013). Towards the circular economy: Economic and business rationale for an accelerated transition.

Ellen MacArthur Foundation. (2021). Lighting as a service: A circular economy business model. Retrieved from https://ellenmacarthurfoundation.org

European Commission. (2020). Smart grids and beyond: EU initiatives and projects.

Everledger. (2021). How Everledger supports sustainability with blockchain technology. Retrieved from https://everledger.io

Geels, F. W. (2002). Technological transitions as evolutionary reconfiguration processes: A multi-level perspective and a case study. *Research Policy*, *31*(8-9), 1257–1274. DOI: 10.1016/S0048-7333(02)00062-8

Geissdoerfer, M., Savaget, P., Bocken, N. M. P., & Hultink, E. J. (2017). The circular economy – A new sustainability paradigm? *Journal of Cleaner Production*, *143*, 757–768. DOI: 10.1016/j.jclepro.2016.12.048

Hilty, L. M., & Aebischer, B. (2015). ICT for sustainability: An emerging research field. In Hilty, L. M., & Aebischer, B. (Eds.), *ICT innovations for sustainability* (pp. 3–36). Springer., DOI: 10.1007/978-3-319-09228-7_1

IBM. (2021). IBM Food Trust blockchain solutions.

Kagermann, H., Wahlster, W., & Helbig, J. (2013). Recommendations for implementing the strategic initiative INDUSTRIE 4.0. Acatech. Müller, M., Kohn, L., & Stark, W. (2020). Digitalization and sustainability: Enabling innovations for achieving the SDGs. Springer.

Kouhizadeh, M., Sarkis, J., & Zhu, Q. (2020). Blockchain technology and the sustainable supply chain: Theoretically exploring adoption barriers. *International Journal of Production Economics, 231*, 107831. DOI: 10.1016/j.ijpe.2020.107831

Microsoft. (2020). AI for Earth: Empowering people and organizations to solve global environmental challenges. Retrieved from https://www.microsoft.com/en-us/ai/ai-for-earth

OECD. (2019). Digital innovation and the circular economy.

Parag, Y., & Sovacool, B. K. (2016). Electricity market design for the prosumer era. *Nature Energy, 1*(4), 1–6. DOI: 10.1038/nenergy.2016.32

Parida, V., Sjödin, D. R., Wincent, J., & Kohtamäki, M. (2019). Digitalization and servitization: A review, synthesis and future research directions. *Industrial Marketing Management, 83*, 302–315. DOI: 10.1016/j.indmarman.2019.05.011

Wolski, P. (2018). The Cape Town "Day Zero" drought: Past, present, and future. *South African Journal of Science, 114*(9/10). Advance online publication. DOI: 10.17159/sajs.2018/a0275

Yildirim, S. (2020). The consumer role for sustainable development: how consumers contribute sustainable development goals. In Chkoniya, V., Madsen, A. O., & Bukhrashvili, P. (Eds.), *Anthropological Approaches to Understanding Consumption Patterns and Consumer Behavior* (pp. 325–341). IGI Global. DOI: 10.4018/978-1-7998-3115-0.ch018

Yildirim, S., Demirtas, I., & Yildirim, D. C. (2022). A Review of Alternative Economic Approaches to Achieve Sustainable Development: The Rising Digitalization and Degrowth Post COVID-19. In Castanho, R. (Ed.), *Handbook of Research on Sustainable Development Goals, Climate Change, and Digitalization* (pp. 288–307). IGI Global. DOI: 10.4018/978-1-7998-8482-8.ch018

Yildirim, S., & Yildirim, D. Ç. (2020). Achieving Sustainable Development Through a Green Economy Approach. In Patti, S., & Trizzino, G. (Eds.), *Advanced Integrated Approaches to Environmental Economics and Policy: Emerging Research and Opportunities* (pp. 1–22). IGI Global. DOI: 10.4018/978-1-5225-9562-5.ch001

Zhou, K., Fu, C., & Yang, S. (2021). Big data driven smart energy management: From big data to big insights. *Renewable & Sustainable Energy Reviews, 68*, 550–565. DOI: 10.1016/j.rser.2021.110640

Zhou, K., Fu, C., & Yang, S. (2021). Big data driven smart energy management: From big data to big insights. *Renewable & Sustainable Energy Reviews, 68*, 550–565. DOI: 10.1016/j.rser.2021.110640

Chapter 2
Digitalization, Sustainability, and Strategy:
An Analytical Model of Green Investment Dynamics

Hrishikesh Desai
https://orcid.org/0000-0002-7945-3744
Arkansas State University, USA

ABSTRACT

This chapter explores the dynamics of subsidies, digitalization, and stakeholder behavior in the green investment ecosystem through a game-theoretic framework. Governments optimize subsidies and penalties, firms decide on green technology investments, and consumers make adoption choices based on costs, benefits, and digital awareness. The model incorporates digitalization as a key variable influencing costs, efficiency, and decision-making. The numerical experiment reveals that digitalization amplifies the effects of subsidies, increasing consumer adoption and firm investments while enhancing government policy efficiency. Key findings include: (1) consumer adoption grows synergistically with subsidies and digitalization; (2) firm investments increase with financial and technological incentives but show diminishing returns at higher subsidies; and (3) government payoffs are optimized at moderate subsidy levels coupled with digital tools. Thus, this chapter highlights how policymakers can design strategies that integrate financial incentives with digital innovation.

DOI: 10.4018/979-8-3693-9909-5.ch002

INTRODUCTION

The push towards sustainability globally has made green investments a focal point of economic, environmental, and social progress. Governments, firms, and consumers play interdependent roles in driving the adoption of green technologies and practices. Governments incentivize or regulate adoption through subsidies and penalties, firms respond by investing in green innovations, and consumers make purchasing decisions influenced by costs, environmental awareness, and personal preferences (Chen & Ma, 2021; Liu et al., 2023). However, the process is far from straightforward due to the complex interplay of incentives, costs, and behavior across these stakeholders (Yan et al., 2023; Rao et al., 2024). In this chapter, I propose a game-theoretic framework to explore these interactions, with a particular focus on the transformative role of digitalization. I also test this framework through a numerical experiment following guidance from Anand et al. (2019) and Knox (2020).

Digital tools such as artificial intelligence (AI), blockchain, and consumer-facing applications are influencing the dynamics of green investment and adoption. By reducing costs, improving information dissemination, and influencing preferences, digitalization plays a key role in accelerating the transition toward sustainable economies (Hao et al., 2023;Ning et al., 2023). Though the significance of digitalization in reshaping sustainable practices is well-documented, the interplay between digitalization and traditional policy levers such as subsidies and penalties remains underexplored. This chapter aims to fill this gap by investigating how digitalization alters the cost structures and strategic behaviors of governments, firms, and consumers, ultimately influencing the adoption of green technologies. Specifically, it seeks to answer the following key questions:

1. How does digitalization influence the costs, incentives, and decision-making processes of key stakeholders in green investments?
2. What equilibrium strategies emerge when governments balance subsidies and penalties, firms optimize their investments, and consumers make adoption decisions influenced by digital tools?
3. How can policymakers and businesses utilize digital tools to maximize the effectiveness of sustainability policies and practices?

By addressing these questions, the chapter aims to bridge the gap between theoretical economic modeling and real-world policy implementation in the context of sustainability and digitalization. It does this by first constructing an analytical mathematical model to identify equilibrium strategies and capture the strategic interactions between governments, firms, and consumers in the context of green investments. This model is then tested through a numerical experiment.

This study builds on extant literature that highlights the effectiveness of subsidies in reducing financial barriers to green investments (Mohammadi et al., 2024; Liu et al., 2020) and the role of penalties in ensuring compliance with environmental standards (Wang et al., 2015). However, prior research often treats digitalization as a secondary variable and overlooks its potential to magnify the effects of subsidies and penalties. Also, existing models largely focus on isolated stakeholder behaviors and fail to capture the interdependencies within the green investment ecosystem.

To explore these dynamics, this chapter employs a game-theoretic framework, modeling the strategic interactions among governments, firms, and consumers. Governments optimize subsidies and penalties to maximize environmental and fiscal payoffs, firms decide on investment levels based on profitability and compliance incentives, and consumers weigh product costs against environmental benefits and digital awareness. The model incorporates digitalization as a key variable that influences costs, awareness, and policy efficiency. The theoretical model is complemented by a numerical experiment that simulates stakeholder interactions across various subsidy and digitalization scenarios. The simulation provides empirical insights into optimal strategies and highlights the synergistic effects of subsidies and digitalization. Key outcomes, including consumer adoption rates, firm investment levels, and government payoffs, are analyzed to identify patterns and policy implications.

The findings reveal several important insights. First, subsidies alone are effective in reducing cost barriers, but their impact is significantly amplified by digitalization, which increases consumer awareness and firm efficiency. Second, firm investments respond positively to subsidies and digitalization, though diminishing returns emerge at higher subsidy levels, which indicates the need for balanced policy designs. Finally, digitalization increases the cost-effectiveness of government interventions, improving subsidy targeting and compliance monitoring while maximizing environmental benefits. These results highlight the key role of digitalization as a multiplier of traditional policy instruments.

This study makes several significant contributions to the literature on sustainability and digital transformation. It extends traditional game-theoretic models by incorporating digitalization as an endogenous factor and provides a more comprehensive framework for analyzing stakeholder interactions. The numerical experiment provides actionable insights for designing policies that balance subsidies with digital incentives, maximizing environmental and economic outcomes. Moreover, the findings highlight the need for interdisciplinary approaches that combine economic modeling with digital tools, to address complex sustainability challenges

The chapter is organized as follows: Section 2 presents the theoretical framework, detailing the roles, strategy spaces, and payoff functions of the stakeholders. Section 3 analyzes the equilibrium outcomes and comparative statics that emphasize the impact of digitalization on stakeholder strategies. Section 4 describes the design and results of the numerical experiment. Section 5 discusses the limitations of the study and suggests directions for future research. Finally, Section 6 concludes by summarizing the findings and their contributions to the field.

THEORY AND MODEL

- **Players**

The strategic interactions in the adoption of green technologies involve three primary stakeholders - government, firms, and consumers. Each player operates with distinct objectives, constraints, and strategies, yet their decisions are deeply interconnected. Understanding their roles and interactions is important for developing a robust analytical framework to examine the dynamics of green investments in a digitalized economy.

The government (G) acts as a central regulator and facilitator by aiming to steer the economy toward sustainability while balancing fiscal and social responsibilities. Governments play a dual role by offering subsidies (s) to incentivize green technology adoption and imposing penalties (p) to deter non-compliance with environmental standards (Wang et al., 2015; Liu et al., 2020; Mohammadi et al., 2024). Subsidies reduce the financial burden on firms and consumers by encouraging investment and adoption of environmentally friendly products and practices. On the other hand, penalties serve as a deterrent mechanism by having firms adhere to green standards. The government's payoff is influenced by the costs of subsidies, revenue from penalties, and the broader environmental and social benefits derived from increased adoption of green technologies. Governments must also consider the role of digitalization in enhancing policy efficiency, as tools like AI can improve the targeting and administration of subsidies while monitoring compliance (Valle-Cruz et al., 2022).

Firms (F) are involved in the development and deployment of green technologies. Their primary decision revolves around the investment level (I) in such technologies, which is aimed at maximizing profits while minimizing costs and penalties. The cost structure of investment is influenced by multiple factors, including the availability of subsidies, regulatory penalties for non-compliance, and the operational efficiencies enabled by digitalization (Xiumei et al., 2023; Chen & Chen, 2024; Wu et al., 2024). Firms benefit from adopting digital tools, such as AI and IoT, which reduce investment costs by optimizing processes and improving resource efficiency. However, these investments are not without risks, as firms can be affected by uncertainties in market demand, policy changes, and consumer preferences. Ultimately, a firm's strategy involves balancing the upfront costs of green investments with the long-term benefits of increased consumer demand for sustainable products and the avoidance of regulatory penalties (Agrawal et al., 2024).

Consumers (C) are the final decision-makers in the green investment ecosystem. They decide whether to adopt green products based on a combination of factors, including the cost of the product (C_g), their digital awareness (D), and the perceived environmental benefits (B_e) of the product. The cost of green products is influenced by the subsidies provided by the government, which can make environmentally friendly choices more accessible (Li et al., 2020; Yi et al., 2022). Digital awareness plays an important role in shaping consumer behavior, as digital platforms can disseminate information about the environmental impact of products and nudge consumers toward sustainable choices (Alamsyah et al., 2020; Armutcu et al., 2024; Nekmahmud et al., 2022). The perceived environmental benefits further influence adoption decisions, as consumers weigh the tangible and intangible benefits of aligning their purchases with their values (Gomes et al., 2023). Consumers' adoption behavior not only determines the success of green products in the market but also feeds back into the decisions of firms and governments, which creates a dynamic interaction between the stakeholders.

In this ecosystem, the actions of each player are interdependent and thus form a strategic interaction that is influenced by digitalization. Governments set the regulatory and economic environment, firms respond by making investment decisions within that framework, and consumers ultimately determine the success of green technologies through their adoption choices. These interactions also highlight the importance of integrating digitalization into the analysis, as it alters the incentives and decision-making processes of all stakeholders.

- **Strategy Space**

The decision-making framework for each stakeholder in the green investment ecosystem is defined by their respective strategy spaces. These strategies determine how each player, government, firms, and consumers, try to achieve their objectives while responding to the actions of others. The strategy spaces are inherently dynamic and shaped by economic incentives, regulatory frameworks, and the role of digitalization in influencing decision-making processes.

For the government, the strategy space involves selecting the optimal combination of subsidies (s) and penalties (p) to promote the adoption of green technologies. Subsidies ($s \geq 0$) are financial incentives provided to reduce the cost burden on firms and consumers, thereby encouraging investment and consumption of environmentally friendly products. Conversely, penalties ($p \geq 0$) are mechanisms to deter non-compliance with environmental regulations by imposing economic costs on firms that fail to meet established green standards. The government's strategy is constrained by fiscal limits and societal goals, which requires a balance between incentivizing sustainable practices and maintaining economic stability. Moreover, the integration of digital tools allows for more targeted and efficient implementation of these strategies, such as real-time monitoring of compliance and dynamic adjustment of subsidies or penalties based on observed outcomes.

Firms operate within a strategy space defined by their investment levels in green technologies ($I \in [0,\infty)$). Firms must decide how much capital to allocate toward green innovations. This involves balancing the upfront costs of investment against the potential long-term benefits. The level of investment is influenced by the availability of government subsidies, the threat of penalties, and the market demand for sustainable products. Moreover, digitalization plays an important role in shaping firm strategies by reducing operational costs through AI-driven process optimization, improved transparency via blockchain, and providing data-driven insights for strategic decision-making (Awan et al., 2021; Desai, 2023a, 2023b; Kulkov et al., 2024). The strategy space for firms is also dynamic, as they must adapt to evolving consumer preferences, technological advancements, and regulatory changes. Ultimately, firms aim to maximize their profitability by aligning their investment decisions with the incentives and penalties set by the government and the purchasing behavior of consumers.

For consumers, the strategy space is binary, as they decide whether to adopt ($a=1$) or not adopt ($a=0$) green products. This decision is influenced by several factors, including the cost of green products (C_g), the perceived environmental benefits (B_e), and their digital awareness (D). Subsidies provided by the government can lower the effective cost of green products and make them more accessible to a broader segment of consumers. Digital platforms further influence the strategy space by enhancing consumer awareness through targeted campaigns, educational content, and real-time feedback on the environmental impact of their purchasing decisions. Consumers also weigh the perceived benefits of adopting green products, which may include intangible factors such as personal satisfaction from contributing to sustainability or tangible benefits like reduced energy costs (Papista & Dimitriadis, 2019; Zhao & Chen, 2021). Thus, their strategy is not only a reflection of individual preferences but also a response to the broader ecosystem of subsidies, penalties, and firm offerings.

Together, the strategy spaces of these stakeholders create a complex and interdependent decision-making environment. Governments design policies to incentivize and regulate, firms respond by investing in green technologies, and consumers make adoption choices that shape market demand. The interplay of these strategies forms the foundation for the game-theoretical analysis, which allows me to explore equilibrium outcomes and their implications for sustainability in the context of digitalization.

- **Payoff Functions**

The strategic interactions among governments, firms, and consumers in the green investment ecosystem are driven by their respective payoff functions. Each player aims to maximize their utility while responding to the strategies of others. The payoff functions incorporate economic incentives, costs, and benefits, and highlight the complex trade-offs that underpin decisions in this dynamic environment. As explained earlier, digitalization also plays a role in shaping these payoffs by altering cost structures, improving efficiency, and influencing behavioral outcomes.

The government's payoff function (U_G) highlights its role as both a regulator and a facilitator of green investments. The government incurs costs associated with subsidies (s) provided to incentivize the adoption of green technologies. This cost is proportional to the quantity of green products adopted (Q_a), represented as $-s \bullet Q_a$. On the other hand, the government generates revenue from penalties (P_t) imposed on firms that fail to comply with environmental regulations. Moreover, the government benefits from a reduction in environmental damage costs (E_c) as green technologies are adopted and carbon footprints are minimized. The payoff function is thus expressed as follows:

$$U_G = -s \cdot Q_a + P_t - E_c$$

This formulation captures the trade-offs the government faces in balancing fiscal expenditures on subsidies with the benefits of reduced environmental damage and penalty revenues. The integration of digital tools further enhances the government's ability to monitor compliance and efficiently allocate subsidies.

The firm's payoff function (U_F) indicates the profit-driven nature of businesses investing in green technologies. Firms generate revenue (R) from the sale of green products, which is influenced by consumer demand and market conditions. However, they incur costs associated with green technology investments ($C(I,D)$), which depend on the investment level (I) and are reduced through digital adoption (D). Digital tools, such as AI and blockchain, help firms optimize their operations, improve supply chain transparency, and lower the cost of compliance with environmental standards. Firms also face penalties (P_f) for failing to meet regulatory requirements. The payoff function is given as follows:

$$U_F = R - C(I,D) - P_f$$

This function captures the interplay between revenue generation, investment costs, and regulatory compliance. Firms aim to maximize their profits by aligning their strategies with government incentives and consumer preferences, while utilizing digital tools to reduce costs and enhance competitiveness.

The consumer's payoff function (U_C) highlights the utility derived from adopting green products. Consumers benefit from the environmental impact of their purchases, represented by the perceived environmental benefit (B_e). However, they also incur the cost of purchasing green products (C_g), which is influenced by government subsidies that lower prices. Digital awareness (D) plays an important role in shaping consumer preferences and adoption behavior by providing information about the environmental impact of their choices and creating nudges toward sustainable consumption. The consumer's payoff function is expressed as follows:

$$U_C = B_e - C_g + D$$

This function highlights the trade-offs consumers face between the costs of green products and the perceived benefits of contributing to environmental sustainability. The increasing availability of digital platforms and tools amplifies consumer awareness and increases their utility.

Together, these payoff functions provide a comprehensive framework for analyzing strategic interactions among governments, firms, and consumers.

Dynamics and Digital Influence

The dynamics of green investment adoption are profoundly influenced by the increasing penetration of digitalization (D) across economic and societal processes. Digitalization not only alters the cost structures and efficiencies of firms but also reshapes consumer awareness and behavior (Rachinger et al., 2019). This creates a transformative effect on the decision-making of all stakeholders. By integrating advanced digital tools such as AI, blockchain, and data-driven platforms, digitalization introduces new opportunities to accelerate the transition toward sustainable practices.

For firms, digitalization significantly reduces the costs associated with green technology investments ($C(I,D)$). Technologies such as AI enable firms to optimize operations, enhance resource efficiency, and streamline supply chains, thereby lowering the financial barriers to adopting green innovations. For example, AI-powered analytics can identify energy-saving opportunities in production processes, while IoT devices enable real-time monitoring of resource consumption and waste generation (Alijoyo, 2024; Olatunde et al., 2024; Tyagi et al., 2024). Blockchain technology further enhances transparency and trust in green investments by providing immutable records of environmental compliance and product life cycles (Desai, 2023a, 2023b;

Rani et al., 2024). Collectively, these digital tools decrease the marginal costs of investment and make green technologies more accessible and profitable for firms. This reduction in costs not only encourages higher investment levels but also enhances firms' ability to compete in increasingly sustainability-focused markets.

Digitalization also exerts a powerful influence on consumer behavior by increasing awareness and the likelihood of adopting green products. Digital platforms provide consumers with easy access to information about the environmental impacts of their purchasing decisions, empowering them to make more informed choices (Parth et al., 2021; Mariani & Borghi, 2023). Tools such as carbon footprint calculators, mobile applications promoting sustainable practices, and personalized recommendations based on consumption patterns nudge consumers toward greener alternatives (Berger et al., 2023; Olatomiwa et al., 2023; von Zahn et al., 2024; Hoffmann et al., 2024). Moreover, social media and digital campaigns amplify the visibility of sustainability issues and create a cultural shift where environmentally conscious behavior is increasingly valued and rewarded (Dwivedi et al., 2021). As a result, consumers become more likely to adopt green products, especially when digital tools increase their understanding of the environmental benefits (B_e) and reduce the perceived cost (C_g) through subsidies or other incentives.

Thus, the interplay between digitalization and stakeholder dynamics creates a feedback loop that accelerates the adoption of green investments. Firms that adopt digital tools not only lower their investment costs but also increase the appeal of their products to digitally aware consumers. In turn, as consumer demand for green products rises, governments are incentivized to further invest in digital infrastructure and policies that support sustainable practices. This cyclical relationship explains the role of digitalization as a catalyst for aligning economic and environmental objectives across the green investment ecosystem. In this context, digitalization emerges not merely as a supportive tool but as a driving force that reshapes the strategic interactions among governments, firms, and consumers.

EQUILIBRIUM ANALYSIS

Nash Equilibrium

The Nash Equilibrium in the context of green investments represents a stable set of strategies for the government, firms, and consumers, where no player can unilaterally alter their strategy to achieve a higher payoff. This equilibrium indicates the strategic interdependence among the stakeholders, as each player's decisions directly impact the outcomes and payoffs of the others. In this analysis, the Nash Equilibrium is represented as (s^*,I^*,a^*), where s^* and p^* denote the government's optimal subsidy and penalty levels, I^* represents the firm's optimal investment in green technologies, and a^* indicates the consumer's adoption decision.

While the government chooses both subsidies (s) and penalties (p), the equilibrium representation (s^*,I^*,a^*) prioritizes the variables that are most directly aligned with decision-making and outcomes. Subsidies (s^*) are typically a more direct incentive for encouraging investment and adoption and including them provides a clearer representation of how the government actively supports green technologies. Penalties (p^*), while equally important, are secondary or complementary to subsidies in this context. The equilibrium representation emphasizes positive incentives (subsidies) rather than deterrents (penalties). This framing aligns with the narrative that governments primarily encourage sustainable practices through financial support rather than enforcement measures. Penalties (p) only come into play if there is non-compliance. If the equilibrium assumes that firms and consumers respond adequately to subsidies and digitalization, which leads to widespread adoption of green technologies, penalties may be redundant. In such cases, the optimal p^* would not directly impact the equilibrium outcomes and could be excluded from the representation.[1]

For the government, achieving s^* and p^* involves determining the optimal balance between subsidies and penalties that maximizes its payoff function. At equilibrium, the government's strategy ensures that the combined effect of subsidies and penalties maximizes the environmental and economic benefits while minimizing fiscal expenditures. This balance depends on the responsiveness of firms and consumers to the incentives provided, as well as the influence of digitalization in improving policy efficiency.

For firms, the Nash Equilibrium investment level I^* indicates the point at which the marginal benefits of investing in green technologies equal the marginal costs. This would take into account subsidies, penalties, and the cost-reduction effects of digitalization. Firms aim to maximize their payoffs by aligning their strategies with government incentives and consumer demand. At equilibrium, the investment level I^* is such that any deviation (either an increase or a decrease in investment) would lead to a lower payoff for the firm. This optimal investment decision is influenced by digital tools, such as AI and IoT, which lower operational costs and improve the return on green investments.

For consumers, the adoption decision a^* is determined by their utility function, which balances the perceived environmental benefits (B_e), the cost of green products (C_g), and their digital awareness (D). At equilibrium, consumers adopt green products ($a^*=1$) if the benefits outweigh the costs, including the influence of subsidies that reduce product prices and digital platforms that increase awareness. Conversely, if the costs outweigh the benefits, consumers choose not to adopt ($a^*=0$). The equilibrium adoption decision depends on the interplay of government subsidies, firm strategies, and digital tools that shape consumer preferences and perceptions.

Equilibrium Derivation

The conditions for equilibrium are derived by analyzing the decision rules for each player and solving for their optimal responses. For consumers, the adoption decision (a^*) depends on their payoff function, which balances the perceived environmental benefits (B_e), the cost of green products (C_g), and the influence of digital awareness (D). The consumer decision rule is expressed as follows:

$$a^* = \begin{cases} 1, \, if \; B_e - C_g + D > 0 \\ 0, \, otherwise \end{cases}$$

This rule implies that consumers will adopt green products ($a^*=1$) if the utility derived from environmental benefits and digital awareness exceeds the cost. If this condition is not met, consumers will choose not to adopt ($a^*=0$). The adoption decision is sensitive to government subsidies that lower C_g, as well as digital platforms that improve D by increasing awareness and nudging behavior. At equilibrium, the consumer decision indicates the combined influence of government policies, firm strategies, and digitalization.

For firms, the equilibrium investment level (I^*) is determined by optimizing their payoff function. Firms invest in green technologies to maximize profits, balancing revenues (R) against investment costs ($C(I,D)$) and penalties for non-compliance (P_f). The optimal investment condition is derived from the first-order condition of the firm's payoff function:

$$\frac{\partial U_F}{\partial I} = \frac{\partial R}{\partial I} - \frac{\partial C(I,D)}{\partial I} - \frac{\partial P_f}{\partial I} = 0$$

This equation states that at equilibrium, the marginal benefit of investment (represented by $\frac{\partial R}{\partial I}$) equals the marginal costs, which include the direct investment cost ($\frac{\partial C(I,D)}{\partial I}$) and the penalty gradient ($\frac{\partial P_f}{\partial I}$). Digitalization ($D$) reduces the marginal investment cost by improving firms' profitability and incentivizing higher levels of green technology adoption. The equilibrium investment strategy depends on government subsidies that offset costs, the threat of penalties, and consumer demand influenced by digital awareness.

To maximize U_F, the firm needs to find the best investment level (I) that makes U_F as large as possible, i.e., the point where the firm's payoff (profit) is maximized with respect to its investment decision. To maximize U_F, we need to look for a point where making a small change in I no longer increases U_F. This happens at the point where the slope of U_F with respect to I is zero. At this point, $\frac{\partial U_F}{\partial I} = 0$. This equation is the first-order condition for maximization since it tells us that the firm's payoff is at its peak, i.e., the marginal benefit of increasing investment equals the marginal cost, and the firm cannot increase its payoff by changing I.

For the government, the optimal levels of subsidies (s^*) and penalties (p^*) are derived by maximizing its payoff function. The government aims to balance the fiscal cost of subsidies, the revenue from penalties, and the environmental benefits of green technology adoption. The optimization condition is given by:

$$\frac{\partial U_G}{\partial s} = -\frac{\partial Q_a}{\partial s} + \frac{\partial P_t}{\partial s} - \frac{\partial E_c}{\partial s} = 0$$

This condition implies that at equilibrium, the marginal cost of providing subsidies ($-\frac{\partial Q_a}{\partial s}$) is equal to the combined marginal benefits, which include the penalty revenue ($\frac{\partial P_t}{\partial s}$) and the reduction in environmental damage costs ($-\frac{\partial E_c}{\partial s}$). The optimal subsidy level s^* indicates a balance between encouraging green technology adoption and minimizing fiscal and environmental costs. Digital tools further increase the government's ability to target subsidies efficiently, monitor compliance, and evaluate the impact of policies in real time.

The government wants to choose s to maximize U_G, i.e., it needs to find the point (the optimal level of s) where any small change in s does not increase U_G. The first-order condition helps identify the subsidy level (s^*) at which the government's payoff is maximized. This happens at the point where the slope of the payoff function with respect to s is zero ($\frac{\partial U_G}{\partial s} = 0$). To reiterate, the subsidy (s) is the government's decision variable, i.e., it directly controls the level of financial support offered to encourage adoption of green technology.

To sum it up, the equilibrium conditions derived for each player illustrate the strategic interdependence in the green investment ecosystem. Consumers' adoption decisions influence firms' investment strategies, which in turn are shaped by government policies. Conversely, the government's optimization relies on the responses of firms and consumers to subsidies, penalties, and digital interventions. These interdependencies create a dynamic and iterative process, where equilibrium is achieved when no player can unilaterally improve their payoff.

Comparative Statics Analysis

The comparative statics analysis explores how changes in digitalization (D) influence the strategic decisions of firms, consumers, and governments in the green investment ecosystem. As explained earlier, digitalization introduces technological advancements, such as AI, IoT, and blockchain, which reduce costs, enhance information flow, and improve operational efficiencies. This analysis investigates the implications of increasing digital adoption on firm investments (I), consumer adoption (Q_a), and government policy effectiveness.

Digitalization directly affects the cost structure of firms by reducing the marginal cost of investments in green technologies. This effect is captured in the firm's cost function, $C(I,D)$, where higher D leads to a lower $\frac{\partial C(I,D)}{\partial I}$, the marginal cost of investment. As discussed earlier, technologies such as AI optimize resource allocation, streamline production processes, and improve energy efficiency, thus making green investments more economically attractive. At equilibrium, firms choose their investment level (I^*) based on the condition:

$$\frac{\partial U_F}{\partial I} = \frac{\partial R}{\partial I} - \frac{\partial C(I,D)}{\partial I} - \frac{\partial P_f}{\partial I} = 0$$

As D increases, $\frac{\partial C(I,D)}{\partial I}$ decreases, which allows firms to invest more in green technologies without sacrificing profitability. This leads to a higher I^*, which suggests that digitalization encourages firms to allocate more resources toward sustainable practices.

As discussed earlier, digitalization increases consumer awareness and influences adoption behavior by improving access to information and providing tools that emphasize the benefits of green products. Digital platforms also disseminate knowledge about the environmental impact of consumer choices, while applications such as carbon footprint calculators and gamified sustainability apps create incentives for green behavior. These tools increase D, the digital awareness parameter in the consumer utility function: $U_C= B_e - C_g+D$. As D rises, the net utility of adopting green products increases and leads to higher adoption rates (Q_a). Additionally, digital platforms can reduce C_g by increasing price transparency and enabling consumers to access subsidies more efficiently. This dual effect, i.e., raising D and lowering C_g, creates a positive feedback loop that drives consumer adoption. As a result, Q_a increases, thus shaping consumer behavior and expanding the market for green products.

Digitalization also increases the effectiveness of government policies by improving targeting, monitoring, and enforcement mechanisms. For instance, blockchain technology can track the life cycle of green investments and create an audit trail ensuring that subsidies are utilized appropriately, and compliance is accurately recorded. AI-powered analytics enable governments to design more efficient subsidy programs by identifying areas with the highest potential for green adoption. In the government's payoff function, $U_G= -s \cdot Q_a+P_t - E_C$, digitalization affects all three components. By increasing consumer adoption (Q_a) and firm investments (I), digital tools increase the environmental benefits ($-E_C$) while reducing the need for penalties (P_t). Moreover, digital platforms can streamline subsidy distribution so that administrative costs are minimized and subsidies (s) achieve maximum impact. As a result, higher D allows governments to achieve sustainability goals with greater efficiency and reduced fiscal burden.

To sum it up, these dynamics illustrate how increasing D creates synergies among stakeholders and accelerates the transition to a sustainable and digitally enabled economy. For firms, digitalization lowers costs and encourages higher investment levels (I^*). For consumers, it increases awareness and reduces barriers to adoption, which drives up Q_a. For governments, digital tools increase the effectiveness and efficiency of policy interventions and thus enable more targeted and impactful programs.

NUMERICAL EXPERIMENT

Design

The numerical experiment is designed to simulate the strategic interactions among governments, firms, and consumers in the green investment ecosystem. It incorporates the effects of digitalization and government subsidies while examining their influence on consumer adoption, firm investment, and government payoffs. The experiment aims to capture the dynamic interplay between these stakeholders and provides insights into optimal strategies and policy outcomes.

Consumers are modeled as rational decision-makers who choose to adopt green products if the perceived environmental benefits and digital awareness outweigh the costs. Subsidies reduce product costs, while digitalization increases consumer awareness and adoption likelihood. Firms are simulated as profit-maximizing entities that invest in green technologies based on subsidies, penalties, and reduced costs enabled by digitalization. Government acts as a regulator and facilitator that optimizes subsidies and penalties to balance fiscal costs, environmental benefits, and compliance incentives. Digitalization (D) influences consumer awareness and firm cost structures by increasing adoption and investment. Subsidies (s) reduce the financial burden on consumers and firms by encouraging adoption and green investments.

Consumer payoff considers environmental benefits, costs, and digital awareness. Firm payoff balances revenues, investment costs, and penalties. Government payoff evaluates the trade-off between fiscal costs (subsidies), penalty revenue, and environmental damage reduction. The experiment spans a range of subsidy levels (s=0 to s=10) and digitalization impacts (D=0 to D=10) to analyze their combined effects.

Simulation Framework

The simulation framework operationalizes the experimental design and incorporates stakeholder decision rules and payoff functions. Subsidy Levels (s) are defined as a range from 0 to 10 in 100 evenly spaced values to capture varying degrees of government support. Digitalization Levels (D) are represented as integers from 0 to 10 to reflect incremental improvements in technology adoption. Following from the discussion in Section 3, Consumer Adoption (Decision) is modeled as follows: $Q_a = max(0, B_e - C_g + D)$, where B_e is the perceived environmental benefit, C_g is the cost of green products (reduced by subsidies), and D is digital awareness. Firm Investment is derived from the following firm payoff function: $I = max\left(0, \frac{\partial R}{\partial I} - \frac{\partial C(I,D)}{\partial I} - \frac{\partial P_f}{\partial I}\right)$, where $\frac{\partial R}{\partial I}$ represents the marginal revenue gained from additional investment in green technologies, $\frac{\partial C(I,D)}{\partial I}$ denotes the marginal investment cost reduced by digitalization, and $\frac{\partial P_f}{\partial I}$ indicates the marginal penalty incurred for non-compliance with green standards. Government Payoff is determined as follows: $U_G = P_t + E_C - (s \bullet Q_a)$, where P_t is penalty revenue, E_C is environmental damage reduction, and $s \bullet Q_a$ represents subsidy costs.

The simulation iterates through all combinations of subsidy and digitalization levels. For each combination:

i. For each combination, consumer adoption (Q_a) is calculated based on environmental benefits, cost reductions from subsidies, and digital awareness.
ii. Investment decisions (I) are determined by balancing revenues, costs, and penalties, and incorporating digitalization effects.
iii. The government's payoff (U_G) is computed using penalty revenues, environmental benefits, and subsidy costs.

Each simulation result, including Q_a, I, and U_G, is stored in a structured array for analysis and visualization. Three key outcomes are plotted as functions of subsidy and digitalization levels. In Figure 1, Consumer Adoption (Q_a) highlights the combined effects of subsidies and digitalization on adoption rates. In Figure 2, Firm Investment (I) demonstrates how firms respond to cost reductions and increased demand. In Figure 3, Government Payoff (U_G) captures the efficiency of policy interventions under varying scenarios.

The Appendix provides the detailed Python code use to run the simulation for the numerical experiment.

RESULTS

The results from the numerical experiment, as summarized in Table 1 and Figures 1 to 3, provide key insights into the interplay of subsidies, digitalization, and stakeholder behavior in the green investment ecosystem.

From Figure 1, we can see that consumer adoption (Q_a) exhibits a positive relationship with both subsidy and digitalization levels. Subsidies reduce the effective cost of green products, thus lowering financial barriers and incentivizing adoption. Digitalization amplifies this effect by increasing consumer awareness of environmental benefits through increased information dissemination and personalized nudges. At higher levels of digitalization (e.g., D=10), adoption increases significantly, particularly at moderate subsidy levels, indicating the synergistic role of subsidies and digital tools in shaping consumer behavior. This tells us that policies that combine financial incentives (subsidies) with investments in digital platforms can accelerate the adoption of green products. Specifically, digitalization serves as a multiplier, particularly in mid-range subsidy scenarios, as it maximizes consumer participation in sustainability efforts.

Figure 2 demonstrates that firm investment (I) in green technologies increases steadily with subsidy levels and is further enhanced by digitalization. Higher subsidies offset investment costs, while digital tools reduce operational and technological expenses, which increases profitability. However, results from Table 1 suggest that the relationship plateaus at higher subsidy levels, indicating diminishing returns as firms reach optimal investment thresholds. For example, with D=10, firm investments are notably higher across all subsidy levels compared to scenarios with no digitalization (D=0). Thus, combining subsidies with digitalization incentives helps firms to achieve cost efficiency and align with sustainability objectives. Policymakers should consider the diminishing returns of subsidies beyond a certain threshold and focus on promoting digital adoption to maintain investment momentum.

Figure 3 reveals a declining trend in government payoff (U_G) as subsidy levels increase. This is expected, as fiscal costs associated with subsidies rise. However, digitalization mitigates these losses by improving policy efficiency by increasing both consumer adoption and firm investments, which lead to greater environmental benefits ($-E_C$) and penalty revenues (P_t). Moderate subsidies combined with high digitalization levels yield the most balanced outcomes for governments, as they achieve substantial adoption and investment without incurring excessive costs. Thus, digitalization increases the cost-effectiveness of government policies by improving subsidy targeting and compliance monitoring. Governments should focus on moderate subsidy levels supplemented by digital interventions to maximize environmental benefits while minimizing fiscal burdens.

Table 1 provides a subsample of the results from the simulation and the first 20 from a total of 121 simulation results are presented for parsimony. The results from Table 1 show a clear trend in consumer adoption (Q_a) as both subsidies and digitalization levels increase. For example, at a subsidy level of 0 and no digitalization (D=0), consumer adoption is 30 units. With the same subsidy but maximum digitalization (D=10), adoption rises to 80 units. This progression demonstrates the amplifying effect of digitalization on subsidies. Thus, consumer adoption increases more steeply when digitalization is present, particularly at moderate subsidy levels. For example, in untabulated results, moving from D=0 to D=10 at a subsidy of 1 lead to a 39-unit increase in adoption (from 31 to 70 units).

The results from Table 1 further indicate that firm investment (I) also benefits significantly from both subsidies and digitalization. For example, at s=0 and D=0, firm investment is 58.5 units. At s=0 and D=10, investment rises to 159.5 units. Similarly, with s=1 and D=10 (untabulated), firm investment increases further to 181.2 units. This pattern aligns with Figure 2, where digitalization consistently raises investment levels across all subsidy scenarios. Notably, the incremental effect of digitalization diminishes at higher subsidy levels, reflecting diminishing returns as firms reach optimal investment thresholds.

The results from Table 1 also indicate that Government payoff (U_G), as shown in Table 1, decreases with rising subsidy levels due to higher fiscal costs. For example, at s=0 and D=0, the payoff is −197.5. At s=0 and D=10, the payoff improves slightly to −152.5, due to digitalization's positive impact on adoption and investment. At s=1, the payoff declines further, highlighting the rising cost burden of subsidies even when digitalization offsets some losses. Figure 3 corroborates this trend as we can see that the government payoff decreases more steeply at higher subsidy levels but stabilizes somewhat with increasing digitalization. This shows that while subsidies are costly, digitalization can mitigate the fiscal burden by improving efficiency and policy effectiveness.

To sum it up, the numerical experiment highlights the interdependence of subsidies and digitalization. While subsidies alone encourage green adoption and investment, their effectiveness is significantly amplified by digitalization. Digital tools reduce firm costs, increase consumer awareness, and optimize policy implementation, which creates a feedback loop that benefits all stakeholders. A balanced approach that integrates financial incentives with digitalization strategies can maximize stakeholder participation and sustainability outcomes. Governments should prioritize investments in digital infrastructure to enhance the efficacy of subsidy programs and accelerate the transition to a green economy.

LIMITATIONS AND FUTURE RESEARCH AGENDA

While the proposed analytical model and numerical experiment provide valuable insights into the strategic interplay of subsidies, digitalization, and green investment dynamics, there are several limitations that must be acknowledged.

First, the model assumes rational decision-making by all stakeholders (government, firms, and consumers), which may not fully capture the complexity of real-world behaviors. For example, the adoption decision assumes that consumers are fully aware of subsidies, digital benefits, and environmental impacts. However, real-world adoption may be influenced by behavioral biases, lack of information, or socio-cultural factors that were not incorporated into the model. The investment strategy is modeled as profit-maximizing, yet firms may face external constraints such as capital availability, organizational inertia, or risk aversion that could alter investment levels. The payoff function for the government assumes efficient allocation of subsidies and penalties that ignore potential administrative inefficiencies or political constraints that may reduce the effectiveness of these policies. While these assumptions simplify behavior, they serve as a baseline for understanding interactions in the green investment ecosystem. The numerical experiment also incorporates sensitivity analyses and explores how variations in parameters like subsidies and digitalization levels influence outcomes. This approach provides robustness to the findings and highlights key trends despite simplified behavioral assumptions.

The model and numerical experiment are static, which represents a snapshot of stakeholder interactions at specific subsidy and digitalization levels. This approach overlooks dynamic adjustments and feedback loops over time. Green technology markets evolve with changes in consumer preferences, technological innovations, and regulatory frameworks, which are not captured in the static analysis. Governments often adjust subsidy and penalty schemes based on observed outcomes; an aspect not accounted for in the static equilibrium framework. Firms may shift investment strategies in response to competitive pressures or long-term market trends, which are beyond the scope of the current model. Static nature is a purposeful simplification made to achieve analytical clarity and computational feasibility. A dynamic model, while richer, would require additional assumptions and increase complexity without necessarily yielding more actionable insights for the scope of this chapter. Incorporating bounded rationality or behavioral biases to expand the model could be an avenue for future research.

While the model incorporates digitalization as a key parameter, its treatment is generalized and may not reflect the diverse ways in which digital tools influence stakeholders. For example, different industries and firm types may experience varying levels of cost reductions and efficiency gains from digitalization, which are not differentiated in the model. The model assumes uniform adoption of digital tools, whereas real-world adoption is often hindered by factors such as technological infrastructure, digital literacy, or access to financing. Digitalization is also represented as an integer scale (0–10), which simplifies its multi-dimensional nature (e.g., AI adoption, blockchain integration, IoT implementation) into a single metric. The simplification of digitalization into a single parameter ensures mathematical and computational tractability. A more granular treatment of digitalization would require significantly more data and assumptions, potentially complicating the model without proportionate benefits. Future research can incorporate temporal adjustments in iterations, such as modeling learning effects, policy feedback loops, and market evolution.

Certain critical parameters, such as the penalty rate and environmental benefit, are treated as exogenous, which limits the model's ability to explore their endogenous relationships with other variables. For example, the penalty rate is fixed and does not respond to compliance levels or firm behaviors, which may oversimplify its role in incentivizing green investments. Treating key parameters as exogenous is a necessary simplification to isolate the primary relationships of interest. Endogenizing these parameters would increase model complexity and reduce its interpretability for policy purposes.

The findings from the numerical experiment may not be directly generalizable to all contexts due to several factors. The model does not differentiate between developed and developing economies, where differences in digital infrastructure, regulatory capacity, and market maturity may alter the dynamics of green investments. The model assumes a homogenous green investment ecosystem, whereas different sectors (e.g., renewable energy, electric vehicles, sustainable agriculture) exhibit distinct cost structures and adoption patterns. Political and institutional constraints, such as lobbying by incumbent industries or limited fiscal capacity, are not captured in the model. However, my focus on a generalized framework is intentional to ensure broad applicability. Tailoring the model to specific contexts would reduce its universality and limit its relevance to a wider audience.

Finally, while the numerical experiment provides theoretical insights, the model lacks empirical validation using real-world data. This limits the ability to test the robustness of the findings and identify potential discrepancies between theoretical predictions and observed outcomes. Future research conducting empirical studies can calibrate the model and validate its predictions against real-world data to strengthen its credibility and practical utility.

CONCLUSION

This chapter explores the complex interplay of subsidies, digitalization, and stakeholder behavior in the green investment ecosystem. It provides a robust analytical and numerical experimental framework for understanding these dynamics. By utilizing a game-theoretic model and conducting a detailed numerical experiment, the study addresses important questions about how governments, firms, and consumers interact to drive sustainable outcomes in a digitalized economy.

The central questions guiding this research were: How does digitalization influence the costs, incentives, and decision-making processes in green investments? What equilibrium strategies emerge among governments, firms, and consumers under different subsidies and digitalization levels? How can policymakers optimize these dynamics to maximize environmental and economic benefits? This study bridges the gap between theoretical modeling and actionable policy recommendations by addressing these core issues.

My findings suggest that subsidies significantly lower financial barriers, while digitalization improves consumer awareness, which creates a synergistic effect that increases adoption rates. Firms respond to subsidies by increasing green investments, with digitalization amplifying this effect through cost reductions and operational efficiencies. However, diminishing returns emerge at higher subsidy levels. While subsidies incur fiscal costs, digitalization offsets these burdens by improving policy efficiency and maximizing environmental benefits. Moderate subsidies combined with high digitalization levels provide the most balanced outcomes.

Thus, this chapter provides important insights into how digital tools can transform green investment dynamics. The findings suggest that subsidies alone are insufficient. Policymakers must invest in digital infrastructure to amplify the impact of financial incentives. Firms can optimize investments by utilizing digitalization to reduce costs and align with evolving consumer preferences for sustainable products. Digital platforms increase transparency and awareness, which encourages environmentally responsible behavior and market participation.

While the study offers several important insights, it has certain limitations. The model assumes rational decision-making, which may not fully capture real-world complexities such as behavioral biases or socio-cultural influences. The equilibrium analysis provides a snapshot rather than a dynamic perspective, which overlooks temporal adjustments and feedback loops. Digitalization is treated as a uniform parameter, which simplifies its multi-dimensional impacts across different sectors and regions. Key factors such as penalty rates and environmental benefits are treated as fixed, which limits the exploration of their endogenous relationships. However, these limitations are acknowledged as necessary trade-offs for analytical clarity and computational feasibility, with opportunities for future research to address them.

During the preparation of this work the author used Anthropic's Claude AI in order to perform spelling and grammar checks and debug Python code. After using this tool/service, the author reviewed and edited the content as needed and takes full responsibility for the content of the publication.

Figure 1. Impact of subsidy and digitalization on consumer adoption

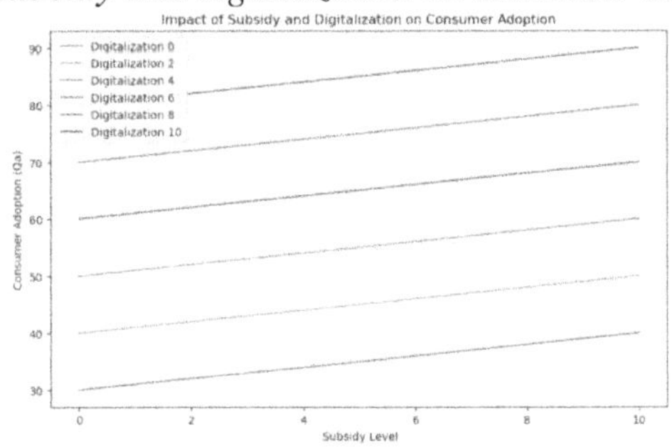

Source: Created by the author.

Figure 1 illustrates the combined impact of subsidy and digitalization levels on consumer adoption (Q_a) of green technologies. Each line represents a specific level of digitalization, ranging from 0 (no digitalization) to 10 (maximum digitalization). The x-axis indicates the subsidy level (s), while the y-axis shows the corresponding consumer adoption rates.

Figure 2. Impact of subsidy and digitalization on firm investment

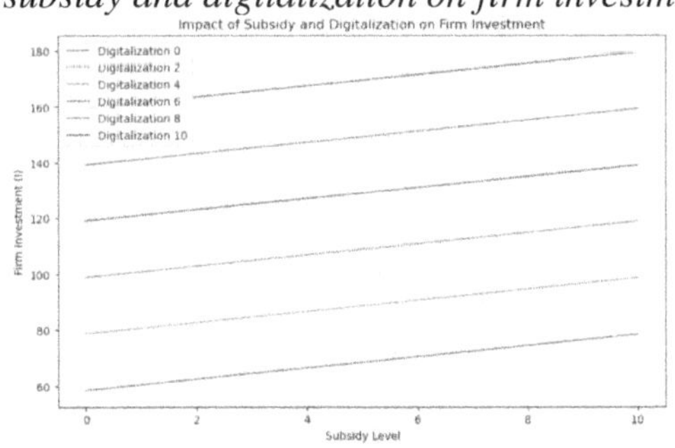

Source: Created by the author.

Figure 2 illustrates the combined impact of subsidy and digitalization levels on firm investment (I) in green technologies. Each line represents a specific digitalization level, ranging from 0 (no digitalization) to 10 (maximum digitalization).

The x-axis displays the subsidy level (s), while the y-axis shows the corresponding firm investment levels.

Figure 3. Impact of subsidy and digitalization on government payoff

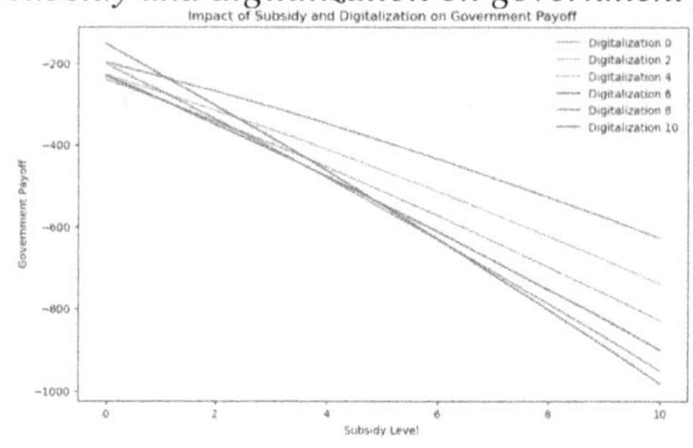

Source: Created by the author.

Figure 3 illustrates the combined impact of subsidy and digitalization levels on government payoff (U_G). Each line represents a specific digitalization level, ranging from 0 (no digitalization) to 10 (maximum digitalization). The x-axis indicates the subsidy level (s), while the y-axis shows the corresponding government payoff values.

Table 1. Sample numerical results from the simulation of green investment ecosystem

Simulation #	Subsidy	Digitalization	Firm Investment	Consumer Adoption	Government Payoff
1	0	0	58.5	30	-197.5
2	0	1	68.6	35	-215.5
3	0	2	78.7	40	-228.5
4	0	3	88.8	45	-236.5
5	0	4	98.9	50	-239.5
6	0	5	109	55	-237.5
7	0	6	119.1	60	-230.5
8	0	7	129.2	65	-218.5
9	0	8	139.3	70	-201.5
10	0	9	149.4	75	-179.5
11	1	10	159.5	80	-152.5
12	1	0	60.5	31	-232.4
13	1	1	70.6	36	-254.4
14	1	2	80.7	41	-271.4

continued on following page

Table 1. Continued

Simulation #	Subsidy	Digitalization	Firm Investment	Consumer Adoption	Government Payoff
15	1	3	90.8	46	-283.4
16	1	4	100.9	51	-290.4
17	1	5	111	56	-292.4
18	1	6	121.1	61	-289.4
19	1	7	131.2	66	-281.4
20	1	8	141.3	71	-268.4

Source: *Created by the author.*

Table 1 presents a sample of the numerical results from the simulation of the green investment ecosystem, and it illustrates the impact of subsidy and digitalization levels on firm investment, consumer adoption, and government payoff. Each row corresponds to a unique simulation, defined by a specific combination of subsidy and digitalization levels. This table includes only the first 20 records out of a total of 121 simulations for brevity and clarity.

REFERENCES

Agrawal, R., Agrawal, S., Samadhiya, A., Kumar, A., Luthra, S., & Jain, V. (2024). Adoption of green finance and green innovation for achieving circularity: An exploratory review and future directions. *Geoscience Frontiers, 15*(4), 101669. DOI: 10.1016/j.gsf.2023.101669

Alamsyah, D., Othman, N., & Mohammed, H. (2020). The awareness of environmentally friendly products: The impact of green advertising and green brand image. *Management Science Letters, 10*(9), 1961–1968. DOI: 10.5267/j.msl.2020.2.017

Alijoyo, F. A. (2024). AI-powered deep learning for sustainable industry 4.0 and internet of things: Enhancing energy management in smart buildings. *Alexandria Engineering Journal, 104*, 409–422. DOI: 10.1016/j.aej.2024.07.110

Anand, V. V., Balakrishnan, R., & Labro, E. (2019). A framework for conducting numerical experiments on cost system design. *Journal of Management Accounting Research, 31*(1), 41–61. DOI: 10.2308/jmar-52057

Armutcu, B., Zuferi, R., & Tan, A. (2024). Green product consumption behaviour, green economic growth and sustainable development: Unveiling the main determinants. *Journal of Enterprising Communities: People and Places in the Global Economy, 18*(4), 798–819. DOI: 10.1108/JEC-05-2023-0074

Awan, U., Shamim, S., Khan, Z., Zia, N. U., Shariq, S. M., & Khan, M. N. (2021). Big data analytics capability and decision-making: The role of data-driven insight on circular economy performance. *Technological Forecasting and Social Change, 168*, 120766. DOI: 10.1016/j.techfore.2021.120766

Berger, M., Lange, T., & Stahl, B. (2022). A digital push with real impact–Mapping effective digital nudging elements to contexts to promote environmentally sustainable behavior. *Journal of Cleaner Production, 380*, 134716. DOI: 10.1016/j.jclepro.2022.134716

Chen, J., & Chen, Z. (2024). How does environmental regulations affect digital green innovation of high-pollution enterprises? Empirical evidence from China. *Heliyon, 10*(13), e33725. Advance online publication. DOI: 10.1016/j.heliyon.2024.e33725 PMID: 39071563

Chen, Y., & Ma, Y. (2021). Does green investment improve energy firm performance? *Energy Policy, 153*, 112252. DOI: 10.1016/j.enpol.2021.112252

Desai, H. (2023a). Infusing Blockchain in accounting curricula and practice: Expectations, challenges, and strategies. *The International Journal of Digital Accounting Research, 23*, 97–135. DOI: 10.4192/1577-8517-v23_5 PMID: 20485478

Desai, H. (2023b). An Auditor's Perspective on Smart Contracts and DAOs. [Retrieved from Link.]. *The CPA Journal, 93*(7/8), 60–63.

Dwivedi, Y. K., Ismagilova, E., Hughes, D. L., Carlson, J., Filieri, R., Jacobson, J., Jain, V., Karjaluoto, H., Kefi, H., Krishen, A. S., Kumar, V., Rahman, M. M., Raman, R., Rauschnabel, P. A., Rowley, J., Salo, J., Tran, G. A., & Wang, Y. (2021). Setting the future of digital and social media marketing research: Perspectives and research propositions. *International Journal of Information Management, 59*, 102168. DOI: 10.1016/j.ijinfomgt.2020.102168

Gomes, S., Lopes, J. M., & Nogueira, S. (2023). Willingness to pay more for green products: A critical challenge for Gen Z. *Journal of Cleaner Production, 390*, 136092. DOI: 10.1016/j.jclepro.2023.136092

Hao, X., Li, Y., Ren, S., Wu, H., & Hao, Y. (2023). The role of digitalization on green economic growth: Does industrial structure optimization and green innovation matter? *Journal of Environmental Management, 325*, 116504. DOI: 10.1016/j.jenvman.2022.116504 PMID: 36272290

Hoffmann, S., Lasarov, W., Reimers, H., & Trabandt, M. (2024). Carbon footprint tracking apps. Does feedback help reduce carbon emissions? *Journal of Cleaner Production, 434*, 139981. DOI: 10.1016/j.jclepro.2023.139981

Knox, B. D. (2020). Semantic processing theory could help explain the effectiveness of creativity incentives: A numerical experiment. *Journal of Management Accounting Research, 32*(3), 155–178. DOI: 10.2308/jmar-18-051

Kulkov, I., Kulkova, J., Rohrbeck, R., Menvielle, L., Kaartemo, V., & Makkonen, H. (2024). Artificial intelligence-driven sustainable development: Examining organizational, technical, and processing approaches to achieving global goals. *Sustainable Development (Bradford), 32*(3), 2253–2267. DOI: 10.1002/sd.2773

Li, Y., Tong, Y., Ye, F., & Song, J. (2020). The choice of the government green subsidy scheme: Innovation subsidy vs. product subsidy. *International Journal of Production Research, 58*(16), 4932–4946. DOI: 10.1080/00207543.2020.1730466

Liu, J., Zhao, M., & Wang, Y. (2020). Impacts of government subsidies and environmental regulations on green process innovation: A nonlinear approach. *Technology in Society, 63*, 101417. DOI: 10.1016/j.techsoc.2020.101417

Liu, L., Wang, Z., Xu, J., & Zhang, Z. (2023). Green baton: How government interventions advance green technological innovation. *Environment, Development and Sustainability*, *25*(10), 11121–11152. DOI: 10.1007/s10668-022-02520-2

Mariani, M., & Borghi, M. (2023). Exploring environmental concerns on digital platforms through big data: The effect of online consumers' environmental discourse on online review ratings. *Journal of Sustainable Tourism*, *31*(11), 2592–2611. DOI: 10.1080/09669582.2022.2033982

Mohammadi, M., Sebastian Heese, H., & Kraft, T. (2024). Environmental Regulation Design: Motivating Firms' Clean Technology Investments With Penalties and Subsidies. *Production and Operations Management*, *33*(1), 32–47. DOI: 10.1177/10591478231224911

Nekmahmud, M., Naz, F., Ramkissoon, H., & Fekete-Farkas, M. (2022). Transforming consumers' intention to purchase green products: Role of social media. *Technological Forecasting and Social Change*, *185*, 122067. DOI: 10.1016/j.techfore.2022.122067

Ning, J., Jiang, X., & Luo, J. (2023). Relationship between enterprise digitalization and green innovation: A mediated moderation model. *Journal of Innovation & Knowledge*, *8*(1), 100326. DOI: 10.1016/j.jik.2023.100326

Olatomiwa, L., Ambafi, J. G., Dauda, U. S., Longe, O. M., Jack, K. E., Ayoade, I. A., Abubakar, I. N., & Sanusi, A. K. (2023). A Review of Internet of Things-Based Visualisation Platforms for Tracking Household Carbon Footprints. *Sustainability (Basel)*, *15*(20), 15016. DOI: 10.3390/su152015016

Olatunde, T. M., Okwandu, A. C., Akande, D. O., & Sikhakhane, Z. Q. (2024). Reviewing the role of artificial intelligence in energy efficiency optimization. *Engineering Science & Technology Journal*, *5*(4), 1243–1256. DOI: 10.51594/estj.v5i4.1015

Papista, E., & Dimitriadis, S. (2019). Consumer–green brand relationships: Revisiting benefits, relationship quality and outcomes. *Journal of Product and Brand Management*, *28*(2), 166–187. DOI: 10.1108/JPBM-09-2016-1316

Parth, S., Manoharan, B., Parthiban, R., Qureshi, I., Bhatt, B., & Rakshit, K. (2021). Digital technology-enabled transformative consumer responsibilisation: A case study. *European Journal of Marketing*, *55*(9), 2538–2565. DOI: 10.1108/EJM-02-2020-0139

Rachinger, M., Rauter, R., Müller, C., Vorraber, W., & Schirgi, E. (2019). Digitalization and its influence on business model innovation. *Journal of Manufacturing Technology Management*, *30*(8), 1143–1160. DOI: 10.1108/JMTM-01-2018-0020

Rani, P., Sharma, P., & Gupta, I. (2024). Toward a greener future: A survey on sustainable blockchain applications and impact. *Journal of Environmental Management, 354*, 120273. DOI: 10.1016/j.jenvman.2024.120273 PMID: 38350276

Rao, Z., Li, P., & Bai, C. (2024). Unraveling the impact of carbon markets on corporate green technology investment behavior: An evolutionary game approach from the lens of competitive manufacturers. *Journal of the Knowledge Economy, 15*(3), 11397–11429. DOI: 10.1007/s13132-023-01544-x

Tyagi, A. K., Bhatt, P., Chidambaram, N., & Kumari, S. (2024). Artificial Intelligence Empowered Smart Manufacturing for Modern Society: A Review. *Artificial Intelligence-Enabled Digital Twin for Smart Manufacturing,* 55-83. DOI: 10.1002/9781394303601.ch4

Valle-Cruz, D., Fernandez-Cortez, V., & Gil-Garcia, J. R. (2022). From E-budgeting to smart budgeting: Exploring the potential of artificial intelligence in government decision-making for resource allocation. *Government Information Quarterly, 39*(2), 101644. DOI: 10.1016/j.giq.2021.101644

von Zahn, M., Bauer, K., Mihale-Wilson, C., Jagow, J., Speicher, M., & Hinz, O. (2024). The smart green nudge: Reducing product returns through enriched digital footprints & causal machine learning. Marketing Science, ahead-of-print. DOI: 10.1287/mksc.2022.0393

Wang, S., Fan, J., Zhao, D., & Wu, Y. (2015). The impact of government subsidies or penalties for new-energy vehicles a static and evolutionary game model analysis. [JTEP]. *Journal of Transport Economics and Policy, 49*(1), 98–114.

Wu, S., Cheng, P., & Yang, F. (2024). Study on the impact of digital transformation on green competitive advantage: The role of green innovation and government regulation. *PLoS One, 19*(8), e0306603. DOI: 10.1371/journal.pone.0306603 PMID: 39088476

Xiumei, X., Ruolan, J., Shahzad, U., & Xiao, F. (2023). Sustainable innovation in small and medium-sized enterprises: Environmental regulations and digitalization as catalyst. *Journal of Environment & Development, 32*(4), 413–443. DOI: 10.1177/10704965231211585

Yan, H., Qamruzzaman, M., & Kor, S. (2023). Nexus between green investment, fiscal policy, environmental tax, energy price, natural resources, and clean energy—A step towards sustainable development by fostering clean energy inclusion. *Sustainability (Basel), 15*(18), 13591. DOI: 10.3390/su151813591

Yi, Y., Wang, Y., Fu, C., & Li, Y. (2022). Taxes or subsidies to promote investment in green technologies for a supply chain considering consumer preferences for green products. *Computers & Industrial Engineering*, *171*, 108371. DOI: 10.1016/j.cie.2022.108371

Zhao, S., & Chen, L. (2021). Exploring residents' purchase intention of green housings in China: An extended perspective of perceived value. *International Journal of Environmental Research and Public Health*, *18*(8), 4074. DOI: 10.3390/ijerph18084074 PMID: 33924311

KEY TERMS AND DEFINITIONS

Consumer Adoption: The decision-making process by which consumers opt to purchase and use green products based on cost, environmental benefits, and digital awareness.

Digitalization: The integration of digital technologies such as AI, blockchain, and IoT to improve operational efficiencies and drive sustainable practices.

Game-Theoretic Framework: A mathematical model that analyzes the strategic interactions and decision-making among players (here, governments, firms, and consumers) in an ecosystem.

Green Investment: The allocation of resources toward technologies and practices that minimize environmental impact and promote sustainability.

Subsidy: The financial support provided by governments to reduce cost burden on firms and consumers and encourage the adoption of green technologies.

ENDNOTE

1 Another relevant argument to support my framing of the equilibrium is that the penalty level p^* may be considered a fixed or externally determined parameter rather than a strategic variable optimized in tandem with s^*. If the model treats penalties as exogenous or assumes that they are set optimally outside the scope of the equilibrium, they would not appear in the equilibrium tuple.

APPENDIX

Figure 4. Python Code for Numerical Experiment Simulation

```python
# Importing necessary libraries
import numpy as np
import matplotlib.pyplot as plt

# Defining parameters for the government, firms, and consumers
# Government parameters
# The range of subsidies provided by the government, ranging from 0 to 10 as integers
subsidy_range = np.arange(0, 11, 1)
# Fixed penalty rate imposed by the government for non-compliance with green standards
penalty_rate = 5
# Environmental damage reduction as a function of consumer adoption (Qa), modeled quadratically
environmental_damage_reduction = lambda Qa: 0.1 * Qa**2

# Firm parameters
# Cost of investment, which decreases with higher levels of digitalization (D)
investment_cost_function = lambda I, D: 100 + 50 * I - 10 * D
# Revenue function dependent on firm investment (I) and consumer adoption (Qa)
revenue_function = lambda I, Qa: 200 * I * Qa
# Penalty for non-compliance, proportional to the shortfall in firm investment
penalty_function = lambda I: penalty_rate * (1 - I)

# Consumer parameters
# Perceived environmental benefit as a function of consumer adoption
environmental_benefit_function = lambda Qa: 50 * Qa
# Effective cost to consumers, reduced by government subsidies
cost_function = lambda Cg, subsidy: Cg - subsidy
# Consumer awareness of green products, increasing with digitalization (D)
digital_awareness = lambda D: 5 * D

# Simulation function
def simulate_game(subsidy, digitalization):
    """
    Simulates the interaction between the government, firms, and consumers, incorporating digitalization.

    Parameters:
    subsidy (float): The level of subsidy provided by the government.
    digitalization (int): The level of digitalization, ranging from 0 to 10.

    Returns:
    tuple: Firm investment (I), consumer adoption (Qa), and government payoff.
    """
    # Consumers decide on adoption based on perceived benefits, costs, and digital awareness
    Qa = max(0, environmental_benefit_function(1) - cost_function(20, subsidy) + digital_awareness(digitalization))

    # Firms decide their investment level to maximize payoff, accounting for costs and revenues
    I = max(0, (revenue_function(1, Qa) - investment_cost_function(1, digitalization)) / 100)

    # Government calculates its payoff based on adoption, penalties, and subsidy costs
    environmental_benefit = environmental_damage_reduction(Qa)
    penalty_revenue = penalty_function(I)
    subsidy_cost = subsidy * Qa
    government_payoff = penalty_revenue + environmental_benefit - subsidy_cost

    return I, Qa, government_payoff

# Running the simulation for various subsidy and digitalization levels
results = []
for subsidy in subsidy_range:
    for digitalization in range(0, 11):
        # Simulate the game for each combination of subsidy and digitalization levels
        I, Qa, government_payoff = simulate_game(subsidy, digitalization)
        results.append((subsidy, digitalization, I, Qa, government_payoff))

# Convert results to a numpy array for easier analysis
results = np.array(results)
```

Chapter 3
Digital Transformation and Sustainability Artificial Intelligence and Sustainable Development

Manju Shree Raman
https://orcid.org/0000-0002-2719-3549
Christ University, India

Vinayak Anil Bhat
https://orcid.org/0000-0001-6546-6037
CMS-B School, Jain University (Deemed), India

Krishna B. Koppa
https://orcid.org/0000-0002-8601-2429
CMS-B School, Jain University (Deemed), India

S. Guru Basava Aradhya
https://orcid.org/0000-0002-3335-2681
Chandigarh Group of Colleges, Jhanjeri, India

V. Chithambar Gupta
https://orcid.org/0009-0002-1364-5425
Presidency University, India

P. Selvakumar
https://orcid.org/0000-0002-3650-4548
Department of Science and Humanities, Nehru Institute of Technology, Coimbatore, India

ABSTRACT

Digital transformation and sustainability are increasingly intertwined in today's rapidly evolving landscape, where technological advancements and environmental concerns converge to shape the future of industries, communities, and societies. Operations, strategies and This transformation can enhance efficiency, improve customer experiences, and foster innovation. However, it also carries the responsibility of addressing the pressing sustainability social inequities. The relationship between digital transformation and sustainability is complex, yet synergistic. On

DOI: 10.4018/979-8-3693-9909-5.ch003

one hand, blockchain—provides powerful tools for enhancing sustainability efforts across various sectors. For example, can consumption be smart while improving efficiencies reducing waste? Additionally, big data analytics can help organizations assess their environmental impact, enabling informed decision-making that supports sustainable practices. This data-driven approach allows companies to track their carbon footprints, manage resources more effectively, and identify areas for improvement.

INTRODUCTION

Digital transformation and sustainability are increasingly intertwined in today's rapidly evolving landscape, where technological advancements and environmental concerns converge to shape the future of industries, communities, and societies. Operations, strategies and This transformation can enhance efficiency, improve customer experiences, and foster innovation. However, it also carries the responsibility of addressing the pressing sustainability social inequities. The relationship between digital transformation and sustainability is complex, yet synergistic. On one hand, blockchain—provides powerful tools for enhancing sustainability efforts across various sectors. For example, can consumption be smart while improving efficiencies reducing waste? Additionally, big data analytics can help organizations assess their environmental impact, enabling informed decision-making that supports sustainable practices. This data-driven approach allows companies to track their carbon footprints, manage resources more effectively, and identify areas for improvement, aligning their operations with sustainability goals. On the other hand, the pursuit of digital transformation must also consider its environmental implications. The digital infrastructure required to support transformation, including data centers and network systems, consumes substantial energy and resources. As organizations invest in digital technologies, they must adopt strategies that mitigate their sources for powering data centers, implement energy-efficient technologies, and reduce electronic By prioritizing sustainability within their digital initiatives, organizations can ensure that their transformation efforts contribute positively to environmental goals rather than exacerbating existing challenges. Furthermore, stakeholder engagement is critical in aligning digital transformation with sustainability objectives.

Companies are increasingly held accountable by consumers, investors, and regulatory bodies to demonstrate their commitment to sustainable practices. As a result, organizations must integrate sustainability into their digital strategies, not just as a compliance measure, but as a core component of their value proposition. Engaging stakeholders—ranging from employees and customers to suppliers and local communities—can foster collaboration and sustainability solutions. By actively involving diverse perspectives, organizations can identify and address sustainability challenges while leveraging digital tools to enhance Education capacity building for equipping organizations navigate intersection of transformation and sustainability. A workforce that is proficient in both digital skills and sustainability practices will be crucial. Training programs and educational initiatives can help individuals understand how to leverage digital technologies to drive sustainability outcomes while promoting responsibility. By fostering a mindset of both digital transformation and sustainability, organizations can cultivate leaders who are prepared to tackle the challenges of the future. Incorporating sustainability into the digital transformation process also requires a shift in organizational culture. Companies must foster an environment that encourages experimentation, collaboration, and a long-term perspective on value creation. This cultural shift can facilitate the adoption of innovative solutions that prioritize sustainability while harnessing the benefits of digital technologies. Leadership commitment is vital in championing this transformation, as it sets the tone for the organization's values and priorities. By embedding sustainability into the corporate ethos, organizations can drive meaningful change that resonates with employees, customers, and other stakeholders.

SUSTAINABLE DEVELOPMENT GOALS (SDGS) AND THE ROLE OF TECHNOLOGY

The established represents a global blueprint for achieving prosperous, sustainable Technology catalyzing innovation, efficiency, and collaboration across sectors. With digital tools and emerging technologies, stakeholders develop solutions that promote sustainable practices, enhance resilience, and drive inclusive growth. Technology to the SDGs facilitates collection, analysis, and dissemination. Reliable data policy implementation. Satellite imagery, big enables governments, NGOs, and researchers to monitor progress, assess challenges, and identify opportunities for improvement across various sectors. For instance, real-time data on agricultural productivity can help policymakers understand food security challenges and implement targeted interventions to support sustainable farming practices (Anwar A., et al, 2022). By providing timely insights, stakeholders' evidence-based aligns with the SDGs. Moreover, technology enhances access to education and information, which is fundamental to achieving several SDGs, particularly Goal 4 (Quality Education). E-learning platforms, mobile applications, and open educational resources are transforming traditional educational models, making learning more accessible to diverse populations. For example, remote and underserved communities can benefit from online courses and digital resources that provide quality education and skills training. Additionally, technology can support lifelong learning initiatives, helping individuals adapt to rapidly changing job markets and contributing to social inclusion. In context health, the potential to significantly impact Telemedicine, mobile health applications, and health information systems to improve rural and marginalized These technologies enable health monitoring, and data management, effectiveness During the COVID-19 pandemic, for instance, telehealth services surged, demonstrating how technology can facilitate healthcare access during crises while maintaining continuity of care.

Sustainable cities and communities, outlined in Goal 11, also benefit from technological innovations. Smart city technologies, enable urban planners' use, improve public transportation, and enhance infrastructure resilience. By integrating technology into urban management, cities can reduce their carbon footprints, and manage waste more effectively, living. For example, enhance efficiency monitoring integrating ultimately to climate action under Goal 13. The technology advancing as emphasized Goal cannot be overlooked. Digital technologies facilitate entrepreneurship, and innovation, and allow businesses to reach global customers, while digital payment systems simplify transactions, making it easier for entrepreneurs to thrive (Al Emran M., et al, 2023). Additionally, technology can support skills development competencies digital economy. By fostering an inclusive and dynamic economic environment, technology can drive job creation and reduce inequalities. Environmental sustainability, a critical component of the SDGs, is also greatly influenced by technological advancements. Innovations in wind, and bioenergy technologies, Goal by providing sustainable alternatives to fossil fuels. Furthermore, technologies for waste management, recycling, and support aligning with Goal routes improve recycling rates, reducing environmental impact and promoting circular economy principles. Despite the vast potential of technology to advance the SDGs, challenges remain. Digital divides persist, particularly in low-income and rural areas, limiting access to technological resources and hindering equitable progress. Ensuring that technology is inclusive and accessible to all is essential for realizing the SDGs. Stakeholders must prioritize investments in digital infrastructure, capacity building, and education to bridge these divides and empower marginalized communities. Additionally, ethical considerations surrounding technology deployment must be addressed. The environmental impact of tech production requires scrutiny. Ensuring that technological solutions are developed and implemented responsibly is crucial for fostering trust and accountability (Hajikhani A., et al, 2022).

ARTIFICIAL INTELLIGENCE (AI) AND MACHINE LEARNING (ML) FUNDAMENTALS

Historical roots trace back to, when theorists retained a foundational concept of AI, examining rule-based systems, relying on predetermined logic to perform tasks. However, as computational power increased and data became more abundant, researchers began to explore probabilistic models and learning-based approaches, paving the way for modern AI and ML. At the heart of ML are systems of three types: datasets, map inputs to outputs based on examples. Analyzing involves training to allow it to identify itself. Techniques like dimensionality reduction fall under this category (Rodríguez García J D., et al, 2020). Reinforcement learning differs by focusing on training agents to make decisions, akin to how humans learn from experience. Deep learning, a specialized area within ML, has gained prominence in handling dataset tasks. Learning models, particularly mimic comprising interconnected layers of data These models excel as demonstrated by AI systems like the success which enables faster computations and the handling of massive datasets. AI and ML have far-reaching applications across diverse sectors. In healthcare, personalize treatment plans, patient care operational efficiencies. In finance, trading the retail industry leverages AI for automation through chatbots. Moreover, in transportation, AI powers autonomous vehicles, enhancing safety and efficiency in logistics and urban mobility. Despite the numerous benefits, the integration job displacement warrants careful consideration (Rodriguez P., et al, 2024). The applications such as or. Ensuring fairness, and accountability, is essential. Furthermore, as AI systems increasingly influence daily life, questions regarding data privacy and consent become paramount, necessitating robust regulatory frameworks to protect individuals' rights. ML also calls for interdisciplinary collaboration. Researchers, policymakers, ethicists, best practices AI deployment. Educational initiatives aimed at equipping the workforce with AI and ML skills are vital to prepare individuals in these technologies play various fields. This includes continuous adaptation and upskilling. Looking ahead, innovation. Emerging trends such as explainable AI, which seeks to make AI decision-making processes more interpretable, and to learn from while preserving privacy, highlight the ongoing evolution of these fields. Enhancing efficiency, security, and interoperability across systems (Blasch E., et al, 2021).

AI for Climate Action: Mitigation, Adaptation, and Resilience

Artificial Intelligence (AI) is increasingly recognized as providing both efforts. As the world grapples with the urgent need for sustainable practices, AI technologies can optimize resource use, and promote resilience in vulnerable communities. Intersection AI and climate action are multifaceted, encompassing data analysis, predictive modeling, and automation sectors, energy, disaster climate action diverse Climate change generates complex datasets—ranging from satellite imagery and weather patterns to socio-economic data—that require advanced analytical tools for effective interpretation. These correlations, and anomalies readily. For example, AI be used to enhance climate models by integrating historical weather data with real-time observations, allowing scientists to predict climate phenomena more accurately potential and human systems (Leal Filho W., et al, 2022). Critical direct mitigation efforts. Smart grids powered by AI can dynamically manage electricity more efficiently. Predictive analytics can optimize energy storage and distribution, ensuring that renewable energy is utilized effectively while minimizing reliance on fossil fuels. Additionally, AI-driven energy management systems can monitor usage patterns in buildings, allowing for targeted energy-saving measures that reduce overall agriculture, sustainability, and resilience promoting precision farming techniques. Satellite imagery crop management. For instance, AI can predict weather conditions and recommend irrigation schedules tailored to specific crops, reducing water waste and enhancing yield improving agriculture, AI can challenge while reducing farming operations. In terms of adaptation, AI is invaluable for enhancing resilience in communities vulnerable to climate impacts. For instance, AI for extreme weather events, to floods, hurricanes, and heatwaves. These systems can communities, that risk and save lives. In urban planning, AI can analyze climate risks and inform the design of resilient infrastructure, ensuring that cities are better equipped to withstand the impacts of climate change. Disaster management is another area where AI can enhance resilience (Kaack L H., et al, 2022). Disaster, AI improves response strategies for natural disasters. Models specific disasters occurring in certain regions, helping governments and organizations allocate resources effectively. Additionally, AI can optimize logistics during disaster response, ensuring that aid reaches affected areas quickly and efficiently, ultimately saving lives and reducing recovery times.

Promise climate action, The implementation of AI solutions requires substantial investment in infrastructure, data collection, and capacity building. Moreover, the effectiveness of AI in addressing climate issues datasets. Ensuring that is accessible, representative, and ethically sourced is critical for developing robust AI models that provide reliable insights. Ethical considerations are paramount when deploying AI for climate action. The potential for biases in algorithms must be addressed to prevent disproportionately negative impacts on marginalized communities. Moreover, engaging locals can be considered, as collaboration. Additionally, the environmental footprint of AI technologies themselves cannot be overlooked. Often require that can negate some of the environmental benefits achieved through their application (Chen L., et al, 2023). To mitigate this, researchers and practitioners must prioritize energy-efficient algorithms and leverage renewable energy sources for powering data centers. In conclusion, AI holds immense potential for advancing climate action through both mitigation and adaptation strategies. Predictive modeling, automation, AI energy use, enhance agricultural practices, improve transportation systems, and bolster community resilience. Commitment practices, investment in data infrastructure, and collaboration across sectors. Urgent climate offers that by embracing innovative technologies, stakeholders can work collectively to address climate issues, ensuring not only livable but thriving. Through targeted investments and collaborative efforts, we can harness AI as a transformative force turning challenges into opportunities for sustainable development.

AI for Sustainable Resource Management: Water, Energy, and Food

Artificial Intelligence (AI) is revolutionizing sustainable resource management across critical sectors such as water, and food. Global and resources escalate, and effective management is essential to ensure sustainability and minimize environmental impact. Provide usage, and reduce waste, ultimately contributing to more sustainable practices in resource management. Energy management is another sector where AI has a transformative impact. AI technologies can enhance by predicting historical energy output forecast production levels, enabling effectively (Xiang X., et al, 2021). Furthermore, AI can industries through smart management. These systems analyze usage patterns and automate controls for lighting, heating, and cooling, leading to substantial energy savings. By promoting energy efficiency, AI contributes not only to lower operational costs but also to reducing environmental impact. In agriculture, AI is instrumental in advancing sustainable food production practices. While minimizing environmental degradation. AI technologies facilitate precision agriculture, which uses data-driven approaches to optimize farming practices. For instance, AI-powered drones and sensors collect crop health and pest populations. Information on when to irrigate, fertilize, or apply pesticides, thus reducing input waste and minimizing chemical runoff into waterways. Additionally, logistics, while minimizing spoilage and waste. Sustainable also extends to enhancing circular economy practices. AI technologies can analyze waste streams to identify opportunities for recycling and resource recovery (Challoumis C., 2024). For example, algorithms effectively Optimize resource use and promote the reuse of materials, AI contributes to the implementation of resource management availability quality systems. In many regions, especially in developing countries, hinder applications. Investments in data infrastructure, sensor networks, and monitoring systems are essential to support technologies. Moreover, to be inclusive and accessible is vital, as bear resource mismanagement.

Significant deployment resource management. Use systems must prioritize transparency, accountability, and fairness to avoid unintended consequences. For instance, algorithms that allocate water resources must be carefully designed to consider social equity, ensuring that vulnerable populations are not disproportionately affected. Engaging local communities in AI fosters that diverse perspectives are considered, enhancing the effectiveness of resource management efforts. The potential for sustainability is vast. Emerging technologies, such as AI-driven climate models and blockchain, can resource supply chains. Integrating AI with technology creates decentralized systems for tracking resource use and sustainability metrics, empowering consumers and businesses to make informed choices (Chamas Z., et al, 2021). Enable public sentiment, and inform policy decisions and community engagement strategies. Poised transformative sustainable resource management across water, energy, and food sectors. By leveraging advanced analytics, and automation, enhance efficiency, and reduce waste, availability, inclusivity, and ethical considerations. As stakeholders across sectors collaborate, they can not only meet the growing demand for resources but also. Embracing AI as a vital tool in resource management will be essential for achieving sustainability goals and fostering resilience in an increasingly complex and interconnected world. Through responsible and equitable implementation, AI is created.

AI for Social Good: Healthcare, Education, and Inclusive Growth

Increasingly recognized powerful social good, with the potential to transform sectors like healthcare, education, and inclusive growth. As societies face complex challenges—ranging from health disparities and educational inequalities to economic exclusion that can quality of life and promote equity. Automating processes, improving service delivery, empowering marginalized communities, and fostering sustainable development. In treatment, and patient care. Advanced imaging, sequences, that may be AI-driven tools such as enabling that can save lives. Predictive analytics can also help healthcare providers forecast patient needs, optimize resource allocation, and reduce wait times, ultimately enhancing the quality of care (Tomašev N., et al, 2020). Additionally, by facilitating continuous monitoring, these technologies help the education sector transform teaching and learning students' styles, strengths, and weaknesses to tailor instructional strategies personalized not only enhances engagement but also improves learning outcomes, particularly for students who may struggle in traditional classroom settings. Furthermore, AI-driven analytics help at-risk enabling early interventions and support to promote retention and success.

The potential of AI for inclusive growth is particularly significant in addressing economic disparities and promoting sustainable development. AI technologies can AI-powered mobile applications can offer financial literacy training, job matching, and entrepreneurship support, helping individuals develop skills and access opportunities for economic advancement (Alami H., et al, 2020). In agriculture, AI can assist smallholder farmers with precision farming techniques, optimizing resource use. Conditions, weather patterns, and market demand, AI provides actionable insights that enhance productivity and income ultimately food security resilience. Moreover, fostering inclusive governance and civic engagement. Natural language processing (NLP) technologies can analyze public sentiment and feedback from diverse communities, enabling governments to better understand the needs and priorities of their constituents. By leveraging AI from surveys, a community that reflects the perspectives of marginalized populations. AI can also facilitate participatory budgeting processes, allowing citizens to engage in decision-making regarding resource allocation in their communities, thus accountability (Cowls J., et al, 2021).

DIGITALIZATION AND SUSTAINABILITY IN SUPPLY CHAIN MANAGEMENT

Digitalization is fundamentally reshaping supply chain management (SCM), presenting new opportunities for enhancing sustainability while optimizing operations. Such as for that align with their operational goals. SCM not only streamlines processes but also fosters transparency, efficiency, and accountability, which are crucial for sustainable development. One of the primary ways digitalization enhances sustainability in supply chains is through improved visibility and traceability. Blockchain, organizations monitor their supply chain inventory levels, monitor equipment performance, and measure environmental conditions throughout the supply chain (Sarkis J., et al, 2021). This wastes and optimizes resource use. For example, real-time tracking of shipments can minimize delays blockchain technology enhances traceability by providing companies with origin-sourcing practices. Digitalization optimizing resource leveraging analytics, companies can analyze historical performance and more predictive capability to adjust their production and inventory strategies, minimizing overproduction and waste. Patterns to optimize stock levels, ensuring that products are available when needed without excess inventory. Storage lessens manufacturing surplus goods. Additionally, digital tools facilitate collaboration among supply chain partners, enabling them to share information and coordinate efforts to minimize resource use across the entire chain. Moreover, digitalization supports the transition to circular economy practices within supply chains. By integrating digital technologies, companies can implement systems that prioritize resource recovery, reuse, and recycling. For example, advanced analytics can identify opportunities for reusing materials in production processes, while IoT devices can monitor the condition of products throughout their lifecycle, enabling timely repairs and maintenance (Dong T., et al, 2023).

Sustainability in supply chain management also necessitates a focus on social responsibility, and digitalization can enhance labor practices and community engagement. Digital tools can facilitate better communication with suppliers, ensuring that labor standards and ethical practices are upheld throughout the supply chain. For instance, AI-driven risk assessment tools can identify suppliers with potential labor violations or environmental issues, enabling companies. Moreover, digital platforms can foster collaboration with local communities, and information that empowers participation practices. Engaging communities in the supply chain not only enhances social responsibility but also builds brand loyalty and strengthens stakeholder relationships. Despite the myriad benefits of digitalization in promoting sustainability, challenges persist. The initial investment required for digital technologies is significant, for such systems. Additionally, can create barriers to adoption, as organizations' tools and. To overcome these challenges, businesses can seek partnerships with technology providers and industry associations that can offer support and resources for digital transformation. Collaborative initiatives can also foster knowledge sharing and best practices among companies, driving broader industry-wide adoption of sustainable digital practices (Zekhnini K., et al, 2022). Other supply chains become more interconnected and reliant on data sharing, so protecting sensitive information becomes paramount. Companies must invest in robust cybersecurity measures and establish clear. By prioritizing data security, and integrity in their digital initiatives.

INDUSTRY 4.0 AND THE CIRCULAR ECONOMY: OPPORTUNITIES AND CHALLENGES

Represent transformative paradigms that are reshaping manufacturing and production processes, aiming for increased efficiency, sustainability, and resource conservation. Automation is a driving wave. Simultaneously, the circular economy seeks to redefine traditional linear economic models by promoting resource reuse, recycling, and regeneration, aiming to minimize waste and environmental impact. The intersection of these two concepts significant formidable business, policymakers, and society at large. One of the key opportunities presented by the convergence is enhanced efficiency. Advanced technologies enable real-time monitoring and data collection throughout the production lifecycle (Abdul Hamid A Q., et al, 2020). Use materials, energy consumption, and emissions, providing valuable insights that help companies identify inefficiencies and areas for improvement. For example, manufacturers analytics minimize usage, aligning operational practices with circular economy principles. And address issues proactively, thereby of machinery resources. Moreover, facilitating circular supply chains is essential for implementing circular economy principles. By integrating digital platforms and blockchain technology, companies can achieve greater transparency and traceability in their supply chains. This transparency enables organizations to verify the sustainability of materials and ensure ethical sourcing, ultimately fostering trust among consumers and stakeholders., can the entire lifecycle of products, helping businesses assess their environmental impact and optimize designs for reuse or recycling. Such innovations are crucial for transitioning to a circular model, where products are designed with their end-of-life in mind. Another opportunity lies in the creation of new business models that promote circularity. Shift from traditional ownership models to service-based. This shift encourages manufacturers to design products that are durable, repairable, and upgradable, as their profitability increasingly relies on long-term customer relationships rather than one-time sales. For instance, companies in the electronics sector can offer consumers the option to lease products instead of purchasing them outright, allowing for easier refurbishment and recycling when the product reaches the end of its life (Awan U., et al, 2021). Such models waste economics by fostering innovation and collaboration among stakeholders. However, the primary need for substantial in technology and infrastructure. Businesses to invest in new processes, equipment, and technologies that support sustainable practices. For many financial burdens adopting can be daunting. This challenge, industry grants, and support programs are sustainable.

Additionally, there is a critical need for a skilled workforce capable of navigating the complexities as digital technologies become more integrated into manufacturing processes, workers must acquire new skills to Upskill reskilling initiatives are essential to prepare the workforce for the future, ensuring that employees have the competencies needed to drive innovation and sustainability. Moreover, circular a cultural shift within organizations. Traditional business models and practices often Change this mindset requires leadership commitment and a willingness to embrace innovation and sustainability as core values (Kumar P., et al, 2021). Organizations must have sustainability that encourages resource use and waste reduction. By integrating sustainability into their corporate strategies and values, companies can align their operations to drive meaningful change. In conclusion, the interplay is significant for businesses and society. Integration of advanced technologies can enhance resource efficiency, create new business models, and promote circular supply chains, ultimately driving sustainability and reducing environmental impact. However, realizing these benefits requires overcoming challenges related to investment, workforce development, supply chain complexity, and cultural change.

By fostering collaboration among stakeholders and prioritizing sustainability as a core value, organizations can navigate the complexities of this transformation. Embracing the convergence offers a pathway to a more sustainable future, where resources are utilized responsibly, waste is minimized, and economic growth is decoupled from environmental degradation. As we move toward a more sustainable industrial landscape, businesses, governments, and communities must work together to harness the potential of these transformative paradigms, ensuring a resilient and equitable future for all.

SMART CITIES AND COMMUNITIES: LEVERAGING AI FOR SUSTAINABLE DEVELOPMENT

Smart cities and communities represent a forward-thinking approach to urban development, urbanization accelerates— and faces pressing challenges related to resource management, transportation, environmental sustainability, and social equity. By harnessing AI, smart cities can address these challenges while fostering inclusive growth and enhancing the most significant opportunities presented by optimization resource-driven informed decision-making. For instance, usage in real-time, allows cities to implement demand-response strategies that reduce energy waste during peak periods (Kalusivalingam A K., et al, 2021). Smart grids equipped with AI capabilities can balance supply and demand more efficiently, this optimization not only lowers energy costs but contributes, to aligning urban development with climate goals. Water-critical AI Urban areas often face challenges related to water scarcity and quality. AI technologies distribution to minimize waste. Water demand develops sustainable management practices. By ensuring efficient water use and reducing contamination, AI helps cities this vital resource for future generations. Transportation is a domain ripe for AI innovation in smart cities. Traffic congestion, air pollution, and inefficient public transit systems are major urban challenges. Predict congestion, optimize traffic improvements, and reduce travel times. Autonomous vehicles, guided by AI, promise to revolutionize urban transport by reducing accidents, enhancing mobility for the elderly and disabled, and lowering emissions through efficient route planning. Additionally, AI can be based on demand, integrating various modes of transport into seamless travel experiences. This not only promotes the use of public transit but reduces contributing to cleaner air and less traffic congestion. Fostering urban extends to waste management as well as AI predicting waste generation trends. By analyzing data from sensors placed in waste bins, cities can implement more efficient collection waste collection (Alahi M E E., et al, 2023).

Moreover, citizen engagement and participation are integral to the success of smart cities. AI technologies can facilitate communication between city officials and residents, enabling more effective feedback mechanisms. Platforms powered by AI can analyze public sentiment expressed through social media and other channels, helping city planners understand community needs and preferences. Engaging citizens in the decision-making process not only enhances transparency but also accountability By leveraging AI participatory governance, cities can ensure that development aligns with the values and aspirations of their residents. Despite the numerous benefits of integrating AI into smart cities, several challenges must be addressed. These are paramount, as ethical questions regarding consent and surveillance to maintaining public trust. Prioritize transparency, privacy protection, and accountability, residents have control over their personal information. Another challenge lies in technology and digital resources. To create truly inclusive smart cities, this training and resources for underserved communities. Ensuring that all residents can benefit from AI-driven initiatives is crucial for promoting equity and preventing further marginalization (Kasinathan P., et al, 2022). Additionally, into planning interdisciplinary among society. Effective partnerships are essential for designing and implementing that diverse urban population. Collaborative approaches can also drive innovation and facilitate knowledge sharing, helping cities conclude, that integration communities offer significant opportunities for sustainable development.

AI-POWERED SUSTAINABLE AGRICULTURE: PRECISION FARMING AND YIELD OPTIMIZATION

AI-powered sustainable agriculture is revolutionizing farming practices by integrating IoT precision and optimizing crop yields. Rise creates an urgent need for sustainable agricultural practices that not only maximize productivity but also minimize environmental impact. AI offers a transformative approach, allowing farmers to leverage data-driven insights for better decision-making and resource management. For instance, AI practices moisture forecasts. Smart irrigation systems powered by AI can automatically adjust water application conditions, optimal without resources help maintain soil structure, and reduce the risk of erosion and nutrient leaching. Additionally, precision irrigation can enhance crop resilience to drought and extreme weather conditions, making farms. Fertilization practices also benefit significantly from AI-powered solutions. Traditional fertilization methods often rely on blanket applications that can lead to excess nutrient runoff, contributing to water pollution and harming aquatic ecosystems (Jack W., et al, 2024). AI-driven precision agriculture techniques allow farmers to analyze soil nutrient levels and apply fertilizers only where and when they are needed. By using remote sensing technology, farmers can identify nutrient deficiencies in specific areas of their fields and target those areas with the appropriate type and amount of fertilizer. This targeted approach over-fertilization, promotes healthier soils and ecosystems. Pest and disease management is another area where AI excels, enhancing sustainable agricultural practices. Preventing reducing computer vision analysis to signs of pest infestation disease outbreaks. Specific different pests and diseases, allowing for rapid identification and targeted interventions. By using AI for pest management, farmers can implement that minimize harm to beneficial insects, and support biodiversity.

Moreover, AI technologies can assist selection planning and are vital to maximizing yields over time. Yield, and climate characteristics, determine specific environments. Crop rotation and diversification practices are essential for preventing soil degradation and enhancing resilience against pests and diseases. AI-driven insights can guide farmers in making informed decisions about which crops to plant and when ensuring optimal productivity while promoting sustainability. Also extends to supply chain optimization, which is critical for reducing waste and improving the efficiency of food distribution. Market demand, production forecast crop yields and optimize harvest schedules (Elbasi E., et al, 2024). This allows farmers to align their production with market needs, reducing surplus and minimizing food waste. Additionally, AI can enhance logistics by optimizing transportation routes and schedules, ensuring that fresh produce reaches consumers more efficiently. This not only contributes to sustainability by reducing carbon emissions associated with transportation but also helps improve food security by minimizing losses in the supply chain. Manpower agriculture fully realizes its potential. One major challenge is the accessibility of technology, particularly for smallholder farmers in developing regions. Many farmers' resources or driven to bridge this gap, and affordable user-friendly can be easily adopted by all farmers. Training programs and workshops can also skills needed to leverage AI in their practices, promoting inclusive agricultural development. These are critical considerations well as vast amounts of agricultural data ownership and the potential for misuse. Regarding data sharing usage to ensure that farmers' rights are protected while still enabling innovation in agricultural practices. Building trust among will be essential encouraging collaboration and sharing. Additionally, the potential for job displacement due to automation and AI technologies must be considered. While AI can change labor dynamics within the agricultural sector. Policymakers and industry leaders should prioritize reskilling agriculture, which requires advanced technological skills (Akintuyi O B., et al, 2024).

DIGITAL HEALTH AND AI-DRIVEN MEDICAL RESEARCH

Digital health and AI-driven medical research are revolutionizing healthcare patient outcomes, and clinical workflows, and accelerating therapies. As grapple with the challenges posed by rising costs, digital technologies promising pathways to transform healthcare delivery and research practices. At the heart of this transformation is the proliferation of applications, and telemedicine platforms. These tools enable patients to manage chronic conditions proactively and promote healthier lifestyles. And physical activity can inform treatment decisions. This data-driven approach not only charges but also allows clinicians to, ultimately leading to improved health outcomes. Critical enhancing capabilities digital health technologies (Ahmadi A., et al, 2024). Health to observers. The ability to extract actionable insights from complex datasets is particularly valuable in medical research, where therapeutic targets, optimizing, for instance, researchers in development that address the underlying causes of conditions such as cancer. Traditionally, patient recruitment for clinical trials has been a significant bottleneck, with many trials failing to meet enrollment targets. Electronic health data sources are eligible more efficiently, streamlining the recruitment process (Bell S., et al, 2023). Furthermore, AI can enhance patient engagement throughout the trial by providing personalized information and support, helping to improve retention rates and the overall quality of trial data. Another critical application of AI in digital health is in, which, assess. By analyzing large datasets identify potential health risks and enable timely responses. For instance, driven models are used to predict transmission patterns, demonstrating the potential of AI in managing public health crises.

While AI-driven medical research is substantial, Primary concerns health important ethical questions regarding patient consent and safeguarding. Essential robust governance frameworks that prioritize transparency, patient rights, and security to build trust among patients and stakeholders. Interoperability is another significant challenge. Many digital health solutions operate in silos, leading to fragmented data that can hinder comprehensive analysis and decision-making. Developing standardized protocols and frameworks is crucial. Policymakers, technology developers, and healthcare providers must collaborate to create interoperable solutions. Additionally, driven healthcare is a skilled workforce capable of navigating the complexities of digital health. Training programs and educational initiatives should focus on equipping healthcare professionals with. Furthermore, fostering interdisciplinary collaboration between data scientists, clinicians, and researchers is vital to ensure that AI applications are clinically relevant and address real-world challenges (Amaro I., et al, 2024). Moreover, regulatory frameworks must evolve to health and AI. Ensuring patient safety and promoting innovation is a delicate balance that requires ongoing dialogue between regulators, industry stakeholders, and healthcare providers.

AI-DRIVEN DISASTER RESPONSE AND RECOVERY: LESSONS FROM PRACTICE

AI-driven disaster response and recovery represent a significant advancement in how societies can be naturally humanitarian. As climate change intensifies the frequency and severity of such events, integrating artificial intelligence (AI) into disaster management has become essential for enhancing situational awareness, optimizing resource allocation, and improving overall resilience. Lessons learned from recent scenarios highlight both the transformative potential of this technology and the challenges that remain. Primary in disaster response rapidly, providing critical insights that inform decision-making. During emergencies, timely information is crucial for coordinating response social media, and sensor networks, to identify affected areas, assess damage, and monitor real-time conditions (Sun L., et al, 2024). For instance, during Hurricane Harvey in 2017, AI was employed to analyze social media posts and satellite images to map flood extents, enabling emergency responders to prioritize their efforts more effectively. By integrating multiple data streams, AI provides a comprehensive view of the disaster landscape, enhancing situational awareness for both responders and decision-makers. Furthermore, AI can enhance predictive analytics, enabling better preparation for disasters. Machine learning models forecast the likelihood of specific floods, earthquakes, and wildfires. For example, AI has been used to predict the paths of hurricanes and wildfires, allowing communities to implement preemptive measures, such as evacuations and resource mobilization. These predictive capabilities can significantly reduce the impacts of disasters by enabling more effective risk mitigation. Also, during efforts. When resources are limited, determining how to allocate them effectively can be challenging. AI various severity disaster, population density, and the availability of resources, to suggest optimal deployment strategies. For instance, was model spreads the virus and predicts healthcare resource needs, enabling health officials to allocate ventilators, PPE, and staff more efficiently to areas with the highest demand. This capability not only saves lives but also, maximizes the overall effectiveness of the response. In the recovery phase, AI can assist in assessing damage and facilitating rebuilding efforts. AI surveys affected areas, capturing high-resolution images that are analyzed to evaluate damage. This data can inform recovery planning by identifying critical infrastructure that requires immediate attention (Kankanamge N., et al, 2021).

Interoperability between different systems and agencies is another critical challenge. Effective disaster response often among, NGOs, ensuring that AI systems can communicate and share data seamlessly across different platforms is vital for coordinated efforts. This necessitates the development of standardized protocols and frameworks that facilitate data exchange and integration among diverse entities involved in disaster management. Training and capacity building are also crucial for the successful implementation of AI-driven disaster response initiatives. While AI technologies offer significant advantages, their effectiveness depends on the expertise of the personnel operating these systems. Training programs must be developed to ensure that responders are equipped not only with skills but also an understanding of how to interpret AI-generated insights and integrate them into decision-making processes. Engaging frontline workers and deployment systems are tailored to real-world challenges. Moreover, addressing equity in AI applications these advancements (Gupta S., et al, 2022). Marginalized groups, often face the greatest risks during disasters and may lack access to the resources needed for effective response and recovery. It is critical to consider the unique needs of diverse populations. This involves actively involving affected communities-driven initiatives and solutions accessible and equitable. Finally, use in disaster management carefully. Decision-making processes, of over-reliance on technology, may overlook the human aspects of disaster response. Essential to maintain a human-centered approach that values empathy, local knowledge, and community engagement.

BLOCKCHAIN FOR SUSTAINABLE DEVELOPMENT: TRANSPARENCY, ACCOUNTABILITY, AND TRUST

Blockchain technology holds transformative potential for sustainable development by enhancing transparency, accountability, and trust across various sectors, including finance, supply chains, healthcare, and governance. As the global community strives, the integration of blockchain can related to resource management, ethical practices, and equitable access to services. By leveraging the unique characteristics of blockchain—decentralization, immutability, and transparency—stakeholders can create systems that promote sustainability and social responsibility. Blockchain provides a transparent tamper-proof record of transactions and interactions. This transparency is particularly crucial in supply chains, where traceability is essential for ensuring that products are sourced sustainably and ethically (Gomez Trujillo A M., et al, 2021). For instance, in the food industry, blockchain can verify the origins of their food. Initiatives like IBM's Food Trust leverage blockchain to chains, identify inefficiencies, reduce waste, and respond swiftly to food safety concerns. This level of transparency not only builds consumer trust but also producers, knowing that actions are subject to scrutiny. In addition to supply chain transparency, blockchain can enhance accountability in financial transactions and aid in reducing corruption. Traditional financial systems often suffer from difficulty tracing the flow of funds and ensuring that resources are allocated effectively. Blockchain's decentralized ledger allows all participants in a transaction to access and verify the same information, reducing the risk of fraud and mismanagement. This has significant implications for international aid and development funding, where misallocation of resources can hinder progress toward the SDGs. Based platforms flow donations in real-time, ensuring that funds reach their intended recipients and are used for their designated purposes. This not only enhances accountability but also increases donor confidence, potentially leading to more robust funding for sustainable development initiatives. Moreover, blockchain technology can facilitate inclusive, particularly systems credit histories, leaving millions without access to essential financial services. Blockchain finance (defi) solutions provide access to banking services via mobile devices, using blockchain's secure and transparent infrastructure. This can empower individuals to save, invest, and access credit, fostering economic growth and reducing poverty (Jiang S., et al, 2022). For instance, projects like Banque utilize blockchain to create economic identities for marginalized individuals, enabling them vital resources. Blockchain data management patient privacy, fostering more equitable access to medical services. Electronic health records (EHRs) often suffer from fragmentation, comprehensive. Blockchain can store health, allowing patients to control their information and grant access to providers as needed. Care promotes trust in the system, encouraging individuals to seek necessary medical attention without fear of data breaches or misuse. Additionally, blockchain can facilitate transparent reporting of health data, which is essential for tracking the spread of diseases and responding to public health crises.

The integration of blockchain in governance can also enhance democratic processes and citizen engagement. Can the administration reduce bureaucratic inefficiencies? Voting systems, where every vote is securely recorded on a tamper-proof ledger. This not only enhances trust in electoral processes but also encourages higher voter participation, as citizens can be confident that their voices are accurately represented. Moreover, blockchain can enable participatory governance models, where citizens can engage directly, fostering community accountability. Despite the immense potential of blockchain for sustainable development, significant concern is. The environmental impact of blockchain operations has raised questions about itself. Other innovative solutions are essential for aligning blockchain applications with sustainability goals. Moreover, as blockchain technology continues to develop platforms minimize environmental impacts while maximizing social benefits. Another challenge lies in the regulatory landscape surrounding blockchain technology.

The rapid pace of innovation often outstrips existing regulatory frameworks, creating uncertainty for businesses and developers. Clear regulations and guidelines are necessary to foster responsible blockchain development while protecting users and stakeholders. Collaboration leaders Furthermore, achieving widespread adoption of blockchain technologies education. Many regions, particularly in technical infrastructure or blockchain solutions are affected. Building local capacity through training programs and partnerships with technology providers can empower communities to leverage blockchain for sustainable development (Schulz K A., et al, 2020). Additionally, ensuring that blockchain applications are user-friendly and accessible to all demographics is crucial for fostering inclusive participation in these initiatives. Lastly, addressing issues of equity and inclusivity is paramount as blockchain technology continues to develop. There is a risk that the benefits of blockchain could disproportionately favor those who are already digitally literate or economically advantaged. To prevent this, it is essential to prioritize initiatives that specifically target marginalized communities, ensuring that they have access to the opportunities provided by blockchain. Engaging these communities in the design and implementation of blockchain solutions ensures the technology serves their needs. In conclusion, blockchain technology offers a promising avenue for advancing sustainable development by enhancing transparency, accountability, and trust across various sectors. By facilitating traceability in supply chains, promoting financial inclusion, improving healthcare data management, and enhancing governance processes, blockchain is equitable and, addresses challenges related to energy consumption, regulatory frameworks, infrastructure, and inclusivity blockchain supporting. As continue to explore innovative solutions to pressing challenges, embracing blockchain as a tool for positive change will be essential in building a more sustainable and just world for all.

ETHICS AND GOVERNANCE OF AI FOR SUSTAINABLE DEVELOPMENT

The ethics and governance for is crucial discourse seeks to leverage AI technologies, economic, and environmental. As AI becomes increasingly integrated into education agriculture and urban planning—ensuring that its deployment aligns with ethical principles and essential. This involves a multifaceted approach that addresses the implications of technologies and fosters inclusive governance frameworks that prioritize the needs of marginalized communities. Primary deployment is often relied on to make decisions, and if these datasets contain biases, the AI existing can lead to inequitable employment outcomes. To mitigate these risks, it is essential to adopt data auditing and inclusive dataset development. By ensuring that datasets reflect diverse populations and experiences, fairness AI contributes to more equitable outcomes in sustainable development initiatives (Truby J., 2020). Principle the governance of black box challenging, which can lead to distrust among users affected communities. Transparency in AI systems involves clear communication about how algorithms function, are used, and criteria for. This can be achieved through, which. By fostering transparency, stakeholders can build trust in AI technologies, enabling communities to understand and engage with these systems more effectively. Accountability in AI governance is paramount to ensure that stakeholders. This includes developers, deployers, and users. In cases where AI systems cause harm—such as in automated surveillance or decision-making that leads to negative social outcomes—it is vital to have mechanisms in place for redress and accountability. This can involve creating regulatory frameworks that outline responsibilities, as well as establishing independent oversight bodies to monitor AI deployment and its impacts on sustainable development. Inclusive for ensuring implemented reflection.

This means actively involving marginalized and underrepresented communities in discussions about AI deployment, ensuring that their voices are heard Participatory approaches, community consultations, and stakeholder engagement, risks benefit AI applications, and equitable. Moreover, fostering interdisciplinary collaboration among ethicists, technologists, policymakers, and community representatives can facilitate a holistic approach to AI governance that prioritizes sustainable development (Larsson S., 2020). The environmental implications of AI must also be considered in discussions of ethics and governance. While AI has the potential to enhance sustainability efforts—such as optimizing energy use and improving resource management—it can also contribute to environmental degradation if not managed responsibly. Large, for instance, the carbon footprint of these technologies. Stakeholders must prioritize efficient algorithms to promote the use of renewable in AI operations to mitigate these impacts. Additionally, AI should be harnessed to support environmental sustainability initiatives, such as climate monitoring and biodiversity conservation, reinforcing its role in achieving the SDGs.

Education and capacity building are critical components of ethical AI governance. As AI technologies evolve, ensuring that individuals, organizations, and communities engage with these is essential. Educational initiatives should focus on raising awareness of the ethical implications of AI and equipping stakeholders with the tools for AI deployment. Includes fostering literacy, promoting critical thinking about AI applications, and encouraging ethical considerations in technology design and implementation. By empowering individuals with knowledge, societies can be better for sustainable development. International cooperation and regulatory harmonization are vital for establishing a coherent global framework for AI governance.

As AI technologies transcend national borders, there is a need for the collaborative development of shared ethical standards. Initiatives (GPAI)'s Principles on Artificial Intelligence provide platforms for dialogue and cooperation among governments, industry, and civil society. Collaboration, best practices in AI governance, and algorithmic accountability on a global scale. Furthermore, potential risks exacerbate existing inequalities (Dara R., et al, 2022). AI technologies are deployed across various sectors, there is a risk that their benefits may disproportionately favor affluent communities while leaving marginalized populations behind. This digital divide can perpetuate socio-economic hindrances. To counter this trend, it is essential to prioritize inclusive policies that promote equitable access to AI technologies, so communities can benefit from their potential. This includes education and training, and fostering an inclusive innovation ecosystem that empowers diverse voices. In conclusion, the ethics and governance of AI for sustainable development are paramount as societies increasingly rely on these technologies to tackle complex global challenges. By prioritizing, stakeholders can create sustainable development. Addressing the implications of interdisciplinary collaboration, and fostering international cooperation will be critical for harnessing. Ultimately, a comprehensive approach to AI governance that emphasizes ethical considerations and inclusive practices will be essential for ensuring that, contributing to a sustainable and equitable world for all. As we navigate this rapidly evolving landscape, the commitment to ethical AI governance will be a vital determinant of our success in achieving sustainable development for future generations.

FUTURE DIRECTIONS FOR DIGITAL TRANSFORMATION AND SUSTAINABILITY

Increasingly dual digital transformation and sustainability, organizations are recognizing the need to harmonize these two domains. Advanced has revolutionized the business landscape, providing unprecedented opportunities for enhancing, reducing, and fostering innovation. Integrating digital tools into sustainability initiatives is not merely an option but a company seeking a rapid One of the key future directions for digital transformation in the context of sustainability is the development of smart, frameworks. Analytics resource consumption patterns, enabling them to identify inefficiencies and optimize their operations. For instance, energy-intensive industries can utilize energy usage, allowing immediate adjustments that reduce waste. This real-time monitoring can extend beyond energy to water usage, waste management, and supply chain processes, creating a holistic view of resource utilization that supports sustainability goals. Furthermore, AI is poised to crucial foresee environmental impacts and can predict model scenarios, helping companies to assess the potential outcomes of their sustainability initiatives (Feroz A K., et al, 2021). Enhance precision farming techniques by predicting crop yields based on weather patterns and soil conditions, thereby minimizing resource use while maximizing output. This predictive capability not only boosts sustainability efforts but also contributes to food security, a pressing global concern. Blockchain technology also presents significant opportunities for the provenance of materials, ensuring ethical sourcing and reducing the carbon footprint associated with transportation and logistics. This transparency is crucial for consumers increasingly concerned about the sustainability practices of the brands they support. Future applications may include smart contracts that automatically enforce sustainability standards, ensuring compliance and fostering a culture of responsibility throughout the supply chain. As digital transformation continues to evolve, the concept of circular economy will likely gain traction. This model emphasizes resource efficiency and recycling. This creates platforms for sharing resources, tracking product lifecycles, and optimizing recycling processes. For instance, digital marketplaces can connect businesses with surplus materials, encouraging collaboration and minimizing waste. By embedding circular economy principles into their digital strategies, organizations can not only enhance sustainability but also unlock new revenue streams and drive innovation.

Collaboration will be another essential pillar of future digital transformation initiatives. No single organization can tackle the complexities of sustainability alone. Cross-industry partnerships, public-private collaborations, and stakeholder engagement will be vital for sharing best practices, resources, and technologies. Digital platforms can facilitate these collaborations, enabling knowledge sharing and fostering innovation networks. For instance, companies in different sectors can come together to co-develop sustainable technologies, leveraging their diverse expertise to address shared challenges. Moreover, consumer engagement is set to digital transformation sustainability. As consumers become more environmentally conscious, they are demanding transparency and accountability from brands. Digital tools such as social media, mobile apps, and gamification can empower informed sustainability initiatives (Nadkarni S., et al, 2021). Brands that harness these technologies to engage their customers in sustainability efforts will not only enhance their reputation but also drive loyalty and advocacy. Education and skill development are critical components of this transformative journey. Organizations must invest in upskilling their workforce employees' digital competencies intersection of technology and sustainability. This involves innovation think critically about how digital tools can be leveraged to achieve sustainability goals. Collaborative learning platforms can facilitate knowledge sharing and ensure that employees are well-versed in the latest technological advancements and sustainability practices. Regulatory frameworks will also need digital transformation sustainability. Bodies have guidelines that incentivize sustainable practices while fostering innovation.

Policymakers in green technologies and digital infrastructure by implementing supportive regulations, tax incentives, and funding initiatives. This collaboration will be essential in driving systemic change and scaling sustainable solutions. Finally, as organizations embark on their digital transformation journeys, they must embrace a mindset of continuous improvement and adaptability. The landscape of technology and sustainability is dynamic, requiring businesses emerging trends and challenges. By fostering innovation and sustainability efforts (Jones M D., et al, 2021). In conclusion, the future directions for digital transformation and sustainability are inextricably linked, presenting both challenges and opportunities for organizations across industries. Advanced, engaging consumers, investing in skill development, companies. As the digital landscape continues to evolve, those organizations that strategically align their digital transformation efforts with sustainability objectives will not only enhance their operational efficiency but contribute to creating a more equitable world. The journey toward digital transformation and sustainability is ongoing, and organizations must lead the charge in this critical endeavor.

CONCLUSION

An era where digital transformation and sustainability are increasingly interwoven emerges cornerstone driving meaningful advancements in sustainable development. The convergence of these fields presents a unique opportunity for organizations' sustainability initiatives, optimizing resource utilization, and fostering innovation. AI, streamlined operations create adaptive strategies that the pressing challenges, resource scarcity, and social equity. AI's that significantly reduce their environmental impact. Forecast demands, allow businesses to optimize supply chains and minimize waste. In agriculture, AI-driven tools can enhance precision farming, ensuring that resources such as water and fertilizers are used efficiently while maximizing crop yields. These applications sustainable practices across various sectors, ultimately contributing to the broader goals. Moreover, sustainability efforts promote transparency and accountability. Blockchain technology, when combined with AI, can enhance. This dual approach fosters consumer trust and encourages businesses to adopt responsible practices, as consumers increasingly seek brands that align with their values. The ability to track and verify sustainable practices through AI and blockchain not only boosts corporate responsibility but also drives competition toward greener solutions. Collaboration harnessing for sustainable development. Complex global challenges require multi-stakeholder engagement, where partnerships between businesses, governments, and civil society can lead to innovative solutions. AI can facilitate these collaborations by providing platforms for data sharing and collective problem-solving. Shared various organizations identify patterns and propose joint initiatives that address sustainability challenges, such as reducing carbon footprints or managing waste more effectively. Furthermore, the role of education and capacity building cannot be understated. As organizations integrate AI into their sustainability efforts, there is a pressing need to upskill the workforce. Investing in education and training ensures the leverage of AI technologies. A well-informed workforce can drive innovation and implementation of sustainable practices, creating improvement. Organizations that prioritize education will have operational capabilities for the development of the economy. Also, sustainability. Policymakers must create environments where ethical considerations and sustainability principles are upheld. By establishing promote responsible sustainability efforts, governments can incentivize businesses to invest in green collaboration between the public and private sectors is essential for scaling solutions that address urgent environmental challenges. As we look towards the future, to drive digital transformation sustainability. However, this potential must be harnessed with a sense of responsibility and ethical consideration. Organizations must ensure do not exacerbate existing inequalities. By prioritizing ethical AI-inclusive solutions that benefit all members of society. In digital transformation, artificial intelligence represents a transformative frontier for organizations seeking to embrace AI as a catalyst for sustainable development, companies can optimize their operations, enhance transparency, and drive collaboration across sectors. The future of sustainable development will be shaped by those who leverage technology responsibly, invest in education, and foster partnerships that prioritize the planet and its people. As we move forward, it is imperative that organizations not only recognize the opportunities presented by AI but also commit to using these technologies equitable future. Through intentional action and innovative thinking, we can turn the challenges of today into the sustainable solutions of tomorrow, ensuring that digital transformation is pursued.

REFERENCES

Abdul-Hamid, A. Q., Ali, M. H., Tseng, M. L., & Lan, S. (2020), "Impeding challenges on industry 4.0 in circular economy: Palm oil industry in Malaysia." Computers & Operations Research, Elsevier.

Ahmadi, A. (2024). *"Digital health transformation: leveraging ai for monitoring and disease management." International Journal of BioLife Sciences*. IJBLS.

Akintuyi, O. B. (2024). *Adaptive AI in precision agriculture: a review: investigating the use of self-learning algorithms in optimizing farm operations based on real-time data*. Research Journal of Multidisciplinary Sciences.

Al-Emran, M., & Griffy-Brown, C. (2023). The role of technology adoption in sustainable development: Overview, opportunities, challenges, and future research agendas. *Technology in Society*, *70*, 102344. DOI: 10.1016/j.techsoc.2023.102240

Alahi, M. E. E., Sukkuea, A., Tina, F. W., Nag, A., Kurdthongmee, W., Suwannarat, K., & Mukhopadhyay, S. C. (2023). Integration of IoT-enabled technologies and artificial intelligence (AI) for smart city scenario: Recent advancements and future trends. *Sensors (Basel)*, *23*(11), 5206. DOI: 10.3390/s23115206 PMID: 37299934

Alami, H., Rivard, L., Lehoux, P., & Hoffman, S. J. (2020), "Artificial intelligence in health care: laying the Foundation for Responsible, sustainable, and inclusive innovation in low-and middle-income countries." Globalization and Health, Springer.

Amaro, I., Citarella, A. A., & De Marco, F. (2024). AI-driven technologies in Digital Health & Well Being: Early detection and intervention strategies. *CEUR Workshop Proceedings*.

Anwar, A., Malik, S., & Ahmad, P. (2022). Cogitating the role of technological innovation and institutional quality in formulating the sustainable development goal policies for E7 countries: Evidence from .... *Global Business Review*, *23*(3), 1–15. DOI: 10.1177/09721509211072657

Awan, U., Sroufe, R., & Shahbaz, M. (2021), "Industry 4.0 and the circular economy: A literature review and recommendations for future research." Business Strategy and the Environment, Wiley Online Library.

Bell, S., Lawrence, C. D., & Dobrin, S. (2023). Near-Term Digital Health Predictions: A Glimpse into Tomorrow's AI-driven Healthcare. *Telehealth and Medicine Today*.

Blasch, E., Pham, T., Chong, C. Y., Koch, W., Leung, H., Braines, D., & Abdelzaher, T. (2021). Machine learning/artificial intelligence for sensor data fusion–opportunities and challenges. *IEEE Aerospace and Electronic Systems Magazine*, *36*(10), 26–39. DOI: 10.1109/MAES.2020.3049030

Challoumis, C. (2024), "BUILDING A SUSTAINABLE ECONOMY-HOW AI CAN OPTIMIZE RESOURCE ALLOCATION." *XVI International Scientific Conference.*

Chamas, Z., Abou Najm, M., Al-Hindi, M., & Yassine, A. (2021). Sustainable resource optimization under water-energy-food-carbon nexus. *Journal of Cleaner Production*, *296*, 126947. DOI: 10.1016/j.jclepro.2021.126947

Chen, L., Chen, Z., Zhang, Y., Liu, Y., & Osman, A. I. (2023). Artificial intelligence-based solutions for climate change: A review. *Environmental Science and Pollution Research International*, *30*(10), 24155–24173. DOI: 10.1007/s11356-022-23717-6

Cowls, J., Tsamados, A., Taddeo, M., & Floridi, L. (2021). A Definition, benchmark and database of AI for social good initiatives. *Nature Machine Intelligence*, *3*(2), 111–115. DOI: 10.1038/s42256-021-00296-0

Dara, R., Hazrati Fard, S. M., & Kaur, J. (2022). Recommendations for ethical and responsible use of artificial intelligence in digital agriculture. *Frontiers in Artificial Intelligence*, *5*, 884192. DOI: 10.3389/frai.2022.884192 PMID: 35968036

Dong, T., Yin, S., & Zhang, N. (2023). The interaction mechanism and dynamic evolution of digital green innovation in the integrated green building supply chain. *Systems*, *11*(3), 122. DOI: 10.3390/systems11030122

Elbasi, E., Mostafa, N., Zaki, C., AlArnaout, Z., Topcu, A. E., & Saker, L. (2024). Optimizing Agricultural Data Analysis Techniques through AI-Powered Decision-Making Processes. *Applied Sciences (Basel, Switzerland)*, *14*(17), 8018. DOI: 10.3390/app14178018

Feroz, A. K., Zo, H., & Chiravuri, A. (2021). Digital transformation and environmental sustainability: A review and research agenda. *Sustainability (Basel)*, *13*(3), 1530. DOI: 10.3390/su13031530

Gomez-Trujillo, A. M., & Velez-Ocampo, J. (2021), "Trust, transparency, and technology: blockchain and its relevance in the context of the 2030 agenda." Corporate Sustainability Initiatives and Corporate Social Responsibility, Springer.

Gupta S, Modgil S, Kumar A, Sivarajah U, (2022), "Artificial intelligence and cloud-based Collaborative Platforms for Managing Disaster, extreme weather and emergency operations." International Journal of Disaster Risk Reduction, Elsevier.

Hajikhani, A., & Suominen, A. (2022). Mapping the sustainable development goals (SDGs) in science, technology and innovation: Application of machine learning in SDG-oriented artefact detection. *Scientometrics, 127*(1), 233–256. DOI: 10.1007/s11192-022-04358-x

Jack, W., & Bagh, S. (2024). *Revolutionizing Agriculture: AI-Powered Crop Yield Forecasting and Precision Farming for Optimal Harvests*. EasyChair.

Jiang, S., Jakobsen, K., Bueie, J., Li, J., & Haro, P. H. (2022). A tertiary review on blockchain and sustainability with focus on Sustainable Development Goals. *IEEE Access : Practical Innovations, Open Solutions, 10*, 114975–115006. DOI: 10.1109/ACCESS.2022.3217683

Jones M D, Hutcheson S, Camba J D, (2021), "Past, present, and future barriers to digital transformation in manufacturing: A review." Journal of Manufacturing Systems, Elsevier.

Kaack, L. H., Donti, P. L., Strubell, E., Kamiya, G., Creutzig, F., & Rolnick, D. (2022). Aligning artificial intelligence with climate change mitigation. *Nature Climate Change, 12*(3), 257–265. DOI: 10.1038/s41558-022-01377-7

Kalusivalingam A K, Sharma A, (2021), "Enhancing Smart City Development with AI: Leveraging Machine Learning Algorithms and IoT-Driven Data Analytics." Journal of AI and Robotics.

Kankanamge, N., Yigitcanlar, T., & Goonetilleke, A. (2021), "Public perceptions on artificial intelligence driven disaster management: Evidence from Sydney, Melbourne and Brisbane." Telematics and Informatics, Elsevier.

Kasinathan, P., Pugazhendhi, R., Elavarasan, R. M., Ramachandaramurthy, V. K., Ramanathan, V., Subramanian, S., Kumar, S., Nandhagopal, K., Raghavan, R. R. V., Rangasamy, S., Devendiran, R., & Alsharif, M. H. (2022). Realization of sustainable development goals with disruptive technologies by integrating industry 5.0, society 5.0, smart cities and villages. *Sustainability (Basel), 14*(22), 15258. DOI: 10.3390/su142215258

Kumar P, Singh R K, Kumar V, (2021), "Managing supply chains for sustainable operations in the era of industry 4.0 and circular economy: Analysis of barriers." Resources, Conservation and Recycling, Elsevier.

Larsson, S. (2020). On the governance of artificial intelligence through ethics guidelines. *Asian Journal of Law and Society, 7*(3), 437–451. DOI: 10.1017/als.2020.19

Leal Filho, W., Wall, T., Mucova, S A R., & Nagy, G. J. (2022). Deploying artificial intelligence for climate change adaptation. *Environmental Science and Pollution Research International*, *29*(4), 5561–5573. DOI: 10.1007/s11356-021-17222-1

Nadkarni, S., & Prügl, R. (2021), "Digital transformation: a review, synthesis and opportunities for future research." Management Review Quarterly, Springer.

Rodriguez, P., & Costa, I. (2024). Artificial Intelligence and Machine Learning for Predictive Threat Intelligence in Government Networks. *Advances in Computer Sciences*, *1*(1), 1–12.

Rodríguez-García, J. D., & Moreno-León, J. (2020), "Introducing artificial intelligence fundamentals with LearningML: Artificial intelligence made easy." *Eighth International Conference on Learning Analytics and Knowledge*, 135-144. Doi: DOI: 10.1145/3434780.3436705

Sarkis J, Kouhizadeh M, Zhu Q S, (2021), "Digitalization and the greening of supply chains." Management & Data Systems.

Schulz, K. A., Gstrein, O. J., & Zwitter, A. J. (2020), "Exploring the governance and implementation of sustainable development initiatives through blockchain technology." Futures, Elsevier. DOI: 10.1016/j.futures.2020.102611

Sun, L., Li, H., Nagel, J., & Yang, S. (2024). Convergence of AI and urban emergency responses: Emerging pathway toward resilient and equitable communities. *Applied Sciences (Basel, Switzerland)*, *14*(17), 7949. DOI: 10.3390/app14177949

Tomašev, N., Cornebise, J., Hutter, F., & Mohamed, S. (2020). AI for social good: Unlocking the opportunity for positive impact. *Nature Machine Intelligence*, *2*(10), 522–526. DOI: 10.1038/s42256-020-00271-8 PMID: 32424119

Truby, J. (2020), "Governing artificial intelligence to benefit the UN sustainable development goals." Sustainable Development, Wiley Online Library. DOI: 10.1002/sd.2048

Xiang, X., Li, Q., Khan, S., & Khalaf, O. I. (2021). Urban water resource management for sustainable environment planning using artificial intelligence techniques. *Environmental Impact Assessment Review*, *87*, 106567. DOI: 10.1016/j.eiar.2020.106515

Zekhnini K, Cherrafi A, Bouhaddou I, (2022), "A model integrating lean and green practices for viable, sustainable, and digital supply chain performance." Journal of Production Research, Taylor & Francis.

Chapter 4
The Intersection of Digitalization and Circular Economy: Advancing Sustainable Development and Climate Change Mitigation

Qamar Ali
https://orcid.org/0009-0004-1306-5820
Virtual University of Pakistan, Pakistan

Muhammad Aamir Aamir
https://orcid.org/0009-0007-4452-2627
Virtual University of Pakistan, Pakistan

ABSTRACT

Digitalization and circular economy are slowly being adopted by industries and are starting to reshape sustainable development and climate change mitigation. Technologies such as Artificial Intelligence (AI), Internet of Things (IoT) and blockchain are presenting new methodologies in handling resources, waste, and sustainability systems. At the same time, Circular Economy principles of reduce, reuse, and recycle are guiding sustainable production and consumption. In combination, these forces serve to shape proper strategies in the face of environmental threats, as well as contribute toward the achievement of the Sustainable Development Goals set by the United Nations. In this book, the link between digitalization and circular economy is explored with regards to the ways in which the two paradigms can help the development of sustainable solutions to address resource management

DOI: 10.4018/979-8-3693-9909-5.ch004

and climate change.

INTRODUCTION

Recently the world has adopted technology and circular economy models in the last few years. This is not only shifting industries but also shift the way how sustainability and climate change mitigation is being considered. Digitalization and circular economy have been identified as two main drivers of change expected to foster innovation and sustainable development. To understand these forces and how they coexist and will shape a more sustainable future, it is necessary to understand how digital technologies have developed and how they can support circularity. Based on the analysis of the effects of digitalization, and the principles of circular economy, in the context of sustainability, and management of challenges in the world today, this section appears(Chauhan et al., 2022).

Most industries have been transformed by the phenomenon of digitalization, which has posed a new challenge to the traditional operating models. The advancement in the use of the digital technology has been fast, and it affects industries from the dawn of computer and automation, using artificial intelligence, big data, IoTs and cloud computing. These innovations have changed the approaches, methods, and business models in areas such as production, healthcare, banking, and transportation. For instance, the application of technologies that define Industry 4.0 like smart factories and automation of the manufacturing process has boosted the manufacturing process due to time efficiency in the consumption of resources. Digitization is not just about technology change. It is a useful instrument to address the current global issues and those concerning sustainable development and climate change. Technological innovations are now being applied in the management of energy efficiency, resources, and emissions. Other key innovations include supply chain management through blockchain and energy management through smart grids in the cities. In this regard, while digitalization is an enabler of economic growth, it is also an essential tool for the formulation and adoption of good polices and behaviors(Visvizi et al., 2018).

The circular economy is an economic system that strives to cut waste and to continuously use products, materials, and resources. Different from the linear economy that products are produced, consumed, and then discarded, the circular economy aims to redesign products for recycling, reuse, and regeneration. These are upstream thinking, stopping pollution at its source, improving resource efficiency, and nourishing nature. These principles look to minimize environmental effects and conservation of scarce resources to enhance sustainability. While the circular economy is a new term, the principles have been around in one form or another for centuries as societies have sought to be more sustainable. The circular economy is a model that is relevant to sustainability because it can help to disentangle economic growth from the consumption of resources in a way that minimizes environmental harm. It provides an approach for industries to shift towards ecologically friendly production and consumption systems as part of the global agenda, including the UN's SDGs(Bengtsson et al., 2018).

Relation between Digitalization and Circular Economy

The integration of digital innovation in circular economy can be considered as a potential format for the sustainable development agenda. Digital technologies enable monitoring and control of resources from the beginning of the circular economy to its end. For example, IoT devices can check how much and in what condition the product is being used, which allows for repair before a breakdown and, therefore, extending the product's life cycle. Similarly, the application of blockchain technology increases the shipment transparency in the supply chain and provides assurance that the materials will be recycled or repurposed. Intersections between circular economy and digital innovation are being understood more and more. Digital tools are the backbone of circular practices as they enable real time data gathering and analysis, which in turn increases resource productivity. For example, platforms that encourage sharing, leasing, or reusing products are already disrupting fashion and electronics industries and shifting consumers' behavior towards circular consumption. In addition, digital twining, which involves creating virtual models of physical systems, has been applied to model and improve production processes with the aim of enhancing resource productivity(Borangiu et al., 2020).

Digitization and circular economy principles create new possibilities for companies to innovate. Through this, organizations can use the circular economy that not only helps in minimizing waste but also in making new business models. For instance, organizations can apply data analysis to decide how to repurpose material or apply automation to enhance the recycling process. In this way, digitalization can be seen as a trigger for the circular economy that allows it to be more easily implemented and at a larger scale. Therefore, the integration of digital technologies and the circular economy has the potential for making a positive impact on sustainability. Using digital technologies, industries can speed up the shift to circular systems and help to address environmental issues and climate change goals. This link between two of the most significant drivers of change, namely digitalization and circular economy, is instrumental in the establishment of sound policies that will support system resilience over the long term(Iacovidou et al., 2021).

Sustainable Policy and Governance

Sustainable policy is one of the main trends of the modern world that developed in the 21st century, as it is impossible to deny the global challenges in the spheres of environment, society, and economy. States, international organizations, and businesses have understood that classical economic models, which are based on the rapid economic development that harms the environment, cannot be sustained. Therefore, the concept of sustainable policies is gradually becoming the key means to make the development process environmentally friendly, economically possible, and socially fair. This section defines sustainable policy in the 21st century, the part that governments and international organizations play in the formulation of these policies and aspects that hinder their implementation(Mbaka & Mugambi, 2014).

SUSTAINABLE POLICY: 21ST CENTURY CONTEXT AND CONCEPTUAL FOUNDATION

The sustainable policy in the 21st century, in its turn, can be described by the following major principles: environmentalism, economic efficacy, and social justice. These principles are congruent with generic sustainability concept, which refers to the delivery of needs in the present generation without limiting the potential of future generations to meet their needs. In the last two decades the world has seen a shift towards sustainable development as a principle for policy formulation and implementation. This change is due to increased sensitivity of the negative impact of human activities on the environment such as climate change, biological diversity, and resource use. Thus, sustainable policies are meant to solve these problems and foster such approaches that can minimize adverse effects on the environment, increase the share of renewable energy sources, and enhance the effective use of resources(Panwar et al., 2011).

That is why the integration of policies with the sustainable development agenda is critical in delivering sustainable environmental and social returns. The UN Sustainable Development Goals and the Paris Agreement on climate change have now become central to the policy agendas of governments and intergovernmental organizations. These sets of goals provide a direction for harmonizing national strategies of different countries with the global sustainable development agenda to achieve the common goal of sustainable development. According to these goals, governments can develop policies that will foster a stronger, fairer, and sustainable society(Globalization, 2004).

Functions of Governments and International Bodies

States have a critically important function in the development of sustainable policy through the formulation and enforcement of laws, rules, and public spending. Over the past few years, several governments have embraced policies that encourage the world to go digital as a way of pursuing sustainable development. To mitigate energy consumption, enhance resource management and cut wastage, smart grids, artificial intelligence, and IoT are being put into use. For instance, smart cities provide data analysis and IoT implementation for energy management, air quality control, and enhanced waste disposal to enhance the quality of the urban environment. Thus, governments can promote the shift towards sustainable economic growth by instituting policies that support the use of digital technology. These include, but not limited to, the United Nations, European Union, and World Economic Forum, which also have the responsibility of ensuring global policies are sustainable. Such organizations afford countries a forum for cooperation in sustainability and set up global guidelines which inform national policies. A most visible case is the European Green Deal that outlines plans for the EU free from emissions, orbital energy, and circular economy. The Green Deal acts as a reference for other areas and shows how a partnership between countries can help make the world a better place. Global frameworks of this nature are important for the purpose of encouraging international cooperation and coherence in the implementation of sustainable initiatives(Nilsson et al., 2012).

Problems of Policy Implementation

Although the work in sustainable policy advancement has been commendable, the issue of how these policies can be effectively put into practice stays a challenge. Some of the main barriers include the limited ability to align technology adoption and circular economy with current economic paradigms and legal systems. In this regard, most traditional industries are skeptical towards innovation and are afraid of losing money when implementing measures that would help them shift towards the concept of sustainability. Lastly, there are also the issues concerning technologies in terms of accessibility and cost, especially in the world's Third World countries where there is almost no infrastructure. Another challenge is the legal framework within which organizations may not encourage the uptake of sustainable measures. Unfortunately, the circular economy is often not backed up by an adequate legal framework and there may be conflicts between national laws and international regulations. For instance, some countries have done a lot to ensure they embrace the use of renewable energy sources while others have not, which results in a partisan effort in achieving sustainability(Ansolabehere & Konisky, 2016).

To address these challenges, the governments and the international organizations need to assess the policy deficits and design the responses. Examples of successful policies are drawn from countries including Germany which has managed to encourage use of renewable energy and circle economy. The German's Energiewende policy is a clear example that favorable government policies and regulations can accelerate deployment of renewable energy technologies, cut CO2 emissions, and advance sustainability. It is hoped that, by learning from these examples, other countries can find their own policy weaknesses and set up plans to help achieve the shift towards a sustainable future. Therefore, sustainable policy and governance are crucial for solving the problems of the 21st century. Policies coordinated with sustainable development aims, the use of digital tools and international cooperation will help to achieve a sustainable shift in the future. Nevertheless, the achievement of these policies poses several challenges such as, opposition from conventional sectors, lack of technology, and legal issues. These barriers can be met through collaboration and innovation to help strengthen and build a more sustainable society at the global level(Reficco et al., 2018).

MEASURES TO COMBAT CHANGE IN CLIMATE

With the increasing cases of global climate change, it is expected to produce effective measures to reduce and prevent future environmental damage. A range of strategies are being proposed and piloted, from the regulatory framework to the use of technological solutions. This section will detail the main climate change mitigation measures with emphasis on how digital technologies and circular economy can decrease carbon footprint(Khan et al., 2022).

Background information on Climate Change

Climate change is one of the biggest issues of the 21st century. Some of the consequences of climate change that are felt today include high temperatures, receding polar ice caps, and adverse weather conditions. These changes affect the world's ecosystems and cause biodiversity loss, changes in agricultural yields, sea level rise and effects on human health. It also worsens the conditions of social injustice, because the most affected groups of people have limited ability to cope with the crisis. The main problem of this concern is the fact that human beings have a great contribution. Pollution, Industrialization, Deforestation and Burning of Fossil fuels have therefore been cited as some of the major causes of greenhouse gases accumulation in the atmosphere. The world's ever-growing need for energy and natural resources has elicited more harm than worsens climate change. The next stage is not only that of emissions cuts but also of ecosystem rehabilitation, and overall human impact minimization(Cameron et al., 2017).

Digital Solutions In Climate Change Mitigation

When it comes to climate change, technology presents great solutions for the modern world. AI, IoT, and Blockchain technology can help to reduce the impact of climate change and its effects through digital means. These technologies provide real time data capturing and analysis to increase the probability of correct risk evaluation and prompt disaster management. AI can analyze climate data on weather, disasters, and emissions, in a way that humans cannot. It also has an energy management function that enhances power grid efficiency, minimizes energy loss, and encourages the uptake of clean energy. Likewise, IoT devices allow for the tracking of environmental conditions on a continuous basis. For example, sensors installed in forests can measure changes in temperature and humidity, which may be used to prevent fires or track forest loss. Iot system in agriculture makes the best use of water and minimizes the use of chemical fertilizers which are emitting pollutants and degrading the environment(Chen et al., 2018). With the help of blockchain technology, an added level of openness and responsibility in the fight against climate change is achieved. As a result of decentralization and security, blockchain makes it possible to track sustainability programs like carbon credit and resource management. An example of this is when organizations use blockchain to check the carbon footprint of goods within the supply chain, to guarantee companies achieve their sustainability standards. Digital technologies also play a significant role in the creation of smart cities, where the design of the cities involves efficient and renewable energy resources, efficient waste disposal systems, and energy efficient structures to reduce adverse effects on the environment. These innovations are important in cities given that population concentration coupled with increased production and economic activities are key sources of greenhouse gases(Romero-Lankao & Dodman, 2011).

Roles of Circular Economy in Minimizing Greenhouse Gas Emission

As a concept, the circular economy introduces a complete change in the use and disposal of resources through recycling and repurpose. This contrasts with the conventional linear model of production, consumption, and disposal of products, which results in a lot of wastage to the environment. Through closing the loop, the circular economy reduces the need for raw materials, perfects waste production, and reduces greenhouse gases. The concepts of circular economy include designing products to be long lasting, making it possible to use materials again, and using waste in new and innovative ways. These actions contribute to the reduction of the demand for new raw materials: Its procurement is linked with increased energy use and adverse effects on the environment. For instance, the recycling of aluminum preserves up to 95% of the energy that would otherwise be used to produce aluminum from scrap, which will also reduce energy and emissions. Also, the circular economy model has resource efficiency as one of its main agendas as well. Currently, organizations are adopting strategies such as material recycling, efficient packaging, and product rejuvenation in their quest to reduce emission of greenhouse gases. They are thus beneficial in sectors like fashion, electronics, and construction, which usually create a lot of waste and discharge when producing their products. With the help of circular economy these industries can minimize their adverse impacts on the environment as well as drive economic growth(Dantas et al., 2021).

Circular economy also supports climate change mitigation through encouraging the use of renewable energy technologies for electricity generation. Environmental management also helps in minimizing energy use in the production processes by controlling the production of wastage hence promoting the change from the use of fossils energy sources. These collaborative efforts make the circular economy effective in avoiding emissions, enhancing resource productivity, and mitigating climate change effects. Thus, one should accept digital technologies and circular economy to combat climate change. Digital technologies enhance the capability of environmental management and enable innovation, while the circular economy provides practical solutions to address emissions and sustainable resource management. Individually, they offer a view of a possible future in which the world is less precarious and more secure(Lorey, 2015).

ROLE OF DIGITALIZATION AND CIRCULAR ECONOMY IN KEY SECTORS INTEGRATION

Energy Sector

It should also be pointed out that at present there are significant trends in the energy sector, related to digitalization and the circular economy. One of the most revolutionary is the function of smart grids that have transformed energy supply and demand beyond recognition. These grids use information technologies to manage energy delivery conduits in a way that reduces energy losses, conserves energy, and taps on renewable energy. It can include more renewable energy sources such as solar and wind power into the present electrical power network for the betterment of the power industry. This integration has the advantages of reducing the fossil energy use, reducing the greenhouse gases emissions, and improving the energy system reliability. Furthermore, circular economy systems are increasingly being practiced in the energy production and consumption processes. Energy resources, energy materials and energy services are consumed, extracted, and recycled in a closed loop for waste reduction in circular economy. For example, renewable energy as a product is more circular than traditional energy as the resources used to produce it are renewable. Energy storage technologies and battery recycling are increasingly being deployed to create a closed loop system for batteries to be used and reclaimed multiple times to reduce the environmental impact of energy usage. About this, aspects of digitalization and circular economy concepts are not only enhancing the sustainability but also fueling advancements in energy production and consumption within the sector(Dantas et al., 2021).

Manufacturing and Production

Digitalization and circular economic concepts are two strategic areas that are already transforming the manufacturing and production processes. Technologies such as 3D printing, Internet of Things (IoT), and advanced robotics are also creating better ways of practicing digital manufacturing. These technologies enable products with zero waste, best material input, and creation of products that can be easily disassembled for recycling at the end of their life. Through the application of these technologies, manufacturers are slowly shifting away from the linear processes of resource extraction and consumption and their disposal, which is the linear production model, to the circular production model that focuses on reducing resource consumption. Several case studies show the positive experience of the circular economy in the manufacturing sector. For instance, some of the companies are using technology to track the life cycle of their products to find that they are disposed of or used in other production. In other cases, manufacturers are changing products to be modular where they can interchange or replace parts of the product without having to replace the whole product. These approaches are not only efficient in minimizing waste but also contribute to the reduction of costs and lengthening of product life. Integrating digital technologies and circular economy strategies is now allowing manufacturers to develop and implement more sustainable, efficient, and profitable production networks(Turner et al., 2019).

Waste Management and Resource Efficiency

Digital technologies are effectively being used to support circular economy in waste management and resource efficiency. Technological advancement in waste management is changing the way waste is collected, treated, and recycled. For example, smart waste collection systems employ sensors and big data to decide the most efficient routes for waste collection so that resources are collected effectively and at the right time. This reduces the carbon footprint of waste collection, as well as ensuring that all waste that is recyclable is correctly sorted for recycling. Moreover, digital tools are used to check the waste streams and materials flows, as well as potential for resource recovery. This level of openness and data analytical thinking is crucial for improving resource productivity in circular economy systems. For example, in implementing the smart manufacturing system, firms are using digital technologies to find where wastes exist so that waste can be reduced, or materials can be recovered. This not only helps in minimizing the wastage generated but also goes a long way in increasing the efficiency of the resources used. However, the circular concept of waste management is not the only area where the concept of improving resource efficiency is being introduced in many industries. Companies are changing the supply chain management to minimize the demand for materials and enhance the effectiveness of material reuse. These are Recycling where the product is either mended or disassembled and used in the manufacturing of new products, and the closed loop system where waste produced by one company forms an input to another company. These practices are being augmented by DTs in that they provide the tools to verify products, manage resources, and dispose of the waste in the most effective manner throughout the production network(Van Engeland et al., 2020).

Consequently, the energy sector, manufacturing, and waste management industries are changing their strategies by implementing the principles of digitalization and circular economy. Technologies like the digital are now being used to enhance the productivity of resources, reduce wastage, and promote circular economies in the respective industries. These technologies also stay useful in the future and will be helpful in addressing some of the world's greatest problems that are currently happening and have a great impact on the world such as the environment, resources(Panel et al., 2011).

GLOBAL CASE STUDIES AND BEST PRACTICES

As the climate change fight continues to progress, the world has turned to digital technologies and circular economy strategies for developing sound climate policies. Countries have also presented measures of how they have been able to incorporate technology while at the same time promoting sustainable economic practices that others can emulate. Such examples along with global case studies are extremely helpful in designing future strategies and policies for combating climate change and encouraging sustainable development(ChANGE, 2008).

Digitalization in Climate-Responsive Policies

Digitalization is already being embraced as a key factor when developing and implementing climate sensitive policies. Several countries have effectively incorporated digital technologies into their climate plans, providing a preview of how technology can enhance and support better management of the climate. One of the best examples is Finland where the Climate Watch was launched as the national initiative to track emissions in real time using digital tools. This platform aids policy makers to produce better and more precise assessment of climate policies and their effects by providing prompt and up to date environmental information. Likewise, Germany has integrated digital technologies into its Energiewende (Energy Turn) policy to regulate renewable energy systems and decrease greenhouse gas emissions in various industries. One of the more interesting examples is China, which is already using AI and IoT in its smart city initiatives. These tools ease efficient energy management in the urban environment as well as waste management through enhanced supervision and use of resource. One such example of how digitalization is helping in the improvement of environment is China's 'Blue Sky' campaign through which big data is used to check and minimizing pollution. It is not only individual countries that have success stories to tell about digitalization but also international cooperation. The European Union's Green Deal is a bold agenda to achieve climate neutrality in Europe by 2050 and digital technologies are at the heart of this strategy. The Green Deal uses the least cost optimization algorithm with advanced monitoring systems, digital twins, and predictive modelling for tracking the emissions and adjusting achieve the desired climate goals. These examples show how the integration of digitalization into climate polities leads to a great improvement of the sustainability and climate change mitigation(Mondejar et al., 2021).

Comparisons Between Developed and Developing Countries and Circular Economy Models

Circular economy that focuses on use, reuse, and recovery of various products is also a key approach to sustainability. However, despite the necessity to apply circular economy models, they are practiced selectively in the developed and developing countries. The circular economy is more easily adopted in developed countries such as the Netherlands and Sweden owing to better infrastructure, effective governance, and the concern of environmental decay. For instance, The Netherlands has set a vision to achieve circular economy by 2050 and has formulated policies that do not allow generation of waste and enhance the flow of materials. However, circular economy models are not flexible enough to be adopted in developing countries. While there are countries such as India and South Africa that are slowly integrating circularity in their waste management and recycling systems, they face problems such as inadequate facilities and funds together with lack of technological innovations. For instance, circular economy strategies are hindered by the fact that in India, waste management is controlled by informal sector. Even though there are some improvements made in tackling electronic waste recycling in the country, there are a lot more concerning aspects that are either related to scalability or formality. Similarly, in African countries, Nigeria and Kenya, there is a slow transition to circular economy because of challenges in infrastructure and logistics through which these countries cannot fully embrace circular systems. Nonetheless, the circular economy is on the rise in developing countries and backed by international actors and alliances. The United Nations Industrial Development Organization (UNIDO) and the African Circular Economy Alliance (ACEA) have signed memorandum of understanding in support of circular economy systems in Africa. These measures are important for bridging the gap between the developed and developing world but much more needs to be done for the promotion of fair growth across the various world regions(Bourguignon et al., 2008).

Partnership for Sustainability: Public Private Partnerships

Digitalization and circular economy are two of the biggest concepts that rely on PPPs to meet their aims. These partnerships therefore are effective in promoting innovation, enhancing the effectiveness of prevention of climate change when governments and private organizations collaborate. There is one that was formed between the World Economic Forum and companies like Microsoft and IBM to build applications that may be able to use AI and data to address environmental issues. It enhances digital technologies but also helps to manage the interactions between the sectors to address climate change challenges. A good example of a successful public private partnership is the Danish Energy Sector where the government has collaborated with other private companies to invest in offshore wind farms making Denmark one of the Denmark has been able to reduce CO2 emissions because of strengthened collaboration between the government agencies, energy companies and tech providers in the country. In the developing countries alone, public private partnership has also been realized to have their share. As a result, in Kenya, the government has partnered with the private sector to put up solar energy projects that provide power to individuals who cannot access power from the national grid. This partnership has not only led to CO2 emission reductions but also has ensured provision of energy for rural population which has in turn boosted social and economic development. Similarly, in India, public private partnerships have taken a front seat in enhancing the accessibility to clean energy and that includes solar energy and electric vehicles(Doukas & Ballesteros, 2015).

These main partners are also playing a role in moving towards more circular economy. South Korea has one of the most effective recycling and waste management systems in the world which combines government and private efforts, technology, and public awareness. This partnership has brought one of the highest recycling rates in the world and a model that other countries should emulate by following. From successful public-private partnerships in sustainability, it is noted that the willingness to collaborate is the key factor that spurs the digital and circular economy trends. These partnerships are expected to play an increasingly important role in responding to climate change as more governments and private companies see the value of collaboration. In addition, the capacity of these collaborations to replicate models that have worked in one region to other regions in the developed and developing world will be important in realizing the potential of these innovations. Therefore, global case studies and best practices reveal that the digitalization and circular economy are critical to the formulation of climate policies. These strategies can be amazingly effective; perhaps because they require cooperation between different stakeholders or countries. These examples serve as useful examples for other countries as they try to create new and effective ways of dealing with climate change. The next step will be to make sure that these innovations will be a benefit for the whole world especially for the third world countries for the better future(Yusuf, 2003).

FINAL REMARKS AND FUTURE WORK

The analysis of the current trends of modern society shows that digitalization and sustainability are the concepts that are actively integrated into the global development. The potential for digital innovation to contribute to environmental conservation and the mitigation of some of the world's greatest challenges, such as climate change, resource scarcity, and increasing energy needs is enormous. This final section finds the trends in the digitalization of sustainability, presents the policy action to boost climate change mitigation and concludes that the digitalization and circular economy offer the opportunity to envision how sustainability can be advanced(Burmaoglu et al., 2023).

Emerging Trends in Digitalization for Sustainability

Over the last few years, digitalization has become a driving force for sustainability and innovation, with technologies like artificial intelligence, quantum computing or the Internet of Things setting the pace. AI has the greatest potential to address the challenge of best resource use, greenhouse gas emissions, and the effectiveness of RE technologies. Others include real time analysis of big data for energy consumption prediction, enhanced control of power networks and the concept of the smart city. Other functions include the identification and control of the environment including the quality of air and water so that early signs of potential danger to the population can be detected. Other emerging tech like quantum computing is also being used in the sustainability space and can really be effective. Because quantum computing allows for analyzing large sets of data in a short amount of time, it will be valuable for enhancing energy storage and advanced materials and developing more correct climate models. Renewable energy is another field predicted to see noticeable developments through quantum computing because the technology potentially makes solar panels and energy storage batteries far more effective. Furthermore, it can serve to refine the numerical climate prediction hence aid decision makers estimate the amount of damage that may be inflicted by climate change(Hallegatte et al., 2012).

Innovation is thus an important component in the growth of the circular economy. This is because technology has slowly been moving the world from the linear economy where products are bought, used, and then disposed of to the circular economy where goods and services are repurposed, recycled and regenerated. For instance, through blockchain, organizations can check the origin and use of materials to improve accountability in the production process and support sustainable production. Likewise, digital technologies are supporting the sharing economy, where goods and services are exchanged in a better way and there is less needed to produce new products and, thus, produce waste. Such innovations are a radical change in the structure of economies and present potential solutions for the improvement of production and consumption processes(Markard & Truffer, 2006).

POLICY RECOMMENDATIONS AND GLOBAL COLLABORATION

Although technological advancements form an important part of the sustainability agenda, it is crucial that there are major changes in policy, and strong collaboration on the international front. To speed up the climate change mitigation process, it is crucial that governments advance policies that encourage the use of environmentally friendly technologies and practices by all sectors of the economy. For instance, carbon pricing may help companies to minimize their CO2 emissions by putting a price on carbon, while subsidy for renewable energy may boost the development of clean energy technologies. Also, measures such as energy conservation in Buildings, Transport and Productions are equally important in addressing emissions and reaching the global Climate Change goals(Röck et al., 2020).

While national policies are important, international collaboration is important given that climate change is a transboundary issue. Frameworks like the Paris Agreement provide a platform for international cooperation, but it is still important that more is done to encourage countries to act on their commitments. Technology transfer that is the provision of sustainable technologies from the developed world to the underdeveloped world is one area where cooperation can have a big dent. Offering clean energy, digital technologies, and circular economy solutions, the developed countries can give Developing Countries a chance to avoid the unenviable path of the industrialization that led to the current crisis. Furthermore, international cooperation between states, business and civic organizations is inevitable in sustainable development. Collective efforts to combine funding and knowledge can help to spur innovation that enables change and introduces sustainable solutions on a large scale. For instance, PPPs have been used to finance large scale renewable energy projects, while international organizations can promote the exchange of information on the most effective way to build climate change resilience. This means that the targets said in the international agreements will be difficult to meet without the collective international action needed to address climate change(Schenck, 2008).

The opportunities from digitalization and circular economy are vast and hold the potential to help to solve some of the greatest challenges facing society today. Advancements in AI, quantum computing, and blockchain will help advance the efficient use of resources and decrease greenhouse gas emissions while building more robust and sustainable systems. There is, therefore, a need to extend the integration of these technologies in the circular economy to ensure that resources are used efficiently, and waste is minimized. However, this potential can only be realized effectively by the technologies that define such polices and frameworks. Governments must play an initiative-taking role in shaping policies that incentivize the adoption of sustainable technologies, while international cooperation is necessary to ensure that these innovations help all countries, particularly those most vulnerable to the impacts of climate change. The importance of global collaboration in the promotion of sustainability cannot be overemphasized because it will define the extent to which humanity can respond to the challenges of environment sustainability. Finally, it is crucial to note that the digital era combined with the circular economy is a great tool to promote sustainability in the 21st century. Therefore, different societies can harness the potential of new technologies and work together to change for the better with sustainable development and environmental conservation, economic growth, and social development. Looking into the future, there are many challenges as well as opportunities and it is critically important to have proper political measures and cooperation to achieve the vision of sustainable development. Digitalization and the circular economy will remain the two significant drivers in defining the future that we are likely to experience as we seek a new and better world(Chauhan et al., 2022).

CONCLUSION

The final part of the document restates the opportunities for digitalization and circular economy as the key drivers for change in response to the challenges of the world, with a focus on sustainable development and climate change. This book finds how circular economy concepts can be combined with emerging technologies such as AI, IoT, and blockchain to limit resource and waste, increase efficiency, and minimize climate change. However, the various benefits cannot be realized without enforcement from both the governments and private sectors through policies and international relations. According to the document, it will be a challenge to reach sustainable development goals globally with special focus on technology transfer and policy measures. From here on, digitization and circular economy will remain two of the most important trends in the pursuit of sustainability. It is important to recognize that policies and partnerships will be central to the success of these efforts so that technology can be used to advance ecological conservation, economic welfare, and human well-being.

REFERENCES

Ansolabehere, S., & Konisky, D. M. (2016). *Cheap and clean: how Americans think about energy in the age of global warming*. Mit Press.

Bengtsson, M., Alfredsson, E., Cohen, M., Lorek, S., & Schroeder, P. (2018). Transforming systems of consumption and production for achieving the sustainable development goals: Moving beyond efficiency. *Sustainability Science, 13*(6), 1533–1547. DOI: 10.1007/s11625-018-0582-1 PMID: 30546486

Borangiu, T., Morariu, O., Răileanu, S., Trentesaux, D., Leitão, P., & Barata, J. (2020). Digital transformation of manufacturing. Industry of the future with cyber-physical production systems. *Romanian Journal of Information Science and Technology, 23*(1), 3–37.

Bourguignon, F., Bénassy-Quéré, A., Dercon, S., Estache, A., Gunning, J. W., Kanbur, R., Klasen, S., Maxwell, S., Platteau, J.-P., & Spadaro, A. (2008). *Millennium Development Goals at Midpoint: Where do we stand and where do we need to go?* EC.

Burmaoglu, S., Ozdemir Gungor, D., Kirbac, A., & Saritas, O. (2023). Future research avenues at the nexus of circular economy and digitalization. *International Journal of Productivity and Performance Management, 72*(8), 2247–2269. DOI: 10.1108/IJPPM-01-2021-0026

Cameron, D. R., Marvin, D. C., Remucal, J. M., & Passero, M. C. (2017). Ecosystem management and land conservation can substantially contribute to California's climate mitigation goals. *Proceedings of the National Academy of Sciences of the United States of America, 114*(48), 12833–12838. DOI: 10.1073/pnas.1707811114 PMID: 29133408

Chauhan, C., Parida, V., & Dhir, A. (2022). Linking circular economy and digitalisation technologies: A systematic literature review of past achievements and future promises. *Technological Forecasting and Social Change, 177*, 121508. DOI: 10.1016/j.techfore.2022.121508

Chen, J., Lü, S., Zhang, Z., Zhao, X., Li, X., Ning, P., & Liu, M. (2018). Environmentally friendly fertilizers: A review of materials used and their effects on the environment. *The Science of the Total Environment, 613*, 829–839. DOI: 10.1016/j.scitotenv.2017.09.186 PMID: 28942316

Dantas, T. E. T., de-Souza, E. D., Destro, I. R., Hammes, G., Rodriguez, C. M. T., & Soares, S. R. (2021). How the combination of Circular Economy and Industry 4.0 can contribute towards achieving the Sustainable Development Goals. *Sustainable Production and Consumption, 26*, 213–227. DOI: 10.1016/j.spc.2020.10.005

Doukas, A., & Ballesteros, A. (2015). Clean energy access in developing countries: perspectives on policy and regulation.

Globalization, W. C. S. D. o. (2004). *A fair globalization: Creating opportunities for all.* International Labour Organization.

Hallegatte, S., Shah, A., Brown, C., Lempert, R., & Gill, S. (2012). Investment decision making under deep uncertainty--application to climate change. *World Bank Policy Research Working Paper*(6193).

Iacovidou, E., Hahladakis, J. N., & Purnell, P. (2021). A systems thinking approach to understanding the challenges of achieving the circular economy. *Environmental Science and Pollution Research International, 28*(19), 24785–24806. DOI: 10.1007/s11356-020-11725-9 PMID: 33289042

Khan, S. A. R., Piprani, A. Z., & Yu, Z. (2022). Digital technology and circular economy practices: Future of supply chains. *Operations Management Research : Advancing Practice Through Research, 15*(3), 676–688. DOI: 10.1007/s12063-021-00247-3

Lorey, I. (2015). *State of insecurity: Government of the precarious.* Verso Books.

Markard, J., & Truffer, B. (2006). Innovation processes in large technical systems: Market liberalization as a driver for radical change? *Research Policy, 35*(5), 609–625. DOI: 10.1016/j.respol.2006.02.008

Mbaka, R. M., & Mugambi, F. (2014). Factors affecting successful strategy implementation in the Water Sector in Kenya. *Journal of Business and Management, 16*(7), 61–68.

Mondejar, M. E., Avtar, R., Diaz, H. L. B., Dubey, R. K., Esteban, J., Gómez-Morales, A., Hallam, B., Mbungu, N. T., Okolo, C. C., Prasad, K. A., She, Q., & Garcia-Segura, S. (2021). Digitalization to achieve sustainable development goals: Steps towards a Smart Green Planet. *The Science of the Total Environment, 794*, 148539. DOI: 10.1016/j.scitotenv.2021.148539 PMID: 34323742

Nilsson, M., Zamparutti, T., Petersen, J. E., Nykvist, B., Rudberg, P., & McGuinn, J. (2012). Understanding policy coherence: Analytical framework and examples of sector–environment policy interactions in the EU. *Environmental Policy and Governance, 22*(6), 395–423. DOI: 10.1002/eet.1589

Panel, U. N. E. P. I. R., Consumption, U. N. E. P. S., & Branch, P. (2011). *Decoupling natural resource use and environmental impacts from economic growth.* UNEP/Earthprint.

Panwar, N. L., Kaushik, S. C., & Kothari, S. (2011). Role of renewable energy sources in environmental protection: A review. *Renewable & Sustainable Energy Reviews*, *15*(3), 1513–1524. DOI: 10.1016/j.rser.2010.11.037

Reficco, E., Gutiérrez, R., Jaén, M. H., & Auletta, N. (2018). Collaboration mechanisms for sustainable innovation. *Journal of Cleaner Production*, *203*, 1170–1186. DOI: 10.1016/j.jclepro.2018.08.043

Röck, M., Saade, M. R. M., Balouktsi, M., Rasmussen, F. N., Birgisdottir, H., Frischknecht, R., Habert, G., Lützkendorf, T., & Passer, A. (2020). Embodied GHG emissions of buildings–The hidden challenge for effective climate change mitigation. *Applied Energy*, *258*, 114107. DOI: 10.1016/j.apenergy.2019.114107

Romero-Lankao, P., & Dodman, D. (2011). Cities in transition: transforming urban centers from hotbeds of GHG emissions and vulnerability to seedbeds of sustainability and resilience: Introduction and Editorial overview. *Current Opinion in Environmental Sustainability*, *3*(3), 113–120. DOI: 10.1016/j.cosust.2011.02.002

Schenck, L. (2008). Climate change crisis-struggling for worldwide collective action. *Colo. J. Int'l Envtl. L. & Pol'y*, *19*, 319.

Turner, C., Moreno, M., Mondini, L., Salonitis, K., Charnley, F., Tiwari, A., & Hutabarat, W. (2019). Sustainable production in a circular economy: A business model for re-distributed manufacturing. *Sustainability (Basel)*, *11*(16), 4291. DOI: 10.3390/su11164291

Van Engeland, J., Beliën, J., De Boeck, L., & De Jaeger, S. (2020). Literature review: Strategic network optimization models in waste reverse supply chains. *Omega*, *91*, 102012. DOI: 10.1016/j.omega.2018.12.001

Visvizi, A., Lytras, M. D., Damiani, E., & Mathkour, H. (2018). Policy making for smart cities: Innovation and social inclusive economic growth for sustainability. *Journal of Science and Technology Policy Management*, *9*(2), 126–133. DOI: 10.1108/JSTPM-07-2018-079

Yusuf, S. (2003). *Innovative East Asia: the future of growth*. World Bank Publications. DOI: 10.1596/0-8213-5356-X

Chapter 5

Digital Transformation as a Catalyst for Sustainability for Advancing Circular Economy and Environmental Innovation:

Harnessing Emerging Technologies for Sustainable Growth and Green Innovation

Priyanka Gupta
https://orcid.org/0009-0006-5341-8314
Graphic Era University, India

Sudhanshu Joshi
https://orcid.org/0000-0003-4748-5001
Doon University, India

Manu Sharma
Graphic Era University, India

Kapil Ahalawat
Graphic Era University, India

Navneet Rawat
Graphic Era University, India

Janmejai Shah
Graphic Era University, India

ABSTRACT

As the world adapts to modern norms which recognises sustainability as key to the

DOI: 10.4018/979-8-3693-9909-5.ch005

progress of the world, digital transformation has consequently become a vitality requisite for ruining sustainability in companies. This chapter examines the relationship between digital technologies including artificial intelligence, blockchain, internet of things and big data to sustainable development in relation to environmental problems including climate change and resource optimization as well as waste management. By extracting sustainable consumption and production patterns in the business environment, enterprise and policymakers are exploring new options on using digital approaches to advance the 2030 Sustainable Development Goals (SDGs). This chapter aims at exploring the potential the relationship between digital transformation and advancing circular economy. It shows where and how digital technology improves productivity, minimises waste and brings innovation to industries like energy, production, logistics and supply chains.

INTRODUCTION

Digitalization or digital business transformation is the process through which organizations incorporate digital technologies across every facet of their operations and into interactions with clients and customers. It covers all these; cloud, AI, blockchain technology, IoT and big data. It can be seen that these technologies allow an organization to redesign business processes, to increase efficiency and to encourage organizational innovation (Vial, 2021).

While sustainable development concerns the enhancement of the standard of living to the present population without detriment to the potential of future generations to improve theirs. It includes three main pillars: of economic, environmental and social sustainability. Economic sustainability relates to activities which enhance long term sustainability without damaging social, environmental and cultural polluting years aspects of the community. Environmental sustainability addresses issues to waste, use of natural resources and impacts on the environment. Static sustainability raises social sustainability to the quality and well-being of people and the society in which they live, work, and engage in the process of establishing operations (Elkington, 2018).

The factors of digital transformation and sustainability are allied in ways that make them even more important. Technology has shown incredible potential to facilitate improvements in sustainability by increasing efficiency, using energy more effectively, minimizing wastage and integrating sustainability in business. For instance, IoT device is capable of optimizing the use of energy at individual household and organizational levels through monitoring in real-time resulting into significant conservation of energy that leads into reduced greenhouse gas emissions (Bibri, 2018). In the same way, blockchain can improve supply chain visibility and guarantee that materials used, goods manufactured and services delivered are done through sustainability (Saberi et al., 2019).

Figure 1. Overview of digital transformation and sustainability

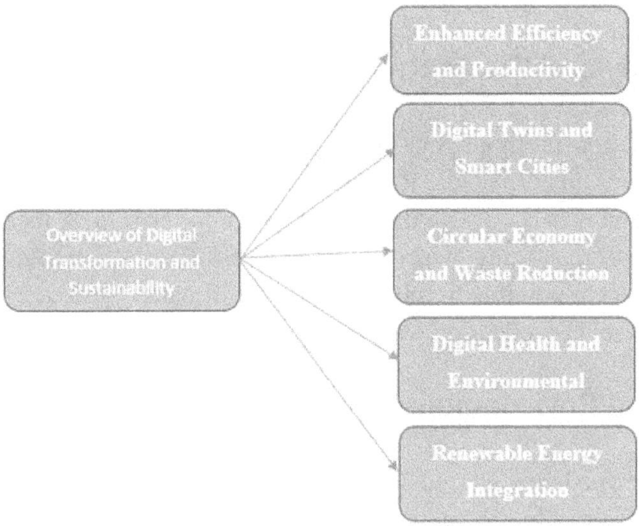

Enhanced Efficiency and Productivity

One of the most significant benefits of digital transformation is the enhancement of operational efficiency and productivity. Through automation and process improvement, calamities have an opportunity to become green thereby cutting down costs. For instance, through using AI and IoT sensors in predicting the appropriate time of maintaining the machineries, a lot of time and resources used in replacing weak or damaged machinery is avoided (Pivoto et al., 2021).

Digital Twins and Smart Cities

Digital twin a digital model of a physical asset – is widely used in several industries for enhancing sustainability in cities. Civilization digital twins allow decision makers to test performance and find a way of arranging traffic, using less energy to power buildings, and improving transport. This approach enhances the social well being of urban dwellers while also resulting in numerous environmental gains

(Dembski, et al., 2020). Smart cities are the urban habitats that integrate information and communications technologies to enhance productivity, maintainability, and habitation quality. IoT sensors, data analytics and AI enable better management of resources such as water and energy. For instance, smart grid can help manage the electric power consumption in real-time promoting effective integration of renewable energy and reducing the use of fossil fuel (Deren et al., 2021.)

Circular Economy and Waste Reduction

Digital transformation promotes the principles of the circular economy that calls for the reuse of resources in continuous production processes. Some emerging technologies for instance, IoT's & Big data enable them to monitor & record the lifecycle of the products thus being in a position to recycle & reuse materials. Such an approach also makes it possible to minimize the use of natural resources apart from minimizing wastes in production and consumption (Vilariño, et al., 2017).

Digital Health and Environmental Impact

It is found that telemedicine and the use of wearable health devices can lower the environmental effects of healthcare solutions. These tools help the doctors and patients avoid some in-person meetings which in their essence cause travel and hence emissions. Furthermore, digital electronic health records eliminate the necessity to use paper documentation which showcased to preserve resources (Keesara et al., 2020).

Renewable Energy Integration

The use of renewable energy sources in the electricity supply mix is another aspect where it is almost impossible to mention digital transformation. Sophisticated data analysis and Artificial Intelligence can predict the energy production by the REs like Solar and wind mills and thus embed them optimally into the grid. which means having a steady and sustainable source of energy, with enhanced usage of clean energy and less dependence on the fossil-based resources (Zhang et al., 2020).

RELEVANCE TO 2030 SUSTAINABLE DEVELOPMENT GOALS (SDGS)

The United Nations' 2030 list of Sustainable Development Goals (SDGs) are 17 targets aiming at improving the quality of life and development for all. They tackle wide subject areas such as poverty and wealth disparity, climate change, environmental conservation, peace and justice to mention but a few (United Nations, 2020). They argue that digital transformation is central to the achievement of these goals because it contributes to finding unique solutions to questions of sustainability as shown in Figure 2.

Figure 2. Relevant 2030 Sustainable Development Goals (SDGs)

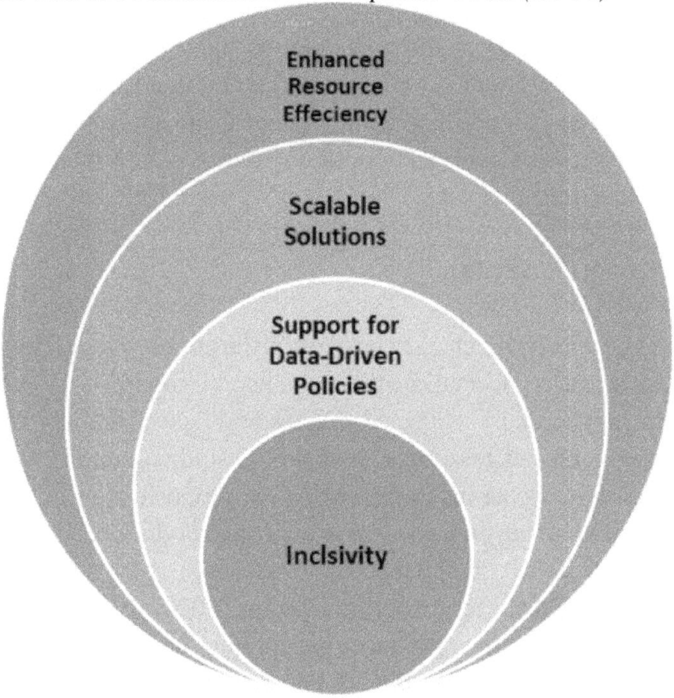

a. SDG 7: Affordable and Clean Energy - Intelligent networks and Interconnected objects can enable efficient utilization of energy and enhance the integration of clean energy resources and reduce the intensity of fossil energy (IEA, 2020).
b. SDG 9: Industry, Innovation and Infrastructure – Industry, Innovation and Infrastructure – The use of digital enables enhancement of innovation and capacity in industries while nurturing sustainable infrastructure. For example, AI improves the predictive maintenance concept in manufacturing, minimising potential interruptions and waste of material (Lee et al., 2015).
c. SDG 12: Responsible Consumption and Production - We use the term Responsible Consumption and Production whereby blockchain can provide accountability for consumption patterns, and encourage manufacturers to take responsibility for their actions. Also, digital platforms can support the sharing economy thus

decreasing demand for new goods, and encouraging reuse or recycling of materials (Gasper, et al., 2019).

d. SDG 13: Climate Action - Digital technologies will help enhance the practice of collecting data and making improvements within climate action initiatives. For instance, AI can process data inputs to forecast the climate change and offer ways through which the change may be dealt with (Rolnick et al., 2022).

Besides the above objectives, the application of digital transformation in ensuring sustainability is not only about attaining the targets outlined above individually, but also promoting the overall concept of integrated sustainable development. Digital technologies offer organisations an opportunity to co-create value in a sustainable, economic and social manner

Objective of the Chapter

Given this, the aim of this chapter is to reveal the possibilities of digital transformation that can lead to the formation of the circular economy. The circular economy is an economic model that is designed to have no disposal and continuous reuse, cycle, and regeneration of resources. It stands in stark comparison with a linear economy, which has a take-make-dispose value proposition. Currently, circular economy aims at eliminating the concept of waste and pollution, built on the remodelling of products and their components in closed loops, and restoring the health of the natural systems (Ellen MacArthur Foundation, 2020). It is within this context that this chapter seeks to offer a detailed insight into the way that adoptive digital technologies can facilitate sustainable behaviours while enabling the attainment of the SDGs. The specific objectives are shown in figure 3:

Figure 3. Objectives

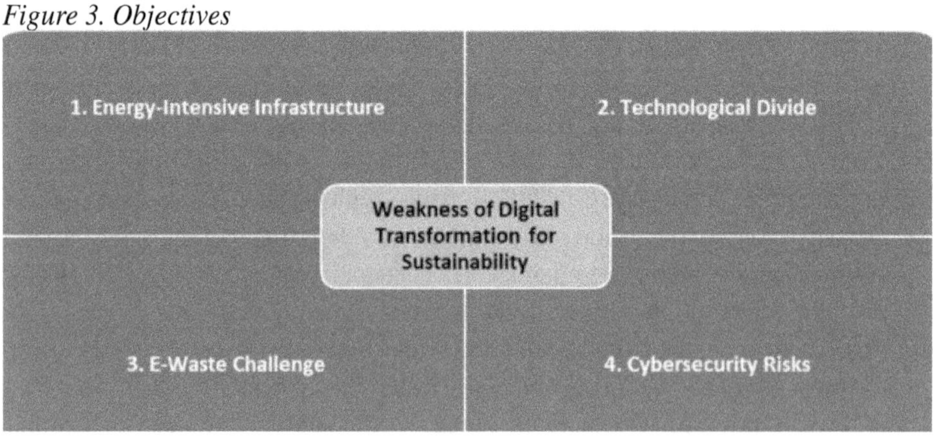

a. Digital Technologies Address Environmental Issues: In the chapter, the author will write on how AI, blockchain, IoT, and other technologies will solve climate change, resource scarcity, and pollution. For instance, IoT products can improve resource efficiency through facilitating efficient use of energy through supply of timely data hence reducing carbon footprints (Dwivedi et al., 2022)
b. Digital Innovations in Organizations: In the chapter, examples of organizations that have implemented the use of digital technologies to support sustainability outcomes will be discussed. For example, through blockchain technology, IBM creates solutions for supply chain sustainably and responsibly, avoiding the use of unlawful materials, as well as minimizing wastes (IBM, 2021).
c. Digital Technologies Enable Circular Economy: The chapter will also seek to describe how digital innovations help build the circular economy through sustainable manufacturing and consumption. For instance, applications based in digital environments can help to catalyse the product-service models so that consumers use goods as services, rather than acquiring ownership of products; thereby preserving products' lifespans and curtailing the production of unwanted objects (Bressanelli et al., 2018).
d. Strategies for Digital Transformation and Sustainability: Many recommendations and suggestions that will be provided at the end of the chapter will help organizations that are planning to digital their sustainability strategies. This entails the identification of solutions to implement existing solutions including the high cost of implementing processes, problems relating to technology application, and issues of non-compliance with change management (Sharma et al., 2022).

THEORETICAL FRAMEWORK

Digital Sustainability is hence the use of digital technologies and innovations in sustainable development. It describes the adoption of technologies to minimize impacts on the Earth's natural capital assets while ensuring social benefits enhancement. The idea is based on the belief that digitalisation remains capable of promoting the sustainable development goals (SDGs) by reducing costs for resources and promoting sustainability and equality (Sparviero et al., 2021). Digital sustainability can be defined as the application of information and communication technologies (ICTs) into various sectors including energy, transport and agriculture for sustainable purpose. For example, advanced grids through application of digital technologies in management and exploitation of electricity expending, and in the incorporation of fresh generation resources, decrease the result of greenhouse impact gases (IEA, 2020). Also, by giving farmers timely and accurate information that includes weather, soil, or crops' characteristics digital platforms will let enhance productivity while at the same time diminishing the effects of lasting rural cultivation on the environment (Lin et al., 2023).

Key components of digital sustainability include as shown in Figure 4:

Figure 4. Key components of digital sustainability

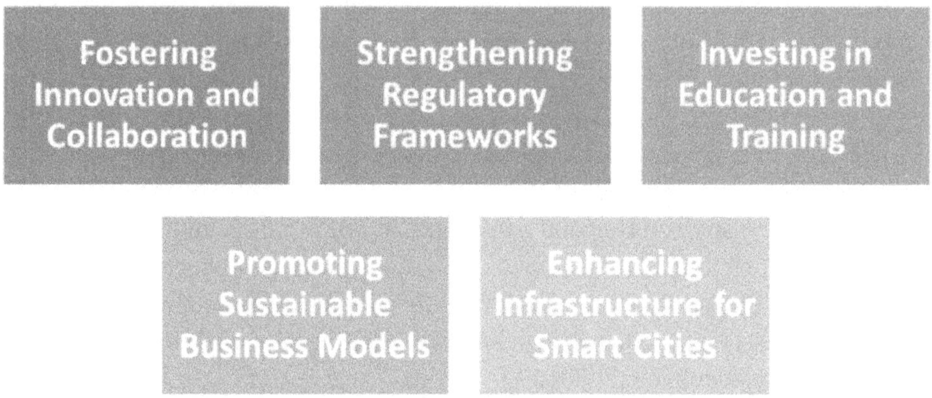

a. Digital Innovation: Creating and designing new technologies, web applications that work in favor of sustainable production and consumption.
b. Digital Infrastructure: Creating strong and scalable digital capabilities that can withstand wear and that supports sustainability work.
c. Digital Literacy: Making sure that everyone especially institutions has the skills that will enable them to use the digital technologies in sustainability.
d. Policy and Governance: Adoption of policies and governance that will promote the use of ICT for sustainability addresses the final agenda (World Economic Forum, 2020)

Concept of the Circular Economy

The circular economy is a concept linked to the idea of a developing an economy for the purpose of reducing waste in the society. This is not like the linear economy that the civilised world has been consuming, which is the 'take make dispose' manufacturing model. Circular Economy is about the idea of abolishing waste and pollution; using up products to employ them as infrastructure; and the rejuvenation of natural ecosystems (Morcillo-Bellido, et al., 2022)

Key principles of the circular economy include:

a. Design for Longevity: Designing products that are so robust that they can be reconditioned, remanufactured, and upgraded in a way that lengthens their operational life cycle.
b. Resource Efficiency: About resource utilisation in the sense where it is used in a more enhanced way and where as much as possible is gotten out of them before they are disposed.
c. Recycling and Reuse: Promoting recycling and reuse, in a manner which helps to ensure not only that items are used for as long as possible, but so that they can be reused or recycled before being used.
d. Systemic Thinking: Noting that this perspective takes into account all stages in the utilisation of products and systems, including their manufacturing, use and end of life (Iacovidou et al., 2021).

There are economic, environmental and social benefits of a transition to circular business models such as increased resource efficiency and reduced environmental impacts and more economic growth. For example, companies can support the ideas of material supply chain, and this can help companies to save material costs in general and enjoy advantages of reuse, such as decreasing the amount of waste (Stahel, 2019).

Overview of Green Technology that Drive Sustainability

Sustainable technologies are also called clean technologies as they are innovations that help to design products with less input from the natural world. These technologies play a crucial role in addressing global challenges such as climate change, resource depletion, and pollution. Figure 5. depicts key green technologies driving sustainability include:

Figure 5. Key green technologies driving sustainability include

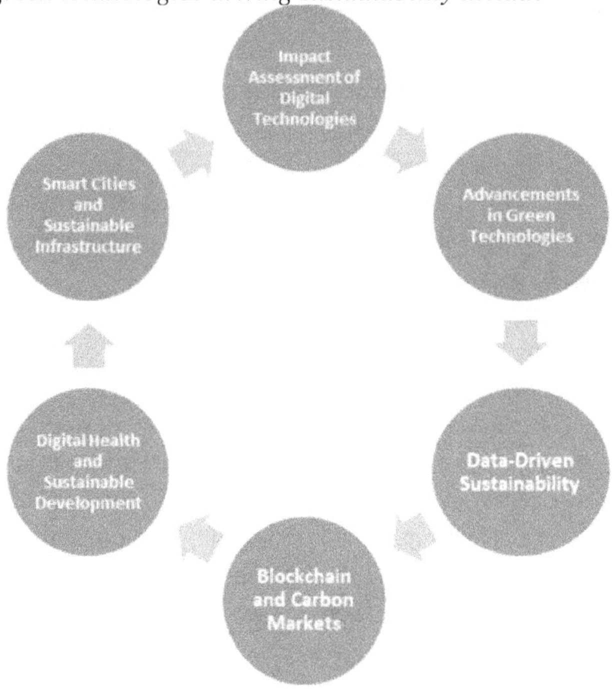

a. Renewable Energy Technologies: The relevant renewable power technologies include solar, wind, hydro and geothermal energy which offer the benefits of low fossil fuel utilization and low carbon emission (IRENA, 2020).
b. Energy Efficiency Technologies: Techno products including LED lights, smart thermostats and energy conserving home appliances decrease energy utilization and increase the efficiency of utilizing energy- (IEA, 2020).
c. Sustainable Agriculture Technologies: Smart farming, modern methods of climatically controlled farming like vertical farming and purely natural farming all increase efficiency of farming with as low effect on the environment which is a role played by the farming industry today.
d. Waste Management Technologies: Better technologies for recycling, composting, and efficient treatment of waste and use of waste energy as a source decrease the utilization of landfills (UNEP, 2020).
e. Water Conservation Technologies: Techniques like drip irrigation, rain water harvesting for crops and water recycled system help save water and water usage tendency (World Bank, 2020). It materializes sustainable development through

increasing resource efficiency by promoting clean and renewable technological applications that minimize the releases of greenhouse gases.

f. Green Building Technologies: Energy efficient construction practices, smart building systems, and sustainable materials like recycled steel and bamboo minimize the carbon footprint of infrastructure.

Triple Bottom Line and Sustainable Business Models

Elkington proposed the idea of TBL concept in 1994, TBL stands for triple bottom line, the concept of sustainability that enables the measurement of a business beyond its financial bottom line of profit. It encompasses three dimensions: as designated in Economic, Environmental and Social (Triple) Bottom Line by Elkington (2018). The three dimensions of Triple Bottom Line as shown in Figure 6:

Figure 6. Three dimensions of Triple Bottom Line

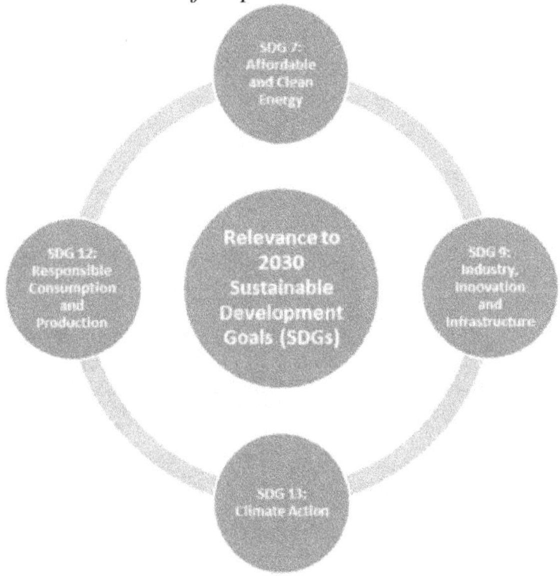

a. Economic Performance: This dimension bears the ability to measure the financial solvency and economic utility of the business. It encompasses profitability, revenues generated as well as cost reduction realised through environmental conservation (Norman & MacDonald, 2020).

b. Environmental Performance: Responsible for evaluating the effects of business operations on the physical world. These are carbon footprint, energy, waste, and resource efficiency indicators that should be captured and reported (Schaltegger & Wagner, 2017).
c. Social Performance: This was considered in this dimension to determine the effects of business activities on society. It encompasses workplace relations, involvement of the community, staff satisfaction and, general stewardship of people's rights (Dyllick & Muff, 2016).

TBL Business Models

TBL framework is incorporated into sustainable business strategies for the creation of economic, social and environmental value. These models are greatly different from such a freestyle business model as they are aimed at a future rather than at immediate financial gain. Examples of sustainable business models included as shown in Figure 7:

Figure 7. Sustainable business models

a. Circular Business Models: Companies integrate circularity through; product design and build for durability, ease of disassembly for recycling, as well as the use of renewable resources (Bocken et al., 2016).

b. Social Enterprises: Soscial missions go hand in hand with financial gains and allocate themselves to solving social challenges starting from poverty to education and healthcare issues (Tate et al., 2018.).
c. Green Businesses: Many organisations concentrate on lowering their environmental impact as a result of implementing green strategies, procurement of electricity from renewable sources, recycling, and green marketing (Hart & Dowell, 2011)

DIGITAL TRANSFORMATION AS A DRIVER OF SUSTAINABILITY

Sustainability is effectively promoted through digital transformation since technology helps to solve global problems in terms of the environment, society and economy. The use of IT such as, artificial intelligence, the internet of things, big data, and blockchain can enhance utilization of resources, energy, and materials therefore the carbon emulsions enhancing sustainable development. This is also possible through the Digital Transformation to a circular economy, where commonly used products and materials are recycled and used over and over. For instance, innovation in waste management are usually smart and the benefits include sorting and recycling thereby reducing landfill. In addition, this digital transformation drives sustainable decision-making through an analysis of available data. Analyzing environmental concerns, companies and governments can work out beneficial policies based on the principles of sustainability. Social media and other digital tools also support remote employment, thus decreasing emission due to commuting and promoting work-life balance.Digital transformation here ensures that digital will be new drivers of change, profit and SDGs, linked to sustainability practice. There is a prospect of this association as a catalyst for sustainability enhancing a more sustainable future as shown in figure 8.

Figure 8. Digital transformation as a driver of sustainability

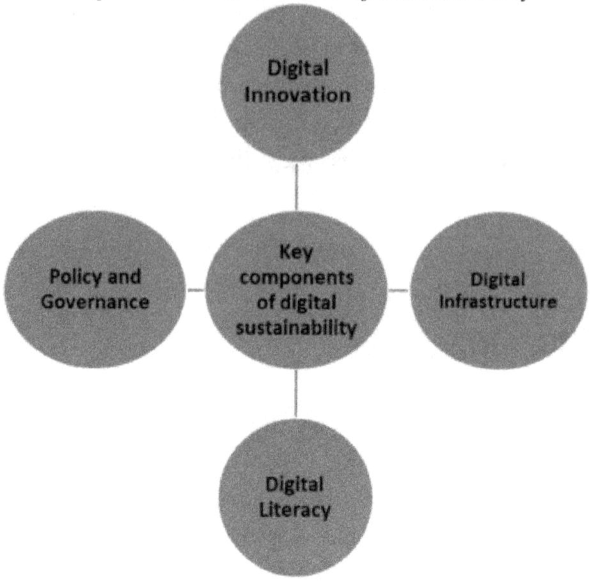

Optimizing Resource Use

Digital transformation means that firms make good use of resources by utilizing other technologies so that optimum results can be obtained when offering their products. The IoT, AI and big data technologies are critical to this optimization. For example, IoT devices can be used to continuously observe resources and efficiently control them; the received data can be analyzed and optimized (Mamudu et al., 2024).

By using AI algorithms, maintenance requirements in industrial equipment can be foreseen, thus decreasing the amount of time that such equipment spends in the workshop and the wastage of resources (Park et al., 2023). In agriculture, precision farming technologies take data from sensor, drones and satellite imagery to improve the usage of water, fertilizer and pesticides with higher crop yields and less harm to the environment (Liakos et al., 2018). In the manufacturing industry, smart factories IoT and AI control production, optimize the usage of energy, and enhance the utilization of raw material (Abubakr et al., 2020).

Digital Solutions for Sustainable Development Goals (SDGs)

Digital transformation fosters the success of SDGs due to the formulation of solutions through the creation of technical advancements in social, economic and environmental fields. For instance, digital health technology seeks to address the unmet need for the healthcare services, increase the quality of health care and decrease the overall health care costs thus supporting the third SDG (Good Health and Well-being) (WHO, 2020). The accessibility of e-learning platforms and all the different forms of education applications on digital based platforms, falls under the support of the UN sustainable development goal number four, which is on quality education (UNESCO, 2020).

Mobile banking for example improves poverty reduction and that, besides, encouraging the utilization of blockchain technology advances SDG 1 on eliminating absolute poverty and SDG 8 on decent work and economic growth. Moreover, the digitalisation platform drives resource efficiency and green consumption which is in alignment with the tenth target of the developed SDG12.

Reduction of Carbon Footprint

Digital technologies significantly contribute to reducing the carbon footprint of various industries. For instance, smart grids and energy management systems optimize energy usage, integrate renewable energy sources, and reduce greenhouse gas emissions (IEA, 2020). Electric vehicles (EVs) and intelligent transportation systems decrease reliance on fossil fuels and reduce emissions from the transportation sector (IEA, 2021).

Telecommuting and virtual collaboration tools reduce the need for physical office space and business travel, leading to lower carbon emissions associated with commuting and air travel (Hook et al., 2020). Additionally, digital platforms for monitoring and managing carbon emissions help organizations track their environmental impact and implement strategies to achieve carbon neutrality (Gholami et al., 2021).

Enabling Circular Economy

Interestingly, digital augmentation fosters the circular economy because digital tools help to design, produce, and consume products in the fashion that optimizes resource usage. Digital platforms and IoT devices also use lifecycle data on the cycle of a product, including how it is used, maintained and when it can be recycled (Pagoropoulos et al., 2017). This data also serves to assist the industry to create products that are easy to maintain, repairable and recyclable, and thus keeping material in use for as long as possible. Blockchain technology improves supply chain transparency and traceability, making sustainable supply chain a possibility while minimizing wastage and fraud (Saberi et al., 2019). It concerns the fact that online marketplaces and sharing economy platforms effectively increase the durability of products by offering opportunities for reuse, repair, and recycling of the received products, and, thus, exclude the need for new resources (Ellen MacArthur Foundation, 2020).

Integration with Environmental Management

Management with Traditional Management Systems' April, 22, 2001 Environmental management systems connect with the digital transformation process that affects environmental monitoring and reporting processes along with regulatory impacts elimination. GIS, remote sensing help to obtain highly detailed information about conditions in environment and are helpful in preserving natural resources and developing effective branching strategies (Asefa et al., 20222). ENVIRONMENTAL APPLICATIONS involves the use of sensors and IoT devices in monitoring quality of air and water pollution, detection as well as recommending measures to be taken (Andries et al., 2022).

Environmental management software solutions automate the process of responding to environmental requirements, monitoring environmental performance and creating reports, thus contributing to corporate reporting and responsibility (Truong et al., 2022). Also, digital twin, the virtual copy of physical assets, also makes it possible to monitor and improve environmental conditions in real-time at manufacturing facilities, power generation and construction sectors among others (Fuller et al., 2020).

CHALLENGES AND BARRIERS

Integration Challenges

One of the primary challenges in leveraging digital transformation for sustainability is the integration of new technologies with existing systems. There are always challenges resulting from such issues as the integration of an organization's old working systems, which may not be compatible with any digital solutions in existence. Integration issues tend to pose negative impacts on business operations such as disruptions, create data silos and lead to costs of running a business. For example, investments in IoT technology to connect embedded systems to existing manufacturing equipment are high in terms of physical infrastructure and human capital (Bowersox et al., 2020). Furthermore, moving data between different platforms and systems is still challenging, and requires well-defined reference architectures and protocols for interoperability (Chen et al., 2021).

Technological Barriers

Critically reviewed and governed constraints and limitations such as, technological factors like restricted access to the advanced technologies, and poor digital infrastructure are disruptive challenges to digital transformation. In many parts of the world, especially the developing world, the digital divide persists as a major factor that hinder the successful advancement of digitization (Ragnedda & Ruiu, 2020). Also, the adoption of advanced technologies including AI, blockchain, and IoT is expensive for those firms especially for SMEs (Zhang et al., 2021). Besides, legal issues emerge as a threat to the implementation of digital technologies, because organizations may be worried about cyber threats and losses of customers' data (Sulich et al., 2021).

Environmental and Social Trade-offs

While digital transformation could provide wide range of sustainability advantages, this occurs with contingent environmental and social trade-offs. The use of technology and the disposal of technology products generate e-waste alike, which has major enviromental impacts (Forti et al., 2020). Moreover, the use of data centres and blockchain processing may extremely increase energy consumption thus becoming the major factor threatening the reduction of greenhouse gases (Jones et al., 2021). From a social perspective, the rate of digitalisation can widen the gap in that particular society since some societies do not possess the knowledge, skills, and probably hardware to go digital (Hilbert, 2020).

Regulatory and Policy Issues

It was, therefore, found that the regulatory and policy environment play a major role in determining the success of the digital change for sustainability (Heath et al.,2019) pointed that sometimes, the original and insufficient regulations do not consider the features of the new media technologies, resulting in increased ambiguity and reduced innovation (Mandel et al., 2020). Furthermore, it is argued that two different policies in different regions make the execution of international digital strategies challenging, and different legal frameworks have to be adhered to by organizations (Deloitte, 2020). Preserving legal data protection means, like the General Data Protection Regulation (GDPR) in Europe, while being crucial from the privacy perspective, come with constraints for data sharing and exchange required by digital transformation processes (de Hert et al., 2020).

Skills and Workforce Challenges

As seen above, effective Digital Transformation for sustainable development needs a workforce that has adequate digital literacy skills and characteristics. However, there is still a large skills deficit, employees often do not have enough experience to effectively operate with sophisticated digital tools. Such a gap can only be closed by training and upskilling programs and these take time and resources, which organisations may not have. Secondly, there is always a demand for new position and new job title that will facilitate the new digital activities which can give rise to shift in employment of workers (Bessen, 2019). These adjustments involve the need for huge investment on the part of the public and private sectors in education and training that will mobilize workforce for a new period of adjustment to the digital world.

Emerging Trends

Trends emerging in the post modern decades of techno-organic development are related with changes and innovations of industries and societies. These trends open the new form of interventions to respond to the urgent environmental problems, to improve the performance and to establish the sustainable environment as seen in Figure 9.

Figure 9. Emerging trends

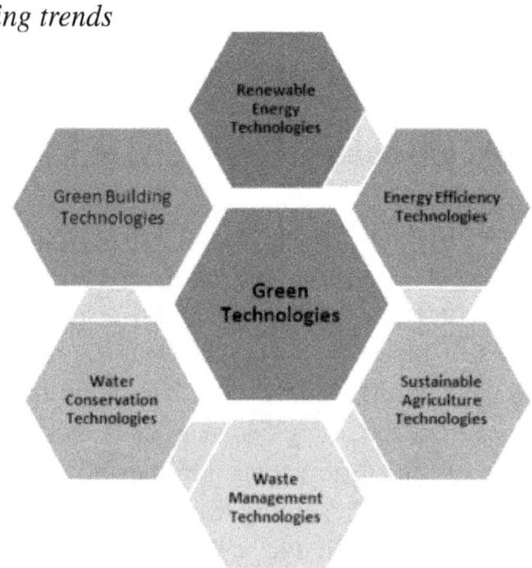

Smart Cities and Infrastructure

The term smart cities is transforming the ways cities are designed and operated through the application of digital systems that boosts sustainability as well as quality of life. IoT, AI, and big data help smart cities concentrate resources, minimize the unnecessary utilization of resources and enhance power utilization effectiveness. For example, smart grids can shift the extent of electrical energy supply according to consumers' actual need or smart water systems can help in identification of leakage and usage of water efficiently (Allam & Dhunny, 2019). In addition, things like smart transit and several other ITS applications play an essential role in minimizing concerns like congestion and emissions (Zhang et al., 2021). As cities continue to grow, the integration of smart technologies will be crucial for achieving sustainable urban development.

Blockchain and Carbon Markets

Blockchain can be used to improve carbon credits system by providing greater transparency and security of trades in carbon markets. Thus, on the one hand, blockchain technology is capable of reducing the presence of false data and cases of double-counting in the processes of carbon trading, improving people's confidence in these markets (Hua et al., 2020). Also, blockchain may help to monitor emissions throughout supply chains more effectively and support environmentally sustainable management, and compliance with legislation (Schulz et al., 2021). Thus, this study examines how blockchain applications are set to become critical as carbon markets develop.

Digital Health and Sustainability

Telemedicine, wearables, and AI-driven diagnosis are expanding as new forms of treatment delivery and as agents of sustainability. A decrease in the number of actual patient visits means that telemedicine also leads to a decrease in emission generated through transportation and infrastructure of healthcare facilities (Hong et al., 2020). Wearable technology helps monitor patient's health conditions on a more frequent basis hence cuts down the frequency of hospital visits, and use of remote monitoring devices ease the burden of the health systems. Diagnostic technologies can contribute positively to healthcare quality and cost-containment since artificial intelligence defines excellent diagnostic approaches (Keesara et al., 2020). Digital health technologies are expected to remain a critical component in the construction of robust, adaptive care delivery systems.

AI and Predictive Analytics

Artificial intelligence (AI) and predictive analytics are driving significant advancements in sustainability by enabling more efficient and informed decision-making. AI algorithms can analyze vast amounts of data to identify patterns and predict outcomes, helping organizations optimize resource use and reduce waste (Sürmeli et al., 2020). For example, AI can predict equipment failures in manufacturing, allowing for proactive maintenance and reducing downtime. In agriculture, AI-driven predictive analytics can optimize irrigation and fertilization schedules, leading to more sustainable farming practices (Kamilaris et al., 2021). As AI technology continues to evolve, its applications in sustainability will expand, offering new opportunities for efficiency and innovation.

Green Innovation Ecosystems

AI and predictive analysis play a pivotal role in making enhanced sustainable improvements in social and management practices as more optimized solutions are pursued. AI computing can process large volumes of information and produce accurate trends and forecasts; organisations can then manage their resources prudently and efficiently, and avoid lavish spending (Sürmeli et al., 2020). For instance, AI can determine equipment failure in production lines, and thus prevents replacement before such failure occurs. In the agricultural sector, AI-driven predictive analysis can enhance companies' policies of watering the crops and feeding the plants in order to save the environment (Kamilaris et al., 2021). The uses of knowledge in sustainability by AI are projected to increase as more development uncovers more possibilities in the system.

Sustainability in Cloud Computing

Green innovation system comprises of several organizations that are in a position to facilitate, generate, disseminate and adopt sustainable improvement throughout the technological value chain. These ecosystems facilitate knowledge exchange, cooperation in research, and venture in green technologies needed for product development (Cohen & Muñoz, 2021). For example, cleantech clusters connect entrepreneurial companies in the sector for renewable energy, energy efficiency, and waste management in order to drive innovation and the implementation of sustainable technologies faster (Bergman et al., 2019). Green Innovation Ecosystems will therefore be pivotal in supporting sustainable development by encouraging collaboration an Innovation.

Strength of Digital Transformation for Sustainability

Digitalisation is the application of digital tools to different sectors transforming how companies, administrations and citizens' work. Especially it is critical in achieving sustainability by encouraging adoption of new ideas, optimizing resources use, and informing better choices. In linking digital transformation with SDGs, organizations use their best approach to solve environmental, social and economic problems. Some of the strengths as shown in figure 10.

Figure 10. Strength of digital transformation for sustainability

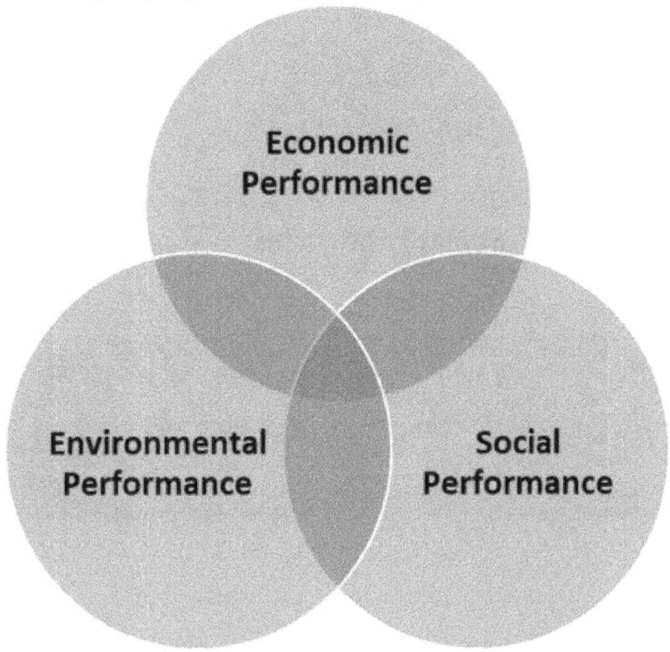

a. Enhanced Resource Efficiency: Some examples of such technologies are the Internet of Things and Artificial Intelligence, which can contribute all round towards reducing wastage and energy utilities.
b. Scalable Solutions: Such challenges, including energy and the circular supply chain, are indeed solved using digital platforms at scale.
c. Support for Data-Driven Policies: This kind of analysis provides greater detail to help create good environmental policies.
d. Inclusivity: Telemedicine and e-learning self-service models break down barriers in healthcare services and learning environment.

Weakness of Digital Transformation for Sustainability

There are various weaknesses noted regarding the same. These challenges can prevent it from achieving its cardinal goals of supporting sustainable development. Improving these weaknesses is crucial in order that digital transformation can be positive on environmental, economical and social perspectives. Some of the weaknesses as shown in Figure 11.

Figure 11. Weakness of digital transformation for sustainability

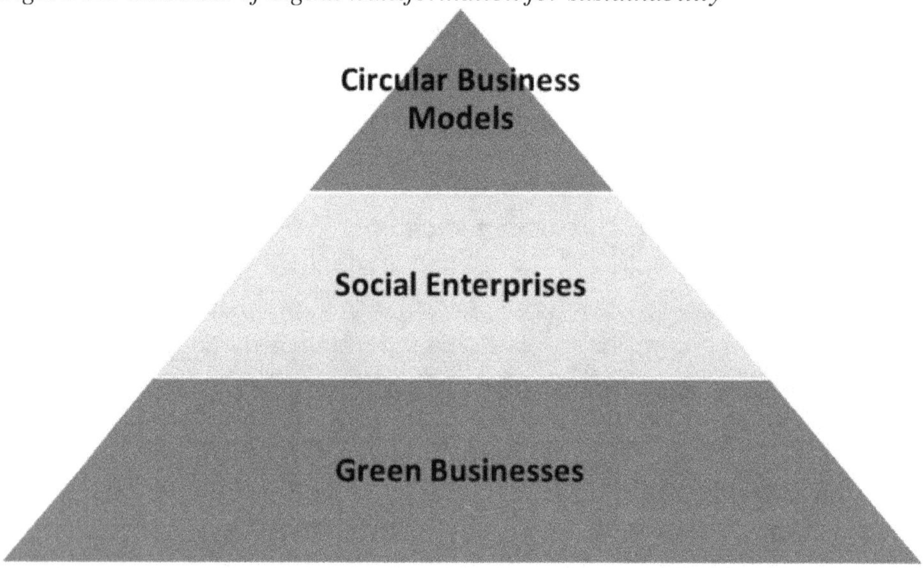

a. Energy-Intensive Infrastructure: That is why the energy consumption by data centers and blockchain can have negative impacts on the environment.
b. Technological Divide: Lack of access to digital technologies in developing areas may be an agent of increasing the gap that separates the haves from the have nots in economic terms.
c. E-Waste Challenge: This makes the generation of the electronic waste to be on the higher side through the enhanced use of the digital devices.
d. Cybersecurity Risks: Thus, there are risks inherent in digital infrastructure that may become dangerous if their impact on sustainable development is taken advantage of.

Recommendations for Practitioners and Policymakers

To effectively harness digital transformation for sustainability, practitioners and policymakers must prioritize several key areas as shown in figure 12:

Figure 12. Recommendations for practitioners and policymakers

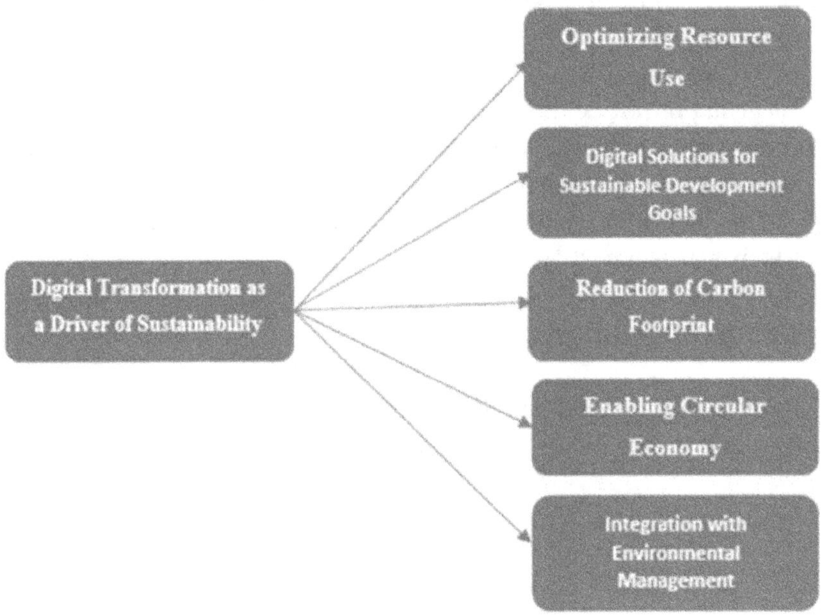

a. Fostering Innovation and Collaboration: Encouraging collaboration between private and public sectors, research institutions, and non-governmental organizations is crucial. Innovation ecosystems that support the development and deployment of green technologies can accelerate progress toward SDGs (Cohen & Muñoz, 2021).
b. Strengthening Regulatory Frameworks: Developing robust regulatory frameworks that support sustainable practices and technologies is essential. Policies should incentivize the adoption of green technologies, promote transparency in carbon markets, and ensure data privacy and security in digital health applications (Howson et al., 2019).

c. Investing in Education and Training: Addressing the skills gap is critical for the successful integration of digital technologies. Investments in education and training programs can equip the workforce with the necessary skills to navigate the digital economy and drive sustainable innovation (Kamilaris et al., 2021).
d. Promoting Sustainable Business Models: Businesses should adopt sustainable practices that align with the Triple Bottom Line, considering the environmental, social, and economic impacts of their operations. Implementing circular economy principles can enhance resource efficiency and reduce waste (Bergman et al., 2019).
e. Enhancing Infrastructure for Smart Cities: Developing infrastructure that supports smart city initiatives can significantly reduce urban environmental impacts. Investments in smart grids, intelligent transportation systems, and sustainable building technologies are essential for creating resilient urban environments (Allam & Dhunny, 2019).

Directions for Future Research

Future research should continue to explore the intersections of digital transformation and sustainability, focusing on several key areas as shown in figure 13:

Figure 13. Directions for future research

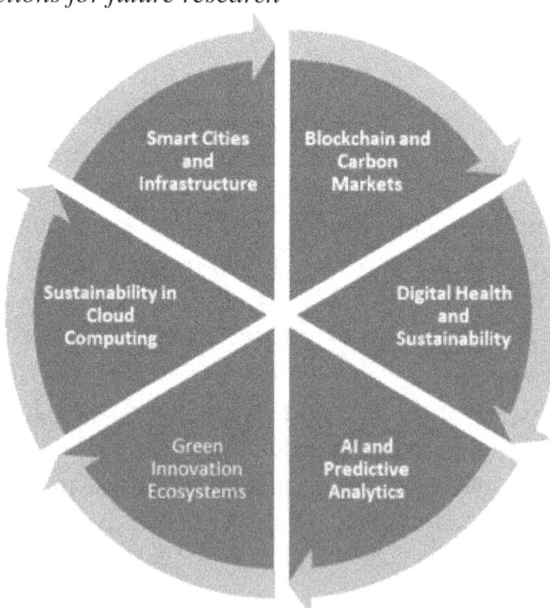

a. Impact Assessment of Digital Technologies: Comprehensive studies assessing the long-term environmental, economic, and social impacts of digital technologies are needed. Such research can inform policy decisions and help identify best practices for sustainable digital transformation (Jiang et al., 2021).
b. Advancements in Green Technologies: Continued research into green technologies, including renewable energy, sustainable agriculture, and waste management, is crucial. Innovations in these areas can significantly contribute to achieving SDGs and mitigating climate change (Sürmeli et al., 2020).
c. Data-Driven Sustainability: Leveraging big data and AI for sustainability requires further exploration. Research should focus on developing predictive analytics and decision-making tools that can optimize resource use and reduce environmental impacts (Zhang et al., 2021).
d. Blockchain and Carbon Markets: Investigating the potential of blockchain to enhance the transparency and efficiency of carbon markets is a promising area for future research. Studies should evaluate the scalability and practical applications of blockchain in various industries (Hong et al., 2020).
e. Digital Health and Sustainable Development: The role of digital health technologies in promoting sustainability and improving healthcare outcomes warrants further investigation. Research should examine the effectiveness of telemedicine, wearable devices, and AI-driven diagnostics in different healthcare settings (Keesara et al., 2020).
f. Smart Cities and Sustainable Infrastructure: Future research should explore the development and implementation of smart city technologies, focusing on their potential to enhance urban sustainability. Studies should evaluate the impacts of smart grids, intelligent transportation systems, and other smart infrastructure on energy use and emissions (Baliga et al., 2021)

CONCLUSION

This chapter presents a deeper and more effective analysis of an interdependence between digitalization and sustainability. It describes how IoT, blockchain, AI and big data are not only making better use of the world's resources but are also creating radical system shifts towards circularity. The issues highlighted hereby point to growing need to ensure that technological breakthroughs are properly complimented by ethical, environmental and social factors. It is crucial for the mutually inclusive relation between digitalization and sustainability, as the long-term development trends, that the concepts should be based on the principles of social and fair distribution of resources, as well as on the prioritization of efficient governance solutions and further advancement of green technologies. Governments and organizations need to join hands with each other to bring about the necessary change needed to achieve sustainable development for the improved use and diffusion of the internet.

The journey through digital transformation and sustainability reveals a nuanced landscape where technology intersects sustainable development goals (SDGs) to foster a more resilient and equitable future. Digital transformation serves as a critical driver for sustainability by optimizing resource use, reducing carbon footprints, and enabling the circular economy. Technologies such as IoT, AI, and blockchain are pivotal in advancing these objectives, offering innovative solutions for energy efficiency, smart cities, and transparent carbon markets.

The theoretical frameworks explored, including the concepts of digital sustainability, the circular economy, and the Triple Bottom Line, provide a robust foundation for understanding how digital innovations can support sustainable business models. Green technologies, ranging from renewable energy to sustainable agriculture, play a vital role in this ecosystem, demonstrating significant potential to mitigate environmental impacts while promoting economic growth.

However, the integration of these technologies is not without challenges. Barriers such as technological limitations, regulatory hurdles, and the need for a skilled workforce present significant obstacle. Environmental and social trade-offs also highlight the complexity of achieving true sustainability, requiring a balanced approach that considers the long-term impacts of digital innovations.

REFERENCES

Abubakr, M., Abbas, A. T., Tomaz, I., Soliman, M. S., Luqman, M., & Hegab, H. (2020). Sustainable and smart manufacturing: An integrated approach. *Sustainability (Basel)*, *12*(6), 2280. DOI: 10.3390/su12062280

Allam, Z., & Dhunny, Z. A. (2019). On big data, artificial intelligence and smart cities. *Cities (London, England)*, *89*, 80–91. DOI: 10.1016/j.cities.2019.01.032

Andries, A., Morse, S., Murphy, R. J., Lynch, J., & Woolliams, E. R. (2022). Using data from earth observation to support sustainable development indicators: An analysis of the literature and challenges for the future. *Sustainability (Basel)*, *14*(3), 1191. DOI: 10.3390/su14031191

Asefa, E. M., Barasa, K. B., & Mengistu, D. A. (2022). Application of geographic information system in solid waste management. In *Geographic information systems and applications in coastal studies*. IntechOpen. DOI: 10.5772/intechopen.103773

Baliga, J., Ayre, R., Hinton, K., & Tucker, R. S. (2021). Green cloud computing: Balancing energy in processing, storage, and transport. *Proceedings of the IEEE*, *99*(1), 149–167. DOI: 10.1109/JPROC.2010.2060451

Bergman, J. P., Hajikhani, A., & Blomqvist, K. (2019). Emergence and development of the cleantech industry: A cognitive construction approach. *Journal of Cleaner Production*, *233*, 1170–1181. DOI: 10.1016/j.jclepro.2019.05.369

Bessen, J. (2019). Artificial intelligence and jobs. The economics of artificial intelligence: an agenda, 291.

Bibri, S. E. (2018). The IoT for smart sustainable cities of the future: An analytical framework for sensor-based big data applications for environmental sustainability. *Sustainable Cities and Society*, *61*, 102328. DOI: 10.1016/j.scs.2017.12.034

Bocken, N. M. P., de Pauw, I., Bakker, C., & van der Grinten, B. (2016). Product design and business model strategies for a circular economy. *Journal of Industrial and Production Engineering*, *33*(5), 308–320. DOI: 10.1080/21681015.2016.1172124

Bressanelli, G., Adrodegari, F., Perona, M., & Saccani, N. (2018). Exploring how usage-focused business models enable circular economy through digital technologies. *Sustainability (Basel)*, *10*(3), 639. DOI: 10.3390/su10030639

Chen, X., Zheng, Y., & Zhu, K. (2021). Integration of IoT with legacy systems in manufacturing: An application to machine fault diagnosis. *Journal of Manufacturing Systems*, *59*, 207–216.

Cohen, B., & Muñoz, P. (2021). Sharing cities and sustainable consumption and production: Towards an integrated framework. *Journal of Cleaner Production, 280*, 124347.

Cricelli, L., & Strazzullo, S. (2021). The economic aspect of digital sustainability: A systematic review. *Sustainability (Basel), 13*(15), 8241. DOI: 10.3390/su13158241

De Hert, P., Papakonstantinou, V., Malgieri, G., Beslay, L., & Sanchez, I. (2020). The right to data portability in the GDPR: Towards user-centric interoperability of digital services. *Computer Law & Security Review, 34*(2), 193–203. DOI: 10.1016/j.clsr.2017.10.003

Deloitte. (2020). [Innovating in the digital era. Deloitte Insights.]. *TechTrends*, •••, 2020.

Dembski, F., Wössner, U., Letzgus, M., Ruddat, M., & Yamu, C. (2020). Urban digital twins for smart cities and citizens: The case study of Herrenberg, Germany. *Sustainability (Basel), 12*(6), 2307. DOI: 10.3390/su12062307

Deren, L., Wenbo, Y., & Zhenfeng, S. (2021). Smart city based on digital twins. *Computational Urban Science, 1*(1), 1–11. DOI: 10.1007/s43762-021-00005-y

Dwivedi, Y. K., Hughes, L., Kar, A. K., Baabdullah, A. M., Grover, P., Abbas, R., Andreini, D., Abumoghli, I., Barlette, Y., Bunker, D., Chandra Kruse, L., Constantiou, I., Davison, R. M., De', R., Dubey, R., Fenby-Taylor, H., Gupta, B., He, W., Kodama, M., & Wade, M. (2022). Climate change and COP26: Are digital technologies and information management part of the problem or the solution? An editorial reflection and call to action. *International Journal of Information Management, 63*, 102456. DOI: 10.1016/j.ijinfomgt.2021.102456

Dyllick, T., & Muff, K. (2016). Clarifying the meaning of sustainable business: Introducing a typology from business-as-usual to true business sustainability. *Organization & Environment, 29*(2), 156–174. DOI: 10.1177/1086026615575176

Elkington, J. (2018). 25 years ago I coined the phrase "triple bottom line." Here's why it's time to rethink it. *Harvard Business Review*.

Forti, V., Baldé, C. P., Kuehr, R., & Bel, G. (2020). *The Global E-waste Monitor 2020: Quantities, flows and the circular economy potential. United Nations University*. UNU.

Fuller, A., Fan, Z., Day, C., & Barlow, C. (2020). Digital twin: Enabling technologies, challenges and open research. *IEEE Access : Practical Innovations, Open Solutions, 8*, 108952–108971. DOI: 10.1109/ACCESS.2020.2998358

Gasper, D., Shah, A., & Tankha, S. (2019). The framing of sustainable consumption and production in SDG 12. *Global Policy*, *10*(S1), 83–95. DOI: 10.1111/1758-5899.12592

Gholami, R., Watson, R. T., Hasan, H., Molla, A., & Bjorn-Andersen, N. (2021). Information systems solutions for environmental sustainability: How can we do more? *Journal of the Association for Information Systems*, *21*(1), 8–21.

Hart, S. L., & Dowell, G. (2011). Invited editorial: A natural-resource-based view of the firm. *Journal of Management*, *37*(5), 1464–1479. DOI: 10.1177/0149206310390219

Havard, V., Sahnoun, M. H., Bettayeb, B., Duval, F., & Baudry, D. (2021). Data architecture and model design for Industry 4.0 components integration in cyber-physical production systems. *Proceedings of the Institution of Mechanical Engineers. Part B, Journal of Engineering Manufacture*, *235*(14), 2338–2349. DOI: 10.1177/0954405420979463

Hilbert, M. (2020). Digital technology and social change: The digital divide. In *Oxford Research Encyclopedia of Communication*. Oxford University Press.

Hoefflinger, B. (2011). The energy crisis. In *Chips 2020: A Guide to the Future of Nanoelectronics* (pp. 421–427). Springer Berlin Heidelberg. DOI: 10.1007/978-3-642-23096-7_20

Hong, Z., Li, N., Li, D., & Shen, J. (2020). Telemedicine during the COVID-19 pandemic: Experiences from Western China. *Journal of Medical Internet Research*, *22*(5), e19577. DOI: 10.2196/19577 PMID: 32349962

Hook, A., Court, V., Sovacool, B. K., & Sorrell, S. (2020). A systematic review of the energy and climate impacts of teleworking. *Environmental Research Letters*, *15*(9), 093003. DOI: 10.1088/1748-9326/ab8a84

Hua, W., Jiang, J., Sun, H., & Wu, J. (2020). A blockchain based peer-to-peer trading framework integrating energy and carbon markets. *Applied Energy*, *279*, 115539. DOI: 10.1016/j.apenergy.2020.115539

Iacovidou, E., Hahladakis, J. N., & Purnell, P. (2021). A systems thinking approach to understanding the challenges of achieving the circular economy. *Environmental Science and Pollution Research International*, *28*(19), 24785–24806. DOI: 10.1007/s11356-020-11725-9 PMID: 33289042

IBM. (2021). Blockchain for supply chain transparency. IBM Supply Chain

IEA. (2020). *World Energy Outlook 2020*. International Energy Agency.

IEA. (2021). *Global EV Outlook 2021: Accelerating ambitions despite the pandemic.* International Energy Agency.

IRENA. (2020). *Renewable capacity statistics 2020.* International Renewable Energy Agency.

Jiang, W., Jia, R., & Yu, X. (2021). Blockchain-based carbon trading: A case study of carbon trading. *Journal of Cleaner Production*, *277*, 123249.

Jones, N., Castellano, A., & Alison, H. (2021). Data centres and the energy use crisis. *Nature Reviews. Earth & Environment*, *2*(9), 435–446.

Kain, R., & Verma, A. (2018). Logistics management in supply chain–an overview. *Materials Today: Proceedings*, *5*(2), 3811–3816. DOI: 10.1016/j.matpr.2017.11.634

Kamilaris, A., Kartakoullis, A., & Prenafeta-Boldú, F. X. (2021). A review on the practice of big data analysis in agriculture. *Computers and Electronics in Agriculture*, *143*, 23–37. DOI: 10.1016/j.compag.2017.09.037

Keesara, S., Jonas, A., & Schulman, K. (2020). Covid-19 and health care's digital revolution. *The New England Journal of Medicine*, *382*(23), e82. DOI: 10.1056/NEJMp2005835 PMID: 32240581

Lee, J., Bagheri, B., & Kao, H. A. (2015). A cyber-physical systems architecture for industry 4.0-based manufacturing systems. *Manufacturing Letters*, *3*, 18–23. DOI: 10.1016/j.mfglet.2014.12.001

Liakos, K. G., Busato, P., Moshou, D., Pearson, S., & Bochtis, D. (2018). Machine learning in agriculture: A review. *Sensors (Basel)*, *18*(8), 2674. DOI: 10.3390/s18082674 PMID: 30110960

Lin, Y., & Li, C. (2023). Towards sustainable development: Research on the green growth effect of digital agriculture in China. *Environmental Science and Pollution Research International*, •••, 1–14. DOI: 10.1007/s11356-023-28367-2 PMID: 37391557

Mamudu, U. U., Obasi, C. D., Awuye, S. K., Danso, H., Ayodele, P., & Akinyemi, P. (2024). Circular economy in the manufacturing sector: Digital transformation and sustainable practices. *International Journal of Science and Research Archive*, *12*(2), 129–141. DOI: 10.30574/ijsra.2024.12.2.1217

Mandel, G. N. (2020). Regulating emerging technologies. In *Emerging Technologies* (pp. 361–378). Routledge.

Masanet, E., Shehabi, A., Lei, N., Smith, S., & Koomey, J. (2020). Recalibrating global data center energy-use estimates. *Science*, *367*(6481), 984–986. DOI: 10.1126/science.aba3758 PMID: 32108103

Morcillo-Bellido, J., Santos-Iscoa, E., & Isasi-Sánchez, L. (2022). Circular Economy Analysis Applying Ellen MacArthur Model: Spanish Glass Sector Case. In *Ensuring Sustainability: New Challenges for Organizational Engineering* (pp. 13–21). Springer International Publishing. DOI: 10.1007/978-3-030-95967-8_2

Pagoropoulos, A., Pigosso, D. C. A., & McAloone, T. C. (2017). The emergent role of digital technologies in the circular economy: A review. *Procedia CIRP*, *64*, 19–24. DOI: 10.1016/j.procir.2017.02.047

Park, M., & Singh, N. P. (2023). Predicting supply chain risks through big data analytics: Role of risk alert tool in mitigating business disruption. *Benchmarking*, *30*(5), 1457–1484. DOI: 10.1108/BIJ-03-2022-0169

Pivoto, D. G., De Almeida, L. F., da Rosa Righi, R., Rodrigues, J. J., Lugli, A. B., & Alberti, A. M. (2021). Cyber-physical systems architectures for industrial internet of things applications in Industry 4.0: A literature review. *Journal of Manufacturing Systems*, *58*, 176–192. DOI: 10.1016/j.jmsy.2020.11.017

Ragnedda, M., & Ruiu, M. L. (2020). *Digital capital: A Bourdieusian perspective on the digital divide*. Emerald Publishing Limited. DOI: 10.1108/9781839095504

Rolnick, D., Donti, P. L., Kaack, L. H., Kochanski, K., Lacoste, A., Sankaran, K., Ross, A. S., Milojevic-Dupont, N., Jaques, N., Waldman-Brown, A., Luccioni, A. S., Maharaj, T., Sherwin, E. D., Mukkavilli, S. K., Kording, K. P., Gomes, C. P., Ng, A. Y., Hassabis, D., Platt, J. C., & Bengio, Y. (2022). Tackling climate change with machine learning. *ACM Computing Surveys*, *55*(2), 1–96. DOI: 10.1145/3485128

Saberi, S., Kouhizadeh, M., Sarkis, J., & Shen, L. (2019). Blockchain technology and its relationships to sustainable supply chain management. *International Journal of Production Research*, *57*(7), 2117–2135. DOI: 10.1080/00207543.2018.1533261

Schaltegger, S., & Wagner, M. (2017). *Managing the business case for sustainability: The integration of social, environmental and economic performance*. Routledge. DOI: 10.4324/9781351280525

Schulz, K., & Feist, M. (2021). Leveraging blockchain technology for innovative climate finance under the Green Climate Fund. *Earth System Governance*, *7*, 100084. DOI: 10.1016/j.esg.2020.100084

Sharma, V., Singh, A., & Rai, S. S. (2022). Disruptions in sourcing and distribution practices of supply chains due to COVID-19 pandemic: A sustainability paradigm. *Journal of Global Operations and Strategic Sourcing*, *15*(2), 235–261. DOI: 10.1108/JGOSS-02-2021-0020

Sparviero, S., & Ragnedda, M. (2021). Towards digital sustainability: The long journey to the sustainable development goals 2030. Digital Policy. *Regulation & Governance*, *23*(3), 216–228. DOI: 10.1108/DPRG-01-2021-0015

Stuermer, M., Abu-Tayeh, G., & Myrach, T. (2017). Digital sustainability: Basic conditions for sustainable digital artifacts and their ecosystems. *Sustainability Science*, *12*(2), 247–262. DOI: 10.1007/s11625-016-0412-2 PMID: 30174752

Sulich, A., Rutkowska, M., Krawczyk-Jezierska, A., Jezierski, J., & Zema, T. (2021). Cybersecurity and sustainable development. *Procedia Computer Science*, *192*, 20–28. DOI: 10.1016/j.procs.2021.08.003

Sürmeli, B. G., & Tümer, M. B. (2020). Multivariate time series clustering and its application in industrial systems. *Cybernetics and Systems*, *51*(3), 315–334. DOI: 10.1080/01969722.2019.1691851

Tate, W. L., & Bals, L. (2018). Achieving shared triple bottom line (TBL) value creation: Toward a social resource-based view (SRBV) of the firm. *Journal of Business Ethics*, *152*(3), 803–826. DOI: 10.1007/s10551-016-3344-y

Truong, T. C. (2022). The impact of digital transformation on environmental sustainability. *Advances in Multimedia*, *2022*(1), 6324325. DOI: 10.1155/2022/6324325

UNEP. (2020). *Global waste management outlook*. United Nations Environment Programme.

UNESCO. (2020). *Global Education Monitoring Report 2020: Inclusion and Education: All Means All*. United Nations Educational, Scientific and Cultural Organization.

United Nations. (2020). The Sustainable Development Goals Report 2020.

Vial, G. (2021). Understanding digital transformation: A review and a research agenda. Managing digital transformation, 13-66.

Vilariño, M. V., Franco, C., & Quarrington, C. (2017). Food loss and waste reduction as an integral part of a circular economy. *Frontiers in Environmental Science*, *5*, 21. DOI: 10.3389/fenvs.2017.00021

WHO. (2020). *Digital health and COVID-19*. World Health Organization.

Willcocks, L. (2020). Robo-Apocalypse cancelled? Reframing the automation and future of work debate. *Journal of Information Technology*, *35*(4), 286–302. DOI: 10.1177/0268396220925830

World Bank. (2020). *Water in agriculture*. The World Bank.

World Economic Forum. (2020). The Global Risks Report 2020. World Economic Forum.

Zhang, J., Long, J., & Von Schaewen, A. M. E. (2021). How does digital transformation improve organizational resilience?—Findings from PLS-SEM and fsQCA. *Sustainability (Basel)*, *13*(20), 11487. DOI: 10.3390/su132011487

Chapter 6
Digitalization and Circular Economy:
Pioneering Sustainable Policies for Climate Change Mitigation and 2030 Sustainable Development Goals

P. Aranganathan
https://orcid.org/0000-0002-4132-5704
Hallmark Business School, India

R. V. Suresh
Gnanam School of Business, India

S. Manikandan
https://orcid.org/0000-0002-1815-410X
Gnanam School of Business, India

S. Christina Sheela
Gnanam School of Business, India

R. Sivarethinamohan
https://orcid.org/0000-0003-1918-8931
Symbiosis Centre for Management Studies, Bengaluru, India

Seda Yildirim
Tekirdag Namik Kemal University, Turkey

ABSTRACT

The 21st century has driven rapid changes in technology, economy, and environmental policies. Addressing climate change, resource depletion, and inequality requires innovations like digitalization and the circular economy. This study examines how digital technologies support sustainability and advance the circular economy and SDGs. Tools like AI, blockchain, and IoT transform industries by reducing energy use, waste, and enabling predictive maintenance. Solutions like product passports boost recycling, while green finance, such as bonds and carbon credits, supports climate action. Initiatives like smart cities, precision farming, and digital health-

DOI: 10.4018/979-8-3693-9909-5.ch006

care lower emissions and improve resource use. Challenges include data privacy, energy consumption, and e-waste. This study aids policymakers, researchers, and practitioners in harnessing digitalization for sustainable growth.

INTRODUCTION

In the twenty-first century, technological development has reached a different level, and environmental issues have come to the forefront. While all the world economies are becoming more integrated and reliant on the digital space, it has also become possible to solve some of the most important problems the world is facing at the present time—climate, resource, and social issues. The economy is experiencing transformation in its structures and processes, and as a result, the concept of "digitisation," which is pervasive in business and society, is becoming crucial to fostering sustainability. Along with the aims of the circular economy, digitisation provides the opportunity to revolutionise waste management, consumption, and production processes in a way that is consistent with the Sustainable Development Goals (SDGs) designed by the United Nations for 2030 (SSD).

Creating alternative methods and strategies to govern the world is imperative in addressing global problems such as climate change. Geissdoerfer et al. (2017) argue that due to the limited resources offered by the planet, the linear economic model, which rests on the ideologies of project planning, action, consumption, and abandonment of resources, is unrealistic and untenable. With an emphasis on resource efficiency and waste management, as well as the rehabilitation of ecosystems, the circular economy provides a feasible way to reduce environmental degradation whilst sustaining economic growth (Kirchherr, Reike, & Hekkert, 2017). The concepts of the circular economy comply with the concepts of the circular economy principles. The combination of decentralization with circular investments can accelerate circular economy operations, as decentralization allows for improved management of material arrays, increased transparency within the supply chains, and innovation in product lifecycle control (Bocken et al., 2016). This is so because optimization of resource flows is made possible by digital technologies.

The awareness of how digital tools can assist to enhance social and economic resilience during quiet emergencies is evident from the COVID-19 outbreak. Remote patient monitoring and telemedicine, among others, are considered digital health technologies that have changed the healthcare delivery system. Such technologies ensure that equal footing in medical treatment is observed and alleviate the stress on the physical facilities. Also, e-governance and delivery of public services in a digital form have also enabled speedy enacting of laws and improved accountability and participation of people in environmental decision-making (Janssen et al., 2020).

From these illustrations, the remarkable digitization and what it has the potential to achieve in solving complex global problems is highlighted.

While digitalisation is a great opportunity, it also entails new risks and challenges. Concerns have been raised regarding the environmental and social sustainability of digital technologies due to the energy usage associated with data centres, the creation of e-waste, and ethical concerns surrounding data and artificial intelligence (AI). Research by Van Dijk, (2020) suggests that disparities in accessing digital infrastructure might reinforce existing economic and social inequalities, thus creating a detrimental "digital divide" that compromises the sustainability agenda's inclusiveness. In order to effectively address these challenges, a more sophisticated understanding of the relationship of digitalisation, the principles of the circular economy, and the environmental objectives is needed.

The objective of this chapter is to investigate how digitisation together with the circular economy would facilitate sustainable development by reducing the impact of climate change and therefore assist in achieving Now under investigation are how AI, blockchain, and IoT can transform industries with sustainable practices, optimize resource use, and accelerate the transition toward a circular economy. Besides this, it studies the socio-political and governance impacts of digitalisation, thus emphasising the opportunities and challenges of integrating digitalisation in the SD strategies.

Digitalization as a Driver of Sustainability

The impact of digital technology is notable and transformational, given that the 21st century digitisation is evolving and revolutionizing not only economies and cultures but also addressing some of the most environmental challenges of our time. Sustainability could, however, be achieved through digital technology that optimises resource usage, enhances performance evaluation, and supports a data-driven model for executive management. With the rise of digitally driven technologies such as AI, blockchain, and IoT, organisations are being reformed to embrace sustainable development and promote the use of circular economies.

Overview of Digital Technologies in Economic and Environmental Transformation

Admits the globalization, recent advancements in digital technologies, including AI, blockchain, and IoT extend their scope in tackling the global problems in sustainability and climate change. They significantly reduce resource consumption and emissions and support waste management through process optimization and big data analytics, thereby improving the efficiency, transparency, and sustainability of operations.

Digital transformation will accompany the shift of old business models into modernized categories with the aim of incorporating green practices in numerous sectors, including transport, manufacturing, and agriculture. AI is transformed into a great disruptive innovation that drives efficiency and sustainability across the whole society beyond being a utilitarian tool (Sorrell, 2015).

- **Contributions of AI, Blockchain, and IoT to Sustainable Practices Artificial Intelligence (AI):** The role that artificial intelligence plays in enhancing the state of sustainability cannot be ignored, as it allows for more precise forecasting and optimisation solutions across various domains. There is a system powered with artificial intelligence that has the ability to monitor and project the trends within the energy industry and consequently optimize the grid's performance and reduce energy losses (Li et al, 2020). Huang et al. (2017) observe that machine learning methods are able to work with large-scale datasets from multiple sources in order to recognize patterns and improve their operational efficiency. Such parameters could include reducing water and energy consumption during production processes or predicting the breakdown of machinery that would extend the production life of the products. Furthermore, Artificial Intelligence technologies are being applied in the monitoring of the environment, for instance, the evaluation of pollution levels or the assessment of the effects of climate change. This facilitates better policy decisions and resource allocation (Jain et al., 2017).

Moreover, it is interesting how artificial intelligence contributes to enhancing the circular economy's approaches in which the business undertakes to redesign the entire production process. A business may design the products in such a way that fewer materials are wasted through optimised materials usage. This is made possible by the use of machine learning algorithms. In their study, Kirchherr et al. (2017) stated that artificial intelligence advanced technologies enable the development of intelligent recycling systems. Such systems autonomously sort and process waste streams based on their composition, thereby increasing recycling rates and decreasing the volume of waste disposed of in landfills. The predictive capabilities of the artificial intelligence systems may also be employed in supply chain management systems and assist companies to curb wastage, overproduction, and plastics' lifetimes of their products.

- **Blockchain:** It is better known for its association with cryptocurrencies, but blockchain technology has great potential for increasing accountability and transparency in environmental governance and sustainability-related activities. Due to the distributed and non-revisable ledger system of the blockchain, it is possible to trace the origin of raw materials to the product's end-of-life stage, ensuring transparency in supply chains (Tapscott & Tapscott, 2016). For example, blockchain technology can help in efficiently sourcing materials. ethically and also ensure the non-depletion of natural resources while meeting the required ecological standards. Records of proper conduct based on verifiable standards of sustainability may allow for the advocacy of responsible consumption, the advancement of corporate social responsibility initiatives, and the provision of transparent histories of products to consumers. Besides the transparency of the supply chains, blockchain technology can also be utilized to implement digital product passports that contain specific data about the environmental impact of a given product, such as its recyclability, materials used, and the energy consumption during its lifecycle.

According to Zhu et al. (2019), the consumers can make more informed decisions as these passports motivate manufacturers to produce goods that are more environmentally sustainable and easy to recycle. In addition, through the security of the blockchain, distributed energy trading is made possible, where individuals can sell surplus renewable energy use, thus endorsing integrated energy systems and the use of clean energy sources. Table 1 lists some of the most important technologies and explains their contribution to the promotion of sustainability.

Table 1. Key digital technologies and their contributions to sustainability

Technology	Contribution	Example Application	Impact on Sustainability
Artificial Intelligence (AI)	Energy optimization, predictive analytics, waste reduction	AI in smart grids for energy efficiency	Reduced carbon emissions and energy use
Blockchain	Supply chain transparency, green finance, tracking renewable energy certificates	Blockchain for ethical sourcing	Enhanced accountability and trust
Internet of Things (IoT)	Real-time monitoring of resources, environmental data collection, predictive maintenance	IoT-enabled precision agriculture	Resource efficiency and lower environmental impact
Digital Twins	Virtual modelling of systems for optimization	Digital twins in urban planning	Reduced waste, better energy and resource use
Big Data Analytics	Climate modelling, policy impact assessment	Climate data visualization platforms	Improved decision-making in sustainability policies

Source: Created by authors

- **Internet of Things (IoT):** The Internet of Things (IoT) is a vast network of interconnected devices that can aggregate and exchange information that allows the current monitoring of multiple systems and their functioning. Many benefits accrue from IoT particularly with regard to sustainability initiatives or conservation targets, as IoT technologies help in resource management and waste minimization. For instance, smart buildings integrate IoT sensors to automatically turn on and off HVAC and lighting systems depending on the patterns of usage, which reduces energy loss (Alwan et al., 2017). Through the collection of data on soil moisture levels, air temperature, and vegetation health, IoT devices enable farms to reduce water and pesticide consumption effectively, thereby facilitating advanced farming technology in agriculture. As for the IoT metrics dealing with ecological education, it includes the intelligent use of water and waste disposal practices. Smart waste management systems utilize IoT sensors to monitor the amount of waste in bins and optimize movement patterns for collection routes, thus reducing the amount of diesel fuel consumed by garbage trucks. Water systems using IoT devices can also quickly repair leaks, thus reducing wastage of water (Gao et al., 2019). These schemes also help municipalities and businesses save considerable amounts of money while at the same time strengthening resource management.

Digitalization's Role in Enhancing Resource Efficiency

Digitization allows for more effective management of resources, helps reduction of waste, and enables the increasing the lifespan of goods and services rendering the operations of industries more efficient. More efficient production processes in the industry are made possible by predictive maintenance systems and advanced robots capable of more effective utilization of resources whilst reducing energy requirements (Sajjad et al., 2021). Moreover, the use of digital technology also ensures that supply chains and logistics are effective, thereby ensuring that resources are distributed in a manner that has a lower impact on carbon emissions from transport.

One of the most significant ways the resource economy is supported through digitalisation is by the development of 'smart' infrastructure such as buildings and highways. For instance, smart cities deploy IoT, sensors, and AI to monitor and manage urban resources such as waste, water, and electricity. Cities can analyse home-to-other-places data in a comprehensive way in order to enhance energy efficiency and cut down on water waste and maintenance of waste treatments, thus reducing expenses and pollution. This, in turn, not only helps cities to be environmentally friendly by reducing pollution but also improves the quality of life for the citizens through better constructive basic service delivery. In recent years, digitalization has been acting as a key driver of sustainability that positively contributes to innovative solutions addressing farmers' emerging challenges. A huge potential of resources optimization, supply chain transparency, and a clear pathway towards the circular economy is offered by artificial intelligence, blockchain, and the internet of things. Such tools not only reduce costs and increase operational efficiency, thus preserving the environment, but also foster economic development. As we all know, many technologies have a great capacity to enhance sustainable practices and help achieve the 2030 Sustainable Development Goals (SDGs), thus, digitalisation is proving to be more crucial for global sustainability in the future as it advances.

CIRCULAR ECONOMY: CONCEPT AND RELEVANCE

The shift towards Circular Economy (CE) focus began in the mid-80s and has remarkably continued to gain momentum in addressing the problematic scenarios of resource depletion, huge waste generation, and environmental issues. Negative linear economy is different, which involves a transport, make, and waste system. The circular economy restores the environment by seeking to create an ecosystem where the product, materials, and resources are utilized for as long as possible. This chapter looks at what the circular economy is, what principles guide it, how societies can evolve from the take-make-waste model to a sustainable business model (the circular system), and how digital technologies help achieve the circular economy objectives.

Definition and Principles of Circular Economy

The circular economy is a production and consumption model focused on sharing, leasing, reusing, repairing, refurbishing, and recycling materials and products to extend their lifecycle. By minimizing waste, it ensures that when a product reaches the end of its life, its materials remain in the economy through recycling. These materials can be reused repeatedly, generating additional value.

- **Definition:** According to Kirchherr et al., (2017), "a circular economy describes an economic system that is based on business models that replace the 'end-of-life' concept with reducing, alternatively reusing, recycling, and recovering materials in production/distribution and consumption processes, thus operating at the micro level (products, companies, consumers), meso level (eco-industrial parks), and macro level (city, region, nation, and beyond), with the aim to accomplish sustainable development, which implies creating environmental quality, economic prosperity, and social equity, to the benefit of current and future generations."

A prerequisite to CE is making goods in such a way that at the initial stages of the life of the product, there is little or no waste and pollution. Here, choosing sustainable materials, designing for longevity, and producing them in such a way that makes them easier to fix, take apart, or recycle Bocken et al. (2016). The expansion of the life of such goods forms the basis of a circular economy where goods and materials are sustained through maintenance, repair, and reuse of such materials in creating new goods. When products approach the end of life, their materials are reclaimed, recycled, and others are retrieved to create new products (Ghisellini et al., 2016).

The CE model focuses on the restoration and restoration of natural ecosystems that have already been disturbed. This principle is promoted by reforestation, the adoption of more eco-friendly farming practices, and the mending of the ecological equilibrium through natural mechanisms (Murray et al. 2017). These concepts aim not only to minimize the harm to the environment but also to create new business opportunities by stimulating innovations, reducing the costs of raw materials, and thereby making production and consumption patterns more eco-friendly. It is, thus, not a surprise that CE has been viewed as an effective approach to achieving sustainability and resolving global challenges such as inequality, resource scarcity, climate change, and others. There are basic changes in the resource use and the business models during the change from a linear to a circular economy, as summarized in Table 2.

Table 2. Comparison between linear and circular economic models

Aspect	Linear Economy	Circular Economy
Resource Use	Extract-Produce-Dispose	Reuse-Recycle-Regenerate
Waste Management	End-of-life disposal	Minimized through recycling and reuse
Environmental Impact	High	Lower
Business Models	One-time sales	Service-based, circular supply chains
Role of Digitalization	Limited	Enables transparency and resource efficiency

Source: Created by authors

Transition from Linear to Circular Economic Models

Moving from a linear to a circular model of the economy is revolutionary in how society and businesses manage their resources. The traditional "use-and-throw" economy is based on the assumption that there are still ample resources available and products do not have to be used again. According to Korhonen et al. (2018), this paradigm is no longer tenable since there is now a persistent increase in waste generation, resource reduction, as well as environmental impacts related to the production and disposal of products. The circular economy is a better alternative as it facilitates the continual use of resources. In reality, this implies that there is a need to redesign production processes, products, and even alter consumption patterns. For instance, they must shift their focus from making products that are single-use to making durable products that can be mended and recycled (Lewansky, 2016). In addition, companies should implement closed-loop systems where materials are collected and reused for production, thereby reducing the need for virgin materials and minimizing waste production. By redesigning their products for ease of disassembly and recyclability, firms operating in the electronics and automobile sectors have also started to integrate circularity into their operations. Circularity within the fashion sector is being fuelled by clothing rental services and upcycling initiatives that utilize these clothes instead of throwing them away (Fletcher & Tham, 2019). Digitalization and the circular economy together make a significant impact on sustainability efforts, as indicated in Figure 1.

Figure 1. The digitalization and sustainability nexus

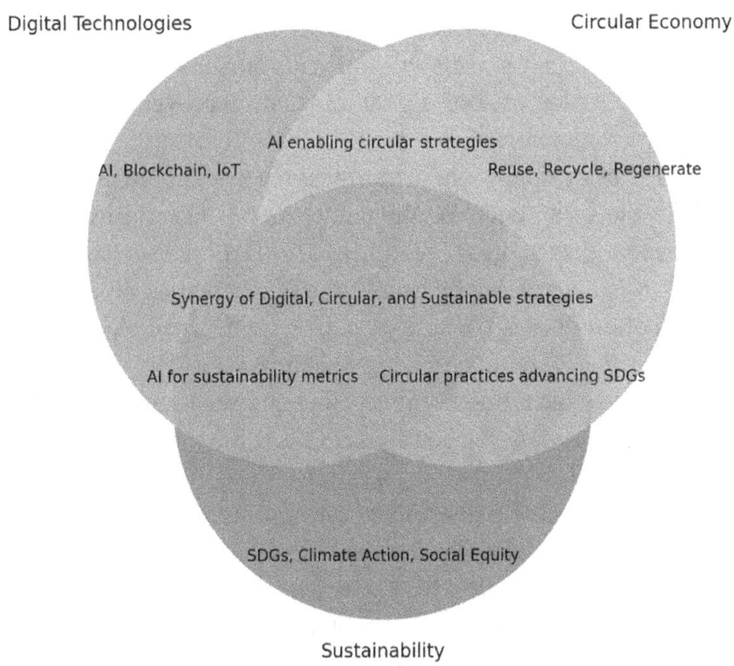

Source: Created by authors

However, the restructuring of the economy into a circular economy is fraught with challenges. There is a need for considerable change in business practices, supply chains, and the customers. Vanderbroek et al. (2020) identify lack of recycling infrastructure and limited availability of secondary raw materials, which are essential for effective closed-loop systems, as one of the main challenges to this transformation. There is also resistance to this transformation from companies that believe in linear systems of mass production and consumption. Embracing the circular economy model in principle should bring such benefits as decreased environmental pollution, reduced dependence on resources, and new business opportunities. The circular economy is more and more integrated into all kinds of market strategies but also in regulations, and therefore its adoption is becoming more and more enforced (Sutherland et al., 2019).

Role of Digitalization in Advancing Circular Economy Goals

Resource efficiency, product lifecycle management, and visibility in material flows can be enabled by advanced technologies to facilitate circular economy objectives. Digitalization constructs logistics that optimize resource movement in circular supply chains for sustainable economic development. Product lifecycle management (PLM) systems provide the company with virtual management of the product, from creation through usage to obsolescence, in such a way that the product is sustainable, repairable, and recyclable—all major attributes for circular economies (Goyal et al., 2021). Other uses of digital tools are improved visualization of the supply chain with effects in resource and process optimization. Technologies like blockchain and IoT bring supply chain transparency by enabling the tracking of the origins and lifecycle of products by the companies; hence, customers trust these products as green products. Digital platforms also facilitate such circular business models as product-as-a-service, which makes leasing or renting possible. This model allows the return of products so that they can be reused, repaired, or refurbished for increasing product life while maximizing resource use (Lacy & Rutqvist, 2015). Digital innovations are including both recycling and waste management under the scope of the circular economy. AI-powered devices automatically and precisely sort recyclables in all steps, leading to a major improvement in efficiency (Geissdoerfer et al., 2017). IoT-based waste management systems are optimizing their collection schedules while monitoring the waste patterns to reduce landfill demand while increasing recycling rates (Gao et al., 2019). Smart cities, as they draw on IoT, AI, and data analytics, play an important role in developing the circular economy. This will help in sustainably managing urban resources, such as water, waste, and electricity. Water leaks can be detected using IoT sensors, reducing waste. Meanwhile, smart grids optimize the use of renewable energy (Alwan et al., 2017).

THE NEXUS OF DIGITALIZATION AND CIRCULAR ECONOMY

As digitisation of the platforms has been integrated into the circular economy concepts, the possibilities of enhancing sustainability across diverse industries have increased significantly. Resource efficiency, waste minimisation, and product lifetime management are fundamental aims of the circular economy, which can be achieved through the application of digital technologies.

Digital Product Passports (DPPs): Enhancing Transparency and Recycling

Digital Product Passports (DPPs) have the potential to generate quite extensive digital documentation of a product's material components, production processes, as well as recyclability. Linking raw materials to design, usage, and end of life management enables consumers, manufacturers, and recyclers to develop tracking, i.e., optimization workings during the lifecycle of said products. Hence making sure they are processed according to their material content and recyclability, improving recycling and resource-out recovery methods. Hence improving the specific recycling and resource recovery processes associated with e-waste brings about a further value towards developing highly improved e-waste managements, bringing huge values in metal recovery from electronic materials, and thus reducing waste and pollution in the same process (Huisman et al., 2017). In areas like electronics and apparel, a combined blockchain and DPP approach would counteract counterfeiting and fraud in the sustainability claims (Bai et al., 2021). Urban areas now going smart will take their initiatives of sustainability by employing the DPPs as one of their most major instruments under which waste is minimized, recycling is made more accurate, and resources are utilized more efficiently.

Table 3. Global smart city initiatives and their sustainability impacts. Source: Created by Authors

City	Initiative	Key Technologies	Impact
Singapore	Virtual Singapore	Digital Twins, IoT	Improved urban planning, energy efficiency
Barcelona	Smart Waste Management	IoT, AI	Reduced waste collection costs, higher recycling rates
Amsterdam	Smart Grid Systems	AI, Big Data Analytics	Decreased energy consumption, reduced CO2

continued on following page

Table 3. Continued

City	Initiative	Key Technologies	Impact
Tokyo	AI-Powered Traffic Management	AI, IoT	Lower fuel consumption, reduced air pollution
Copenhagen	Carbon Neutral Goals by 2025	IoT, Renewable Energy Tech	Significant reduction in greenhouse gas emissions

Data-Driven Innovations for Waste Reduction and Resource Optimization

Innovations that are based on open data assist the circular economy in practices of waste reduction and resource optimization. Big data analytics analyzes material flow in supply chains to help find inefficiencies such as underutilization, waste, and overproduction. This can improve inventory management and support material recovery and reuse toward more effective circular systems (Dufresne et al. 2021). Green IoT should integrate sensors throughout the lifecycle of a product for continuous monitoring and data collection for waste reduction. Geo-fencing IoT technologies have streamlined the entire process since they have real-time tracking of amounts and compositions of waste, as well as volume from their containers; therefore, the collection routes are more efficient, and overflow occurrences are reduced. These technologies create new opportunities for better resource utilization and waste management in the circular economy. IoT systems have the potential to extend product lifespan and thus minimizes the requirement of repairs and maintenance under smart cities, where waste, energy, and water in municipalities are managed by IoT. Smart grids improve energy distribution, while some devices save water in homes. (Alwan et al., 2017).AI further enhances the principles of a circular economy by improving resource efficiency and waste reduction. AI-powered algorithms predict the life span of the products, allowing reductions in design and production costs. The predictive maintenance programs also eliminate equipment failures, thereby conserving raw material (Zhou et al., 2020). AI supports design for disassembly (DfD) and recycling (DfR) to further guarantee that products will be recyclable after their lifecycle (Bocken et al., 2016). Transformational digitization tools such as digital product passports, big data, IoT, and AI are fundamentally altering the circular economy.

DIGITALIZATION IN TACKLING CLIMATE CHANGE

The process of digitalization remains instrumental in dealing with the complex issues raised by climate change. Advanced technologies for environmental management and reduction of carbon emissions in business activities offer opportunities for creating new businesses and meeting the aspirations of global sustainability initiatives. This section looks at the role of digital technologies in contributing to the fight against climate change with special emphasis on carbon emissions, renewable energy solutions, and agricultural technologies.

Role of Technology in Reducing Carbon Emissions

In the industrial sector, AI algorithms can estimate the energy consumption of machinery and equipment to aid its optimization, thus, less power will be consumed in operations. Smart manufacturing means, as well, identifying and eliminating energy inefficiencies, which should be possible in emissions from factories and production lines with them having to be dependent on these ever-increasing measures of energy efficiency (Dufresne et al., 2021). The digitalisation impact of the industrial sectors rests on the use of Figure 2 as indicators of carbon emission reduction.

Figure 2. Global carbon emission reductions enabled by digital solutions

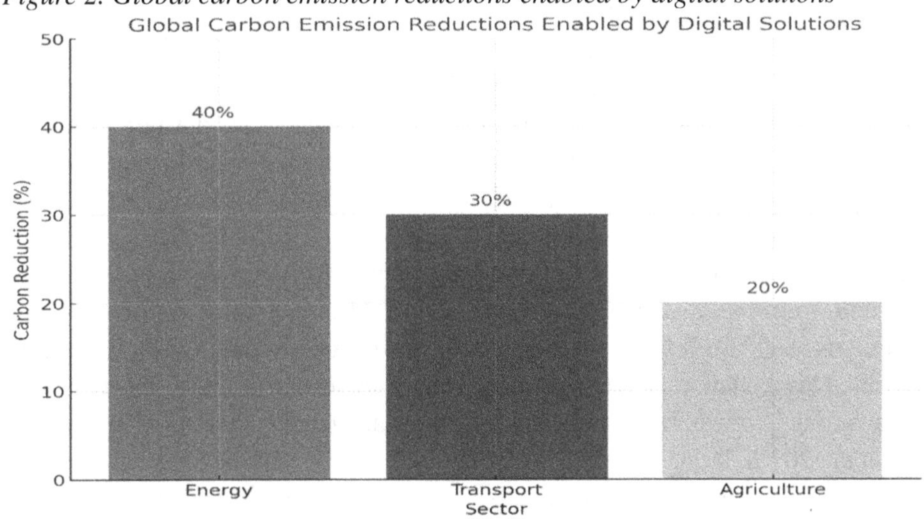

Source: Created by authors

The intelligent energy control schemes adopted in buildings and towns operate according to the real-time requirement such as heating, cooling, and lighting, which account for considerable reduction of energy use and greenhouse gas emissions (Yang et al.,2020). These systems are often linked with smart grids to enhance the control over energy flow consumption to decrease their dependency on non-renewable sources during peak demand. Digital ICTs forecast energy consumption patterns avoid wastages and overuses. Data analytics derive optimization of emissions from hotspot identification and carbon footprint estimation. According to Perlow et al. (2021), big data from satellite images, IoT-like devices, and environmental monitoring collates in sectors like transportation, energy, and buildings to target critical emission hotspots. Such knowledge leads to effective reduction strategies. Smart grids also make it possible for further integration of renewable energy because they balance demand and supply regarding the intermittencies from solar and wind power generation. Smart grids use advanced technologies for energy forecasting, storage, and distribution, thus reducing fossil fuel dependency and carbon emissions (Zhou et al., 2020).

Digital Agriculture: Precision Farming and Sustainable Food Systems

This makes agriculture one of the highest greenhouse gas emitters but potentially the greatest solution to changing climate. Improved productive and sustainable agriculture with minimal environmental impact is embraced by precision farming. This modality provides advanced technologies like satellite imaging, IoT devices, drones, and AI to collect and analyze data about soil, crops, weather, and water resources. Such information is able to improve agricultural practices while conserving resources, reducing emissions, and maximizing efficiency. By application of fertiliser and pesticide, the above-mentioned techniques apply chemicals only where necessary, minimize the chemical application, integrate biodiversity, minimize greenhouse gas emissions from chemical manufacturing, and avoid leaching into the water systems (Kamilaris et al., 2017; Gouveia et al., 2020). IoT smart irrigation systems function on principles of monitoring soil moisture and weather for optimized water usage and avoiding overwatering, thus aiding water management in areas with scarcity (Liu et al., 2020). It takes every food from farm to table and makes it traceable. In this way, blockchain will minimize waste and enable consumers to make sustainable choices to promote green food production (Dufresne et al., 2021).

SOCIO-ECONOMIC IMPACTS OF DIGITALIZATION ON SUSTAINABILITY

Digitalisation not just enables environmental sustainability but also facilitates rapid socio-economic transformation. Digital technology, by driving advances, enabling resource accessibility, and overcoming adverse socio-economic divisions, may address the pressing challenges in public health, finance, and resource distribution. With an emphasis on developments in digital health, the role of fintech in green finance, and the provision of sustainable resources considered within the context of digital inclusion, this section examines the socio-economic impacts of the processes of digitalisation on sustainability.

Digital Health Innovations for Public Health Sustainability

In transforming the healthcare context by allowing equitable availability of healthcare interventions and therefore reducing ecological footprints while enhancing the sustainability of health of the public, the healthcare provision systems are being transformed due to the use of digital health technologies. In particular, the integration of telemedicine, AI-based diagnostics, and health wearables devices has helped changed the way patients are treated in more isolated areas with limited resources. For example, telehealth has become an important strategy for providing care during outbreaks such as COVID-19. With many physical visits now being replaced by video calls, the need for commuting and its associated emissions have reduced (Wosik et al., 2020). Not only does this invention enhance the connectivity to institutions by minimizing the need for infrastructure but it also contributes to environmental objectives. In addition, quicker and more accurate medical evaluations facilitated by the AI-powered diagnostic tools, such as imaging algorithms that assist in confirming detection of diseases, have saved time and resources, hence leading to reduced need for many diagnostic procedures (Topol, 2019). Another significant development is the proliferation of wearable health technologies that monitor heart rate, blood pressure, blood glucose levels, and other related measures continuously, which can help to prevent and detect diseases at an earlier stage. It allows people to take an active role in their well-being, thereby shifting some of the responsibilities from the health system to individuals. Digital health records and interoperable platforms, moreover, contribute to improving the efficiency of healthcare delivery by reducing administrative duplication and unnecessary paper usage in keeping with environmental goals (Wang et al., 2021). Better health outcomes and reduced inequalities also highlight the broader sustainability benefits from the adoption of digital health technologies. Ensuring availability of primary healthcare services for those at the bottom of the socio-economic pyramid is vital to enhancing equity in our society, which is crucial to the Sustainable Development Goals (SDGs).

Fintech and Green Finance for Climate Action

Green finance, as well as fintech, gives up multiple avenues for addressing climate change and sustainability. In fact, fintech services form a significant part of the entire system and include various digital platforms and payment modes, crowdfunding towers, and online blockchains for boosting access to money. They thus provide more finance for green infrastructure projects, renewable energy projects, and mitigation in climate change. Indeed, green bonds are securities reframed for environmental projects that have been issued where digital platforms improve the issuance, tracking, and transparency through which financing flows happen in both clean energy and conservation (Banga, 2019). Blockchain is a technology with which carbon credits can be efficiently traded without a third party, allowing for transparency and safety of carbon trade and, therefore, encouraging carbon-lowering activities (Chen et al., 2020). It will support microfinancing and crowdfunding for small renewable projects, especially in developing areas that do not have access to mainstream financial institutions, public or private. Fintech takes stakeholders at all levels, bringing together academic, private, business, and government sectors to increase the social-economic impact of the climate agenda. Mobile wallets or platforms fuel the financial access of users residing in underprivileged areas where farmers can obtain loans to buy green equipment or renewable energy products (Demirgüç-Kunt et al., 2018).

Digital Inclusion: Democratizing Access to Sustainable Resources

Digital inclusion plays an important role in sustainable development by closing the digital gap and giving disadvantaged parts of the world the room to participate in environmental activities. It fills the gap of socioeconomic inequality and environmental equity by ensuring that there is access to and use of information, education, and other resources for sustainable livelihoods. The growing adoption of societies in sustainable behavior would be aided by knowledge-sharing technologies and platforms for e-learning—all that is possible through digital inclusion: waste management, use of renewable energies, and sustainable farming practice example, for example, access to current market prices, crop tips, and weather forecasts for farmers in rural areas by providing mobile apps. Blockchain technology allows the remote community to sell the surplus solar energy collected in the houses within the neighbour community, thereby making a peer-to-peer renewable energy trading network between members and making energy equity in their community—from that grid system—less dependence on centralized grids (Parag& Sovacool, 2016). Further, water resource management systems that are supported by electronic sensors for Internet usage, measure the amount consumed, thus increasing the efficiency of the resource and providing better accessibility, especially in arid or drier regions. Digital inclusion also extends to making provision for citizens to make proper participatory engagement in such a way that governance looks into environmental sustainability—using digital platforms to create opportunities through which citizens participate in environmental governance, thus making sure more sustainability projects consider input from diverse stakeholders' needs, given in an equitable arrangement.

GOVERNANCE AND POLICY IMPLICATIONS

The current shift towards digital governance has also its advantages regarding the new levels of transparency and sustainability that can be achieved in government operations. With the province of many policies aimed at sustainability, active citizenship among people is encouraged along with accountability, which can be achieved using technology.

E-Governance and Policy Monitoring Platforms

Such understanding of digital technology has already become one of the main features of contemporary government since it enables the effective and clear implementation of sustainability concepts. The use of electronic platforms facilitates the collection, monitoring, and evaluation of real-time data on social and environmental activities.

For example, policy analytics and machine learning classifiers may estimate the success rates of district waste management strategies, deforestation policies, and technology deployment for solar energy. The monitoring of land use change, biodiversity loss, and climate consequences is generally done using remote sensing techniques & GIS. With these systems, lawmakers have the flexibility to adjust their strategies depending on evidence and specific resource application aimed at achieving set goals.

Moreover, the use of digital technologies enables governments to create interactive maps and other applications that allow citizens to monitor changes in the environment – effects of over-logging, air pollution, etc. (Petrova et al., 2021). Furthermore, e-governance has, to a great degree, moved the bulk of administrative functions online, hence alleviating the use of paper-based systems and their associated environmental concerns.

Citizen Engagement through Digital Public Services

This openness has the effect of fostering responsibility and a sense of ownership towards engagement in the environmental initiatives. This participatory governance may be viewed as soliciting active citizens' participation in environmental or civic issues. Some countries have created smartphone applications. For example, in the United States, an app that allows citizens to report illegal picture shows violations like illegal logging or even illegal dumping. American citizens use the "SeeClickFix" system to report environmental or other issues in their community to the relevant authorities. The same holds for people living in India; for instance, the Swachh Bharat App allows citizens to report issues of sanitation, thereby obligating the citizens and promoting the ideal of clean areas in urban environments (Mehta & Gupta, 2020). There is also an improvement in civic awareness as a result of these two apps since people receive news about environmental policies and initiatives as they happen. This knowledge democratization gives power to individuals on how and what to consume, how and whether to manage waste, and how to use energy. For example, apps that enable tracking of personal carbon emissions and thus targeting reduction of them empower users to work towards reducing their carbon footprint. Crowdsourcing for policy creation represents another dimension of digital public service. Citizens take part in policy development through the use of surveys, polls, and forums where policy proposals are made, ensuring that the government has many stakeholders.

Examples of Policy Frameworks Supporting Digital Sustainability

There are benefits resulting from the adoption of these technologies, but there are also downsides. Data protection and privacy, cybersecurity threats, and lack of access to technology are core issues. For example, rural and deprived sections of society may not be able to possess the digital proficiency or the necessary infrastructure prerequisites to get actively involved in the e-governance platforms, there by aggravating existing disparities (van Deursen & van Dijk, 2019). In order to meet these challenges, there is a need for governments to emphasize on digital citizenship programs, equitable infrastructure, as well as security. The policies will also have to embrace measures aimed at protecting the citizens against the abuse of their data and political abuse of the policies made. These challenges notwithstanding, the prospects of enhancing sustainability through digitalized governance are very promising. By using different tools, digital governance creates conditions of inclusiveness, openness and use of evidence in policy making processes to global challenges of climate change, resource use and social inequities.

CHALLENGES AND RISKS OF DIGITALIZATION IN SUSTAINABILITY

The angle of enhancing sustainability is undertone in profiling the role of digitalization. This however is contrary as problems as the digitalization process creates its own challenges and risks. In this regard, this section discusses the negative impact of digital technology on the environment, the risks of ethical issues – data privacy, algorithmic discrimination, and the unequal levels of access to technology, and legal and business approaches to their resolution.

Environmental Costs of Digital Technologies (Energy Use and E-Waste)

The means of the digitalization process, through which great changes are expected to occur, is the outcome of the process of the innovation. In this, two that are energy consumption and e-waste are growing to become major future challenges.

- **Energy Use:** Data centres are energy-intensive buildings that are deployed in the running of cloud-based services. It is reported by Jones et al. (2021) that these structures alone account for 1 percent of global power consumption, a figure that is only going to increase as demand grows in the future. This is further worsened by blockchain and cryptocurrencies and other invasive technologies like these that are used in mining. Nowadays, Bitcoin mining is criticized for its excessive energy use. For example, in 2022 alone, Bitcoin mining consumed more energy annually than the entire country of Argentina (Cambridge Centre for Alternative Finance, 2022). Furthermore, the improper disposal of e-waste also shows grave environmental risks, with examples including soil and water pollution from cadmium, mercury, and lead. In addition, carbon emissions are emitted during the use, manufacture, and disposal of electronic devices to some extent. In the development of new gadgets, neodymium is frequently used because its extraction and processing are energy-consuming and ecologically harmful, and its biodegradable substitute is not known.
- **E-Waste:** Deficient e-waste disposal poses serious threats to the environment through the leaching of cadmium, mercury, and lead into the soils and water bodies. Also, the manufacture and disposal of electronic equipment cause somewhat significant carbon emissions. In the development of new gadgets, neodymium is frequently used because its extraction and processing are energy-consuming and ecologically harmful, and its biodegradable substitute is not known.

Ethical Concerns: Data Privacy, AI Bias, and Digital Divide

Digitalisation is another rapidly growing aspect of the society that brings up serious ethical issues that need to be handled in order to preserve a combination of sustainability and equity in the future. Data Privacy: On several occasions, there are no proper regulations dictating what the data will be used for; governments and businesses tend to hoard excessive amounts of data regarding their customers, clients, and citizens. One key ethical challenge is the tension created in attempts to exploit the collection of data for sustainability whilst maintaining respect for individual privacy.

- **AI Bias:** However, strong ones do include many biases that structures pose. Machine learning algorithms, many times, reflecting the biases in their training set, are discriminative in their outcomes especially in activities such as recruiting, loan approvals, and law enforcement (Obermeyer et al., 2019). It's worth emphasising that such biases lead to sustained discrimination rather than resolution, which runs against the many goals of social sustainability that digitalisation seeks to achieve.
- **Digital Divide:** Although urban populations are helped with the use of modern digital instruments, the ability to do so is not available for many rural and low-income areas and regions (Van Dijk, 2020). This gap constrains the possibilities of the disadvantaged groups to participate in e-governance, access sustainable resources, or benefit from green technologies. Addressing the issue of equitable distribution of the benefits of digitalisation is predicated on overcoming the digital divide.

Reducing the Risks: Industry Methodologies And Policy

Policymakers, business leaders, and civil society working together will help to solve the problems and dangers. connected with digitalisation. These techniques help to reduce some dangers below.

- **Encouraging Renewable Energy and Energy Efficiency:** Energy-efficient solutions must be given top priority by governments and businesses as they move to renewable energy sources. Data centres can use energy-saving strategies, including virtualisation, better cooling systems, and artificial intelligence use to maximise energy utilisation, for instance (Huang et al., 201). Furthermore, including renewable energy into digital infrastructure would help to greatly lower its carbon impact. Already, IT firms are advancing this field. For example, Google Sustainability (2021) states that by 2030 it will be running totally on carbon-free electricity. Such projects show how realistically it is possible to match digital innovation with environmental sustainability.

- **E-waste handling:** Good e-waste management calls for industry-led projects as well as legal structures. Extended producer responsibility (EPR) laws let governments make manufacturers responsible for the end-of-life handling of their goods (Forti et al., 2020). Tech businesses may create goods with sustainability in mind, including modular designs for simpler repair and recycling at the same time.
- **Development of Ethical AI:** Developers of AI must give openness, responsibility, and diversity in AI training datasets a priority in order to solve AI bias. Ethical rules such as the OECD AI Principles offer a structure for making sure AI systems are equitable, inclusive, and consistent with society's ideals (OECD, 2019).
- **Closing the Digital Divide:** Working together, governments and international organisations may offer underprivileged populations subsidised digital equipment, digital literacy initiatives, and reasonably priced internet access (Broadband Commission for Sustainable Development, 2020). Aiming to link millions of people in rural regions to the internet, public-private collaborations like Microsoft's Airband Initiative show the promise for group answers to this problem.

FUTURE DIRECTIONS AND GREEN INNOVATIONS

The junction of digitalisation and sustainability has produced a fresh wave of green ideas meant to solve urgent world problems. This part looks at how smart cities help with waste management and energy efficiency, investigates new developments in digital sustainability, and addresses chances for worldwide cooperation on digital green ideas.

Energy Efficiency and Waste Management: Smart Cities

Leveraging cutting-edge technology to increase energy efficiency, maximise resource utilisation, and boost waste management systems, smart cities epitomise urban digitalisation. These developments not only solve the environmental problems brought on by urbanisation but also raise living standards for city dwellers.

- **Energy Conservation:** The foremost use for smart city projects is energy conservation, particularly through smart grids making use of IoT devices, AI, and big data for the conditional management of supply and demand. Such systems allow the forecasting of energy consumption patterns, energy waste reductions, and the integration of renewable energy sources such as solar and wind into the grid (Siano, 2020). Platforms powered by AI optimize lighting, heating, and cooling systems according to real-time occupancy patterns. A study by Hu et al. (2021) highlights the fact that these AI-based solutions in smart buildings can effectively bring down energy expenditures by as much as 30%, thus significantly contributing toward urban sustainability objectives.

- **Environmental Management of Waste:** AI systems may also examine trash composition and recommend more efficient ways of recycling or reusing. Waste segregation systems powered by artificial intelligence have been used by smart cities such as Tokyo and Singapore, therefore improving resource recovery rates and lowering landfill reliance. For example, emerging patterns in digital sustainability, including digital twins, have produced innovative tools and ideas. pushing the envelope of sustainability in the fast-developing digital world. Among them, digital twins are one of the transforming ideas having uses in several fields.
- **Digital Shadows:** A digital twin is a virtual copy of a real-time data-based physical system, environment, or process used for simulation and prediction. Urban planning, energy management, and manufacturing are just a few of the applications for this technology that help to maximise resources and reduce environmental effect.

AI- POWERED ENVIRONMENTAL SOLUTIONS

Generally, AI and massive data, machine learning, play a great part in environmental monitoring and climate modeling for predicting climate patterns, monitoring deforestation, and analysing conservation outcomes.

- **Blockchain for Sustainability:** Blockchain serves the purpose of tracking green energy credits and verifying sourcing materials ethically. IBM, for example, has created a traceable supply chain via blockchain technology, thus ensuring social and environmental compliance (Hackius & Petersen, 2020).
- **Global Cooperation on Digital Green Solutions:** Digital technologies allow humanity to interact virtually with which everyone can share their resources and best practices for sustainability. Examples including the ‘UN Global Digital Compact’ mention their creation of shared value by using digital solutions for global sustainability (United Nations, 2023).
- **Transborder Technology Collaboration:** Digital technologies have offered directions that permit developing nations to adopt sustainable behav-

iors swiftly. For example, countries with high incomes are now teaming up with digital companies to ensure IoT solutions will be affordable for use in water, energy, and agriculture. One illustration is the Global Cooling Prize, an award supporting innovation and cooperation between all concerned for combating climate change.

RECOMMENDATIONS FOR POLICYMAKERS, RESEARCHERS, AND PRACTITIONERS

To ensure that the hazards associated with digitalization do not outweigh the benefits, stakeholders need to develop appropriate strategies such as:

For Policymakers

1. Create well-coordinated policy regimes: Policies that integrate innovative life-sustaining digital initiatives such as tax subsidies for eco-friendly innovations and controls on electronic waste management should be put in place.
2. Foster collaborative initiatives with the Private Sector: Government, business, and academia partnerships can aid in the rapid implementation of new technologies and promote the fair distribution of technology's benefits.
3. Reinforce existing international commitments: Endorse frameworks that support sustainability, for example, the UN SDGs by embedding country strategies into global sustainable targets.

For Researchers

1. Conduct cross-disciplinary research: Study the relationship between digitization, the circular economy, and sustainability in spheres such as energy, medicine, and urbanism.
2. Examine new paradigms: Research cutting-edge technologies like digital twins, AI-based climate solutions, and blockchain as instruments for environmental governance.
3. Tackle moral issues: Conduct research on the political and social impacts of digitalization and policy practices, including the normative aspects of privacy, AI discrimination, and digital divides.

For Practitioners

1. Utilize green technologies: Enterprises and institutions must embed digital sustainability initiatives into business processes such as smart energy systems and data-centric waste management techniques.
2. Engage in capacity building: Develop the knowledge and skills of operational staff and partners on the sustainable opportunities provided by the innovations.
3. Assess the outcome: Incorporate specific metrics and performance indicators of the outcomes of the digital technologies' implementation within the environment and society, for further improvement of the processes.

CONCLUSION AND RECOMMENDATIONS

Combining sustainability with digitalisation is immensely powerful to address the world's ever-growing challenges in problems like resource scarcity, 'course' disparities, and pollution including climate change. This final section provides a roadmap for integrating digitalisation with green development goals as well as conceptualizing the scope of this empirical research, and presenting concrete recommendations for policymakers, researchers and practitioners in the field.

Summary of Key Findings

Policymakers are faced with competing demands that require rapid and effective solutions, which they view as an opportunity for digitalisation. For key findings, include:

- **Digitalization for Sustainability and its tools as a Service:** Y. Yang et al. (2017) specifies that internet and technology tools, specifically AI technology, digital space, supply systems change, and efficiency. Clearly such tools include blockchain and the Internet of Things, which directly aid the improvement of resource management, minimizing carbon output, switching from traditional to green development models.

These tools are quite significant in monitoring energy use, reducing waste, and making informed decisions based on information gathered through data analytics. Circular Economy Enablement: Emerging technologies, like the digital product passport or blockchain, make the whole supply chain more transparent and accountable, making it easier to adhere to the principles of reuse, recycling, and regeneration.

- **Climate Change Mitigation:** Digital technologies such as smart grids, intensive agriculture, and renewable systems reduce emissions of greenhouse gases and improve the ability to deal with climate change.
- **Socio-Economic Benefits:** Sustainability challenges are also accompanied by advancements in digital inclusion, fintech solutions or health technologies in the fight against socio-economic inequalities.
- **Challenges and Risks:** However damaging impacts of the digitalisation approaches, such as e-waste and energy consumption, as well as non-classical ethical dilemmas related to AI bias or the digital gap have to be addressed in a balanced and inclusive policy.

The perception of digitalization and sustainability as separate activities are more or less over. They have come together and can determine the fate of mankind. The results of this research highlight the revolutionary role of digital technologies in the application of the circular economy, in solving the problem of global warming, and in overcoming socio-economic inequalities. But to make this happen, a holistic and intersectional model responsive to energy consumption, e-waste, and ethical issues is needed. Step by step, there seems to be hope for the international community that governments, scholars, and practitioners will work as one. By integrating technology with sustainable development, the world will be ready for a new technological age capable of addressing challenges including climate change. The two domains, digital transformation and sustainability, should continue to go hand in hand, working for the current generation as well as future generations.

REFERENCES

Alwan, W.. (2017). Smart building and smart cities: A review of recent trends. *Energy and Building*, *143*, 28–38. DOI: 10.1016/j.enbuild.2017.03.058

Bai, Y., Fan, K., Zhang, K., Cheng, X., Li, H., & Yang, Y. (2021). Blockchain-based trust management for agricultural green supply: A game theoretic approach. *Journal of Cleaner Production*, *310*, 127407. DOI: 10.1016/j.jclepro.2021.127407

Banga, J. (2019). The green bond market: A potential source of climate finance for developing countries. *Journal of Sustainable Finance & Investment*, *9*(1), 17–32. DOI: 10.1080/20430795.2018.1498617

Bocken, N. M., De Pauw, I., Bakker, C., & Van Der Grinten, B. (2016). Product design and business model strategies for a circular economy. *Journal of industrial and production engineering, 33*(5), 308-320.

Broadband Commission for Sustainable Development. (2020). The State of Broadband 2020: Tackling digital divides. Retrieved from https://www.broadbandcommission.org

Cambridge Centre for Alternative Finance. (2022). Bitcoin energy consumption index. Retrieved from https://ccaf.io/cbeci/index

Chen, Y.. (2020). Blockchain applications in sustainability: A review of challenges and opportunities. *Journal of Cleaner Production*, *262*, 121057. DOI: 10.1016/j.jclepro.2020.121057

Demirgüç-Kunt, A., Klapper, L., Singer, D., Ansar, S., & Hess, J. (2018). *The Global Findex Database 2017: Measuring financial inclusion and the fintech revolution*. World Bank. DOI: 10.1596/978-1-4648-1259-0

Dufresne, G.. (2021). Big data and the circular economy: Challenges and opportunities. *Journal of Cleaner Production*, *288*, 125617. DOI: 10.1016/j.jclepro.2020.125617

Fletcher, K., & Tham, M. (2019). *Sustainable fashion and textiles: Design journeys*. Earthscan.

Gao, S.. (2019). Internet of things (IoT)-based waste management systems: Opportunities and challenges. *Journal of Cleaner Production*, *234*, 1245–1259. DOI: 10.1016/j.jclepro.2019.06.336

Geissdoerfer, M., Savaget, P., Bocken, N. M. P., & Hultink, E. J. (2017). The circular economy – A new sustainability paradigm? *Journal of Cleaner Production*, *143*, 757–768. DOI: 10.1016/j.jclepro.2016.12.048

Ghisellini, P., Cialani, C., & Ulgiati, S. (2016). A review on circular economy: The expected transition to a balanced interplay of environmental and economic systems. *Journal of Cleaner Production, 114*, 11–32. DOI: 10.1016/j.jclepro.2015.09.007

Google Sustainability. (2021). Google's commitment to a sustainable future. Retrieved from https://sustainability.google

Gouveia, M.. (2020). Precision agriculture and sustainability: A review of emerging trends in digital farming. *Agronomy for Sustainable Development, 40*(4), 1–18. DOI: 10.1007/s13593-020-00634-2

Goyal, S.. (2021). Product lifecycle management: A review of digital technologies and sustainability practices. *Sustainability, 13*(5), 2702. DOI: 10.3390/su13052702

Hackius, N., & Petersen, M. (2020). Blockchain in logistics and supply chain: A review. *Logistics Journal, 15*(4), 12–22.

Hu, L.. (2021). AI-powered energy optimization in smart buildings. *Energy Efficiency Journal, 14*(6), 987–1005. DOI: 10.1007/s12053-021-0987-9

Huang, X.. (2021). Energy-efficient data center cooling: Advances and perspectives. *Energy and Building, 240*, 110986. DOI: 10.1016/j.enbuild.2021.110986

Huisman, J.. (2017). The role of product recovery in a circular economy: A global perspective. *Waste Management & Research, 35*(3), 1–12. DOI: 10.1177/0734242X16670247

Jain, R.. (2017). Artificial intelligence and sustainability: Applications in environmental management. *The Science of the Total Environment, 574*, 1040–1051. DOI: 10.1016/j.scitotenv.2016.09.029

Janssen, M., Konopnicki, D., Snowdon, J., & Wahyudi, A. (2020). Digital government: Building a smart and sustainable government ecosystem. *International Journal of Information Management, 55*, 102141. DOI: 10.1016/j.ijinfomgt.2020.102141

Jones, M. W., Andrew, R. M., Peters, G. P., Janssens-Maenhout, G., De-Gol, A. J., Ciais, P., Patra, P. K., Chevallier, F., & LeQuéré, C. (2021). Gridded fossil CO2emissions and related O2combustion consistent with national inventories 1959–2018. *Scientific Data, 8*(1), 2. DOI: 10.1038/s41597-020-00779-6 PMID: 33414478

Kamilaris, A.. (2017). A survey of the applications of deep learning in agriculture. *Computers and Electronics in Agriculture, 140*, 157–167. DOI: 10.1016/j.compag.2017.06.001

Kirchherr, J., Reike, D., & Hekkert, M. (2017). Conceptualizing the circular economy: An analysis of 114 definitions. *Resources, Conservation and Recycling, 127*, 221–232. DOI: 10.1016/j.resconrec.2017.09.005

Korhonen, J., Nuur, C., Feldmann, A., & Birkie, S. E. (2018). Circular economy as an essentially contested concept. *Journal of Cleaner Production, 175*, 544–552. DOI: 10.1016/j.jclepro.2017.12.111

Lacy, P., & Rutqvist, J. (2015). *Waste to wealth: The circular economy advantage.* Palgrave Macmillan.

Lewandowski, M. (2016). Designing for a circular economy. *Sustainability Science, 11*, 243–257. DOI: 10.1007/s11625-016-0353-1

Li, Z.. (2020). Artificial intelligence in renewable energy: A comprehensive review. *Renewable & Sustainable Energy Reviews, 120*, 109634. DOI: 10.1016/j.rser.2019.109634

Liu, L.. (2020). Internet of Things (IoT)-based smart agriculture: From sensor data to precision farming. *Agronomy (Basel), 10*(12), 1836. DOI: 10.3390/agronomy10121836

Mehta, S., & Gupta, R. (2020). Digital India: Implications for sustainability governance. *Journal of Public Administration, 42*(2), 234–249. DOI: 10.1080/01900692.2020.1718693

Murray, A., Skene, K., & Haynes, K. (2017). The circular economy: An interdisciplinary exploration of the concept and application in a global context. *Journal of Business Ethics, 140*(3), 373–382. DOI: 10.1007/s10551-015-2693-2

Obermeyer, Z., Powers, B., Vogeli, C., & Mullainathan, S. (2019). Dissecting racial bias in an algorithm used to manage the health of populations. *Science, 366*(6464), 447–453. DOI: 10.1126/science.aax2342 PMID: 31649194

OECD. (2019). OECD AI Principles: Recommendations for trustworthy AI. Retrieved from https://www.oecd.org

Parag, Y., & Sovacool, B. K. (2016). Electricity market design for the prosumer era. *Nature Energy, 1*(4), 16032. DOI: 10.1038/nenergy.2016.32

Petrova, D.. (2021). Citizen science and e-governance: Opportunities and challenges for sustainable development. *Environmental Research Letters, 16*(3), 034001. DOI: 10.1088/1748-9326/abde45

Sajjad, M.. (2021). Digital technologies for enhancing the resource efficiency of the manufacturing industry. *Sustainable Production and Consumption, 27*, 272–285. DOI: 10.1016/j.spc.2020.11.022

Sorrell, S. (2015). Reducing energy demand: A review of issues, challenges and approaches. *Energy Policy, 75*, 77–88. DOI: 10.1016/j.enpol.2014.09.005

Sutherland, W.. (2019). Circular economy: A systematic review of the literature. *Journal of Environmental Management, 248*, 109363. DOI: 10.1016/j.jenvman.2019.109363

Swan, M. (2015). *Blockchain: Blueprint for a new economy*. O'Reilly Media.

Tapscott, D., & Tapscott, A. (2016). *Blockchain Revolution: How the Technology Behind Bitcoin and Other Cryptocurrencies is Changing the World*. Penguin.

Topol, E. J. (2019). High-performance medicine: The convergence of human and artificial intelligence. *Nature Medicine, 25*(1), 44–56. DOI: 10.1038/s41591-018-0300-7 PMID: 30617339

United Nations. (2023). Global Digital Compact: Aligning technology with global goals. Retrieved from https://www.un.org

van Deursen, A. J. A. M., & van Dijk, J. A. G. M. (2019). The first-level digital divide shifts from inequalities in physical access to inequalities in material access. *New Media & Society, 21*(11–12), 3542–3561. DOI: 10.1177/1461444818797082 PMID: 30886536

Van Dijk, J. (2020). *The digital divide*. John Wiley & Sons.

Vanderbroek, W.. (2020). Circular economy: Challenges and opportunities. *Sustainable Development, 28*(3), 1626–1639. DOI: 10.1002/sd.2101

Wosik, J., Fudim, M., Cameron, B., Gellad, Z. F., Cho, A., Phinney, D., Curtis, S., Roman, M., Poon, E. G., Ferranti, J., Katz, J. N., & Tcheng, J. (2020). Telemedicine transformation: COVID-19 and the rise of virtual care. *Journal of the American Medical Informatics Association : JAMIA, 27*(6), 957–962. DOI: 10.1093/jamia/ocaa067 PMID: 32311034

Yang, J.. (2020). Smart grid and its applications in energy systems: A review. *Renewable & Sustainable Energy Reviews, 118*, 109541. DOI: 10.1016/j.rser.2019.109541

Zhou, Z., & Wang, W. (2020). Standardization of smart grid technology for the development of global energy interconnection. In *2020 IEEE 3rd Student Conference on Electrical Machines and Systems (SCEMS)* (pp. 569-573). IEEE. DOI: 10.1109/SCEMS48876.2020.9352385

Chapter 7
Leveraging Digital Technologies for Effective SDG Partnerships:
A Case Study Approach

Aliza
https://orcid.org/0000-0003-1087-7239
Banaras Hindu University, India

ABSTRACT

In the face of global challenges, achieving the United Nation's Sustainable Development Goals (SDGs) requires innovative and collaborative efforts across sectors. However, there is a paucity of literature to study the effectiveness of collaboration and initiatives that foster SDG achievement. This study explores the role of digital technologies in enhancing partnerships that are aimed at achieving SDGs. Using a case study approach, the research examines two initiatives (i.e., UNDP's Digital SDG Partnership and WEF's Digital Agenda) where digital tools and platforms have facilitated effective multi-stakeholder collaboration, improved resource mobilization and enhanced transparency and accountability. The findings highlight the transformative potential of digital tools in bridging gaps, fostering innovation, and scaling impact for SDGs. The paper concludes by mentioning significance, limitations, and recommendations for future directions of the study for leveraging digital technologies in future SDG partnerships to maximise their potential for sustainable development.

DOI: 10.4018/979-8-3693-9909-5.ch007

INTRODUCTION

Digitalization transforms companies, economies, and communities by incorporating digital technologies into routine operations. It increases efficiency, facilitates real-time communication, and promotes innovation in industries such as healthcare, education, banking, and manufacturing. Katz et al. (2013) and Katz and Koutroumpis (2013) define digitalization as the societal change brought about by the widespread use of digital technologies for information generation, processing, sharing, and transaction. Digitalization lowers costs, increases productivity, and improves customer experience by automating operations and tasks. It also supports data-driven decision-making using advanced analytics and artificial intelligence, allowing firms to respond swiftly to changing market conditions. The process of transforming analogue data into a digital representation is known as digitization (Katz & Koutroumpis, 2013). As a social process, digitization is the process by which digital communications and applications change the socio-institutional activities and the techno-economic environment and it is built upon the advancements in semiconductor and network access technologies, software and computational engineering, and the ramifications of their use (Katz & Koutroumpis, 2013). Digitalization has made remote work and worldwide cooperation more practical, breaking down geographical constraints and extending prospects for growth and development. Digital innovations are revolutionizing society and business. However, the future path of these revolutions remains unknown (Schneider, 2019).

According to Wu et al. (2023), the quick advancement of digital technology, together with its adaptability and penetrating attributes, has made it possible to improve environmental conditions and energy efficiency, but it is still unknown how digital technology affects energy efficiency and what its underlying logic is. However, digital transformation alone does not always result in better environmental performance; to be successful, it must be integrated with green solutions (Ahmadova et al., 2022). A comprehensive evaluation of the social ramifications of integrating digital technologies and sustainability issues into new twin transition business models is also necessary, considering issues with privacy, human rights, health and safety, and working conditions (Trittin-Ulbrich et al., 2021; West, 2019). Further, for partnerships and collaborations in the digital realm, priority must be given to developing a common vision for digital collaboration and a digital future and strengthening digital cooperation strategies is necessary (Deganis et al., 2021).

The growing global complexities that are accelerated by climate changes have made it difficult for nations and their individuals to feasibly survive and prosper. The condition is worsened with the growing economic inequality among nations. To cope with such problems, the United Nations came up with mutually accepted targets aimed at achieving sustainability known as SDGs, which share close linkage with digitalisation. To achieve the global target regarding a healthy planet and sustainable development, the United Nations came up with a well drafted 17-point Sustainable Development Goals, also known as SDGs. The comprehensive list of SDGs includes goals regarding poverty reduction (SDG 1), no hunger (SDG 2), good health and well-being (SDG 3), quality education (SDG 4), gender equality (SDG 5), clean water and sanitation (SDG 6), affordable and clean energy (SDG 7), decent work and economic growth (SDG 8), industry, innovation and infrastructure (SDG 9), reducing inequalities (SDG 10), sustainable cities and communities (SDG 11), responsible consumption and production (SDG 12), climate action (SDG 13), life below water (SDG 14), life on land (SDG 15), peace justice and strong institutions (SDG 16), and Partnership for the Goals (SDG 17). The 2030 Agenda for Sustainable Development is made up of 17 Sustainable Development Goals (SDGs), and SDG 17 is one of them. SDG 17, also referred to as Partnerships for the goals, is about revitalizing the global partnership for sustainable development. It refers to the cooperation between governments, the private sector, and civil society to take the ambitious targets on the 2030 Agenda. Ciulli and Kolk (2023) offer scholars a fresh perspective on how to bridge the gap between international business, digitalization and sustainable development through the affordance perspective, in order to enable scholars to navigate the complexities of digitalization and sustainable development. Further, based on a micro-approach, to investigate the role of digital technologies in reaching Net Zero Emission and the SDGs, Yadav et al. (2023) adopted a resource-based view for strategically managing firms' tangible and intangible resources based on lean and green philosophies. They also emphasized the necessity of adopting digital technology and implementing lean and green practices since lean-green manufacturing may help reduce the relative cost of reducing pollution and aids in raising awareness of the value of pollution avoidance or reducing the cost of implementing ecological development. In a similar way, by adopting a comprehensive approach to sustainability that considers social, economic, and environmental factors, Okot et al. (2023) aim to create a conceptual framework that outlines a plan for using the Internet of Things to promote sustainable development in Costa Rica. The results of this study indicate that technology has a significant potential to support the nation's sustainable development. However, to completely reap its benefits, a few obstacles must be overcome. There are going number of research to study the drivers of digitalization that targets SDGs. With the advancement in digital technologies and global connectivity, it is likely that the dynamics between digitalization and SDG partnership will evolve in new dimensions. Despite the importance of partnerships for achieving the Sustainable Development Goals (SDGs), many partnerships face challenges in terms of effectiveness, scalability, and sustainability. Digital technologies have the potential to enhance SDG partnerships, but there is limited research on their impact. As ineffective partnerships hinder SDG achievement there is a need for global level integrative efforts of policy makers and practitioners to work in the direction of global sustainability and equity. The objective of the present study is multiple, i.e., firstly to explore digital technologies' potential for SDG partnerships. Secondly, to analyze two pioneering and successful case studies of digital SDG partnerships. Lastly, to identify the best practices and challenges based on the above case studies. In the present study, a case study research approach is used that is a common approach in the field of social science research (Noor, 2008). Based on the case studies, the present study provides an answer to the question: How can digital technologies enhance SDG partnerships? This study contributes to the growing body of knowledge on SDG partnership and digital technologies, providing actionable insights for policymakers, practitioners, and researchers that are seeking to harness the potential of digital innovation for sustainable development.

LITERATURE REVIEW

The current section provides an overview of SDG 17 and partnerships along with reviews of digital technologies for development (e.g., blockchain, AI, IoT). In addition, existing research on digital SDG partnerships is reviewed to identify gaps and limitations.

Introduction to Digital Technologies

Digitalization is a driving force the shapes the future by promoting economic growth, technological innovation, and societal progress. There are various research that deals with the role of digitalization in achieving a better future. Ignatov (2020) argued that while the digital economy generates new economic growth drivers, it also exacerbates global inequality and affects the development of nations worldwide. The author stated that nevertheless, not enough research has been done on how global governance organizations like BRICS can support digital growth. Additionally, there is substantial uncertainty regarding the degree of development of the digital economy in specific nations, and the BRICS nations' digital economy varies in terms of maturity, and their digital infrastructure is generally of poor quality and low cost. By combining digital technology adoption (digitalization and digitization), globalization, and environmental quality, Elfaki and Ahmed (2024) sought to investigate the effects of globalization innovation and digital technology adoption on green, sustainable economic growth in a subset of Asian Pacific nations. The results showed that adoption of digital technology fosters economic growth in these nations. They also discovered that the internalization of globalization and the adoption of technology play a significant role in fostering innovation, knowledge, and technology transfer. This allows businesses to access a variety of markets, promotes direct foreign investment, collaboration, and strategic partnerships, and facilitates the dissemination of knowledge both within and between countries regarding technological innovation, globalization-based technology transfer, and intellectual property development in order to achieve the Sustainable Development Goals (SDGs). Additionally, Dwivedi et al. (2024) stated that digital technology can greatly assist in cnhancing circular supply chain operations and achieving the SDGs. There have been discoveries that integrating digital technology within circular supply chains tremendously increases organizational effectiveness. Organizations hoping to prosper in the age of digital and sustainable transformation must adopt new leadership paradigm, which embodies the values and methods that will characterize success in the dual transition environment, but it can be difficult and resource-intensive to integrate digital technologies into current processes and match them with sustainability goals (Aagaard & Vanhaverbeke, 2024).

Digital Technologies Implementation in Different Sectors

Digitalization has positively impacted several industries in how organizations conduct their business and how they offer services. In the overall span of these industries, digitalization will increase connectivity; increase productivity; and at the same time deliver products that will be of personalized experiences, crucial for keeping up in this constantly changing market landscape. Singh et al. (2024) show that integration of digital technologies with green healthcare practices are revolution perfect. They also underline the importance of enabling policies, leadership as well as stakeholder participation in the realization of sustainable delivery of healthcare. Setyaningrum et al. (2024) confirmed that digital based business training would positively influence the empowerment and digital self-efficacy of MSME players. In addition, MSME participants that are empowered are able to improve their digital creativity and digital self-efficacy. Additionally, digital-based business training on the part of MSME actors can be mediated by their own digital self-efficacy, which shows a very positive effect on digital innovation. Nevertheless, the empowerment of MSME players cannot diminish the impact of corporate digital based training in terms of digital innovation. Earlier, Martínez-Peláez et al (2023) determines how MSMEs' owners or senior management can start a long-term digital transformation initiative and highlighted how critical it is to prioritize cultural change as a key component of the digital transformation process. Furthermore, MSMEs starting their digital transformation journey towards sustainability need to have digital competencies. MSMEs may successfully walk the road to sustainability, open up new prospects, and support their expansion while having a beneficial social and environmental impact by utilizing these competencies and tackling related obstacles.

Zhang and Liu (2014) examined the mechanism between digital orientation and sustainable commitment orientation from the standpoint of digital green capability. Ahmed (2023) examines the relationship between technology and sustainability, emphasizing the ways in which technology-driven solutions can improve sectors like supply chain sustainability, waste management, energy, agriculture, and transportation. Ahmed concluded that emerging technologies have the potential to make a substantial contribution to sustainability initiatives. Green technology, such energy-efficient equipment and renewable energy sources, are become cheaper and easier to use. Furthermore, it is acknowledged that the integration of digital technology and sustainability is essential for spurring future growth in various industries on the global scene as new technologies have enormous potential for sustainability initiatives and utilizing them to their full potential requires creativity and teamwork. Akter (2024) looks at important uses of AI for sustainability and suggested for endeavor to create a more resilient, just, and sustainable future for present and future generations by skillfully utilizing AI technologies.

Digitalization has also been studied in the sectors that indirectly affect SDG achievement in various ways. According to De Giovanni (2023) there are frameworks to analyze the metaverse from the triple bottom line perspective (i.e. environmental, social, and governance [ESG]) or from the point of view of its most impacting business strategies or that from the perspective of sustainable development goals (SDGs). Giovanni also analyses the sustainability of the metaverse technology by adopting a responsible digitalization perspective to drive the transition to Industry 5.0. Further, to achieve the Sustainable Development Goals, Aini et al. (2024) sought to determine whether a recently created learning model—a digital literacy model with a video content project—was effective in raising civic intelligence. In the context of the SDGs, Mishra et al. (2024) assess the connection between digital orientation, digital eco-innovation, and Circular Economy (CE) practices. The results demonstrated that digital eco-innovation is positively impacted by digital orientation and digital capacity and that digital eco-innovation mediates the effects of these factors on both financial and non-financial performance. According to the study, digital eco-innovation and digital orientation have a good link. In the automotive sector, the digital eco-innovation and CE nexus align with SDGs 7, 8, 11, and 12. Utilizing greener energy sources to minimize waste, optimize resources, and minimize pollution is encouraged by digital eco-innovation based on new digital technologies. In order to combat climate change and advance sustainable development, George et al. (2021) examine how digital technologies are assisting in addressing major concerns by developing a digital sustainability strategy based on the imaginative and inventive use of digital technologies and concluded that climate change is the biggest existential concern facing businesses today, and digital transformation is without a doubt one of the most significant trends influencing them. The authors bring forth that the intersection of the sustainability and digital imperatives has led to some of the most innovative research concepts and business endeavors.

Since the triple link between digital technologies, various industry sectors, and SDGs has not been thoroughly examined in the previous scientific literatures, Varriale et al. (2024) set out to create a robust body of knowledge to guide decision-making when adopting digital technologies for sustainability. They offered a framework that links digital technologies to particular industries and sectors and the attainment of the SDGs, improving comprehension among managers, policymakers, and practitioners. (Popkova et al., 2022) confirms that the results of SDG16 were impacted in developing nations by the institutions of economic freedom and human development, which depend on digital technology and infrastructure.

Digital technologies and SDGs

Digitalization is inextricably related to the achievement of the United Nations' Sustainable Development Goals (SDGs), which provide innovative answers to some of the world's most critical concerns. Digital technologies serve as enablers of sustainable development, allowing countries to accomplish their SDG targets while promoting inclusive and equitable progress for all. The Sustainable Development Goals (SDGs) are inextricably linked with economic development since they aim to promote inclusive, equitable, and sustainable growth around the world. By addressing concerns such as poverty, inequality, and environmental sustainability, the SDGs lay the groundwork for long-term economic growth that benefits all sectors of society. Mitra and Prasad (2021) highlighted that modernization, automation, and technology have been major contributors to poverty and inequality in developed nations; however, it's also possible that the same technology can be an enabler if applied to lessen inequity and human suffering as digital technology increases connectedness because in order to achieve SDGs 4, 8, 9, 11, and 12, digital science and technology are essential. The effects of corporate digitization on enterprise carbon emissions are investigated by An and Shi (2023). By differentiating between two kinds of green patents, the authors examine the transmission process in more detail and investigates how green finance might regulate corporate digitization and lower corporate carbon emissions. According to research, companies that go digital significantly lower their carbon footprints. Dionisio et al. (2023) carried out a systematic review to determine how social innovations and digital transformations address social problems and SDGs. They found that the adoption and growth of technologies related to all 17 SDGs, particularly blockchain, IoT, AI, and autonomous robots, are rapidly expanding their role and presence and providing a dramatic evolution in a variety of sectors, including healthcare, smart cities, agriculture, and the fight against poverty and inequality. However, concerns over the effects on the labour force, potential instability, and potential effects on low-skilled/low-paying occupations were among the numerous ethical threats that the authors found, particularly with the growing usage of public data. In order to help microbusinesses realize their full potential and meet the difficulties of the dual transition era (digitalization and sustainability), Astuty et al. (2024) carried out a study to create and forecast a sustainable business model. According to the study, survival, continuity, reorientation, and synergy are all examples of sustainable resilience strategies (SRS), which are created by matching internal resources with responsive tactics. Furthermore, microbusinesses must be highly focused on attaining sustainable development goals in order to meet the challenges of a sustainable business model. Once SRS mediates these goals, sustainable business practices will be implemented and their long-term effect on sustainable business performance can be achieved through sustainable business performance.

Alojail and Khan (2023) examine the matter of the adoption of innovative digital technologies in digital transformations because digital technologies are adopted successfully only if some issues are considered carefully, for example the issues of stakeholder engagement, resource allocation, risk mitigation and the existence of resources and implementation support. The authors maintain that integrating sustainability principles behind digital transforms will amplify the efficacy of the transformation, according to metrics of economic, social as well as environmental performance. Secondly, the organizations' long-term sustainability is also improved when objectives for Digital Transformation are aligned with the Sustainable Development Goals (SDGs). Finally, Clark et al. (2022) further expand upon the 'Digital Ethical Framework' with the idea that four driver: digital capabilities, technology, infrastructure, and governance are key to achieving all 17 SDGs. Moreover, the authors argue that the importance of digital technologies in realizing the SDGs has become more focused in the context of the COVID 19 pandemic and an SDG related to 'Digital connection' would benefit in order to pursue the whole goals as digital inclusion and connectivity are critical to achieving all the targets.

Problems in Achieving SDGs

Approximately, about a dozen of these hurdles include lack of financial resources, especially in low-income countries, and weak governance that create roadblocks for implementation of effective policies. Income, gender and access to services are hurdles that prevent inclusive advancement. Climate change and environmental degradation have an impact on vulnerable groups, challenging goals such as clean water, sustainable agriculture, and climate action. Geopolitical disputes and trade obstacles also impede international cooperation. Stafford-Smith et al. (2016) investigate how the UN's SDGs implementation targets mostly fail to mention the interdependencies and links between the goals. Regarding the UN's categories of means of implementation—finance, technology, capacity building, trade, policy coherence, partnerships, and, lastly, data, monitoring, and accountability—the authors issued seven recommendations to strengthen these connections on a national and international scale. Timely achievement of SDGs requires adequate financing, which is mainly covered under green financing. The goal of green financing, according to Abramova et al. (2024), is to increase the amount of financial flows from the public, private, and non-profit sectors (from banking services, microcredit, insurance, and investment) to the priorities of sustainable development, specifically the efficient management of social and environmental risks and the utilization of opportunities that offer both a respectable rate of return and environmental benefits and a higher degree of accountability. Furthermore, the author highlighted that the current level of investment is insufficient to create a more sustainable world and to develop in a way that is more socially, environmentally, and climate-sustainable.

Few previous research studies explored hurdles in implementations of plans for SDGs, particularly in developing and underdeveloped economies across the globe. In accordance with national and international low-carbon targets and the Sustainable Development Goals (SDGs), Li et al. (2024) argued that Net-Zero Carbon Cities should minimize urban carbon emissions for healthier cities. They discovered that smart city initiatives and data-driven decision-making approaches are essential to reaching net-zero carbon cities. Dzator et al. (2023) examine how access to information and communication technology (ICT) affects poverty reduction in 44 countries in sub-Saharan Africa. They discovered that while internet, broadband, and ICT goods export raise poverty rates across all age and gender groups, telephone, mobile phone, and ICT goods import help reduce poverty. Later, Divi et al. (2024) carried out a study to examine the main obstacles that Indian rural local governments face in accomplishing the Sustainable Development Goals. They discovered that rural India faces significant obstacles in achieving the SDGs, such as fewer financial resources, manpower, technological know-how, and capabilities. However, the bright spot lies in the government's readiness to convert the digital advantages into useful information, raise public awareness, and advocate for women's more effective participation in local administration. Ardi and Putra (2024) examine how different sectors—such as the government, private sector, education, and civil society organizations—affect mental health in Indonesia. They concluded that inter-sectoral cooperation was crucial to enhancing mental health in the digital age, particularly when it came to implementing the SDGs in Indonesia.

Linkage between SDGs and Digital Technologies

Digitalization is a major enabler of the SDGs at the same time it is a catalyst for the sustained progress of many sectors. Digitalization enhances progress towards the SDGs, creating a more inclusive, sustainable and equitable global development framework. There is much research backing the theory that digitalization is the key collaborator on timely and efficient achievement of the SDGs. Popkova et al. (2022) investigate the seemingly interdependent interdependencies of institutions, SDGs and digital technologies to determine their precise interdependencies. It is found that sustainable development factors have conflicting dependence on the development of digital technologies. It also indicates that advancement in digital technology is suitable to deal with the major problems of sustainable development. Also, they uncovered a relationship between SDGs 3 and 17 that depend on digitalized information and the use of digital technology with human development, globalization and innovation institutions in industrialized countries.

According to Mondejar et al. (2021), digitalization serves the way to a smart green planet, offering answers and contributing to sustainable development, since the different advantages of integration artificial intelligence, big data management and Internet of Things clearly stand out. Moreover, the author argued that to ensure that we have a just, ecologically sustainable and healthy society it might have special chances for strategically meeting the issues related to Sustainable Development Goals (SDGs) of the United Nations. Similarly, in their research, Malhotra et al. (2018) aim to showcase a range of advancements concerning big data, data analytics, and its function in governmental procedures aimed at accomplishing the Sustainable Development Goals and discovered that when large data is appropriately analyzed using sound methods and tools, it can generate a lot of ideas and undoubtedly aid in decision-making. However, it is important to note that big data technology-based applications would come with several hazards and data security concerns. ICT development and digitization have a direct impact on attaining SDG 9, and it has been demonstrated that integrating digital technologies into business operations and digitizing the economy's non-IT sectors will help achieve several SDGs, including SDGs 3, 4, 8, 9, 11, and 12, and will spur the accomplishment of other objectives, which will typically have a synergistic effect (Strilchuk et al., 2024).

In the context of digital transformation, Strilchuk et al. (2024) seek to identify the global determinants of sustainable development and found that ICT is one of the key forces behind sustainable development. According to their study, no single SDG can be accomplished alone without the others, as they are all connected and have an impact on one another and can be accomplished in a variety of ways with various instruments and techniques. Further, in order to support global sustainable development, nations that are not sufficiently digitalized must enhance their digital transformation process. Yan et al. (2024) talked about how standardization can be integrated with cutting-edge technologies like big data, artificial intelligence, blockchain, and knowledge graphs to increase standard creation efficiency, application accuracy, and implementation effects. This study examines how national and international standards for urban sustainable development respond to the UN Sustainable Development Goals in order to keep up with current developments.

METHODOLOGY

The present study is qualitative and descriptive in nature. It is based on secondary data based on document analysis (reports, articles, websites) that is used in the present study. Further, two separate case studies of UNDP's Digital SDG Partnerships (i.e., multi-stakeholder partnerships for SDGs achievement) and the World Economic Forum's Digital Agenda (i.e., private-public partnerships for digital transformation) are provided to understand the real-world scenario of digital partnership initiatives. Tight (2017) provides an overview of case study research methods and highlights that case study involves a study of a single or limited number of cases that can be complex and bounded. Furthermore, the author advocated that the particular case is studied in a holistic way within the context of study. Similarly, in previous research, Baxter and Jack (2008) advocated that qualitative case study is a tool to study complex phenomena, and as a research approach, case study is a valuable tool for developing theory, evaluating programs, and developing interventions. There are studies that highlight the need to include multiple case studies within a single research study (Yin, 1993; Noor, 2008). Further, Yin (1993) argues that in the case of multiple-case studies, it should follow replication rather than simple logic. Cavaye (1996) found case study as a highly versatile research strategy that can be used in positivist and interpretative traditions and for building and testing theories. Moreover, the author mentioned the use of qualitative or mixed methods while conducting the case study.

Case Study 1: UNDP's Digital SDG Partnerships

The United Nations Development Programme (UNDP) has long been at the forefront of global development initiatives, with a central focus on achieving the Sustainable Development Goals (SDGs) set by the United Nations in 2015. As we move further into the digital age, the role of technology and digital tools has become increasingly critical in addressing global challenges, improving governance, and fostering innovation. One significant aspect of UNDP's work in the digital space is its Digital SDG Partnerships, an initiative designed to harness digital technologies and innovative partnerships to accelerate progress toward the SDGs. This case study explores how UNDP's Digital SDG Partnerships leverage the power of digital transformation, data, and collaborative efforts to drive sustainable development in various regions of the world.

The SDGs encompass 17 global goals ranging from poverty eradication to environmental sustainability and gender equality to global peace and justice. The ambition of the SDGs is vast, and achieving them requires global collaboration, innovative solutions, and multi-stakeholder partnerships.

In this context, digital technologies—such as artificial intelligence (AI), blockchain, the Internet of Things (IoT), big data, and digital platforms—offer new ways to address these challenges more effectively. UNDP recognized the importance of digital tools in realizing the SDGs and launched the Digital SDG Partnerships to focus on how partnerships between governments, the private sector, and development agencies can use digital tools to achieve these goals.

- **Key Objectives**
    - The Digital SDG Partnerships initiative aims to:
    - *Promote Digital Innovation for Sustainable Development:* By harnessing digital technologies, UNDP seeks to address pressing development challenges such as poverty, inequality, climate change, and access to education.
    - *Foster Multi-Stakeholder Collaboration:* This initiative emphasises collaboration between governments, the private sector, civil society organisations, and international agencies to build solutions for sustainable development.
    - *Scale Proven Solutions:* Through these partnerships, UNDP works to scale successful digital initiatives that can be adapted and implemented in various regions to accelerate progress on the SDGs.
    - *Enhance Data and Monitoring Systems:* Digital solutions improve the ability to collect, analyse, and share data on SDG progress, enabling better monitoring and decision-making.
- **Approach and Strategies**

UNDP's approach to digital SDG partnerships is multi-faceted and designed to create synergies between technology and sustainable development. Here are the core strategies:

- *Partnerships with Tech Companies and Innovators*: UNDP partners with tech giants, startups, and innovation hubs to leverage cutting-edge digital solutions. For instance, collaborations with companies like Microsoft, Google, and IBM have enabled the development of advanced analytics platforms, AI models, and cloud technologies to support SDG-related initiatives. One such partnership is with Microsoft, where they focus on harnessing AI and cloud computing to scale initiatives aimed at promoting financial inclusion and improving education access.

- *Digital Platforms for SDG Solutions*: UNDP has worked with various organisations to develop digital platforms aimed at advancing SDGs. The SDG Acceleration Lab is one such initiative that uses a digital approach to incubate innovative solutions for SDGs in countries like the Maldives, Bangladesh, and Ecuador. These platforms connect entrepreneurs, technologists, and policymakers to address key development challenges.
- *Blockchain for Transparency and Accountability*: UNDP has embraced blockchain technology to ensure transparency, accountability, and efficiency in development programs. For example, UNDP has partnered with organisations to implement blockchain-based systems that track aid distribution, ensuring that resources reach the intended recipients without delays or mismanagement. This is particularly relevant in humanitarian settings and crisis management.
- *Big Data for Monitoring and Evaluation*: Using big data analytics, UNDP has enhanced its monitoring and evaluation systems. By collecting and analysing vast amounts of data from different sectors, UNDP can track progress on SDGs and make real-time adjustments to its strategies. This includes the use of satellite imagery for environmental monitoring, mobile data to track poverty, and digital platforms to gather insights on education access.
- *Public-Private Partnerships (PPP)*: UNDP actively works to foster PPPs, wherein private-sector companies contribute to the development of digital solutions that align with the SDGs. These partnerships enable the deployment of digital solutions at scale, ensuring that technology is used to solve real-world challenges. For example, the partnership with Vodafone Foundation focuses on digital connectivity and financial inclusion, particularly in remote areas of Africa.

- **Notable Projects and Initiatives**

The UNDP's Digital SDG Partnerships encompass a wide range of initiatives that focus on both immediate and long-term issues. Some of the most prominent initiatives include:

- *Digital Innovation Hubs (DIHs)*: UNDP's Digital Innovation Hubs are set up in various countries to incubate digital solutions designed to achieve specific SDGs. These hubs facilitate experimentation, testing, and scaling of innovative digital solutions in sectors such as health, agriculture, education, and governance. In countries like India and the Philippines, DIHs have successfully piloted e-government platforms, digital health systems, and digital learning tools for underserved communities.

- *'Faster Together' Initiative*: In collaboration with the private sector, UNDP launched the Faster Together initiative, which seeks to create new digital partnerships that can speed up the achievement of the SDGs. The initiative focuses on projects that use technology for social good, such as providing digital financial services to rural populations or deploying IoT solutions to monitor and protect natural ecosystems.
- *Blockchain for Impact*: In partnership with the Swiss Agency for Development and Cooperation (SDC), UNDP has explored the potential of blockchain to enhance development impact. One notable project in the Pacific Islands involves using blockchain to ensure the traceability of aid funds, reducing corruption and ensuring that funds are allocated efficiently to areas most in need.
- *AI for Climate Action*: In partnership with various tech organisations, UNDP has been integrating artificial intelligence (AI) into climate action. This includes AI-based systems that predict weather patterns, monitor carbon emissions, and improve disaster response times. The partnership with Google and DataKind on climate data has enabled developing countries to use AI for better disaster preparedness.

- **Challenges and Lessons Learnt**

While the Digital SDG Partnerships initiative has achieved considerable success, it has faced several challenges:

- *Digital Divide*: The gap in access to digital technologies between developed and developing nations remains a significant hurdle. UNDP has worked to ensure that digital solutions are inclusive, but the divide still poses challenges to universal access to technology.
- *Data Privacy and Security*: With the increased use of digital platforms and data collection, ensuring the privacy and security of citizens' data has become a key concern. UNDP's partnerships emphasise the importance of developing ethical guidelines and ensuring the protection of data.
- *Sustainability of Solutions*: While many digital solutions have been successfully implemented, the sustainability of these solutions in the long term remains a concern. UNDP has focused on ensuring that digital solutions can be maintained and scaled beyond the pilot phase, relying on strong local ownership and capacity-building.
- *Regulatory and Policy Frameworks*: In many countries, existing legal and regulatory frameworks are not yet adapted to digital technologies. UNDP

has worked to engage governments in creating policies that foster innovation while protecting citizens' rights.

- **Case Summary**

UNDP's Digital SDG Partnerships represent a powerful approach to accelerating progress toward the SDGs through digital transformation. By leveraging innovative technologies, fostering collaboration, and scaling successful solutions, UNDP is helping countries tackle complex challenges in a more effective and efficient way. While challenges remain—particularly related to inclusivity, data security, and sustainability—the initiative demonstrates the potential of digital tools to drive systemic change. Moving forward, UNDP continued focus on these partnerships will likely play a pivotal role in shaping the future of sustainable development globally. The lessons learnt from these efforts can inform future digital solutions for SDGs and provide a blueprint for using technology for social good on a global scale.

Case Study 2: World Economic Forum's Digital Agenda

The World Economic Forum (WEF), founded in 1971, has evolved into one of the most influential platforms for public-private collaboration, bringing together leaders from business, government, academia, and civil society to address global challenges. As the digital age reshapes economies and societies, the WEF's Digital Agenda has become a crucial framework to navigate the opportunities and challenges posed by digital technologies. The WEF's Digital Agenda seeks to accelerate the responsible adoption of digital technologies, foster innovation, and shape global policies in a way that promotes inclusivity, security, and sustainable growth. This case study examines the core aspects of the WEF's Digital Agenda, its initiatives, partnerships, and the impact of its efforts on global digital transformation. The WEF's Digital Agenda thrives on multi-stakeholder collaborations. It partners with a wide range of organizations, including governments that play a central role in shaping the digital agenda, as they create regulatory frameworks and policies; major tech companies to drive innovation and develop frameworks for responsible technology adoption; academia and think tanks that help the WEF to create data-driven policy recommendations; and civil society and nonprofits to ensure that digital policies prioritize human rights, inclusivity, and equity.

Digital technologies, including artificial intelligence (AI), blockchain, 5G, and the Internet of Things (IoT), are rapidly transforming industries and societies. As these technologies advance, they present unprecedented opportunities for innovation but also raise new challenges in areas such as governance, cybersecurity, and equity. The WEF recognised the importance of digital transformation early on and launched its Digital Agenda to provide a platform for addressing these emerging issues.

The agenda aims to guide the world toward the adoption of digital technologies in a way that benefits all stakeholders, ensuring that technological advancement contributes positively to economic growth, social inclusion, and environmental sustainability.

- **Key Objectives of the Digital Agenda**

The World Economic Forum's Digital Agenda revolves around several key objectives:

- *Promoting Inclusive Digital Transformation*: The WEF aims to ensure that digital technologies are accessible to all, especially in underserved regions, and that the benefits of digital transformation are shared equitably across the global population.
- *Fostering Collaboration on Global Digital Issues*: By providing a neutral platform, the WEF brings together governments, businesses, and civil society organisations to collaborate on the development of digital policies, frameworks, and best practices.
- *Addressing Ethical and Regulatory Challenges*: The WEF works to develop ethical guidelines and regulatory frameworks that ensure the responsible use of emerging technologies, protecting users and promoting transparency, privacy, and trust.
- *Supporting Innovation and Digital Entrepreneurship*: The WEF's Digital Agenda also aims to encourage innovation and the scaling of digital startups and emerging technologies. This is done by providing platforms for dialogue and creating the conditions for digital entrepreneurship to thrive.
- *Promoting Cybersecurity and Digital Governance*: With the increasing reliance on digital platforms, the WEF focuses on strengthening cybersecurity measures and digital governance structures to protect individuals, businesses, and nations from cyber threats.
    - **Approaches and Strategies**

The WEF's Digital Agenda aims at focusing on harnessing the power of technology to drive a positive global change. It enables leaders across sectors, including business, government, academia, and civil society, to collaborate and address the opportunities and challenges posed by digital transformation. The key approaches and strategies that are used by WEF to shape the future of the digital world are as follows:

- *Digital Inclusion and Accessibility*: The WEF advocates for universal access to digital technologies, ensuring that no marginalised communities or developing nations are left behind in digital transformation. Initiatives like 'Internet for All' work to connect the remote communities and underserved populations to the global digital ecosystem. Further, the forum emphasised the need for digital infrastructure and connectivity across each part of the globc.
- *Building Digital Trust and Security*: As trust and security are deemed to be essential in the digital era, WEF places significant focus on building a secure and trustworthy digital environment. This is done by addressing cybersecurity threats, handling data security and privacy concerns, and understanding the ethical implications of digital technologies. WEF facilitates global conversations on establishing norms and standards for cybersecurity, data governance, and ethical AI usage. The 'Centre for Cybersecurity' initiative focuses on fostering collaboration between governments, businesses, and tech providers for safeguarding critical digital infrastructure.
- *Shaping Digital Policy and Governance*: WEF's strategy includes guiding policymakers for creating frameworks that support innovations along with protecting public interest, as digital transformation requires adaptive policies that can address the rapidly evolving nature of technology. WEF informs regulatory approaches and best practices in digital governance via initiatives like Digital Trade and Connectivity Platforms and Global Governance of AI. It also includes fostering collaboration between governments and private sectors for supporting digital innovation and mitigating risks associated with market monopolisation, misinformation, and privacy violations.
- *Advancing Artificial Intelligence and Emerging Technologies*: WEF acknowledges that technologies like AI and IoT have transformative potential, so their development and implementation must be done responsibly and ethically. Through initiatives like 'Shaping the Future of Technology,' the forum develops thought leadership on AI and emerging technologies. Further, WEF fosters discussions on the responsible use of AI, long-term implications, and ethical consideration of AI use on society, labour markets, and global economies. It encourages the creation of the ethical framework, e.g., 'AI for All,' that looks at the ethical development and implementation of AI.
- *Promoting Digital Transformation in Business and Industry*: The WEF works to accelerate the digital transformation of industries and organisations by integrating digital technologies into their operations to improve efficiency, sustainability, and competitiveness. The 'Centre for the Fourth Industrial Revolution' (C4IR Network) supports businesses in adopting cutting-edge technologies like AI, blockchains, and 5G in industries like manufacturing,

healthcare, and finance. The WEF also helps companies in navigating the challenges of digital disruption by providing frameworks for innovation, workforce transformation, and scaling digital technology solutions globally.

- *Ethical Frameworks for Digital Innovation*: The WEF addresses concerns regarding ethics in AI, bias in algorithms, and the impact of automation on employment. WEF is among the pioneer organisations that develop ethical guidelines that ensure digital innovations are aligned with human values and global sustainable goals. The ethical frameworks and governance models for emerging technologies are guided by initiatives like the 'Global AI Council' and the 'Digital Trade and Connectivity Initiative.'. Such initiatives help in mitigating the negative social and economic impact of technologies. It also helps in building confidence in the system for its easy adoption.
- *Data and Privacy Management*: The WEF emphasises the importance of robust data governance systems, focusing on privacy, protection, and interoperability. Initiatives like 'Data for Good' promote responsible data use for social impact and balance privacy with economic and technological advancement.
- *Supporting Digital Innovation Ecosystems*: The WEF fosters digital entrepreneurship and innovation through programs like 'Global Innovators Community' and 'Tech Pioneers' that also aim at promoting a conducive environment for startups, investors, and policymakers in the digital space.
- *Reskilling and Workforce Development*: The WEF is promoting the 'Reskilling Revolution Initiative,' which is a major part of WEF's Digital Agenda, to reskill the global workforce for future job roles in the digital economy. This initiative aims to equip individuals with necessary skills through collaboration between governments, companies, and educational institutions, bridging the digital skills gap and preparing workers for the future.
- *Fostering Digital Sustainability*: The WEF emphasizes the need to integrate digital technologies for economic growth and environmental sustainability. The 'Shaping the Future of Digital Economy and Society' platform promotes the use of AI, blockchain, and IoT to address global environmental challenges, aiming for a more resilient and responsible future.

- **Notable Projects and Key Initiatives under the Digital Agenda**

The WEF's Digital Agenda encompasses a wide range of initiatives that focus on both immediate and long-term digital issues. Some of the most prominent initiatives include:

- *Global Network for Digital Cooperation*: In 2018, the WEF launched the Global Network for Digital Cooperation (GNDG), in partnership with other organisations like the United Nations and the International Telecommunication

Union (ITU). The GNDG aims to promote inclusive, human-centred digital policies that can guide global governance of the digital space. It focuses on connecting stakeholders across sectors to foster dialogue and collaboration on issues like digital equity, privacy, data governance, and internet access.

- *Centre for the Fourth Industrial Revolution*: The Centre for the Fourth Industrial Revolution was established by the WEF to engage governments, businesses, and civil society in creating policies and frameworks for the responsible adoption of emerging technologies such as AI, blockchain, and autonomous systems. The Centre's initiatives focus on areas like AI governance, blockchain policy, IoT regulation, and digital identity.
- *Digital Economy and Society Initiative*: This initiative aims to help countries and businesses adapt to the rapidly changing digital economy. It focuses on driving digital transformation in key industries such as finance, manufacturing, and healthcare while ensuring that the economic benefits of digital technologies are distributed equitably. The initiative brings together industry leaders to develop strategies for digital resilience, digital inclusion, and sustainable economic growth.
- *Shaping the Future of the Internet*: The WEF's Shaping the Future of the Internet initiative brings together stakeholders to discuss and design the future of the global internet. This includes tackling issues like internet governance, digital trade, privacy, and online safety. The initiative focuses on the importance of a multi-stakeholder approach to internet governance that ensures an open, accessible, and secure internet for all.
- *Blockchain and Digital Assets*: The WEF has also focused on exploring the potential of blockchain technology to transform industries, including financial services, supply chains, and governance. Through its blockchain initiatives, the WEF brings together thought leaders and practitioners to discuss best practices and develop regulatory frameworks for blockchain technology and digital assets.

- **Challenges and Lessons Learnt**

Despite its successes, the WEF's Digital Agenda has faced several challenges:

Global Digital Divide: Access to technology remains unequal, particularly in developing countries. While the WEF works to address this gap, digital inclusion remains a key challenge.

- *Cybersecurity Risks*: The rise of cyber threats, data breaches, and the need for secure digital infrastructure has led to ongoing discussions on global standards for cybersecurity. The WEF's efforts to create a unified cybersecurity policy have been hampered by differing national security interests.

- *Ethical Concerns*: As technologies like AI and blockchain advance, ethical dilemmas surrounding data privacy, automation, and digital rights have emerged. The WEF has worked to address these issues but continues to refine its approaches to ethical digital governance.

- **Case Summary**

The World Economic Forum's Digital Agenda is an ambitious and impactful initiative designed to guide the global community toward a more inclusive, secure, and prosperous digital future. By fostering multi-stakeholder collaboration, promoting innovation, and addressing key ethical and regulatory challenges, the WEF has played a pivotal role in shaping the digital landscape. Through its various initiatives, such as the Global Network for Digital Cooperation, the Centre for the Fourth Industrial Revolution, and the Digital Economy and Society Initiative, the WEF has made significant strides in promoting digital transformation across industries and regions. While challenges persist, the WEF's efforts continue to influence global digital policies and practices, ensuring that digital technologies are harnessed for the benefit of all. The lessons learnt from the Digital Agenda will be crucial as the world navigates the next stages of the digital revolution. To achieve the SDGs by 2030, we will need to coordinate global efforts, innovate, and form stronger partnerships.

Significance of the Research

This study contributes to the understanding of digital technologies' role in enhancing SDG partnerships. The analysis of case studies of successful programs that use digital technologies for SDG achievement that involve diverse and multiple stakeholders will inform policymakers, practitioners, and stakeholders on effective strategies for digital SDG partnerships and will also help them in framing more holistic and effective guidelines for policy implications. The present study is significant for several key reasons. First, it provides valuable insights into how digital tools can enhance the effectiveness and scalability of partnerships aimed at achieving SDGs. As global challenges become increasingly complex and interconnected, traditional approaches to development need to be augmented by innovative technological solutions to drive impactful change. This study highlights the potential of digital technologies—such as artificial intelligence, big data, and blockchain—to foster more efficient and transparent partnerships among governments, NGOs, private companies, and other stakeholders.

Second, by focusing on real-world case studies, the research offers practical evidence and lessons learnt from successful digital initiatives. These case studies can serve as models for future projects, helping stakeholders replicate and adapt successful strategies in different contexts. The research also identifies barriers and challenges, such as issues of digital equity, cybersecurity risks, and capacity gaps, providing crucial information for overcoming these obstacles in future collaborations.

Finally, the study is timely, as digital technologies continue to evolve rapidly and their role in global development becomes more pronounced. Understanding how these tools can be harnessed to accelerate progress toward the SDGs will inform policy decisions, shape funding priorities, and guide the development of frameworks that enable more inclusive, impactful, and sustainable partnerships. In essence, this research contributes to the growing body of knowledge on how digital technologies can reshape global development, offering actionable recommendations for enhancing SDG partnerships and ensuring their long-term success.

Limitations Of The Study

While this research provides valuable insights into the role of digital technologies in leveraging SDG partnerships, it has certain limitations. First, the findings of the present research have a lack of generalizability due to the case study approach. Also, due to contextual limitations, the finding might not be easily generalizable, particularly in areas with limited access to digital infrastructure. Second, the digital tools and approach discussed in the research may become outdated or evolve in unexpected ways as the technological innovations have a rapid pace to follow. Third, the case studies relieve publicly available data, which may not always be comprehensive, accurate, or up-to-date. Fourth, there are challenges in measuring the dimensional impact of initiatives, as evaluating the effectiveness of digital technologies in achieving SDGs can be complex, especially for long-term projects where outcomes take time to materialize. Lastly, while the research focuses on multi-stakeholder partnerships, it does not necessarily account for the broader ecosystem of stakeholders involved in SDG implementation, such as grassroots organizations, individual citizens, and local communities, whose perspectives and contributions may not always be captured in partnership models. Moreover, unique contextual factors may limit its applicability when it is seen in comparison to the schemes and programs that are functional on a national or regional level. These limitations highlight the need for ongoing research and further investigation into the dynamic relationship between digital technologies and SDG partnerships, particularly in diverse and evolving global contexts.

RECOMMENDATIONS FOR FUTURE RESEARCH

Future research should include a wider variety of case studies, especially for organizations that have limited digital infrastructure to facilitate SDG partnership initiatives. Further, as digital technologies continue to evolve, research should explore emerging tools like quantum computing and advanced AI that can potentially transform SDG partnerships. Understanding how these new technologies can be integrated into SDG efforts will provide valuable foresight into their long-term impact on global development. Moreover, a deeper quantitative and longitudinal study that tracks the long-term effects of digital technology adoption in SDG partnerships is essential. Further, by including the perspectives of people from grassroots organizations and local communities, the case studies should be enriched, as it will provide a more holistic view of the challenges and benefits at different levels of strategy implementation. Future research should focus on developing metrics that combine insights from allied fields to effectively measure the impact of digital technologies on SDG outcomes. Since the involvement of private sectors can be a key driver in scaling digital solutions for SDGs, research should explore how businesses, startups and tech innovators can collaborate with government and other partners for creating scalable sustainable solutions. Lastly, researchers should investigate the role of governance structures and policy frameworks that are helpful in creating supportive environments to promote digital innovation, protect data privacy, and encourage collaborations across sectors. These recommendations aim to further refine the understanding of digital technologies in SDG partnerships and provide a strategic foundation for future research and practical application in global development initiatives.

CONCLUSIONS

This study demonstrates the transformative potential of digital technologies in fostering effective partnerships for achieving SDGs. The pivotal role of digital technologies in facilitating effective SDG partnership is highlighted in the study. Digital technologies not only facilitate communications and coordination across diverse stakeholders but also provide innovative solutions to overcome challenges in SDG implementation. However, the research also highlights the need for strategic planning, inclusive approaches, and robust digital infrastructure to maximize the impact of these technologies. It is crucial to continue exploring how digital innovations can be harnessed to bridge the gaps, scale solutions, and accelerate progress towards a more sustainable and equitable future. As the global community continues to navigate the complexities of achieving the SDGs, this research emphasized the need for stakeholders to harness the power of digital technologies to accelerate progress, foster collaborative action, and drive meaningful impact. As we move forward, two case studies are presented that are pioneers in SDG partnership. The UNDP's Digital SDG Partnership and WEF's Digital Agenda are discussed as global innovative projects that use digital technology to speed up progress toward the Sustainable Development Goals (SDGs).

REFERENCES

Aagaard, A., & Vanhaverbeke, W. (2024). The Twin Advantage: Leveraging Digital for Sustainability in Business Models. In *Business Model Innovation: Game Changers and Contemporary Issues* (pp. 227–262). Springer International Publishing. DOI: 10.1007/978-3-031-57511-2_8

Abramova, A., Popelo, O., Fedyshyn, M., Popova, L., Shchur, R., & Marych, M. (2024). Digital Green Finance Architecture in the Context of Green Deal Implementation: Modern Fintech Tools and Solutions Under the Sustainable Development Concept. *Journal of Lifestyle and SDGs Review, 4*(2), e02166. DOI: 10.47172/2965-730X.SDGsReview.v4.n02.pe02166

Ahmadova, G., Delgado-Márquez, B. L., Pedauga, L. E., & Leyva-de la Hiz, D. I. (2022). Too good to be true: The inverted U-shaped relationship between home-country digitalization and environmental performance. *Ecological Economics, 196*, 107393. DOI: 10.1016/j.ecolecon.2022.107393

Ahmed, F. (2023). Towards a sustainable era: Leveraging technology for positive impact. *International Research Journal of Modernization in Engineering Technology and Science, 5*(6).

Aini, D. N., Abdulkarim, A., Sapriya, & Rahmat. (2024). Effectiveness of digital literacy learning model with video content project in enhancing sustainable (SDG) civic intelligence. Journal of *Lifestyle and SDG'S Review, 5*. DOI: 10.47172/2965-730X.SDGsReview.v5.n01.pe02616

Akter, M. S. (2024). AI for sustainability: Leveraging technology to address global environmental. *Journal of Artificial Intelligence General science (JAIGS) ISSN: 3006-4023, 3*(1), 40–48.

Alojail, M., & Khan, S. B. (2023). Impact of Digital Transformation toward Sustainable Development. *Sustainability (Basel), 15*(20), 14697. Advance online publication. DOI: 10.3390/su152014697

An, Q., & Shi, Y. (2023). Does enterprise digitization reduce carbon emissions? Evidence from China. *Zhongguo Renkou Ziyuan Yu Huanjing, 21*(4), 219–230. DOI: 10.1016/j.cjpre.2023.11.003

Ardi, Z., & Putra, A. H. (2024). Inter-sectoral collaboration for the Mental Well-being of Indonesians in the digital age: A narrative review of SDGs implementation. *E3S Web of Conferences*, 568. DOI: 10.1051/e3sconf/202456804022

Astuty, E., Sudirman, I. D., & Aryanto, R. (2024). Sustainable resilience strategy: Unleash the micro-businesses's potential in the digitalization and sustainability era. *Cogent Business & Management, 11*(1), 2313672. Advance online publication. DOI: 10.1080/23311975.2024.2313672

Baxter, P., & Jack, S. (2008). Qualitative case study methodology: Study design and implementation for novice researchers. *The qualitative report, 13*(4), 544–559.

Cavaye, A. L. (1996). Case study research: A multi-faceted research approach for IS. *Information Systems Journal, 6*(3), 227–242. DOI: 10.1111/j.1365-2575.1996.tb00015.x

Ciulli, F., & Kolk, A. (2023). International Business, digital technologies and sustainable development: Connecting the dots. *Journal of World Business, 58*(4). *Elsevier Inc.* DOI: 10.1016/j.jwb.2023.101445

Clark, S., Maclachlan, M., Marshall, K., Morahan, N., Carroll, C., Hand, K., Boyle, N., & O'Sullivan, K. (2022). Including Digital Connection in the United Nations Sustainable Development Goals: A Systems Thinking Approach for Achieving the SDGs. *Sustainability (Basel), 14*(3), 1883. Advance online publication. DOI: 10.3390/su14031883

de Giovanni, P. (2023). Sustainability of the Metaverse: A Transition to Industry 5.0. *Sustainability (Basel), 15*(7), 6079. Advance online publication. DOI: 10.3390/su15076079

Deganis, I., Haghian, P. Z., Tagashira, M., & Alberti, A. (2021). *Leveraging digital technologies for social inclusion*. United Nations Department of Economic and Social Affairs.

Dionisio, M., de Souza Junior, S. J., Paula, F., & Pellanda, P. C. (2024). The role of digital social innovations to address SDGs: A systematic review. *Environment, Development and Sustainability, 26*(3), 5709–5734. Springer Science and Business Media B.V. DOI: 10.1007/s10668-023-03038-x

Divi, S., Dash, D. N., & Sahoo, M. K. (2024). Problems and Challenges of Indian Rural Local Governments in Achieving Sustainable Development Goals: An Analysis of the Viable Perspectives. *International Journal of Sustainable Development and Planning, 19*(7), 2455–2464. DOI: 10.18280/ijsdp.190703

Dwivedi, A., Ganguly, A., & Paul, S. K. (2024). Critical success factors for linking digital technologies and circular supply chains. *Business Strategy and the Environment, 33*(8), 8332–8360. Advance online publication. DOI: 10.1002/bse.3907

Dzator, J., Acheampong, A. O., Appiah-Otoo, I., & Dzator, M. (2023). Leveraging digital technology for development: Does ICT contribute to poverty reduction? *Telecommunications Policy*, *47*(4), 102524. Advance online publication. DOI: 10.1016/j.telpol.2023.102524

Elfaki, K. E., & Ahmed, E. M. (2024). Digital technology adoption and globalization innovation implications on Asian Pacific green sustainable economic growth. *Journal of Open Innovation*, *10*(1), 100221. Advance online publication. DOI: 10.1016/j.joitmc.2024.100221

Fuchs, I., Rajasekharan, J., & Cali, Ü. (2024). Decentralization, decarbonization and digitalization in swarm electrification. *Energy for Sustainable Development* (Vol. 81). Elsevier B.V. DOI: 10.1016/j.esd.2024.101489

George, G., Merrill, R. K., & Schillebeeckx, S. J. (2021). Digital sustainability and entrepreneurship: How digital innovations are helping tackle climate change and sustainable development. *Entrepreneurship Theory and Practice*, *45*(5), 999–1027. DOI: 10.1177/1042258719899425

Ignatov, A. (2020). The Digital Economy of BRICS: Prospects for Multilateral Cooperation.

Katz, R., Koutroumpis, P., & Callorda, F. (2014). The Latin American path towards digitization. *Revista de Estudios Para El Desarrollo Social de La Comunicación*, 151–179. DOI: 10.15213/redes.n9.p151

Katz, R. L., & Koutroumpis, P. (2013). Measuring digitization: A growth and welfare multiplier. *Technovation*, *33*(10-11), 314–319. DOI: 10.1016/j.technovation.2013.06.004

Li, J., Shirowzhan, S., Pignatta, G., & Sepasgozar, S. M. E. (2024). Data-Driven Net-Zero Carbon Monitoring: Applications of Geographic Information Systems, Building Information Modelling, Remote Sensing, and Artificial Intelligence for Sustainable and Resilient Cities. *Sustainability (Switzerland)*, *16*(5). Multidisciplinary Digital Publishing Institute (MDPI). DOI: 10.3390/su16156285

Malhotra, C., Anand, R., & Singh, S. (2018). Applying Big Data Analytics in Governance to Achieve Sustainable Development Goals (SDGs) in India. *Studies in Big Data*, *38*, 273–291. Springer Science and Business Media Deutschland GmbH. DOI: 10.1007/978-981-10-7515-5_19

Martínez-Peláez, R., Ochoa-Brust, A., Rivera, S., Félix, V. G., Ostos, R., Brito, H., Félix, R. A., & Mena, L. J. (2023). Role of digital transformation for achieving sustainability: Mediated role of stakeholders, key capabilities, and technology. *Sustainability (Basel), 15*(14), 11221. DOI: 10.3390/su151411221

Mishra, R., Singh, R. K., & Rana, N. P. (2024). Digital orientation, digital eco-innovation and circular economy in the context of sustainable development goals. *Business Strategy and the Environment, 33*(4), 2752–2770. DOI: 10.1002/bse.3619

Mitra, C. K., & Prasad, M. N. V. (2021). Information Explosion in Digital Ecosystems and Their Management. *Handbook of Ecological and Ecosystem Engineering*, 451–467.

Mondejar, M. E., Avtar, R., Diaz, H. L. B., Dubey, R. K., Esteban, J., Gómez-Morales, A., Hallam, B., Mbungu, N. T., Okolo, C. C., Prasad, K. A., She, Q., & Garcia-Segura, S. (2021). Digitalization to achieve sustainable development goals: Steps towards a Smart Green Planet. *Science of the Total Environment, 794*. Elsevier B.V. DOI: 10.1016/j.scitotenv.2021.148539

Noor, K. B. M. (2008). Case study: A strategic research methodology. *American Journal of Applied Sciences, 5*(11), 1602–1604. DOI: 10.3844/ajassp.2008.1602.1604

Okot, T., Madrigal-Mendez, P., & Solorzano-Arias, D. (2023). The Internet of Things (IoT) for Sustainability: A Framework for Costa Rica. *Journal of Technology Management & Innovation, 18*(4), 3–17. DOI: 10.4067/S0718-27242023000400003

Popkova, E. G., de Bernardi, P., Tyurina, Y. G., & Sergi, B. S. (2022). A theory of digital technology advancement to address the grand challenges of sustainable development. *Technology in Society, 68*, 101831. Advance online publication. DOI: 10.1016/j.techsoc.2021.101831

Schneider, S. (2019). The impacts of digital technologies on innovating for sustainability. *Innovation for Sustainability: Business Transformations Towards a Better World*, 415–433.

Setyaningrum, R. P., Rulianti, E., & Susilo, P. (2024). Digitalisation empowerment of business to digital innovation in the achievement of SDGS: A survey on creative MSME actors in Indonesia. *Journal of Lifestyle and SDG'S Review, 5*(1), e02770. Advance online publication. DOI: 10.47172/2965-730X.SDGsReview.v5.n01.pe02770

Singh, M., Srivastava, S., Kumar, M., Singh, D., Anupriya, Sahu, P. K., & Chaturvedi, A. (2024). SDG'S and emerging synergy: green healthcare system for sustainable healthcare delivery in metropolitan cities in India. *Journal of Lifestyle and SDG'S Review, 4*(4). DOI: 10.47172/2965-730X.SDGsReview.v4.n04.pe02319

Stafford-Smith, M., Griggs, D., Gaffney, O., Ullah, F., Reyers, B., Kanie, N., Stigson, B., Shrivastava, P., Leach, M., & O'Connell, D. (2017). Integration: The key to implementing the Sustainable Development Goals. *Sustainability Science, 12*(6), 911–919. DOI: 10.1007/s11625-016-0383-3 PMID: 30147763

Strilchuk, Y., Krasnova, I., Khodakevich, S., Metsger, I., Stryzhak, A., & Dubas, A. (2024). Sustainable development determinants in the context of digital transformation. *Financial and Credit Activity: Problems of Theory and Practice, 3*(56), 293–307. DOI: 10.55643/fcaptp.3.56.2024.4367

Tight, M. (2017). Understanding case study research: Small-scale research with meaning. *Sage (Atlanta, Ga.)*. Advance online publication. DOI: 10.4135/9781473920118

Trittin-Ulbrich, H., Scherer, A. G., Munro, I., & Whelan, G. (2021). Exploring the dark and unexpected sides of digitalization: Toward a critical agenda. *Organization, 28*(1), 8–25. DOI: 10.1177/1350508420968184

Varriale, V., Camilleri, M. A., Cammarano, A., Michelino, F., Müller, J., & Strazzullo, S. (2024). Unleashing digital transformation to achieve the sustainable development goals across multiple sectors. *Sustainable Development*. Advance online publication. DOI: 10.1002/sd.3139

West, S. M. (2019). Data capitalism: Redefining the logics of surveillance and privacy. *Business & Society, 58*(1), 20–41. DOI: 10.1177/0007650317718185

Wu, H., Han, M., & Shen, Y. (2023). Technology-driven energy revolution: The impact of digital technology on energy efficiency and its mechanism. *Frontiers in Energy Research, 11*, 1242580. Advance online publication. DOI: 10.3389/fenrg.2023.1242580

Yadav, S., Samadhiya, A., Kumar, A., Majumdar, A., Garza-Reyes, J. A., & Luthra, S. (2023). Achieving the sustainable development goals through net zero emissions: Innovation-driven strategies for transitioning from incremental to radical lean, green and digital technologies. *Resources, Conservation and Recycling, 197*, 107094. Advance online publication. DOI: 10.1016/j.resconrec.2023.107094

Yan, M., Yang, F., Pan, H., & Li, C. (2024). Research on the Response of Urban Sustainable Development Standards to the United Nations Sustainable Development Goals Based on Knowledge Graphs. *Land (Basel), 13*(11), 1962. Advance online publication. DOI: 10.3390/land13111962

Yin, R. (1993). *Application of Case Study Research.* Sage Publication.

Zhang, J., & Liu, M. (2024). How to leverage digital sustainability orientation to promote environmentally sustainable practices of manufacturing enterprises in China. *Sustainability (Basel), 16*(12), 5112. DOI: 10.3390/su16125112

Chapter 8
Digital Circular Economy for Attaining Sustainable Development Goals:
Technologies for Global Sustainability

Rejaul Karim
https://orcid.org/0000-0001-6175-8825
Varendra University, Bangladesh
INTI International University, Malaysia

Abdul Waaje
https://orcid.org/0009-0004-7207-0750
Michigan Technological University, USA

Md. Mustaqim Roshid
https://orcid.org/0000-0001-8517-6336
University of Rajshahi, Bangladesh

ABSTRACT

Digital technology with circular economy concepts is one of the game-changing strategies for sustainable development goals (SDGs). The digital circular economy presents possibilities and threats, which this section discusses. Concerns about data privacy and security, regulatory and economic hurdles, and the need for a shift in culture and behavior are among the most pressing problems. However, there are many chances to be innovative through working with others, forming partnerships, and expanding into new areas. Stakeholders may make strides towards a more sustainable and resilient economy by responding to these threats and capitalizing on

DOI: 10.4018/979-8-3693-9909-5.ch008

these possibilities. To better understand and implement digital circular practices, future research should focus on creating standardized impact metrics, developing cutting-edge technologies, establishing supportive policy frameworks, gaining insights into consumer behavior, creating innovative business models, conducting sector-specific studies, considering social equity, collaborating globally, and ensuring long-term sustainability.

INTRODUCTION

Embracing the circular economy model and the human-centric integration of Industry 4.0 transformative technologies is becoming essential in today's digital age, where the word sustainability is becoming an absolute need in both the rethinking and reshaping of commercial processes. In response to the targets for sustainable development, the notion of a digital circular economy (DCE) is gaining traction, and with it, there is a shift from the traditional linear to a circular economic framework. Most research on the DCE's sustainability consequences has appeared in the past two years. Whereas there has been an upsurge across this field of learning, a substantial number of it has relied on informal testimony or quoted literature compared to concrete data (Piscicelli, 2023). To achieve SDG targets, specific technologies and the broader area of information systems researchers have already begun focusing on the circular economy (Zeiss et al., 2021; Gazi et al., 2024). The progression from the Fourth Industrial Revolution to Industry 5.0, featuring elevated human-digital interaction and incorporating digitalized circular economy concepts, has greatly influenced the changing dynamics of biodiversity, economic models, and organizational frameworks in contemporary society. By combining circular economy concepts with modern technology, Industry 4.0 and 5.0 bring a creative panorama on manufacturing and using resources while spotlighting the digitization of industries (Voulgaridis et al., 2022). This convergence further amplifies the vital role of the digital circular market in meeting the SDGs.

The primary focus of DCE is to leverage technological advancements such as artificial intelligence (AI), the Internet of Things (IoT), and blockchain to reduce waste, promote efficient resource management, and drive the principles of the circular economy model. Contrary to popular belief, the Sustainable Development Goals (SDGs) are a collection of seventeen global objectives that the UN General Assembly formed to address major challenges such as inequality, poverty, global warming, and environmental degradation by 2030 (United Nations, 2015). Incorporating digital technology into the current circular economic model tends to lower resource utilization and accelerate circular processes (Moreno & Charnley, 2016; Roshid et al., 2025). By utilizing the DCE, built on AI, companies may evaluate customer input in depth to determine demands and create sustainable goods or services. Optimal machine utilization, reduced downtime, and extended asset life expectancy are all possible thanks to digital CE, which also helps to decrease the requirement for resource-consuming upgrades. Simplifying the supply chain is another way an AI-powered DCE might improve sustainability through better waste management.

Similarly, a DCE centered around the IoT can enhance the operational effectiveness of circular economy models by propagating process-specific instantaneous information. Other benefits include realizing the SDGs and DCE and integrating sensor-based IoT technology, which could drastically reduce fossil fuel use in industrial plants and lessen gaseous exposure threats to workers' health (Ahmad et al., 2024). Transparency and the continued traceability of circular operations are essential to a DCE, and blockchain technology can help make this a reality. These cutting-edge DCE frameworks are in direct alignment with the SDGs, which have the objectives of guaranteeing environmentally friendly habits of consumption and production (SDG 12), developing robust infrastructure (SDG 9), and mitigating global warming (SDG 13).

The central goal of a circular economy (CE) is to maximize the economic value of the acquisition compared to supplies throughout a closed-loop cycle of renovation and recycling, thus maximizing monetary and green efficiency (Lieder & Rashid, 2016; Karim et al., 2024; Roshid, Karim, et al., 2024; Roshid, Waaje, et al., 2024). The long-term distribution of circular economy (CE) strategies is favorable for manufacturing entities because it reduces the number of raw materials used and improves their innovation ability (Rahman et al., 2023). The application of DEC in various merchandise types results in numerous benefits witnessed at multiple manufacturing phases. These benefits include optimizing produced goods to improve lifespan and superior quality, documenting utilization phases and fabrication statuses based on stops and circumstances, and giving feedback on merchandise for the formulation of approaches. As a result, the adoption of DCE strengthens the influence of various characteristics in the production and service sector by utilizing process data throughout problem-solving to gather, analyze, and integrate with the entire system. DCE is not only able to capitalize on the monetary advantages of digitalization by implementing industry 4.0/5.0 innovations within the manufacturing industries, but it additionally can assist the service sector, such as healthcare, as well as the government entities by ensuring broadened tracing of inventory, upgrading of government amenities, and effective solutions for sustainable energy management. Several different industries are undergoing paradigm shifts that involve the transition from linear socioeconomic models to digital circular economic models or from the circular economic model that is currently in place to DCE models, which include digital integration of circular economy, digital transformation of circular economy, and smart circular economy. Considering the SDGs, this chapter will investigate how digital technology and circular economy practices might collaborate to create synergies. The presentation will go into specifics about the overview of DCE, with a particular emphasis on digital advances that have already been implemented and their applicability in various industries, and it will showcase successful implementations through current case studies.

Additionally, the chapter will discuss the difficulties and possibilities associated with merging technologically enhanced CE principles with digital solutions to expedite progress toward the SDGs. Because the world is currently struggling to deal with the urgent problems of climate change, the depletion of resources, and the deterioration of the environment, the need for creative and long-term solutions has remained relatively high (Karim et al., 2024). The DCE presents a promising possibility for mitigating these difficulties and can encourage economic development, generate employment opportunities, and advance a more egalitarian and sustainable future. With the right implementation of human-centered advanced technology features of Industry 4.0, also known as Industry 5.0, DCE projects have the potential to play a crucial role in constructing a more sustainable society. This can be accomplished by coordinating themselves with the SDGs.

THE DIGITAL CIRCULAR ECONOMY

The Digital Circular Economy (DCE) denotes an economic framework that amalgamates digital technology with circular economy tenets to improve resource efficiency, sustainability, and economic development. It utilizes sophisticated technologies including the Internet of Things (IoT), Artificial Intelligence (AI), blockchain, big data, and cloud computing to enhance the lifetime of products, materials, and resources. In recent years, the circular economy has emerged as a metaphor for sustainable economic growth, as it envisions products in a way that eliminates their consequences on the environment. To gain sustainability standards in an economically feasible way and coordinate circular ecosystems in general, digitizing and digital technology are becoming crucial aspects of the economy that operates circularly. Due to the dearth of scholarly work on the topic, it could be difficult to understand what the term "digital circular economy" means. The digital ecosystem is an approach to advancing upward the circular economy. It highlights connecting and integrating the economic value chain to develop new ways of transferring it, surveillance, and controlling it within digitalized platforms (Berg et al., 2020; Wilts, 2017; Yıldırım & Yıldırım, 2020). It assists enterprises in overseeing their products and services effectively while determining their practicability, functionality, and end-user (or user) contentment. In order to improve resource utilization, cut down on waste, and encourage the reuse, refurbishing, and recycling of items, it makes use of technology developments that are appearing in Industry 4 or Industry 5 technologies. These technologies include the IoT, blockchain, AI, and big data analytics. The DCE aims to decrease the negative effects on the environment, extend the life cycles of products, and generate value from trash. This is accomplished by enabling current-time surveillance and tracking of resources, improving supply chains, and promoting fresh business models like product-as-a-service and mutually beneficial platforms. The shift to an economic structure that is more resilient and recuperating is supported by this strategy, which not only adds to the survival of the economy but also helps to assist the change. About the manufacturing supply chain, DCE is required to conform to the fundamentals of the circular economy and adopt circular methods for reinvention.

Thus, a DCE integrates digital technologies into the principles of the circular economy to enhance sustainability, resource efficiency, and economic value. The DCE aims to create a sustainable, efficient, and resilient economic system by leveraging digital innovations to enhance the traditional circular economy model.

Key Components of the Digital Circular Economy

The Digital Circular Economy (DCE) combines digital technology with circular economic ideas as an innovative strategy for long-term sustainability. Fundamental to the DCE are novel approaches that use technological developments to reduce waste and increase resource efficiency. One of the main parts is Resource Efficiency and Optimisation, which stresses the need to use resources sustainably using automation and advanced analytics (Vazhenina et al., 2023). Product as a Service (PaaS) encourages durability and decreases environmental impact by moving away from the traditional ownership paradigm and towards service-based solutions (Tseng et al., 2023). Reducing demand for new resources, closed-loop supply chains allow the reuse of commodities indefinitely (Waaje et al., 2024). To maximize material recovery and reutilization, Enhanced Recycling and Waste Management employ state-of-the-art technology (Sarc et al., 2019). Lastly, the foundation of all these parts is data-driven decision-making, which guarantees that educated, real-time insights propel forward the circular economy (Awan, Shamim, et al., 2021). Together, these elements lay the groundwork for an environmentally and socially responsible economic model that can weather any storm.

Optimization and effectiveness of resources are key to the DCE. Companies employ digital technologies to enhance resource consumption and avoid wastage across the manufacturing lifecycle. IoT, AI, and big data analytics assist in tracking and managing the usage of resources in real-time. The IoT sensors in manufacturing facilities may supply continual usage of asset data. These sensors accurately track the flow of materials, power consumption, and industrial outputs, allowing instantaneous modifications to improve productivity and eliminate waste. IoT devices can track energy and material usage, highlighting opportunities for resource conservation or recycling during production (Vazhenina et al., 2023). Machine learning and AI techniques find trends in IoT sensor reports and forecast resource needs. Technology can optimize production plans, manage logistics, and lower downtime, heightening resource utilization. AI-driven preventative care can extend mechanical longevity by predicting faults and planning appropriate interventions, reducing resource waste (Kurniawan et al., 2022). Blockchain technology provides supply chain accountability and traceability, vital to closed-loop systems that recycle and reuse materials. Blockchain may confirm material origins, sustainability, and ecological compliance (Stucki et al., 2023). Transparency promotes confidence with consumers and stakeholders, encouraging sustainable production and consumption. Effective resource utilization requires better disposal of waste and recycling. Intelligent Robots and AI could automate garbage sorting and treatment, enhancing recycling accuracy and efficiency. Intelligent machines remove recyclable materials from trash streams, while AI optimizes recycling procedures by discovering the best ways to process diverse materials. These digital technologies improve the sustainability of resources and help create a circular economy. The DCE promotes economic viability and hazard resilience by optimizing resource use, decreasing junk, and extending product lifecycles.

Product as a Service (PaaS)

One novel approach to conducting business in the emerging DCE is "Product as a Service," or PaaS. It switches a concentration from distributing physical goods to delivering the solutions that those items compose conceivably. Optimized performance, energy savings, and environmental protection are the cornerstones of this paradigm, prioritizing interdependent gadgets and technologies. Businesses in Indonesia's domestic appliance fabrication industry could provide servicing, customization, and efficient use of resources through PaaS models, all while minimizing the environmental effects (Tseng et al., 2023; Karim, et al., 2024). With the help of IoT technology, PaaS keeps tabs on product use, forecasts when maintenance will be needed, and guarantees smooth operation. Devices connected to the IoT gather data in real-time, which allows for preventative services, which in turn increases goods' lifespan, decreases waste, and prevents failures and replacements (Karim et al., 2024; Roshid et al., 2024; Tseng et al., 2023). Furthermore, digital platforms are vital to PaaS since they allow for the smooth exchange of information between vendors and their goods. Cloud computing, which offers a scalable network for storing data and investigation, enables personalized services, and these platforms could further improve it by integrating data from IoT devices to deliver insights into product efficacy and consumer habits. Products are made to last, repairable, and recyclable when circular ideas are incorporated into PaaS models. Modular designs made of recyclable materials help achieve the circular economy's aim of reducing waste and making the most of resources. This approach has ecological and financial advantages because it generates income from services instead of selling products. In this way, PaaS promotes a sustainable future by utilizing cutting-edge technology to increase efficiency and longevity, therefore balancing monetary incentives with ecological advantages.

Closed-Loop Supply Chains

The DCE relies on closed-loop supply chains (CLSCs) to reduce waste and maximize resource efficiency by reusing recycling, and remanufacturing products. Improving CLSCs' operational effectiveness and environmentally friendly operation by integrating digital innovations like the IoT, AI, blockchain, etc. For effective recycling and remanufacturing, the IoT allows for continuous watching over and recording of products at all lifecycle phases via integrated sensors that gather data on use, condition, and location (Ghobakhloo, 2018). In order to decrease downtime and increase product life, AI and machine learning algorithms examine this data to forecast product failures, optimize maintenance schedules, and enhance decisions about lifecycle management (Ondemir & Gupta, 2014; Yildirim, et al., 2022). Blockchain technology, another safeguard for closed-loop systems, creates an immutable record of interactions and commodity identifiers (Kazancoglu et al., 2023). The application of the IoT in the food supply chain, for example, shows great promise for reducing food waste by maintaining ideal conditions for perishable commodities (Kazancoglu et al., 2023). In order to efficiently manage the lifecycles of goods and guarantee ingredient reuse and recycling, companies such as Fairphone utilize IoT and blockchain. Utilizing these digital solutions, businesses may lessen their environmental impact by cutting down on waste and using resources better.

Enhanced Recycling and Waste Management

Advanced technologies improve recycling and trash management in the DCE, maximizing resource recovery and limiting environmental harm. Innovative dumpsters and computerized separating mechanisms are crucial to modern garbage disposal. Smart bins with sensors monitor waste levels and adjust collecting pathways, saving petroleum and reducing greenhouse gases. Sensor-equipped garbage containers and autonomous trucks in Berlin have reduced garbage removal trips, improving efficiency and cutting costs (Cheah et al., 2022). Resource trade and trash management are easier with IoT and digital platforms. Mobile Automation

(MOBA) tracks garbage containers, manages client data, and optimizes waste collection routes, enhancing efficiency (Maiurova et al., 2022). Furthermore, machine learning programs reliably identify and categorize garbage, increasing the rate of recycling and decreasing contamination (Cheah et al., 2022; Wube Dametew, 2015). The elimination of waste is increasingly using robotics. Robotics with enhanced sensors and AI sort recyclables accurately, eliminating manual labor and improving recycling processes. This technology improves waste recovery, helping the circular economy (Cheah et al., 2022; Sarc et al., 2019). Public participation and enforcement of policies depend on technology-based solutions. Self-sweeping cars and AI-based littered characterization improve urban management of waste and citizen engagement in separating garbage and recycling (Cheah et al., 2022). Extension of producer accountability (EPR) policies encourage recycling by forcing producers to be liable for commodity end-of-life governance, supporting an increasingly sustainable waste removal system (Cheah et al., 2022; Sarc et al., 2019). Digitizing recycling and disposal processes improves resource efficiency and sustainability. IoT, machine learning, and autonomous systems are capable of helping cities and industries reduce their ecological footprint and promote equitable development by transitioning to a regenerative economy.

Data-Driven Decision Making

Decision-making informed by data is necessary for the DCE to improve efficiency and sustainability (Awan et al., 2021; Charnley et al., 2019). Technologies such as the IoT, big data analytics, and cyber-physical systems (CPS) enable organizations to collect and analyze data in real-time and monitor and optimally utilize their resources. Incorporating IoT sensors into manufacturing equipment allows for the monitoring of machinery performance, the forecasting of maintenance requirements, and the reduction of downtime, ultimately resulting in the extension of machinery life and the reduction of waste (Awan et al., 2021). According to (Charnley et al., 2019), big data analytics provides an additional interpretation of the huge amount of data created by IoT devices. This interpretation identifies patterns and insights that steer optimal product recycling or refurbishment decision-making. Simulation methods such as "discrete event simulation (DES)" and "system dynamics (SD)" offer valuable foresight by simulating various scenarios. According to (Okorie et al., 2018), these models help forecast remanufacturing processes and reduce costs. A good example of how simulation can help improve efficiency is the Certainty of Product Quality (CPQ) idea (Charnley et al., 2019). This concept demonstrates how simulation can help improve efficiency by forecasting it. To combine economic goals with environmental stewardship, the incorporation of these technologies promotes continuous improvement and sustainability in manufacturing processes (Turner et al., 2016).

Figure 1. The key Components of the DCE

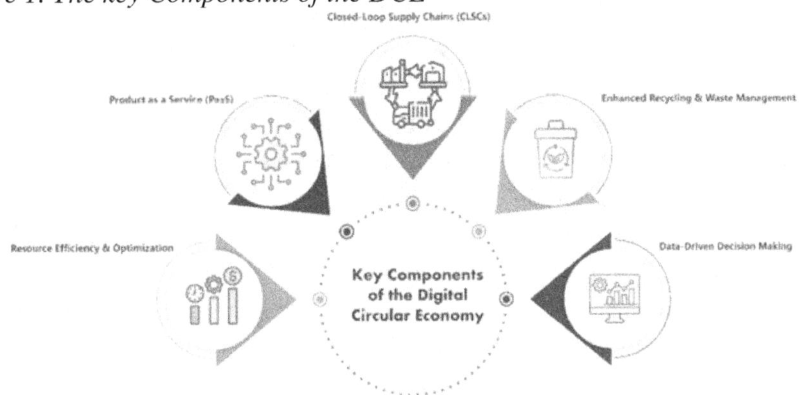

Source: Authors' Illustration

The Sustainable Development Goals: A Brief Overview

The United Nations (UN) ratified the 17-point SDGs in 2015 to tackle various environmental, social, and economic issues. The SDGs foster a balanced approach to sustainable development, which aims to increase wealth while maintaining environmental protection. With the help of all nations, industries, and other interested parties, the 2030 Agenda for Sustainable Development aims to accomplish enormous worldwide progress by 2030. Given that poverty is a significant barrier to achieving sustainable development, the primary Sustainable Development Goal (SDG) is to eradicate poverty in all its forms. In 2017, around 689 million individuals, which accounts for nearly 9.2% of the global population, lived on less than $1.90 per day. This is the case despite significant progress made in reducing extreme poverty. These problems were made worse by the COVID-19 epidemic, which sent millions of people back into poverty and demonstrated the importance of strong social safety nets (Erdogan, 2022; United Nations, 2022, 2024a). The second goal, eradicating hunger, is to accomplish food security, improve nutrition, and promote sustainable agriculture. According to the Food and Agriculture Organisation (FAO), over 690 million individuals suffered from undernourishment in 2019. This trend is projected to persist and escalate to over 840 million by 2030. In order to fight hunger and guarantee food security, the SDGs stress the significance of sustainable farming methods, funding rural infrastructure, and assisting smallholder farmers (Anand et al., 2020; Grosso et al., 2020; United Nations, 2024b). Goal 3 of the Sustainable Development Agenda is to guarantee that all people of all ages enjoy good health and strive to improve their quality of life. Reducing maternity and infant death rates and fighting infectious diseases are a few great achievements in health outcomes. Nevertheless, there are still obstacles to overcome, especially regarding mental health, non-communicable diseases, and attaining universal health care (United Nations, 2024c). Inclusion and equity in high-quality education and possibilities for continuous learning throughout life are emphasized in the fourth SDG on quality education. Despite education's central role in driving development, inequalities in access to education continue, especially in low-income nations. At its height in 2018, the COVID-19 epidemic affected more than 1.6 billion students worldwide, further disrupting educational systems. 258 million children and youth were not in school (United Nations, 2024d). The fifth SDG on gender equality is empowering all girls and women to attain gender equality. There are still huge gender gaps, especially when it comes to economic participation and violence against women, even if we have come a long way in areas like education and political representation. The SDGs aim to eliminate prejudice and violence in all its manifestations, promote gender equality in leadership roles, and guarantee the right to reproductive health care (United Nations, 2024e). The sixth Sustainable Development Goal (SDG) prioritizes the crucial significance of ensuring access to safe and cheap drinking water and adequate sanitation for the promotion of human health and overall well-being. In 2017, a staggering 4.2 billion individuals lacked access to adequate sanitation, while 2.2 billion were deprived of safely regulated drinking water, despite some advancements. The SDGs promote sustainable water management, ecological protection, and infrastructure investment to provide universal access (United Nations, 2024f). Sustainable development and economic growth depend on affordable and clean energy, which is the topic of the seventh SDG. In order to meet fundamental requirements and conduct economic activity, access to energy is vital. Almost 789 million people in 2018, mostly in Sub-Saharan Africa, had no electricity access. To guarantee that everyone has access to contemporary energy services, the SDGs stress the importance of renewable energy, energy efficiency, and sustainable energy technology (United Nations, 2024g). The eighth SDG is Decent Work and Economic Growth, which ensures everyone has access to decent work and full and productive employment. Unemployment, underemployment, and unsafe working conditions are problems that affect the worldwide labor market. Worldwide, the unemployment rate was 6.5% in 2020, according to the International Labour Organisation (ILO), and it was even higher among young people. Enhanced labor standards, universal access to good jobs, and inclusive economic growth are all pushed for by the SDGs (United Nations, 2024h). Recognizing that innovation, inclusive and sustainable industrialization, and resilient infrastructure are critical drivers of economic growth and development, the ninth SDG focuses on these three areas. Productivity, employment, and long-term growth are all aided by technology and physical infrastructure investments. Science, technology, and the promotion of environmentally friendly business practices are all urged upon in the SDGs (United Nations, 2024i). The tenth SDG is to lessen inequality, both domestically and internationally. Disparities in income, wealth, and opportunity exist within and between nations, making inequality a major obstacle. For a more equitable society, the SDGs call for more social protection systems, inclusive policies, and equal opportunities to access resources (United Nations, 2024j). Inclusion, safety, resilience, and sustainability in urban areas are the goals of the eleventh SDG. Congestion, pollution, and insufficient housing are some of the problems urbanization brings, along with the chances it brings for economic progress. By 2030, almost 60% of the global population will reside in metropolitan regions. To guarantee sustainable urbanization, the SDGs promote resilient infrastructure, affordable housing, and sustainable urban planning (United Nations, 2024k). The twelfth SDG aims to promote responsible production and consumption to guarantee long-term sustainability. Degradation of the environment, loss of resources, and climate change result from unsustainable use and production. According to the SDGs, sustainable development can only be accomplished via responsible use of resources, less waste, and ethical corporate practices (United Nations, 2024l). The thirteenth SDG of climate action stresses the critical importance of taking immediate action to halt or mitigate the effects of climate change. The effects of climate change on ecosystems and human populations, including higher average temperatures, more frequent and severe weather events, and higher sea levels, constitute a danger to international progress. According to the SDGs, Reducing greenhouse gas emissions, increasing resilience, and incorporating climate policy into national goals are urgently needed (United Nations, 2024m). Underwater life is the focus of the fourteenth SDG, which seeks to protect and responsibly utilize the world's seas, oceans, and marine resources. Global ecosystems rely on the oceans for various reasons, including food production, economic stability, and climate regulation. Nevertheless, marine biodiversity is in jeopardy due to overfishing, pollution, and climate change. Marine conservation, sustainable fishing techniques, and pollution reduction are all emphasized in the SDGs (United Nations, 2024n). The sixteenth Sustainable Development Goal (SDG) focuses specifically on terrestrial ecosystems and biodiversity. The objectives of this initiative are to halt the decline in biodiversity, implement sustainable forest management practices, combat desertification, and save, restore, and encourage the sustainable utilization of terrestrial ecosystems. Carbon sequestration, water purification, and biodiversity habitat are just a few of the essential services provided by ecosystems like forests and wetlands. Sustainable land management, conservation initiatives, and restoring damaged ecosystems are all highlighted in the SDGs (United Nations, 2024). The 16 Sustainable Development Goal (SDG) encompasses the following aims: promoting peaceful and inclusive societies, guaranteeing universal access to justice, and establishing effective, responsible, and inclusive institutions. War, bloodshed, and ineffective government work together to stymie development initiatives and widen existing income gaps. The SDGs promote inclusive government, human rights, and the rule of law as means to achieve justice and peace (United Nations, 2024o). Last, the seventeenth SDG emphasizes partnerships to achieve the goals, acknowledging that reaching the SDGs calls for cooperation among various sectors, nations, and stakeholders (United Nations, 2024p). In order to achieve sustainable development, the SDGs emphasize the need for increased global collaboration, resource mobilization, capacity building, and partnerships involving multiple stakeholders.

Synergies Between the Digital Circular Economy and SDGs

Integrating digital technologies into the circular economy can significantly contribute to achieving the SDGs. Below are the synergies between the digital circular economy and specific SDGs.

- **SDG1:**By working to end poverty in all its manifestations, the DCE has the potential to be an instrumental force in reaching SDG 1. Economic prospects and new jobs may materialize due to the DCE's advocacy of sustainable manufacturing practices, waste reduction, and resource efficiency. The DCE improves resource management and reduces waste through digital technologies like the IoT, AI, and blockchain, increasing productivity and decreasing costs. Poverty reduction and economic inclusion programs can benefit from these savings (Babu & Srivastava, 2024). One example is how digital platforms may help small-scale producers and entrepreneurs gain access to markets. This will allow them to engage in the economy more efficiently. Renewable energy and sustainable agriculture are two more areas that the DCE backs and both are important for helping low-income people meet their basic needs and improve their standard of living. The goals of SDG 1 and the DCE are complementary; they work together to end poverty and advance sustainable development by promoting inclusive growth and fair allocation of resources.
- **SDG2:**Goal 2 of the SDGs aims to accomplish the objectives of eradicating hunger, attaining food security, enhancing nutrition, and fostering sustainable agriculture. This objective can be made much more effective by incorporating the ideas of a circular economy. With a DCE, reusing, recycling, and repurposing materials is key to reducing waste. Reducing waste and increasing resource efficiency in agriculture is the goal of digital technology-enabled

closed-loop systems that recycle agricultural by-products. For example, soil health and crop productivity can be enhanced by transforming agricultural leftovers into bioenergy or organic fertilizers (Kazancoglu et al., 2023; Ng & Wong, 2024; Schützenhofer et al., 2024).

- **SDG3:**Promoting health and wellbeing for all people of all ages is the third of the SDGs. Incorporating DCE ideas will greatly aid this endeavor by encouraging sustainable production techniques, reducing waste, and using resources most efficiently. Health-related businesses can improve the lifetime of healthcare materials, develop more efficient supply chains, and decrease the environmental imprint of medical items through the DCE (De Laurentis et al., 2024; Moutaoukil et al., 2024). For example, digital platforms can make recycling and repurposing medical equipment easier, which helps to use resources better and lessens the burden of extracting raw materials. We can build a healthcare system that can withstand greater challenges and last longer, which will benefit people worldwide by bringing the DCE in line with Goal 3.
- **SDG4:**In order to promote sustainable development, it is crucial to achieve Quality Education (SDG 4). The DCE provides new possibilities to improve educational opportunities for all. Resource efficiency, waste reduction, and sustainable practices can be inculcated at a young age by incorporating circular economy ideas into the educational system. The proliferation of open-access resources and virtual learning environments made possible by digital platforms can democratize education by cutting down on the need for physical materials and wasted schooling (Shadiev et al., 2024). Educators can ensure that all students have access to high-quality education by using technology to design engaging, personalized lessons that meet the requirements of a wide range of learners. In addition, by incorporating ideas from the circular economy into school curricula, we may raise a new generation of eco-warriors who can propel sustainable practices (Amballoor & Naik, 2023). In this way, the DCE and SDG 4 work hand in hand to improve educational outcomes while teaching students important values for responsible resource management and equipping them to make meaningful contributions to a sustainable future.
- **SDG5:**Gender equality and women's empowerment can be propelled forward by merging the ideas of the DCE with SDG 5. One way the circular economy is helping to level the playing field for women in the workplace is by supporting sustainable business models that use digital technologies (Pant, 2024). Recycling and upcycling are examples of circular behaviors that can empower women by allowing them to start their businesses and earn their own money (Abdelwahed et al., 2024; Vijaya Lakshmi et al., 2024a). In addition to creating a more welcoming economy, this integration promotes gen-

der equality by ensuring everyone shares in the rewards of sustainability and innovation.

- **SDG6:**Achieving universal access to safe drinking water and sanitation through improved infrastructure and long-term planning is the focus of SDG 6. This objective takes on additional significance within the framework of the DCE as a result of ground-breaking methods that improve recycling, water efficiency, and sustainable resource management. In order to make industrial processes and urban water systems more efficient, digital technologies like data analytics and IoT sensors allow for real-time water usage monitoring. Industries can lessen their environmental impact and lessen the likelihood of water scarcity by adopting circular economy principles. These principles stress the importance of minimizing water usage, reusing treated water, and recycling materials (Cairone et al., 2024; Saadatinavaz et al., 2024). This collaboration helps achieve both goals while promoting resilient water management methods in many industries and preserving natural resources.
- **SDG7:**The seventh SDG ensures that everyone can access modern, cheap, dependable, and sustainable energy. Efforts to optimize energy use and promote renewable sources strongly align with Goal 7 within the circular economy context. The DCE uses smart grids, predictive maintenance, and energy management systems to improve energy efficiency. It does this by leveraging sophisticated technologies such as the IoT and AI (Zhang et al., 2024). These innovations enable tracking energy usage in real-time and adjusting as needed, opening the door for more renewable power to enter the grid. The DCE helps slow global warming by encouraging new approaches to storing and distributing energy. Also, it helps with energy decentralization, which means businesses and communities may switch to greener energy sources (Jakubelskas & Skvarciany, 2023, Karim 2020). In order to promote sustainable energy solutions and encourage technological innovation on a global scale, governments, businesses, and communities must work together to achieve Goal 7 through the DCE.
- **SDG8:**The eighth objective of the SDGs is to guarantee universal access to quality jobs, promote sustainable economic growth, and ensure inclusivity and durability. Fostering economic activities that prioritize resource efficiency, waste reduction, and creating green jobs becomes crucial within the context of the circular economy. Goal 8 plays a pivotal role in this. Achieving greater innovation and productivity with less environmental effect is our goal when we apply the circular economy concepts to companies and sectors (Sinkovics et al., 2024). Together, these goals—improving working conditions and providing more job possibilities, particularly for underrepresented groups—and shifting towards more sustainable production and consumption

patterns align with the SDG (Taron & Gebrezgabher, 2024a). We can make great strides towards Goal 8 and a future where economic growth is in harmony with environmental preservation and social wellbeing if we embrace circularity, which strengthens economic resilience, promotes social equity, and helps with environmental stewardship.

- **SDG9:**Goal 9 of the SDGs focuses on developing strong and adaptable infrastructure, promoting industrialization that is fair and sustainable, and fostering innovation. Goal 9 stresses the significance of incorporating sustainable practices into manufacturing procedures and infrastructure development within the circular economy framework. By embracing circular economy ideas, industries can reduce their environmental impacts, resource consumption, and waste generation. These principles include designing products for lifespan, reuse, and recycling (Ahmed et al., 2024). This method encourages new developments in eco-design, waste management technology, and environmentally friendly production methods, all of which contribute to sustainable industrialization. By combining economic growth with environmental responsibility and social inclusion, Goal 9 encourages a paradigm change towards sustainable development.
- **SDG10:**The tenth SDG is to lessen inequality, both domestically and internationally. With the DCE in mind, this objective highlights inclusive growth and helps everyone, especially those on the margins and in vulnerable situations. Society can better address economic inequality if digital behaviors adhere to circular economy principles, which include reducing waste, increasing resource efficiency, and promoting sustainable consumption patterns (Vijaya Lakshmi et al., 2024). Economic empowerment is promoted across diverse socioeconomic categories through the digitalization of circular economy processes, which promotes access to markets and employment prospects. To further promote equitable allocation of resources, digital technologies such as the IoT and AI allow for real-time monitoring and optimization of resource flows (Roshan et al., 2024). This partnership bolsters Goal 10 by advocating for measures that help marginalized communities gain agency, promote long-term economic growth, and help create more inclusive and resilient communities worldwide.
- **SDG11:**The eleventh SDG ensures everyone can live in safe, inclusive, and sustainable urban areas. Sustainable practices that increase resource efficiency, decrease environmental impact, and promote social inclusion within urban contexts are the focus of Goal 11, which is part of the circular economy framework. The circular economy concept aligns with Goal 11 because it promotes closed-loop technologies that minimize pollution and waste by continuously cycling materials back into production (Kalebaila et al., 2024a).

Creating resilient housing, sustainable infrastructure, and efficient resource management methods are all essential for urban sustainability, and this strategy encourages their growth. The circular economy helps build more sustainable, equitable, and climate-resilient cities by encouraging new approaches to waste management, energy efficiency, and city planning (Santos et al., 2024). Governments, businesses, and communities must work together to advance sustainable development worldwide by optimizing resource use, improving infrastructure, and improving urban residents' quality of life to achieve Goal 11 through circular economy principles.

- **SDG12:**SDG 12 focuses on sustainable consumption and production practices. The circular economy plays a crucial role in this endeavor by advocating for the efficient use of resources, minimizing waste, and encouraging the reuse of materials at every stage of a product's lifespan. Society can lessen its negative effects on the environment, save resources, and encourage resource-independent economic growth by shifting from the traditional linear "take-make-dispose" paradigm to a circular economy (Moorthy et al., 2024a). This strategy encourages new business models that emphasize product durability, remanufacturing, and recycling. In addition, the circular economy framework is improved by incorporating digital technologies like AI and the IoT, which allow for smarter resource management and decision-making by increasing material efficiency and traceability. Thus, when matched with circular economy concepts, Goal 12 advances broader sustainable development objectives by contributing to environmental sustainability, fostering economic resilience, and social inclusion (Sylvester, 2024).
- **SDG13:**Combating climate change and its consequences is the primary emphasis of SDG 13, also known as Climate Action. Sustainable consumption and production patterns are promoted by methods that align with Goal 13, which is part of the circular economy. Businesses can do their part to slow global warming by embracing circular economy practices like reusing materials and cutting down on trash (Moorthy et al., 2024b). This method improves resource efficiency and resistance to climate-related risks while simultaneously decreasing emissions of greenhouse gases. By improving transparency throughout supply chains, optimizing resource consumption, and tracking environmental impacts, advancements in digital technologies are crucial in easing these transitions (Burggräf et al., 2024). A more resilient and sustainable future can be achieved by integrating circular economy principles with Goal 13, which tackles climate concerns, creates economic possibilities, improves environmental sustainability, and supports inclusive growth.
- **SDG14:**SDG 14 focuses on the preservation and sustainable utilization of ocean, sea, and marine resources. The circular economy is crucial since it

advocates for production and consumption patterns that reduce pollution and waste. Reducing plastic waste, improving resource efficiency, and enhancing ecosystem resilience can be achieved by adopting circular techniques in marine industries, including fisheries and aquaculture (Ji & Li, 2024). New digital technologies enable real-time monitoring of marine ecosystems, which helps with sustainable management. To prevent illicit fishing and promote sustainable fishing methods, blockchain technology can, for example, monitor fish stocks and guarantee transparency in seafood supply chains. The preservation of ocean biodiversity and the free flow of information are two areas that can greatly benefit from the rise of digital platforms that allow for stakeholder collaboration and the exchange of information (Li & Ji, 2024). Goal 14 and the SDGs' broader environmental sustainability goals can be advanced by integrating digital innovations and circular economy concepts.

- **SDG15:**Goal 15 of the SDGs is to preserve, repair, and encourage the responsible use of terrestrial ecosystems, sustainably manage forests, fight desertification, stop and reverse land degradation, and stop biodiversity loss (Kalebaila et al., 2024b). By increasing resource efficiency, decreasing waste creation, and encouraging sustainable land use practices, implementing circular economy concepts improves the attainment of Goal 15. In contrast to conventional linear industrial models, which degrade ecosystems and cause habitat loss, the circular economy seeks to close material loops while reducing resource extraction. Reducing demands on natural resources and ecosystems is another way circular behaviors like recycling and reusing materials help retain biodiversity (Mathiazhagan & K.R., 2024). Assuring the conservation of terrestrial habitats, protecting endangered species, and promoting sustainable management of forests and land resources are all goals of SDG 15, which can be achieved through circular economy solutions. Together, these factors highlight why embracing circular economy strategies is critical to help achieve the larger objective of protecting terrestrial ecosystems and biodiversity for the benefit of all people.
- **SDG16:**Encouraging peaceful and comprehensive communities for sustainable development is the focus of Goal 16 of the SDGs. This goal also emphasizes the importance of developing effective, responsible, and inclusive institutions at all levels and providing everyone with access to justice. Goal 16 is very important in the circular economy because it encourages governments and businesses to be more open, accountable, and transparent (Hendriks & Babu, 2024). By embracing circular economy concepts, societies can strengthen their ability to tackle systemic issues like corruption, injustice, and inadequate access to justice. These include resource efficiency, waste reduction, and sustainable consumption and production patterns. More

than that, by committing to circularity, stakeholders like governments, corporations, and civil society are encouraged to work together to create laws and practices that protect the environment without compromising fairness and equity in resource access (Taron & Gebrezgabher, 2024b). The necessity of taking a comprehensive approach to sustainability is shown by the alignment between Goal 16 and the circular economy. Such a strategy aims to achieve economic advantages, social cohesion, and environmental stewardship, all of which contribute to the overall achievement of the SDGs.

- **SDG17:**In order to achieve sustainable development, partnerships are crucial, as stated in Sustainable Development Goal 17 of the SDGs. Goal 17 is more significant in the DCE since it stresses the importance of public-private partnerships in utilizing digital technology to promote circular practices (Almeida & Ocon, 2024). Leveraging technology-driven innovations that promote resource efficiency, waste reduction, and sustainable consumption habits requires these partnerships. Countries should promote inclusive partnerships that improve capacity-building, technology transfer, and information sharing by combining the circular economy concepts with Goal 17. In addition to hastening the completion of all SDGs, this coordinated effort guarantees that digitalization's advantages are fairly distributed, which boosts economic growth and protects environmental sustainability (Makhura, 2024). To promote resilient infrastructure, sustainable industrialization, and innovation—all of which are essential for moving towards a better future—Goal 17 acts as a push to incorporate the DCE into worldwide development agendas.

Figure 2. The SDGs and key components of DCE

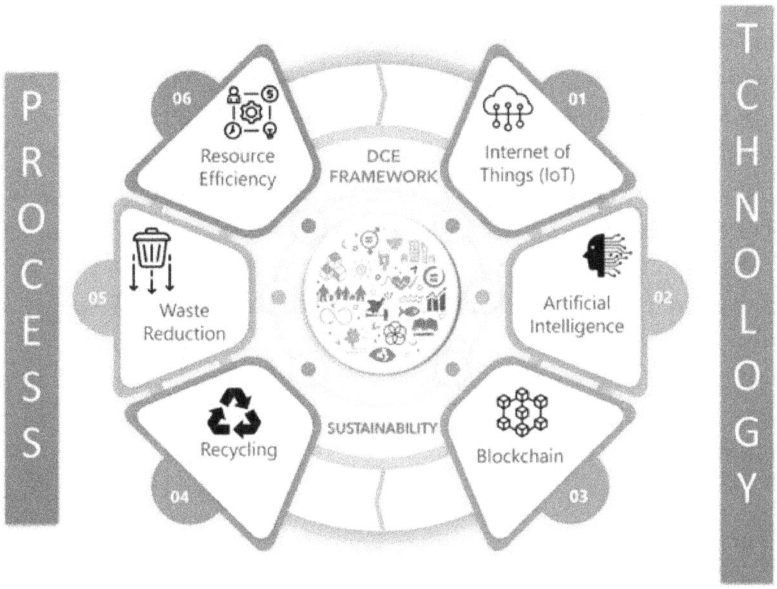

Source: Authors' Illustration

The Future of Digital Circular Economy

The future of the DCE holds transformative potential for global sustainability and economic resilience. As digital technologies continue to advance, integrating these innovations into circular economy principles will become more seamless and impactful. The IoT will play a pivotal role in this evolution, enabling real-time tracking and managing resources, products, and waste. Sensors and connected devices will provide detailed product usage and lifecycle data, facilitating more efficient recycling, remanufacturing, and resource recovery. This granular level of insight will allow businesses to optimize supply chains, reduce waste, and design products with longer lifespans and easier recyclability from the outset.

Cutting-edge technologies like AI and big data analytics will further enhance the capabilities of the DCE by providing advanced predictive and optimization tools. AI algorithms can forecast maintenance needs, reducing downtime and extending product lifecycles. Big data analytics will offer insights into consumption patterns and waste generation, helping policymakers and businesses to develop targeted strategies for reducing environmental impact. Moreover, blockchain technology will ensure transparency and traceability across supply chains, building trust among consumers and stakeholders regarding the sustainability of products. This increased transparency will also facilitate enforcing environmental regulations and standards more effectively.

In the future, digital platforms and marketplaces will enable the proliferation of circular business models, such as product-as-a-service, sharing economies, and peer-to-peer exchanges. These platforms will make it easier for consumers and businesses to access, share, and repurpose products and materials, fostering a more collaborative and resource-efficient economy. As digital literacy and infrastructure improve globally, the benefits of the DCE will become more accessible to emerging markets, driving inclusive and sustainable economic growth. However, achieving this vision will require concerted efforts to address data privacy, security, and regulatory framework challenges. Policymakers, businesses, and communities must work together to develop robust standards and policies that support the ethical and effective use of digital technologies in advancing circular economy practices to ensure future sustainability.

CONCLUSION

By integrating state-of-the-art digital technology with the concepts of the circular economy, the digital circular economy offers tremendous promise for promoting sustainable development. Opportunities for innovation, collaboration, expanding markets, and improving resource efficiency are just as compelling as the existing constraints, which range from data privacy concerns and technology restrictions to economic and regulatory obstacles and the necessity for cultural transformations. All parties involved, including companies, lawmakers, consumers, and researchers, must work together to reap these benefits fully. Building strong evaluation frameworks, making good use of cutting-edge innovation, creating welcoming policy settings, deciphering consumer behavior, and encouraging equitable socioeconomic growth should all be priorities for the next generation of researchers. A digital circular economy that is both inclusive and resilient can help achieve the SDGs and provide a sustainable future for everyone if we follow these lines of inquiry and work together on a global scale.

REFERENCES

Abdelwahed, N. A. A., Bano, S., Al Doghan, M. A., Aljughiman, A. A., Shah, N., & Soomro, B. A. (2024). Empowering women through digital technology: unraveling the nexus between digital enablers, entrepreneurial orientation and innovations. Equality, Diversity and Inclusion, ahead-of-print(ahead-of-print). https://doi.org/ DOI: 10.1108/EDI-02-2024-0072/FULL/XML

Ahmad, S., Saxena, C., Islam, S., & Karim, R. (2024). Impact of Insur-Tech on the Premium Performance of Insurance Business. *SN Computer Science*, *5*(1), 1–7. DOI: 10.1007/s42979-023-02462-0

Ahmed, K., Swain, M. S. S., & Sahoo, M. S. (2024). Assessing the Status of SDG-9 in Indian States. *International Journal for Research in Applied Science and Engineering Technology*, *12*(5), 5897–5901. DOI: 10.22214/ijraset.2024.63027

Almeida, F., & Ocon, E. (2024). Achieving sustainable development goals through digitalization in ports. *Business Strategy and the Environment*, *33*(7), 6737–6747. Advance online publication. DOI: 10.1002/bse.3842

Amballoor, R. G., & Naik, S. B. (2023). Digital technologies and education for sustainable development. Fostering Sustainable Development in the Age of Technologies, 225–237. https://doi.org/DOI: 10.1108/978-1-83753-060-120231016/FULL/EPUB

Anand, P. B., Parra, C., & Weakliam, J. (2020). End Hunger, Achieve Food Security and Improved Nutrition and Promote Sustainable Agriculture. In Mining, Materials, and the Sustainable Development Goals (SDGs) (pp. 7–14). CRC Press. DOI: 10.1201/9780367814960-2

Awan, U., Shamim, S., Khan, Z., Zia, N. U., Shariq, S. M., & Khan, M. N. (2021). Big data analytics capability and decision-making: The role of data-driven insight on circular economy performance. *Technological Forecasting and Social Change*, *168*, 120766. DOI: 10.1016/j.techfore.2021.120766

Awan, U., Sroufe, R., & Shahbaz, M. (2021). Industry 4.0 and the circular economy: A literature review and recommendations for future research. *Business Strategy and the Environment*, *30*(4), 2038–2060. DOI: 10.1002/bse.2731

Babu, S. C., & Srivastava, N. (2024). Sustainable Development Goal 1: Ending poverty. In Hendriks, S. L., & Babu, S. C. (Eds.), *Handbook on Public Policy and Food Security* (pp. 80–89). Edward Elgar Publishing., DOI: 10.4337/9781839105449.00013

Berg, H., Le Blèvennec, K., Kristoffersen, E., Streè, B., Witomski, A., Stein, N., & Vrancken, K. (2020). Digital circular economy as a cornerstone of a sustainable European industry transformation. https://doi.org/http://dx.doi.org/10.13140/RG.2.2.13769.36966

Burggräf, P., Steinberg, F., Becher, A., Sauer, C. R., & Wigger, M. (2024). Towards a Sustainable Industrial Society – Critical Capabilities for the Transformation to a Circular Economy in Manufacturing Companies. Lecture Notes in Production Engineering, Part F1764, 304–315. DOI: 10.1007/978-3-031-47394-4_30

Cairone, S., Hasan, S. W., Choo, K. H., Lekkas, D. F., Fortunato, L., Zorpas, A. A., Korshin, G., Zarra, T., Belgiorno, V., & Naddeo, V. (2024). Revolutionizing wastewater treatment toward circular economy and carbon neutrality goals: Pioneering sustainable and efficient solutions for automation and advanced process control with smart and cutting-edge technologies. *Journal of Water Process Engineering*, *63*, 105486. DOI: 10.1016/j.jwpe.2024.105486

Charnley, F., Tiwari, D., Hutabarat, W., Moreno, M., Okorie, O., & Tiwari, A. (2019). Simulation to Enable a Data-Driven. *Sustainability (Basel)*, *11*(12), 3379. DOI: 10.3390/su11123379

Cheah, C. G., Chia, W. Y., Lai, S. F., Chew, K. W., Chia, S. R., & Show, P. L. (2022). Innovation designs of industry 4.0 based solid waste management: Machinery and digital circular economy. *Environmental Research*, *213*, 113619. DOI: 10.1016/j.envres.2022.113619 PMID: 35700763

De Laurentis, C., Beverley, K., Clifton, N., Bacon, E., Rudd, J., & Walpole, G. (2024). Exploring opportunities for public sector organisations to connect wellbeing to resource loops in a regional circular economy. *Contemporary Social Science*, *19*(1–3), 303–336. DOI: 10.1080/21582041.2024.2356190

Erdogan, T. (2022). Poverty Problem in the Framework of Sustainable Development Goals. *Economics Literature*, *4*(1), 1–21. DOI: 10.22440/elit.4.1.1

Gazi, M. A. I., Islam, H., Islam, M. A., Karim, R., Momo, S. M., & Senathirajah, A. R. B. S. (2024). Unlocking sustainable development in East Asia Pacific and South Asia : An econometric exploration of ESG initiatives. *Sustainable Environment*, *10*(1), 1–15. DOI: 10.1080/27658511.2024.2366558

Ghobakhloo, M. (2018). The future of manufacturing industry: A strategic roadmap toward Industry 4.0. *Journal of Manufacturing Technology Management*, *29*(6), 910–936. DOI: 10.1108/JMTM-02-2018-0057

Grosso, G., Mateo, A., Rangelov, N., Buzeti, T., & Birt, C. (2020). Nutrition in the context of the Sustainable Development Goals. *European Journal of Public Health*, *30*(Supplement_1), i19–i23. DOI: 10.1093/eurpub/ckaa034 PMID: 32391903

Jakubelskas, U., & Skvarciany, V. (2023). Circular economy practices as a tool for sustainable development in the context of renewable energy: What are the opportunities for the EU? *Oeconomia Copernicana*, *14*(3), 833–859. DOI: 10.24136/oc.2023.025

Ji, J., & Li, Y. (2024). Does fishery digitalization matter in the sustainable development of fisheries? Evidence from China. *Sustainable Development (Bradford)*, *32*(6), 7382–7396. Advance online publication. DOI: 10.1002/sd.3091

Kalebaila, N., Kapari, M., Nhamo, L., & Mpandeli, S. (2024a). The circular economy as a catalyst for environmental and human health. Circular and Transformative Economy: Advances towards Sustainable Socio-Economic Transformation, 139–158. DOI: 10.1201/9781003327615-7

Karim, R. (2020). Green Banking Practices in Bangladesh: Present Scenario and Future Challenges. *Journal of Business Studies*, (January), 34–44.

Karim, R., Roshid, M. M., & Waaje, A. (2024). Circular Economy and Supply Chain Sustainability. In F. Mizrak (Ed.), Strategic Innovations for Dynamic Supply Chains (pp. 1-30). IGI Global Scientific Publishing. DOI: 10.4018/979-8-3693-3575-8.ch001

Karim, R., Roshid, M. M., & Waaje, A. (2024). Does the Firm Size Matter for Financial Performance? —Evidence from the Financial Sector of a South Asian Emerging Economy. In Kurnaz, N., & Akıncı, A. (Eds.), *2nd International Congress on Economics Public Finance Business & Social Sciences* (pp. 461–474).

Karim, R., Waaje, A., & Roshid, Md. M., & Yeamin, Md. B. (2024). Turning the Waste Into Wealth: Progressing Toward Global Sustainability Through the Circular Economy in Waste Management. In R. Sucheran (Ed.), Sustainable Waste Management in the Tourism and Hospitality Sectors (pp. 507–552). IGI Global Scientific Publishing. DOI: 10.4018/979-8-3693-6110-8.ch019

Kazancoglu, Y., Ekinci, E., Mangla, S. K., Sezer, M. D., Choudhary, S., & Dora, M. (2023). Smart Closed-Loop Food Supply Chain in Circular Economy. *IEEE Transactions on Engineering Management*. Advance online publication. DOI: 10.1109/TEM.2023.3303272

Kurniawan, T. A., Maiurova, A., Kustikova, M., Bykovskaia, E., Othman, M. H. D., & Goh, H. H. (2022). Accelerating sustainability transition in St. Petersburg (Russia) through digitalization-based circular economy in waste recycling industry: A strategy to promote carbon neutrality in era of Industry 4.0. *Journal of Cleaner Production, 363*, 132452. DOI: 10.1016/j.jclepro.2022.132452

Li, Y., & Ji, J. (2024). The digitalization of Chinese fisheries and its configuration path to empower fishery sustainable development. *Journal of Cleaner Production, 466*, 142807. DOI: 10.1016/j.jclepro.2024.142807

Lieder, M., & Rashid, A. (2016). Towards circular economy implementation: A comprehensive review in context of manufacturing industry. *Journal of Cleaner Production, 115*, 36–51. DOI: 10.1016/j.jclepro.2015.12.042

Makhura, M. N. (2024). Sustainable Development Goal 17: Partnerships for the goals. In Handbook on Public Policy and Food Security (pp. 394–405). Edward Elgar Publishing. DOI: 10.4337/9781839105449.00043

Moorthy, K. S., Balakrishnan, G., Kumar, S. S., Raja, L., & Vijayalakshmi, A. (2024). *Embracing Circular Economy Principles for Sustainable Green Supply Chain Management in Manufacturing Industries*. IGI Global., https://services.igi-global.com/resolvedoi/resolve.aspx?doi=10.4018/979-8-3693-1343-5.ch005, DOI: 10.4018/979-8-3693-1343-5.ch005

Moreno, M., & Charnley, F. (2016). Can Re-distributed Manufacturing and Digital Intelligence Enable a Regenerative Economy? An Integrative Literature Review. Smart Innovation. *Systems and Technologies, 52*, 563–575. DOI: 10.1007/978-3-319-32098-4_48

Moutaoukil, O., Mahfoud, H., & Bouhaddou, I. (2024). A novel conceptual framework for predictive maintenance of medical devices in a circular healthcare supply chain. *2024 International Conference on Control, Automation and Diagnosis (ICCAD)*, 1–6. DOI: 10.1109/ICCAD60883.2024.10553792

Mukombwe, J. S., du Toit, A., & Hendriks, S. L. (2024). Sustainable Development Goal 16: Peace, justice and strong institutions. In Handbook on Public Policy and Food Security (pp. 376–385). Edward Elgar Publishing. DOI: 10.4337/9781839105449.00041

Ng, S.-L., & Wong, F.-M. (2024). Recent Developments in Research on Food Waste and the Circular Economy. *Biomass, 4*(2), 472–489. DOI: 10.3390/biomass4020024

Okorie, O., Salonitis, K., Charnley, F., & Turner, C. (2018). A Systems Dynamics Enabled Real-Time Efficiency for Fuel Cell Data-Driven Remanufacturing. *Journal of Manufacturing and Materials Processing, 2*(4), 77. DOI: 10.3390/jmmp2040077

Ondemir, O., & Gupta, S. M. (2014). Quality management in product recovery using the Internet of Things: An optimization approach. *Computers in Industry*, *65*(3), 491–504. DOI: 10.1016/j.compind.2013.11.006

Pant, D. (2024). The Role of E-Governance in Gender Equality and Sustainable Development. In Gender Roles and Green Concepts: Pathways to Environmental Stainability (Vol. 2, pp. 58–75). Anu Books, Meerut. DOI: 10.31995/Book.AB306-M24.Chapter5

Piscicelli, L. (2023). The sustainability impact of a digital circular economy. *Current Opinion in Environmental Sustainability*, *61*, 101251. DOI: 10.1016/j.cosust.2022.101251

Rahman, Md. H., Yaqot, M., & Menezes, B. C. (2023). Towards Circular Economy in Manufacturing Industries Based on Industry 4.0 Technologies. 2023 IEEE International Conference on Industrial Engineering and Engineering Management (IEEM), 1502–1506. DOI: 10.1109/IEEM58616.2023.10406739

Roshan, R., Balodi, K. C., Datta, S., Kumar, A., & Upadhyay, A. (2024). Circular economy startups and digital entrepreneurial ecosystems. *Business Strategy and the Environment*, *33*(5), 4843–4860. DOI: 10.1002/bse.3727

Roshid, M., Karim, R., Waaje, A., Islam, S., & Bhowmik, R. C. (2025). Mapping the carbon finance research: A bibliometric exploration of innovation in sustainable finance. *Development and Sustainability in Economics and Finance*, *5*, 100048. DOI: 10.1016/j.dsef.2025.100048

Roshid, M. M., Karim, R., & Waaje, A. (2024). Synergizing Circular Economy and H2H Marketing: A Sustainable Approach to Human-Centric Business. In Kurnaz, N., & Akinci, A. (Eds.), *EMIDWORLD 2nd International Congress on Economics Public Finance Business & Social Sciences* (pp. 765–783). Altınbaş University., www.emidworld.com

Roshid, M. M., Waaje, A., & Karim, R. (2024). Industry 6.0 as an Emerging Research Field: A Systematic and Bibliometric Analysis. In N. Kurnaz & A. Akinci (Eds.), EMIDWORLD 2nd International Congress on Economics Public Finance Business & Social Sciences (pp. 784–802). Altinbas University.

Roshid, M. M., Waaje, A., Meem, T. N., & Sarkar, A. (2024). Logistics 4.0: A Comprehensive Literature Review of Technological Integration, Challenges, and Future Prospects of Implementation of Industry 4.0 Technologies. *International Journal of Technology, Knowledge and Society*, *20*(1), 65–85. Advance online publication. DOI: 10.18848/1832-3669/CGP/v20i01/65-85

Saadatinavaz, F., Alomari, M. A., Ali, M., & Saikaly, P. E. (2024). Striking a Balance: Decentralized and Centralized Wastewater Treatment Systems for Advancing Sustainable Development Goal 6. *Advanced Energy and Sustainability Research, 2400097*, 2400097. Advance online publication. DOI: 10.1002/aesr.202400097

Santos, P., Cervantes, G. C., Zaragoza-Benzal, A., Byrne, A., Karaca, F., Ferrández, D., Salles, A., & Bragança, L. (2024). Circular Material Usage Strategies and Principles in Buildings: A Review. Buildings 2024, Vol. 14, Page 281, 14(1), 281. DOI: 10.3390/buildings14010281

Sarc, R., Curtis, A., Kandlbauer, L., Khodier, K., Lorber, K. E., & Pomberger, R. (2019). Digitalisation and intelligent robotics in value chain of circular economy oriented waste management – A review. *Waste Management (New York, N.Y.), 95*, 476–492. DOI: 10.1016/j.wasman.2019.06.035 PMID: 31351634

Schützenhofer, S., Pibal, S., Wieser, A., Bosco, M., Fellner, M., Petrinas, V., & Kovacic, I. (2024). Digital Ecosystem to enable Circular Buildings – The Circular Twin Framework Proposal. Journal of Sustainable Development of Energy. *Water and Environment Systems, 12*(2), 1–20. DOI: 10.13044/j.sdewes.d12.0500

Shadiev, R., Reynolds, B. L., & Li, R. (2024). The Use of Digital Technology for Sustainable Teaching and Learning. *Sustainability (Basel), 16*(13), 5353. DOI: 10.3390/su16135353

Sinkovics, N., Sinkovics, R. R., Boussebaa, M., & Fletcher, M. (2024). *International Business and SDG 8: An Introduction to the Theme*. Palgrave Macmillan., DOI: 10.1007/978-3-031-46802-5

Stucki, T., Woerter, M., & Loumeau, N. (2023). Clearing the fog: How circular economy transition can be measured at the company level. *Journal of Environmental Management, 326*, 116749. DOI: 10.1016/j.jenvman.2022.116749 PMID: 36446223

Sylvester, O. (2024). Sustainable Development Goal 12: Sustainable consumption and production patterns. In Hendriks, S. L., & Babu, S. C. (Eds.), *Handbook on Public Policy and Food Security* (pp. 289–298). Edward Elgar Publishing., DOI: 10.4337/9781839105449.00033

Taron, A., & Gebrezgabher, S. (2024a). Circular bioeconomy: A pathway to sustainable development in an age of global crisis. International Trade, Economic Crisis and the Sustainable Development Goals, 99–117. https://doi.org/DOI: 10.1108/978-1-83753-586-620241007/FULL/XML

Taron, A., & Gebrezgabher, S. (2024b). Circular bioeconomy: A pathway to sustainable development in an age of global crisis. International Trade, Economic Crisis and the Sustainable Development Goals, 99–117. https://doi.org/DOI: 10.1108/978-1-83753-586-620241007/FULL/XML

Tseng, M. L., Sujanto, R. Y., Ardaniah, V., Bui, T. D., Lan, S., Negash, Y. T., & Tseng, J. W. (2023). Product-service system toward circularity: Hierarchical framework in Indonesia's home appliance manufacturing industry under uncertainties. *Annals of Operations Research*, •••, 1–27. DOI: 10.1007/s10479-023-05538-5

Turner, C. J., Hutabarat, W., Oyekan, J., & Tiwari, A. (2016). Discrete Event Simulation and Virtual Reality Use in Industry: New Opportunities and Future Trends. *IEEE Transactions on Human-Machine Systems*, *46*(6), 882–894. DOI: 10.1109/THMS.2016.2596099

United Nations. (2015). The 2030 Agenda for Sustainable Development. "Transforming our world." Dianova. https://www.dianova.org/press-reviews/the-2030-agenda-for-sustainable-development-qtransforming-our-worldq/?gad_source=1&gclid=CjwKCAjwtNi0BhA1EiwAWZaANMxt4GNTdQBcKzKncAgx7OIahXZRvgoQbNGOlp2Esmyr8qEoRx-qIhoCztIQAvD_BwE

United Nations. (2022). *The Sustainable Development Goals Report*. United Nations., DOI: 10.18356/9789210018098

United nations. (2024). Goal 15 | Protect, restore and promote sustainable use of terrestrial ecosystems, sustainably manage forests, combat desertification, and halt and reverse land degradation and halt biodiversity loss. Department of Economic and Social Affairs. https://sdgs.un.org/goals/goal15

United Nations. (2024a). Goal 1 | End poverty in all its forms everywhere. Department of Economic and Social Affairs. https://sdgs.un.org/goals/goal1

United Nations. (2024b). Goal 2 | End hunger, achieve food security and improved nutrition and promote sustainable agriculture. Department of Economic and Social Affairs. https://sdgs.un.org/goals/goal2

United Nations. (2024c). Goal 3 | Ensure healthy lives and promote wellbeing for all at all ages. Department of Economic and Social Affairs. https://sdgs.un.org/goals/goal3

United Nations. (2024d). Goal 4 | Ensure inclusive and equitable quality education and promote lifelong learning opportunities for all. Department of Economic and Social Affairs. https://sdgs.un.org/goals/goal4

United Nations. (2024e). Goal 5 | Achieve gender equality and empower all women and girls. Department of Economic and Social Affairs. https://sdgs.un.org/goals/goal5

United Nations. (2024f). Goal 6 | Ensure availability and sustainable management of water and sanitation for all. Department of Economic and Social Affairs. https://sdgs.un.org/goals/goal6

United Nations. (2024g). Goal 7 | Ensure access to affordable, reliable, sustainable and modern energy for all. Department of Economic and Social Affairs. https://sdgs.un.org/goals/goal7

United Nations. (2024h). Goal 8 | Promote sustained, inclusive and sustainable economic growth, full and productive employment and decent work for all. Department of Economic and Social Affairs. https://sdgs.un.org/goals/goal8

United Nations. (2024i). Goal 9 | Build resilient infrastructure, promote inclusive and sustainable industrialization and foster innovation. Department of Economic and Social Affairs. https://sdgs.un.org/goals/goal9

United Nations. (2024j). Goal 10 | Reduce inequality within and among countries. Department of Economic and Social Affairs. https://sdgs.un.org/goals/goal10

United Nations. (2024k). Goal 11 | Make cities and human settlements inclusive, safe, resilient and sustainable. Department of Economic and Social Affairs. https://sdgs.un.org/goals/goal11

United Nations. (2024l). Goal 12 | Ensure sustainable consumption and production patterns. Department of Economic and Social Affairs. https://sdgs.un.org/goals/goal12

United Nations. (2024m). Goal 13 | Take urgent action to combat climate change and its impacts. Department of Economic and Social Affairs. https://sdgs.un.org/goals/goal13

United Nations. (2024n). Goal 14 | Conserve and sustainably use the oceans, seas and marine resources for sustainable development. Department of Economic and Social Affairs. https://sdgs.un.org/goals/goal14

United Nations. (2024o). Goal 16 | Promote peaceful and inclusive societies for sustainable development, provide access to justice for all and build effective, accountable and inclusive institutions at all levels. Department of Economic and Social Affairs. https://sdgs.un.org/goals/goal16

Vazhenina, L., Magaril, E., & Mayburov, I. (2023). Digital Management of Resource Efficiency of Fuel and Energy Companies in a Circular Economy. Energies 2023, Vol. 16, Page 3498, 16(8), 3498.

Vijaya Lakshmi, V., Mishra, M., Kushwah, J. S., Shajahan, U. S., Mohanasundari, M., & Boopathi, S. (2024). Circular Economy Digital Practices for Ethical Dimensions and Policies for Digital Waste Management. In Harnessing High-Performance Computing and AI for Environmental Sustainability (pp. 166–193). IGI Global. DOI: 10.4018/979-8-3693-1794-5.ch008

Voulgaridis, K., Lagkas, T., Angelopoulos, C. M., & Nikoletseas, S. E. (2022). IoT and digital circular economy: Principles, applications, and challenges. *Computer Networks*, *219*, 109456. DOI: 10.1016/j.comnet.2022.109456

Waaje, A., Karim, R., Roshid, M. M., & Meem, T. N. (2025). Minimizing Waste in the Hospitality Industry: The Opportunities and Challenges of Lean Tourism. In Sucheran, R. (Ed.), *Sustainable Waste Management in the Tourism and Hospitality Sectors* (pp. 125–166). IGI Global Scientific Publishing., DOI: 10.4018/979-8-3693-6110-8.ch006

Wilts, H. (2017). The Digital Circular Economy: Can the Digital Transformation Pave the Way for Resource-Efficient Materials Cycles? *International Journal of Environmental Sciences & Natural Resources*, *7*(5). Advance online publication. DOI: 10.19080/IJESNR.2017.07.555725

Wube Dametew, A. (2015). Problems of Solid Waste Management in Small and Medium Enterprises. *International Journal of Environmental Monitoring and Analysis*, *3*(2), 50. DOI: 10.11648/j.ijema.20150302.14

Yildirim, S. (2020). The consumer role for sustainable development: how consumers contribute sustainable development goals. In Chkoniya, V., Madsen, A. O., & Bukhrashvili, P. (Eds.), *Anthropological Approaches to Understanding Consumption Patterns and Consumer Behavior* (pp. 325–341). IGI Global. DOI: 10.4018/978-1-7998-3115-0.ch018

Yildirim, S., Demirtas, I., & Yildirim, D. C. (2022). A Review of Alternative Economic Approaches to Achieve Sustainable Development: The Rising Digitalization and Degrowth Post COVID-19. In Castanho, R. (Ed.), *Handbook of Research on Sustainable Development Goals, Climate Change, and Digitalization* (pp. 288–307). IGI Global. DOI: 10.4018/978-1-7998-8482-8.ch018

Yildirim, S., & Yildirim, D. Ç. (2020). Achieving Sustainable Development Through a Green Economy Approach. In Patti, S., & Trizzino, G. (Eds.), *Advanced Integrated Approaches to Environmental Economics and Policy: Emerging Research and Opportunities* (pp. 1–22). IGI Global. DOI: 10.4018/978-1-5225-9562-5.ch001

Zeiss, R., Ixmeier, A., Recker, J., & Kranz, J. (2021). Mobilising information systems scholarship for a circular economy: Review, synthesis, and directions for future research. *Information Systems Journal*, *31*(1), 148–183. DOI: 10.1111/isj.12305

Zhang, L., Liu, S., Chiu, Y., Pang, Q., Chen, Q., & Shi, C. (2024, December). ho, Pang, Q., Chen, Q., & Shi, C. (2024). Assessing China's energy-related efficiency towards SDG7 and influencing factors. *Sustainable Development (Bradford)*, *32*(6), 6449–6470. Advance online publication. DOI: 10.1002/sd.3034

Chapter 9
The Role of Circular Economy in Advancing Sustainability

Asheetu Bhatia Sarin
https://orcid.org/0000-0001-5321-1969
VIPS, Guru Gobind Singh Indraprastha University, India

ABSTRACT

This chapter investigates the role of the circular economy (CE) in advancing sustainability, focusing on how circular models can drive environmental, economic, and social benefits by minimizing waste, conserving resources, and promoting regenerative practices. The study reviews core CE strategies such as product lifecycle extension, resource recovery, and sustainable material sourcing and assesses their effectiveness in achieving sustainable development goals (SDGs). Through industry case studies and policy analyses, it identifies key challenges, including financial, regulatory, and cultural barriers that hinder CE adoption. This chapter highlights the need for collaborative frameworks, supportive government policies, and innovative business models to enable a transition from linear to circular systems, emphasizing that CE is not only essential for reducing environmental impact but also crucial for building resilient, sustainable economies. The findings underscore the importance of CE as a fundamental approach to sustainability and climate resilience.

INTRODUCTION

Sustainability refers to meeting present needs without compromising the ability of future generations to meet theirs. It balances environmental, economic, and social considerations to ensure long-term well-being. Environmental sustainability

DOI: 10.4018/979-8-3693-9909-5.ch009

focuses on protecting natural resources and ecosystems by minimizing pollution, conserving biodiversity, and reducing ecological footprints. Economic sustainability seeks long-term economic growth that does not harm social or environmental systems, while social sustainability emphasizes equity, access to resources, and community resilience.

The world faces significant environmental challenges driven largely by unsustainable human practices. Human activities have significantly increased the strain on Earth's natural resources. In 1961, the global ecological footprint was within the planet's capacity to regenerate, but by 2005, it had exceeded this limit, reaching the equivalent of 1.4 planet Earths (Galli et al., 2012). Projections suggest that by 2030, human consumption and environmental impact could require resources equivalent to two Earths (Moore et al., 2012). Meanwhile, research on planetary boundaries highlights that natural ecosystems have a limited ability to withstand stress and regenerate, emphasizing the urgent need for sustainable practices (Rockström et al., 2009).

Climate change, caused by greenhouse gas emissions from fossil fuel use, deforestation, and industrial processes, leads to rising temperatures, sea-level rise, extreme weather, and biodiversity loss. Habitat destruction, pollution, overexploitation, and invasive species contribute to the decline of biodiversity, threatening ecosystem services. Deforestation for agriculture, logging, and urbanization reduces carbon sinks, disrupts water cycles, and destroys habitats. Pollution in air, water, soil, and oceans harms human health, contaminates resources, and degrades ecosystems. Water scarcity, exacerbated by overconsumption, pollution, and climate change, limits access to clean water, reduces agricultural productivity and heightens conflicts. Additionally, waste mismanagement, driven by overreliance on single-use plastics and unsustainable consumption, leads to landfills overflowing and oceans polluted with harmful microplastics. Overconsumption and the depletion of nonrenewable resources strain ecosystems, reduce biodiversity, and foster competition for limited resources.

Addressing these challenges requires adopting sustainable practices. Shifting to renewable energy sources like wind and solar power can greatly decrease greenhouse gas emissions A circular economy minimizes waste through recycling, reuse, and sustainable production practices. Conservation efforts focus on protecting habitats, restoring ecosystems, and ensuring sustainable land use. Effective policies and international agreements, like the Paris Accord, enforce environmental regulations and promote sustainable development. Public education and awareness campaigns encourage behavioral changes, while technological innovations help reduce emissions and enhance resource efficiency. By integrating these strategies, societies can tackle environmental challenges and move toward a sustainable future.

The circular economy is a transformative economic model aimed at redefining how resources are used and managed to minimize waste and create sustainable systems. Unlike the traditional linear economy, which follows a "take-make-dispose" approach where raw materials are extracted, used to develop products, and ultimately discarded the circular economy focuses on closing the loop by keeping resources intact(Geng & Doberstein, 2008). The existing economic model is no longer viable on a planet with limited resources and a finite ability to absorb emissions (Steffen et al., 2015). To ensure long-term sustainability, a shift towards more resource-efficient and environmentally responsible practices is essential. This is achieved through reducing, reusing, repairing, refurbishing, and recycling materials, ensuring they retain their value and functionality throughout multiple cycles.

The foundation of the circular economy lies in creating products and systems that prioritize durability, flexibility, and sustainability. Products are made to be durable and easy to repair or upgrade, reducing the need for frequent replacements. Materials used in production are selected for their ability to be safely recycled or returned to natural systems, forming a regenerative cycle rather than a wasteful one. For example, biodegradable materials might return to the earth as nutrients, while durable materials like metals and plastics are reclaimed and reused in manufacturing. Accenture has estimated that transitioning to a circular economy (CE) could generate $4.5 trillion in global economic growth by 2030, strengthening the resilience of economies worldwide (García-Sanchez et al., 2021). Recognizing the potential of CE, the European Union has integrated it into its recovery strategy following the COVID-19 pandemic. Through the Next Generation EU initiative, the EU has allocated 750 billion euros to various investment areas, with a significant focus on ecological transition and the adoption of a circular economic model as a key priority (García-Sanchez et al., 2021).

This model offers significant environmental, economic, and social benefits. Environmentally, it helps reduce waste, pollution, and greenhouse gas emissions by minimizing the extraction of raw materials and the energy-intensive processes associated with manufacturing. Economically, it decreases reliance on finite resources, lowers production costs, and fosters innovation in sustainable design and materials science. Socially, it creates new business opportunities, generates green jobs, and encourages collaboration among industries, governments, and communities to develop sustainable systems.

By prioritizing resource efficiency and environmental responsibility, the circular economy addresses pressing global challenges such as climate change, resource depletion, and waste management. It represents a shift from the current disposable culture to a more regenerative, sustainable way of thinking, where the value of materials is preserved, ecosystems are protected, and future generations can thrive in a healthier, more resilient world. The purpose of this study is to examine the role of

CE in promoting sustainability, the benefits, the challenges, and its broader impact on society and the environment.

LITERATURE REVIEW

According to Murray et al. (2017), the Circular Economy (CE) is the latest approach to integrating economic activities with environmental and resource sustainability. This concept builds on longstanding principles of resource efficiency while emphasizing the economic benefits of conserving resources. By explicitly linking sustainability efforts with financial gains, the Circular Economy promotes a more balanced and sustainable approach to production and consumption. The circular economy is a financial framework designed to minimize excess, maximize resource efficiency, and establish sustainable systems by keeping materials, products, and resources in use for as long as possible. Unlike the traditional linear economy which follows a "take-make-dispose" approach, extracting finite resources to produce goods that are discarded after use CE seeks to create a closed-loop system. This system minimizes environmental impact, reduces resource extraction, and fosters economic growth while prioritizing sustainability and efficiency.

Several scholars have significantly contributed to the literature on the circular economy, with the work of Pearce & Turner (1990) being widely recognized as a foundational study (Geissdoerfer et al., 2017; Ghisellini et al., 2016; Govindan & Hasanagic, 2018). In their research, Pearce & Turner (1990) describe the circular economy as a closed system, where natural resources such as materials, water, and energy serve as inputs for production. This approach emphasizes the importance of resource efficiency and sustainability by ensuring that materials are continually reused, reducing waste and environmental impact as seen in Fig 1. A linear economy operates in a way that disregards environmental factors, functioning as an open system where resource extraction, waste generation, and recycling are not fully considered (Pearce & Turner, 1990). In contrast, a circular economy is designed as a closed system, where environmental impact is accounted for at every stage. This model ensures that resource extraction is balanced with regeneration and that waste is minimized by being repurposed as raw materials through recycling (Boulding, 1966; McDonough & Braungart, 2010; Pearce & Turner, 1990). The same principles apply to supply chains. A linear supply chain follows an open-system approach, ignoring the effects of resource depletion and waste accumulation. On the other hand, a circular supply chain integrates sustainability by considering resource efficiency, waste reduction, and recycling, making the entire process more environmentally responsible (Batista et al., 2018; De Angelis et al., 2018).

Figure 1. Circular economy model

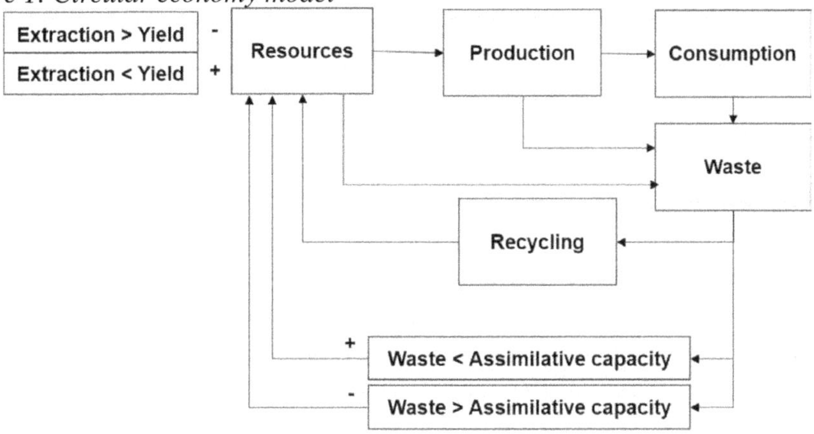

source: Pearce & Turner, 1990

Geissdoerfer et al. (2017) explored the circular economy as an emerging sustainability paradigm and examined how the circular economy differs from traditional linear economic models by emphasizing resource efficiency, waste minimization, and sustainable production processes. The paper discusses the benefits, challenges, and practical applications of circular economy principles, positioning it as a key strategy for achieving long-term environmental and economic sustainability. Ghisellini, Cialani, and Ulgiati (2016) provide a comprehensive review of the circular economy (CE) concept, focusing on its potential to create a harmonious balance between economic growth and environmental sustainability. The study examines various CE models and strategies, including waste minimization, resource efficiency, and sustainable production. It highlights the benefits of CE, such as reduced environmental impact and improved economic performance, while also addressing challenges like regulatory barriers, implementation costs, and the need for behavioral change.

Studies suggest that achieving ambitious circular economy (CE) objectives by 2050 could result in a 55.3% decrease in CO2 emissions (Aguilar-Hernandez et al., 2021). Additionally, shifting from primary to secondary materials on a global scale could reduce the demand for primary resources by 27% for metals and 8% for nonmetallic minerals (OECD, 2021). Embracing the CE approach could lead to significant environmental benefits, such as reduced air, land, and water pollution, while also positively impacting human health (OECD, 2021). Moreover, it could drive job creation and boost a country's GDP (Aguilar-Hernandez et al., 2021), offering a "win–win–win" scenario across economic, social, and environmental dimensions (Palea et al., 2023). Governments worldwide increasingly recognize the circular economy (CE) as a key strategy for combating climate change and enhancing economic stability (Domenech & Bahn-Walkowiak, 2019; Palea et al.,

2023). Japan was a pioneer in this effort, introducing CE-related legislation in 2000 (Ministry of the Environment, 2000). The European Union (EU) followed suit with its "Circular Economy Action Plan" in 2015 (European Commission, 2015). Meanwhile, the United States has yet to launch any dedicated national initiatives in this area (Palea et al., 2023)

Principles of the Circular Economy

1. **Design out Waste and Pollution**: CE emphasizes designing products, processes, and systems that eliminate waste and prevent pollution. This involves creating items with longer lifespans, easier repairability, and recyclability while minimizing the environmental footprint at every stage of production.
2. **Keep Products and Materials in Use**: CE focuses on maximizing the value of materials by reusing, repairing, refurbishing, and recycling products. This principle encourages businesses to shift to models like leasing, sharing, and remanufacturing, ensuring resources remain productive throughout multiple cycles.
3. **Regenerate Natural Systems**: CE goes beyond reducing harm to actively restoring and regenerating ecosystems. It promotes the use of renewable energy, sustainable materials, and regenerative agricultural practices, allowing natural systems to replenish and thrive.

These principles collectively support sustainability by decoupling economic growth from resource depletion and environmental degradation. The circular economy thus represents a transformative approach that benefits businesses, the environment, and society by fostering innovation, reducing waste, and building resilient systems for the future.

The CE aligns closely with the United Nations Sustainable Development Goals (SDGs) by addressing critical challenges related to resource efficiency, environmental protection, and social equity. Rodríguez-Antón et al. (2019) analyzed the relationship between the circular economy (CE) and the Sustainable Development Goals (SDGs) in the European Union (EU). Their study highlights a clear connection between CE practices and SDG achievements, revealing a statistically significant correlation between certain CE indicators and both individual SDGs and overall SDG compliance. These findings suggest that by implementing policies that promote circularity, the EU can enhance its progress toward achieving the targeted SDGs, reinforcing the role of CE in sustainable development. Specifically, CE supports SDG 12, which emphasizes responsible consumption and production, by promoting practices that minimize waste, optimize resource use, and create sustainable production systems. Through its principles of reducing, reusing, and recycling, CE reduces environmental degradation and fosters more efficient use of natural resources, aligning with SDG 13 on climate action and SDG 15 on life on land. By encouraging the adoption of renewable energy, sustainable materials, and regenerative practices, CE reduces the carbon footprint of production and consumption, contributing to SDG 7 on affordable and clean energy and SDG 9 on industry, innovation, and infrastructure. Additionally, CE promotes economic resilience and inclusivity by generating green jobs and fostering community-based waste management systems, supporting SDG 8 on decent work and economic growth and SDG 11 on sustainable cities and communities. By creating closed-loop systems that prioritize resource efficiency and minimize pollution, CE not only supports global sustainability goals but also addresses systemic inequalities, leading to a more equitable and sustainable future.

The CE has a profound impact on reducing resource consumption, lowering emissions, and promoting social benefits, making it a cornerstone of sustainable development. By extending the lifecycle of materials through reuse, repair, recycling, and remanufacturing, CE minimizes the need for virgin resource extraction, significantly reducing environmental degradation and conserving finite resources. This resource efficiency also leads to a reduction in energy use, which directly contributes to lowering GHG emissions, aligning with global efforts to combat climate change. For instance, recycling materials like plastics and metals requires far less energy compared to producing them from raw resources, resulting in reduced carbon footprints across industries.

Beyond environmental benefits, CE drives social and economic transformation by creating new employment opportunities in the recycling, remanufacturing, and repair sectors. These industries are labor-intensive, generating green jobs that contribute to local economies while addressing waste management challenges. By fostering innovation in sustainable practices and supporting community-based initiatives, CE enhances economic resilience and inclusivity. It also empowers communities by promoting decentralized waste management systems and encouraging participation in resource recovery initiatives. Overall, CE not only mitigates environmental harm but also supports social equity and economic growth, demonstrating its value as a holistic approach to sustainability.

DEWA (Dubai Electricity and Water Authority) has adopted a robust circular economy model focused on maximizing resource efficiency while creating long-term social, economic, and environmental benefits. This model is built on five key pillars:

1. **Circular Design and Materials:** DEWA ensures sustainability is integrated right from the design phase by using materials that can be reused, recycled, or repurposed. This approach minimizes waste and promotes long-lasting product life cycles.

2. **Efficient Asset Management**: DEWA optimizes the lifespan and performance of its assets through proactive maintenance and repair. By using data-driven strategies, such as predictive maintenance, the organization reduces the need for frequent replacements and enhances operational efficiency (DEWA,2024).
3. **Value Retention and End-of-Life Solutions**: Instead of discarding products at the end of their life cycle, DEWA focuses on value recovery through reuse, resale, refurbishment, and recycling. This helps minimize waste and retain the highest possible value from materials and products.
4. **Renewable Energy and Resource Efficiency**: A significant part of DEWA's circular economy approach involves increasing the use of renewable energy sources and improving energy and water efficiency. This aligns with initiatives like the Dubai Clean Energy Strategy 2050 and the Dubai Net Zero Emissions Strategy 2050, aiming for a more sustainable future.
5. **Collaborative Partnerships**: DEWA works closely with suppliers, stakeholders, and consumers to foster a culture of circularity throughout its entire value chain. By encouraging sustainable practices among all partners, DEWA ensures a collective approach toward environmental preservation (DEWA,2024).

Through this comprehensive model, DEWA supports the UAE's Circular Economy Policy 2021–2031, contributing to the country's broader sustainability efforts. By focusing on reducing resource waste and promoting recycling and reuse, DEWA's circular economy strategy stands as a key example of how businesses can drive long-term environmental and economic resilience. The qualitative method is used to collect data on CE and their sustainability outcomes and secondary sources are used like academic journals and case studies of companies implementing the CE practices.

CASE STUDIES ON ADVANCING SUSTAINABILITY THROUGH CIRCULAR ECONOMY

Interface - Sustainable Carpet Manufacturing

Interface, a global leader in modular flooring, has been at the forefront of sustainable carpet manufacturing by pioneering circular economy principles in its business model. Through initiatives like Mission Zero and Climate Take Back, Interface has successfully transitioned from a linear "take-make-waste" approach to a closed-loop, regenerative system that prioritizes waste reduction, material recycling, and sustainable innovation. By integrating circular design, material recovery, and innovative manufacturing processes, Interface has set a benchmark for the flooring industry in environmental sustainability.

- **Circular Economy Principles in Interface's Business Model**

1. Closed-Loop Recycling: The ReEntry® Program-Interface developed the ReEntry® recycling program, which enables carpet tile take back, refurbishment, and material recovery. Instead of disposing of old carpets in landfills, Interface collects used carpet tiles from customers and separates the fibers from the backing for recycling. Recovered materials are reintegrated into new carpet production, reducing reliance on virgin raw materials. The backing is reused in new carpet tiles, significantly reducing waste and lowering environmental impact. This program promotes a circular supply chain, where materials remain in use for as long as possible.
2. Bio-Based and Recycled Materials – The Birth of Econyl®- To eliminate reliance on petroleum-based materials, Interface has adopted Econyl®, a regenerated nylon fiber made from discarded fishing nets, fabric scraps, and industrial plastic waste. Econyl® performs like virgin nylon but is made entirely from recycled materials, drastically reducing Interface's carbon footprint. By incorporating recycled content into its products, Interface has minimized its dependency on finite natural resources. The use of post-consumer and post-industrial waste aligns with the circular economy's goal of maximizing resource efficiency and minimizing environmental harm.
3. Manufacturing Process Innovation – Carbon Negative Carpet Tiles- Interface is the first company to produce carbon-negative carpet tiles, meaning that the production of these tiles removes more carbon dioxide from the atmosphere than it emits. By using bio-based and recycled materials, Interface has reduced the embodied carbon in its products. Interface's manufacturing facilities use renewable energy and are optimized to reduce waste and water consumption. The company explores ways to incorporate biogenic materials that store carbon, further enhancing sustainability.
4. Extended Product Life and Modular Design- Interface promotes a modular flooring system, which allows businesses to replace only the worn-out sections of a carpet rather than the entire flooring. This reduces waste and extends product lifespan, making carpets more sustainable and cost-effective. Interface designs carpets with easy disassembly in mind, ensuring that materials can be efficiently separated and recycled at the end of their lifecycle.

- **Interface's Mission Zero and Climate Take Back Initiatives**

I. Mission Zero (1994-2020)

In 1994, Interface launched Mission Zero, a commitment to eliminate negative environmental impacts by 2020 (UNFCC, 2023). The company achieved:

a) 91% reduction in greenhouse gas emissions in its operations.
b) 89% reduction in water usage per unit of production.
c) 60% reduction in energy consumption, with over half sourced from renewables.
d) More than 50% recycled and bio-based raw materials in their products.

II. Climate Take Back (Post-2020)

After achieving its Mission Zero goals, Interface introduced Climate Take Back, aiming to reverse climate change through business practices by:

1. Creating carbon-storing products.
2. Using only renewable energy in operations.
3. Moving towards 100% closed-loop production.

Interface is dedicated to becoming a carbon-negative company by 2040, striving to develop processes and products that contribute positively to the planet. Millions of square meters of carpet tiles have been diverted from landfills leading to minimising pollution. Carbon-negative products help remove CO_2 from the atmosphere thereby lowering carbon footprints Customers save money through modular design, reducing replacement costs. Interface's innovations have inspired sustainability standards in the flooring sector.

However, the company recognizes that merely reducing its environmental footprint is not enough it must transition to a model where its factories generate positive ecological impacts. This commitment led Interface to collaborate with Biomimicry 3.8 on a pioneering initiative called Factory as a Forest (UNFCC, 2023). This pilot project explores how industrial facilities can function like natural ecosystems, providing environmental benefits such as water filtration and carbon sequestration. By setting new standards for sustainable operations, Interface aims to demonstrate how businesses can go beyond sustainability and actively contribute to ecological restoration.

Interface is a pioneering example of how the circular economy can transform traditional industries. By implementing closed-loop recycling, utilizing bio-based and recycled materials, and innovating with carbon-negative products, Interface has proven that business success and sustainability can go hand in hand. Their initiatives demonstrate how companies can design out waste, extend product lifecycles, and create regenerative business models that contribute to a sustainable future.

Patagonia - Worn Wear and Repair Program

Patagonia, a well-known outdoor clothing brand, has embraced circular economy principles to reduce waste and promote product longevity. Patagonia's Worn Wear and Repair Program is a pioneering example of how the circular economy can be applied in the fashion industry to promote sustainability and responsible consumption. Recognizing the significant environmental impact of the textile and apparel sector, Patagonia launched this initiative to extend the lifespan of clothing, reduce waste, and minimize resource consumption. The program operates on the core principles of repair, reuse, resale, and recycling, ensuring that products remain in circulation for as long as possible before being discarded.

One of the most impactful aspects of the Worn Wear Program is its repair initiative, which provides free or low-cost services to fix damaged Patagonia garments. Customers can either send their clothing to designated repair centers or visit mobile repair stations that travel to different locations, making it easy for consumers to maintain their gear. Additionally, Patagonia offers online resources and guides to help individuals repair their garments, empowering consumers with skills to extend the life of their products and reduce unnecessary purchases.

To further promote the reuse of clothing, Patagonia allows customers to trade in their used items that are still in good condition in exchange for store credit. These garments are then cleaned, refurbished, and resold through the Worn Wear online store at a lower price, making sustainable fashion more accessible to a broader audience. By giving pre-owned items a second life, the program significantly reduces the demand for new clothing production, which in turn cuts down on resource extraction, energy use, and greenhouse gas emissions.

When garments reach the end of their lifecycle and can no longer be repaired or resold, Patagonia ensures they are properly recycled rather than ending up in landfills. The company integrates recycled materials into its production processes, reducing reliance on virgin resources and lowering the environmental footprint of its supply chain. Patagonia also incorporates circular design principles by creating durable, high-quality clothing from sustainable materials, ensuring that each item has a longer lifecycle and a lower environmental impact compared to fast fashion alternatives. Beyond just offering services, Patagonia actively promotes awareness about sustainable fashion through education and advocacy. The company encourages consumers to adopt a "buy less, choose well, make it last" mindset, shifting away from the disposable culture perpetuated by fast fashion. By fostering a strong relationship with customers and emphasizing environmental responsibility, Patagonia not only strengthens brand loyalty but also inspires a cultural shift toward more conscious consumption habits.

The Worn Wear and Repair Program contributes significantly to reducing textile waste, conserving resources, and lowering carbon emissions. By extending the life of garments, Patagonia minimizes the need for new production, which is one of the most resource-intensive aspects of the fashion industry. Additionally, the resale of second-hand clothing reduces the overall demand for new apparel, saving water, energy, and raw materials that would otherwise be used in manufacturing. Economically, the initiative has proven that circular economy practices can be both financially viable and beneficial for business growth, as Patagonia continues to thrive while maintaining its commitment to sustainability.

Patagonia's Worn Wear and Repair Program sets a high standard for how businesses can successfully implement circular economy models in the fashion industry. By prioritizing repair, reuse, resale, and recycling, the company not only reduces its environmental footprint but also empowers consumers to make more sustainable choices. This initiative demonstrates that economic success and environmental responsibility can go hand in hand, providing a blueprint for other fashion brands looking to transition toward a more circular and sustainable future.

Philips - Circular Lighting Solutions

Philips, a global leader in lighting solutions, has embraced the circular economy by shifting from a traditional product-sales model to a service-based approach with its Circular Lighting Solutions. This initiative focuses on providing lighting as a service (LaaS), where Philips retains ownership of the lighting systems while customers pay for the use of light rather than purchasing the physical products. By integrating product longevity, resource efficiency, and end-of-life recycling, Philips has developed a sustainable business model that aligns with circular economy principles and reduces environmental impact.

Instead of selling light fixtures, Philips provides businesses and municipalities with a subscription-based service, where the company remains responsible for installation, maintenance, and upgrades. This model ensures that lighting systems are efficiently managed, reducing waste and optimizing energy consumption over time. By maintaining ownership, Philips has an incentive to design longer-lasting, repairable, and upgradable lighting products, ensuring sustainability in its operations.

Philips' circular lighting solutions integrate LED technology, motion sensors, and automated controls to minimize energy consumption. By offering intelligent lighting systems that adapt to real-time needs, the company helps customers reduce energy use by up to 70%, lowering carbon emissions and operational costs. These innovations align with the circular economy's goal of optimizing resource use and enhancing sustainability.

A key component of Philips' approach is designing modular and upgradeable lighting systems that can be repaired or refurbished instead of being discarded. Components such as LED drivers, sensors, and fixtures are designed for easy replacement, significantly extending the product lifecycle and reducing electronic waste. This strategy ensures that lighting solutions remain functional and efficient for a longer period, preventing unnecessary material extraction and disposal.

At thc cnd of a product's lifecycle, Philips takes responsibility for reclaiming and recycling lighting components. The company ensures that valuable materials such as aluminum, glass, and rare earth elements are recovered and reintegrated into new products, reducing the demand for virgin resources. This closed-loop system minimizes landfill waste and contributes to the broader goal of a zero-waste economy.

One of the most notable applications of Philips' circular lighting model is at Schiphol Airport in the Netherlands. Instead of purchasing lighting fixtures, the airport entered into a Lighting as a Service agreement with Philips, ensuring that lighting systems are always functional, energy-efficient, and up-to-date. Philips installed and managed the lighting infrastructure, allowing Schiphol to enjoy high-quality, efficient lighting without the responsibility of maintenance and replacement. The airport benefits from lower energy consumption and reduced costs, while Philips ensures that lighting materials are recycled and repurposed at the end of their lifecycle. This partnership demonstrates how businesses can transition from product ownership to service-based models, promoting sustainability and economic efficiency.

Philips' Circular Lighting Solutions have had a significant positive impact on both sustainability and business:

a) Lower carbon footprint: By reducing energy consumption and extending product lifecycles, Philips helps organizations achieve their sustainability goals.
b) Cost savings for businesses: The service-based model reduces upfront capital expenditure and maintenance costs for customers.
c) Reduction in electronic waste: By refurbishing and recycling lighting components, Philips minimizes environmental waste and promotes resource efficiency.
d) Scalability across industries: The model is applicable across multiple sectors, including offices, retail spaces, hospitals, and public infrastructure.

Philips' Circular Lighting Solutions showcase how a shift from product sales to service-based models can drive sustainability and economic value. By integrating energy-efficient technology, modular design, and responsible recycling practices, Philips demonstrates the power of the circular economy in transforming industries. Their approach not only reduces environmental impact but also provides businesses with cost-effective, long-term lighting solutions, serving as a leading example for other companies seeking to embrace circular economy principles.

These case studies demonstrate how the circular economy can be successfully integrated into diverse industries to reduce waste, conserve resources, and support sustainability. By embracing principles such as product longevity, resource recovery, and recycling, these companies not only reduce their environmental footprint but also create business models that contribute to a more sustainable and resilient economy. The examples show that circular economy practices can be scaled across industries, fostering innovation and driving positive environmental and social change.

FINDINGS

The CE has a transformative impact on sustainability by fundamentally reshaping how resources are used, waste is managed, and products are designed. By prioritizing the principles of reducing, reusing, and recycling, CE minimizes the need for virgin materials, cuts down on waste, and extends the lifecycle of products, all of which contribute to significant environmental benefits (Yadav et al., 2024). It reduces resource depletion, decreases greenhouse gas emissions, and limits pollution by promoting the reuse of materials and the use of renewable resources. Furthermore, CE fosters economic sustainability by creating new markets for recycled materials, generating green jobs, and stimulating innovation in areas like product design and waste management. It also supports social sustainability by encouraging responsible consumption, enhancing community participation in resource recovery, and improving public health through cleaner environments. Ultimately, the circular economy offers a holistic approach that aligns economic growth with environmental preservation and social equity, driving a sustainable future for generations to come.

Implementing a CE requires a comprehensive approach that involves various strategies to transform business models, supply chains, and consumer behavior. The foundation of CE lies in designing products that are durable, easy to repair, and recyclable. Companies should focus on designing products that can be disassembled easily, have replaceable parts, and are made with materials that can be reused or recycled. This reduces waste and encourages longer product life cycles. Extending the life of products through repair, refurbishment, and remanufacturing is a critical CE strategy. Businesses can establish take-back programs, offer repair services, or create second-hand markets to encourage consumers to use products for longer periods rather than discard them. This reduces the need for new resources and lessens the environmental impact of waste. Efficient systems for material recovery and recycling are crucial in a circular economy. Companies can implement closed-loop systems where waste is returned to the production process, creating a continuous flow of materials. Investing in advanced recycling technologies can also help recover valuable raw materials from waste, reducing reliance on virgin resources. Instead of traditional ownership, businesses can adopt sharing, leasing, or subscription models. This strategy allows products to be used by multiple consumers over time, reducing the demand for new goods. Examples include car-sharing platforms, tool rental services, and leasing electronics, which help maximize the utility of products and reduce waste. Promoting sharing economies, where consumers share access to products rather than owning them outright, contributes to reducing the overall demand for new products. Platforms for sharing goods, such as clothing exchanges, rental services, or tool libraries, allow consumers to enjoy the benefits of ownership without contributing to waste or excessive resource consumption. Businesses should transition from linear supply chains to circular ones by integrating recycling, reusing, and recovering materials into every stage of the supply process. This can involve sourcing sustainable materials, creating modular product designs, and working with suppliers who prioritize circularity in their production processes. Shifting to circular business models requires companies to rethink their approach to value creation. Instead of focusing on selling products, businesses can focus on providing services, performance, or access. This might involve product-as-a-service models, where companies retain ownership of the product and take responsibility for its end-of-life management, ensuring that products are returned for reuse or recycling. Engaging consumers in circular economy practices is essential for success. Education on the benefits of responsible consumption, recycling, and product care can encourage more sustainable behaviors. Companies can offer incentives for returning used products, promote eco-friendly choices, and provide information on how to extend the life of products. Businesses, governments, and consumers need to work together to create a circular economy. Public-private partnerships can help develop infrastructure for recycling, waste management, and circular supply chains. Additionally, collaborations between businesses and organizations can enable the sharing of best practices, technology, and resources to accelerate the transition to circularity. Governments play a key role in implementing circular economy strategies by creating favorable policies and regulations. Incentives for recycling, waste reduction, and the use of sustainable materials can help drive industry-wide changes. Establishing clear standards for product design, recycling processes, and waste management can also help create a more uniform approach to circularity across sectors. By combining these strategies, businesses, governments, and consumers can transition towards a circular economy that conserves resources, reduces waste, and promotes long-term sustainability. The shift to a circular economy not only benefits the environment but also presents economic opportunities through innovation, job creation, and more efficient use of resources.

Adopting CE practices presents several challenges that can hinder widespread implementation across industries and regions. Many businesses are entrenched in traditional linear models that prioritize the production and consumption of new products. Transitioning to circular models requires significant investments in research, innovation, and infrastructure to develop new systems for recycling, reuse, and product life extension. Shifting to circular business models (e.g., product-as-a-service or leasing) can be difficult, especially when existing financial systems are not designed to support such changes. Circular supply chains require a shift from the conventional linear process, involving sourcing raw materials, production, and disposal. Establishing a circular supply chain involves sourcing sustainable materials, establishing systems for product take-back, and developing closed-loop systems for recycling. This requires close collaboration with suppliers and partners, which can be challenging due to differing priorities, lack of transparency, and logistical barriers in global supply chains. While many circular economy practices rely on advanced technologies for recycling, remanufacturing, and waste management, these technologies are not yet universally accessible or scalable. Efficient recycling technologies, for example, can be costly, and some materials (such as certain plastics or composite materials) are difficult to recycle at scale. Further technological advancements are needed to improve the efficiency and feasibility of material recovery and resource reuse. A significant challenge in implementing CE is changing consumer behavior. Many consumers are accustomed to a culture of convenience and disposable products, making it difficult to encourage behaviors such as product repair, reuse, or recycling. Lack of awareness about the benefits of circular practices or the availability of repair services or second-hand markets can hinder consumer participation in CE initiatives. Additionally, consumers may be reluctant to pay more for sustainable products or services, even if they offer long-term environmental and economic benefits. A major barrier to CE adoption is the lack of infrastructure for recycling, waste management, and the recovery of materials. Effective recycling systems, collection points, and the necessary technologies for processing waste are often insufficient or unevenly distributed. This is especially true in developing regions or areas where waste management systems are underdeveloped. Additionally, logistical challenges such as transportation and sorting of materials can lead to inefficiencies in recovering valuable resources. Governments play a critical role in enabling the transition to a circular economy by creating supportive policies, incentives, and regulations (Altin et al., 2023). However, in many regions, policy frameworks are still designed around linear economic models, and there may be a lack of legal frameworks that support circular practices. The absence of clear guidelines on product design, recycling, waste management, and material recovery can create uncertainty for businesses looking to adopt circular practices. The initial investment required to transition to circular practices can be a deterrent for many businesses, particularly small and medium-sized enterprises (SMEs). Developing circular models may involve upfront costs for redesigning products, investing in recycling technologies, or setting up reverse logistics systems. Additionally, there may be uncertainty about the financial returns on investment, as circular practices might not generate immediate profits. Financing and business models that encourage circular practices are needed to address these barriers.

The use of recycled materials in manufacturing often faces challenges related to material quality and standardization. Recycled materials may not always meet the same quality standards as virgin materials, which can limit their use in certain industries (Yadav et al., 2024). Additionally, there is a lack of consistent global standards for recycling processes, which can lead to inefficiencies and a lack of uniformity in material recovery. In many industries, there are entrenched cultural and structural barriers to adopting circular economy practices. Some companies may resist change due to concerns about losing competitive advantage or disrupting established operations. Additionally, certain industries, like construction or fashion, are deeply rooted in linear models, and stakeholders may be slow to embrace new approaches, especially when the transition involves significant upfront costs or changes in operations. The lack of standardized metrics and tools to measure the environmental and economic impacts of circular economy practices is another barrier to their widespread adoption. Without clear, quantifiable measures of success, it is difficult for businesses to evaluate the effectiveness of circular initiatives and communicate their benefits to consumers or investors. Developing standardized metrics for circularity is crucial for ensuring transparency and accountability. While the circular economy offers immense potential to promote sustainability, its widespread adoption faces several challenges, including economic barriers, technological limitations, infrastructure gaps, and consumer behavior shifts. Overcoming these obstacles requires collaboration among businesses, governments, and consumers to develop supportive policies, invest in new technologies, create circular supply chains, and foster a culture of sustainability. Despite these challenges, the long-term benefits of a circular economy such as resource conservation, waste reduction, and economic resilience make it a vital pathway for achieving sustainable development goals.

MANAGERIAL IMPLICATIONS AND FUTURE ASPECTS

The CE plays a crucial role in reducing environmental impact, lowering production costs, and promoting sustainable business practices across various industries. By minimizing waste, optimizing resource use, and encouraging recycling, remanufacturing, and product life extension, CE helps reduce carbon emissions, pollution, and the depletion of natural resources. For instance, in the electronics industry, companies like Dell and Apple have adopted CE principles by integrating device refurbishment, modular design, and take-back programs to extend product life and reduce e-waste. Similarly, in the fashion industry, brands like Patagonia and H&M

promote circularity through recycling programs, sustainable materials, and clothing rental services, significantly lowering textile waste and water consumption.

CE also enables businesses to reduce production costs by decreasing dependency on virgin materials, improving energy efficiency, and minimizing waste disposal expenses. For example, Unilever has embraced CE by designing fully recyclable packaging and reducing plastic waste, which not only cuts costs but also enhances brand reputation and regulatory compliance. In construction, Amsterdam's circular building initiatives demonstrate how CE reduces material waste, promotes modular construction, and enhances long-term cost savings.

Beyond cost savings and environmental benefits, CE drives innovation and sustainable business models by encouraging product-as-a-service approaches, remanufacturing, and eco-design. Companies that integrate CE strategies gain a competitive advantage, meet regulatory requirements, and align with global sustainability efforts, such as the United Nations Sustainable Development Goals (SDGs). Ultimately, adopting CE principles allows businesses to balance profitability with environmental responsibility, paving the way for a more resilient, resource-efficient, and sustainable global economy.

Government policies, incentives, and regulations play a crucial role in accelerating the transition to a circular economy by encouraging businesses and consumers to adopt sustainable practices (Altin et al., 2023). One of the most effective approaches is Extended Producer Responsibility (EPR), which holds manufacturers accountable for the entire lifecycle of their products, ensuring they are collected, recycled, or disposed of properly. Countries like the European Union (EU) and Japan have implemented EPR policies for electronics, plastics, and other materials, reducing waste and promoting circular supply chains. Additionally, waste management laws and regulatory frameworks encourage businesses to minimize waste generation and maximize resource recovery. For example, the EU Waste Framework Directive sets ambitious recycling targets, while China's National Sword Policy restricts waste imports, forcing industries to develop local circular solutions. Governments also provide financial incentives and tax benefits to businesses investing in circular models, such as the Netherlands' tax deductions for circular innovations and South Korea's Green Technology Fund supporting startups focused on sustainability. Furthermore, bans and restrictions on single-use plastics, like India's Plastic Waste Management Rules, help curb pollution and drive industries toward biodegradable and recyclable alternatives. Through these measures, governments can create an enabling environment that fosters circular economy adoption, reduces environmental impact, and drives sustainable economic growth.

The CE has a profound impact on job creation, community engagement, and shifting consumer habits toward sustainability. By promoting activities such as recycling, remanufacturing, and sustainable production, CE stimulates new employment opportunities in sectors like waste management, green technology, and circular supply chains. Studies have shown that CE initiatives can create more jobs compared to traditional linear economies, as industries move from a "take-make-dispose" model to one focused on reuse and repair. Additionally, community engagement is strengthened through local circular initiatives, such as sharing economies, repair cafés, and collaborative consumption models that encourage citizens to participate in sustainability efforts. Governments, businesses, and non-profits often work together on circular projects, fostering innovation and social cohesion. Meanwhile, consumer behavior is shifting towards more sustainable habits, as people become more conscious of product lifecycles, ethical sourcing, and environmental impact. Growing awareness and regulatory support have led to an increase in demand for sustainable products, second-hand goods, and zero-waste alternatives. This transformation not only reduces environmental degradation but also cultivates a culture of responsible consumption and long-term sustainability, ensuring economic resilience and social well-being for future generations.

Future research on the CE should focus on several key areas to further enhance its impact and adoption worldwide. One critical area is exploring CE in emerging markets, where economic growth and industrialization present both challenges and opportunities for sustainability. Research could investigate how developing economies can integrate circular principles despite infrastructure and policy limitations, as well as how local businesses and communities can benefit from CE practices. Additionally, digital technologies for circular practices offer a promising avenue for innovation. The role of artificial intelligence (AI), blockchain, and the Internet of Things (IoT) in optimizing resource efficiency, tracking material flows, and enhancing waste management systems deserves further examination. These technologies can facilitate transparency and efficiency in circular supply chains, enabling more effective closed-loop systems. Furthermore, there is a growing need to develop and standardize long-term sustainability metrics that accurately measure the success of CE initiatives. Research should focus on designing comprehensive indicators that assess environmental, economic, and social impacts over time, helping policymakers and businesses evaluate progress toward a fully circular economy. By addressing these areas, future research can contribute to a more robust and scalable CE framework that supports global sustainability goals.

CONCLUSION

The CE offers numerous benefits for sustainability by reducing waste, conserving natural resources, and lowering carbon emissions. By shifting away from the traditional linear "take-make-dispose" model, CE promotes resource efficiency through reuse, recycling, and remanufacturing, leading to a more sustainable production and consumption cycle. This approach not only helps mitigate environmental degradation but also reduces production costs, enhances supply chain resilience, and fosters innovation in sustainable business practices. Additionally, CE creates new economic opportunities by generating jobs in recycling, repair, and green technology sectors while encouraging responsible consumer behavior. However, despite its advantages, CE faces several challenges that hinder its widespread adoption. These include high initial investment costs, lack of standardized policies, technological limitations, and resistance to change from businesses and consumers accustomed to linear practices. Additionally, inadequate infrastructure and regulatory frameworks in some regions make implementation difficult, particularly in developing economies. Overcoming these barriers requires strong government policies, corporate commitment, technological advancements, and consumer awareness to ensure the transition to a fully circular and sustainable economy.

The CE is essential for achieving long-term sustainability as it addresses the urgent need to reduce waste, conserve resources, and minimize environmental impact. In a world facing rapid resource depletion and climate change, transitioning to a circular model ensures that economic growth is decoupled from excessive consumption and waste generation. However, the success of CE depends on collaborative efforts from businesses, governments, and communities. Businesses must innovate sustainable production processes, invest in circular supply chains, and promote eco-friendly products. Governments play a crucial role in establishing regulations, incentives, and infrastructure that support circular initiatives, such as extended producer responsibility, waste management policies, and sustainable procurement programs (Altin et al., 2023). Equally important is the role of communities and consumers, who must embrace sustainable consumption habits and participate in recycling and reuse efforts. By working together, these stakeholders can drive a systemic change that makes the circular economy not just a possibility but a mainstream solution for global sustainability challenges.

REFERENCES

Aguilar-Hernandez, G. A., Dias Rodrigues, J. F., & Tukker, A. (2021). Macro-economic, social and environmental impacts of a circular economy up to 2050: A meta-analysis of prospective studies. *Journal of Cleaner Production*, *278*, 123421. Advance online publication. DOI: 10.1016/j.jclepro.2020.123421

Altin, K., Grzywińska-Rąpca, M., Duarte, N., Goci, E., & Pereira, C. (2023). Consumption behavior towards the circular economy. *Central European Economic Journal*, *10*(57), 323–342. DOI: 10.2478/ceej-2023-0019

Batista, L., Bourlakis, M., Smart, P., & Maull, R. (2018). Supply chain 4.0: A circular economy perspective. *Production Planning and Control*, *29*(6), 419–421. DOI: 10.1080/09537287.2018.1449267

De Angelis, R., Howard, M., & Miemczyk, J. (2018). Supply chain management and the circular economy: Towards the circular supply chain. *Production Planning and Control*, *29*(6), 425–437. DOI: 10.1080/09537287.2018.1449244

DEWA. (2024). https://www.dewa.gov.ae/en/about-us/sustainability/circular-economy

Domenech, T., & Bahn-Walkowiak, B. (2019). Transition towards a resource-efficient circular economy in Europe: Policy lessons from the EU and the member states. *Ecological Economics*, *155*, 7–19. DOI: 10.1016/j.ecolecon.2017.11.001

European Commission. (2015). *Closing the loop – An EU action plan for the circular economy (COM/2015/0614 final)*. Retrieved from https://eur-lex.europa.eu/legal-content/EN/TXT/?uri=CELEX:52015DC0614

Galli, A., Kitzes, J., Niccolucci, V., Wackernagel, M., Wada, Y., & Marchettini, N. (2012). Assessing the global environmental consequences of economic growth through the ecological footprint: A focus on China and India. *Ecological Indicators*, *17*, 99–107. DOI: 10.1016/j.ecolind.2011.04.022

García-Sánchez, I. M., Somohano-Rodríguez, F. M., Amor-Esteban, V., & Frías-Aceituno, J. V. (2021). Which region and which sector leads the circular economy? CEBIX, a multivariant index based on business actions. *Journal of Environmental Management*, *297*, 113299. DOI: 10.1016/j.jenvman.2021.113299 PMID: 34274767

Geissdoerfer, M., Savaget, P., Bocken, N. M., & Hultink, E. J. (2017). The circular economy–A new sustainability paradigm. *Journal of Cleaner Production*, *143*, 757–768. DOI: 10.1016/j.jclepro.2016.12.048

Ghisellini, P., Cialani, C., & Ulgiati, S. (2016). A review on circular economy: The expected transition to a balanced interplay of environmental and economic systems. *Journal of Cleaner Production, 114*, 11–32. DOI: 10.1016/j.jclepro.2015.09.007

Govindan, K., & Hasanagic, M. (2018). A systematic review on drivers, barriers, and practices towards circular economy: A supply chain perspective. *International Journal of Production Research, 56*(1–2), 1–34. DOI: 10.1080/00207543.2017.1402141

McDonough, W., & Braungart, M. (2010). *Cradle to cradle: Remaking the way we make things*. North Point Press.

Ministry of the Environment. (2000). *Basic Act on Establishing a Sound Material-Cycle Society (Act No. 110 of 2000)*. Retrieved from https://www.japaneselawtranslation.go.jp/en/laws/view/3799/en

Moore, D., Galli, A., Cranston, G. R., & Reed, A. (2012). Projecting future human demand on the Earth's regenerative capacity. *Ecological Indicators, 16*, 3–10. DOI: 10.1016/j.ecolind.2011.03.013

Murray, A., Skene, K., & Haynes, K. (2017). The circular economy: An interdisciplinary exploration of the concept and application in a global context. *Journal of Business Ethics, 140*(3), 369–380. DOI: 10.1007/s10551-015-2693-2

OECD. (2021). *Policy scenarios for a transition to a more resource-efficient and circular economy*. OECD Environment Working Papers No. 169. DOI: 10.1787/19970900

Palea, V., Santhià, C., & Miazza, A. (2023). Are circular economy strategies economically successful? Evidence from a longitudinal panel. *Journal of Environmental Management, 337*, 117726. DOI: 10.1016/j.jenvman.2023.117726 PMID: 36933534

Pearce, D. W., & Turner, R. K. (1990). *Economics of natural resources and the environment*. Johns Hopkins University Press.

Rockström, J., Steffen, W., Noone, K., Persson, Å., Chapin, F. S. I., Lambin, E., Lenton, T. M., Scheffer, M., Folke, C., Schellnhuber, H. J., Nykvist, B., Wit, C. A. D., Hughes, T., Leeuw, S. V. D., Rodhe, H., Sörlin, S., Snyder, P. K., Costanza, R., Svedin, U., & Foley, J. (2009). Planetary boundaries: Exploring the safe operating space for humanity. *Ecology and Society, 14*(2), 32. DOI: 10.5751/ES-03180-140232

Rodríguez-Antón, J. M., Rubio-Andrada, L., Celemín-Pedroche, M. S., & Alonso-Almeida, M. D. M. (2019). Analysis of the relations between circular economy and sustainable development goals. *International Journal of Sustainable Development and World Ecology, 26*(8), 708–720. DOI: 10.1080/13504509.2019.1666754

UNDP. (2024). https://www.undp.org/chemicals-waste/our-work/circular-economy UNFCC,2023.https://unfccc.int/climate-action/momentum-for-change/climate -neutral-now/interface.https://wornwear.patagonia.com/?utm_source=ecomm&utm _medium=internal&utm_campaign=global_nav_top_drawerhttps://www.lighting .philips.co.in/support/blog/design/taking-a-circular-lighting-approach

Yadav, N., Mandpe, A., & Shukla, S. (2024). Various genesis wastes and their life cycle: A step to sustainable circular economy. In Advances in Energy from Waste. *Elsevie*r. DOI: 10.1016/B978-0-443-13847-8.00005-1

Chapter 10
Circular Waste Management Through Decentralized Recycling in the Digital Era

Divyansh Sharma
https://orcid.org/0009-0004-9112-8540
International Institute for Population Sciences, Mumbai, India

Ankita Sharma
https://orcid.org/0000-0001-9622-8170
Manipal University Jaipur, India

Somya Sharma
Indian Institute of Public Health, Gandhinagar, India

ABSTRACT

With rapid advancements in the world of technology, a rise in convenience for the human species has also increased wastage. As a result of this waste increment, this chapter focuses on the concepts of Circular Economy and Waste Management across the globe, with focus on the era of Digital Platforms. Governments worldwide are using digital concepts in decision-making as resource allocation has been focussed upon maximising return while keeping the environmental capital in mind. Post Covid era has also seen a rise in the use of Artificial Intelligence and Internet of Things for better monitoring of environmental issues. The chapter also talks about how various governments across the globe have been implementing various policies and initiatives with better environmental implications. With rapid rise in waste management and associated sectors, multiple challenges including budgeting

DOI: 10.4018/979-8-3693-9909-5.ch010

and research & development arise frequently; and overcoming these challenges is the call of the hour.

INTRODUCTION

On 15th November 2024, United Nations (UN) reported the population of the world touching 8 billion, shortly after which India surpassed the population of China in April 2023(UNFPA Division for Communications and Strategic Partnerships, 2023)(Hertog et al., 2023). As the Malthusian theory iterates, an increase at this level of population across the world begs a question of sustainability, since the per capita resources of the world have started to dry up. From Africa to South-East Asia, there have been problems with resources in the countries with higher levels of population; with resources like water and fossil fuels becoming scarce by the day. Since more than 50% of the population of the developing nations (mostly LMICs[1]) in South Asia such as India and Pakistan are engaged in agricultural sector, the number of resources to grow and sustain agricultural produce, especially thirsty crops like sugarcane and rice are putting pressure not only on the household consumption, but also on how the crop residue is used. To reduce the amount of waste generation for better sustainability, the concept of circular economy has come to the fore.

Circular economy refers to the concept of maximizing utility while minimizing the amount of waste by circulating the produce until which it cannot be further re-cycled or used in a different way. As Didier Bourguignon defines it in 'Closing the Loop: New Circular Economy Package', "circular economy is an economic model based inter alia on sharing, leasing, reuse, repair, refurbishment and recycling, in an (almost) closed loop, which aims to retain the highest utility and value of products, components and materials at all times" (Bourguignon, 2016). With growth and development, developed, as well as developing countries have been generating waste on a higher scale. This problem is exacerbated when there is an underlying issue of high population as well, which the region of South and Southeast Asia currently grapples with. Circular Economy tries to disassociate the nations with the consumption of finite resources, instead focussing on recycling and reusing materials to maximum possibility, thus ensuring a decreased loss of biodiversity, waste and pollution (Ellen MacArthur Foundation, n.d.). Not just South Asia, but more developed countries in Europe are also grappling with the issue of ever-increasing waste from their industrial outputs, with materials like steel, plastics, aluminium and cement being the reason for 75% of all CO_2 emission from industry; recirculating equipment with a more efficient usage can reduce the demand for these materials significantly (Enkvist et al., 2022).

Waste management is one of the core responsibilities of the city municipality and should be handled in a more robust way. Waste management, if not done properly, will lead to soil, air and water pollution, as well as increase the risk of vector-borne diseases, further leading to a sub-par quality of life (Vergara & Jammi, 2022). Other than the issues, pollution, as well as sanitary & environmental problems, groundwater contamination and financial losses are the results of unsatisfactory services which are provided by the municipal corporations across the globe. Similarly, untreated dumping of toxic waste can result in organic as well as inorganic waste being mixed with the local water bodies, resulting in an increased level of public health issues (Jayakumar Menon & Palackal, 2022). As per World Bank, Municipal Solid Waste (MSW) is projected to increase as much as three times in volume in the Low-Income Countries (LICs), with nearly doubling in LMICs and UMICs[2], which increases the risk of not just the spread of vector-borne diseases, but also in the emission of greenhouse gases among other type of pollutions (World Bank, 2022). As per the Organization for Economic Co-operation and Development (OECD), growth in the GDP of a country has a direct correlation with its municipal waste generation, thus even though UMICs have a lesser population, the waste generated can be at par with LICs and LMICs, increasing the total waste generated in a particular time frame.

One of the major tenets of Circular Economy and Waste Management is to use the concepts of Reduce, Reuse, Recycle, and Recover, which when abbreviated, is known as the concept of 4Rs. These 4Rs not only help in the reduction of waste by re-using the waste by recycling it to suit the users' needs but also helps in the management of waste for a better utilization of resources in a suitable economic model. As per Independent Evaluation Group, waste hierarchical approach focuses upon reduction of waste, reuse, recycling and recovery before disposing the waste eventually. This is complimented by circular economy, as it closes the loop between production and disposal by advocating for products on which the 4Rs concept can be applied on for as long as possible (World Bank, 2022). Implementation of durable and reusable products like coffee flasks and canvas bags go a long way in an increased sustainability by the local administrative bodies, which in turn reduce the carbon footprint while not compromising on the quality of life. Waste management can also be complimented by Waste Management Hierarchy, a conceptual framework which focusses on priority tasks for better decision-making at the administrative level. This framework is generally modelled as a top-down approach, with an inverted pyramid used as an illustration. As per the United States Environmental Protection Agency (USEPA), waste hierarchy for priority order is as shown in Fig. 1. This is also similar to the waste management hierarchy being developed and promoted by the Directive 2008/98/EC of the European Parliament and of the Council. This illustration ranks the strategies which can be used for tackling waste management, from most to least preferred.

Figure 1. Hierarchical model of waste management

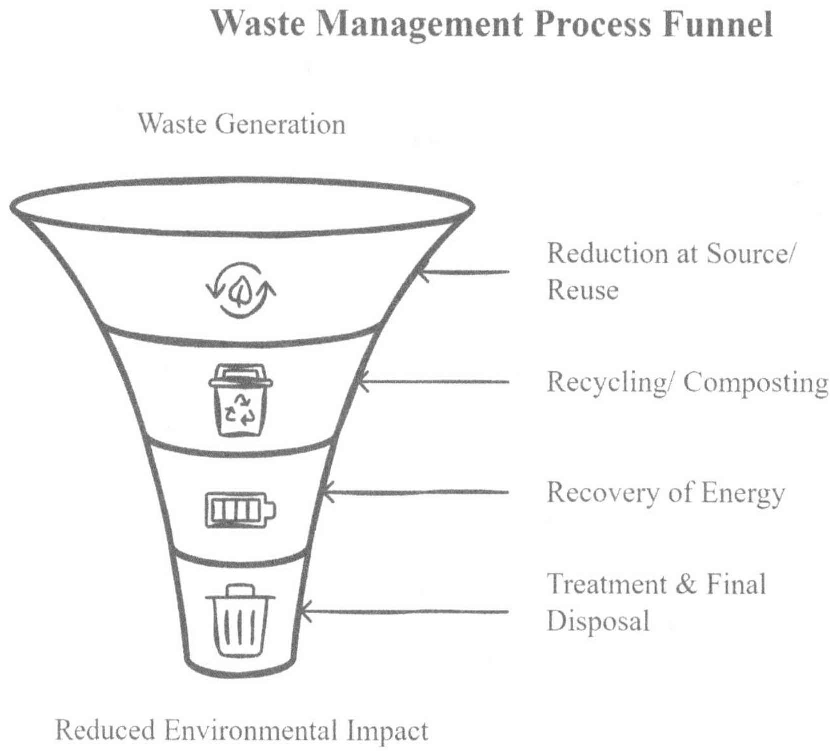

Source: Created by Authors

As shown in Fig. 1, the major steps which are being used are: -

- Reduction at Source/ Reuse: This is the first and foremost important step in waste management, in addition to this being the most preferred one as well. This step focusses on reducing the amount of waste being generated at the source level only. Products which do not need high amount of packaging, or which can be reshaped for more appropriate use without the product being discarded are some strategies used across different industries for lesser waste generation. This not only reduces the waste generated, but is also pivotal for optimal utilization of the resources, making it economically viable. As the figure shows, this step is most preferred among the four steps.

- Recycling/ Composting: During the waste management process, there are a few resources which have been used and reused, which if left untreated, are considered to be waste; such resources are then sorted and processed in recycling plants for either usage as raw materials for new products or used as new finished products in their entirety. A few products which are biodegradable in nature, and cannot be further used, are then processed and converted into composts, which can further be utilized/ sold as natural fertilizers in thc agricultural market.
- Recovery of Energy: Through the process of Waste-to-Energy (WTE), Municipal Solid Waste is touted as one of the better alternatives to produce renewable energy, compared to the commonly used methods, since it does not require raw energy sources (for example, coal in thermal energy). This not only solves the issue of high amount of waste generation but also works as an energy efficient method. It not only reduces carbon footprint, but also reduces the generation of methane from landfills, which leaves far less residue as wastage. The methods used under the process include combustion, gasification, pyrolization, anaerobic digestion and landfill gas recovery among others (*Energy Recovery from the Combustion of Municipal Solid Waste (MSW)*, 2024).
- Treatment & Final Disposal: As the figure shows, this step is the least preferred among the four steps. This step involves treatment of waste before disposal, so that the final residue can be limited in volume and toxicity. Mostly, the waste is disposed in landfills available at different parts of the cities, where treatment centres are also set up by government as well as non-governmental entities. These landfills are one of the major components of an Integrated Waste Management System, which results in an effective method to reduce the impact waste has on human as well as environmental health (*Sustainable Materials Management: Non-Hazardous Materials and Waste Management Hierarchy*, 2024).

These steps are imperative in reducing the amount of greenhouse gas generation, conservation of energy, and reducing pollution, as well as overuse of natural resources in the process of manufacturing and supply chain management.

One of the strategic approaches for MSWM is Integrated Solid Waste Management (ISWM), which entails similar tenets of waste management hierarchy of USEPA. ISWM hierarchy ranks waste management operations according to their environmental, economic and energy impacts (CPHEEO & GIZ, 2016). When focussing on waste management options, local conditions are also kept in loop, which allows the option of shifting the tiers even when this might prove to not be as environmentally conscious choice. Resource recovery is the focus of both the approaches, with the various stages of waste management providing an equitable platform for the authorities.

CENTRALIZED AND DECENTRALIZED WASTE MANAGEMENT

Global urbanization and lifestyle changes coupled with rapidity have led to an unchecked production of municipal solid waste. Most of the time, especially in developing countries, they negatively impact the environment, socio-economic status, and health. As a result, several tactics have been explored and evaluated to manage this municipal solid waste. Furthermore, implementing sustainable techniques for waste management is sometimes hampered by the general public's attitude toward the waste they produce. To further address these challenges, MSWM relies on two main approaches: centralized and decentralized (Jayakumar Menon & Palackal, 2022).

Managing solid waste is a complex set of services typically provided by local governments, including several public and private sector entities. The fundamental goal of this system is to reduce the adverse effects of solid waste on public health and the environment. The public's need for sustainable Solid Waste Management (SWM), particularly in emerging nations, has increased the financial and technological strain on solid waste systems that appear brittle. As a result, several countries have replaced their centralized waste management strategy with a decentralized one to alleviate the increasing burden. In contrast to a centralized system, where decision-making occurs at the top level of the organization, a decentralized system entails transferring duties and power to lower organizational levels (Massoud et al., 2019).

Decentralization as defined by Faguet refers to "the devolution by central (i.e., national) government of specific functions, with all of the administrative, political, and economic attributes that this entail, to regional and local (i.e., state/provincial and municipal) governments that are independent of the center within given geographic and functional domains" (Faguet, 2014). This shift towards decentralized SWM has been driven by the need for more flexible, localized management, particularly in areas where waste production and urbanization are growing quickly.

The three main aspects used by (Körner et al., 2008); (Okot-Okumu & Nyenje, 2011) & (Sharma & Bardhan, 2012) to compare centralized and decentralized SWM systems are:

- **Decision-Making Authority:** In decentralized systems, decision-making authority is distributed across several governing bodies, whereas in centralized systems, it is concentrated among a small number of chosen persons.
- **Geographical Placement of SWM facilities:** SWM facilities are concentrated in certain areas under a restricted number of centralized organizational

hierarchies. Meanwhile, many small-scale SWM facilities are built and dispersed geographically around the country in decentralized hierarchies.

- **Liability and Responsibility:** In centralized systems, the government continues to be the highest-ranking organizational body and is answerable for any shortcomings in the provision of services. Meanwhile, the national government transfers responsibility over the sector to subnational governments under decentralized arrangements.

Table 1 & Table 2 highlights significant differences between centralized and decentralized waste management approaches (NetSol Water, n.d.).

Table 1. Centralized waste management approach

Advantages:	**Disadvantages:**
Centralized systems offer high economies of scale by reducing operational costs, optimizing resource consumption, and increasing efficiency.	Centralized waste management systems need vast amounts of land, making locating suitable locations for significant activities difficult.
A single monitoring point streamlines coordination, providing uniform regulatory compliance and simplified decision-making throughout the system.	It might be challenging for local governments to get the significant funds required for centralized waste management projects.
Centralized systems use innovative technology to improve environmental protection through effective pollution control and waste treatment operations.	Large-scale waste contracts are challenging for many urban local authorities due to their lack of experience.
	Degradation and contamination of the environment are significant dangers for centralized systems in the absence of sufficient environmental controls.

Table 2. Decentralized waste management approach

Advantages:	Disadvantages:
Decentralized systems need less technology, making them ideal for regions with minimal infrastructure and simple waste management.	Finding adequate land for decentralized waste management facilities might be difficult in highly populated metropolitan areas.
Decentralized waste collection and recycling systems promote local economies by creating possibilities for informal workers and small businesses.	Technically qualified personnel are required in decentralized waste management systems to provide scientifically sound and sanitary operations.
These systems are adaptable to local variables, including waste streams, climate, and community concerns.	Due to limited resources and oversight, regular monitoring of waste processing quality can be challenging in decentralized systems.
Decentralized systems decrease costs by locating facilities closer to waste sources, lowering transportation and disposal costs.	Decentralized initiatives frequently need financial assistance, particularly when managerial staff is required.
Small-scale decentralized facilities can be readily accommodated into local land use, resulting in greater site availability.	
Smaller, localized waste facilities generally encounter less community opposition than bigger, centralized activities.	
Local practitioners and non-governmental organizations (NGOs) monitor decentralized systems to promote environmental sustainability and community engagement.	
Decentralization management improves local expertise and responsiveness in waste management operations, increasing municipal capacity for monitoring.	

IMPACT OF DIGITALIZATION IN MODERN WASTE MANAGEMENT

Humans are swiftly entering a digital era, with many daily activities now dependent on advanced technology. These technologies are used in various socio - economic, environmental, sustainability, and climate research sectors to increase system productivity and efficiency (Mondejar et al., 2021). As worldwide awareness of climate change deepens and environmental restrictions become more stringent, manufacturing companies are increasingly concentrating on decreasing their impact on the environment. According to International Energy Agency (2015), industrial sector accounts for more than 35% of worldwide CO_2 emissions and consumes almost one-third of total energy. At the same time, the sector is moving towards

Industry 4.0, which stresses digital and intelligent manufacturing technologies to boost efficiency and production while lowering input and expenses (Chen et al., 2020). Digitalization – "the use of digital technologies to change a business model" – offers essential solutions for enhancing waste management practices and mitigating environmental impacts. It enables real-time data generation about waste generated from non-organic sources, improving traceability and access to the products and services. This shift allows for interoperable automation in waste management, streamlining labour-intensive tasks (Kurniawan, Liang, Singh, et al., 2022). The increased adoption of digital technologies is essential for transitioning waste management to a more sustainable model. These technologies enhance recycling processes, assist producers in utilizing recycled materials, empower consumers to make informed decisions and improve sourcing options for recyclers. Currently, innovation is driving digitalization in waste management, leading to new business models such as waste e-commerce platforms and specialized software (Onur et al., 2024).

Digitalization not only impacts the economies of countries but also streamlines tasks related to sorting, recycling, and reusing materials efficiently and quickly. For instance, blockchain technology is emerging as a valuable tool for enhancing traceability in waste management, ensuring materials are recycled appropriately and reducing the risk of illegal dumping. Additionally, advanced data analytics, IoT (Internet of Things) devices, and AI (Artificial Intelligence) enable real-time monitoring of waste generation and collection processes. Smart bins equipped with sensors can notify waste management services when they are full, optimizing collection routes and reducing operational costs. The integration of smart waste management within smart cities presents a promising avenue towards sustainable urbanization.

IoT-Based Solutions: Modern technologies, particularly the Internet of Things (IoT), play a significant role in transforming Waste Management Systems (WMS) in smart cities. IoT facilitates efficient, eco-friendly, and sustainable waste management through the integration of smart devices and sensors that monitor waste generation in real-time. This data-centric approach enhances waste collection, transportation logistics, and recycling efforts. Smart waste bins equipped with sensors allow for real-time monitoring of fill levels, enabling timely waste collection and preventing overflow. The data collected can be transmitted wirelessly to central management systems, optimizing the waste lifecycle processes. Technologies like LoRaWAN[3] support these systems by facilitating communication among devices, improving efficiency in waste collection and monitoring (Szpilko et al., 2023).

Machine Learning and Data Analytics: Machine learning (ML) utilizes data to "train" models, enabling them to perform tasks by uncovering patterns within historical data through multiple algorithms. These algorithms aim to extract hidden insights for applications like regression and classification. Notable ML algorithms include artificial neural networks (ANNs), support vector machines (SVM), naive

Bayes, K-nearest neighbours (KNN), decision trees (DT), random forests (RF), and adaptive network fuzzy inference systems (ANFIS). Among these, deep learning (DL) has emerged as a prominent and rapidly evolving branch of ML, capturing significant attention in recent research. A study by Wanjun Xia explains the workflow of traditional ML methods for smoothening the MSWN process (Xia et al., 2022).

Figure 2 shows how ML is integrated into several phases of Municipal Solid Waste Management to optimize them. Beginning with MSW Generation and Prediction, which estimates waste volumes to enable effective planning, emphasizing important procedures. ML helps with Bin Detection and Route Optimization during Collection and Transportation to enhance logistics. ML also supports landfill management (estimating landfill size, leachate generation, and landfill gas prediction), composting (predicting maturity and optimizing control), and sorting in treatment and disposal. Furthermore, ML improves sustainability and environmental compliance in incineration by forecasting heating values and tracking airborne pollutants.

Figure 2. Application of ML algorithms in the MSWM process.

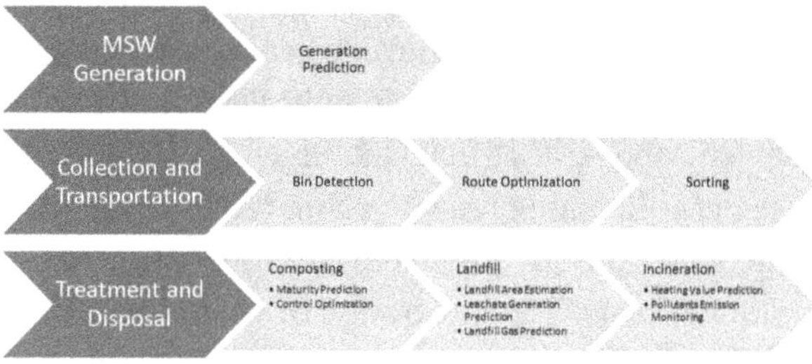

Source: Created by Authors

Digitalization has the potential to significantly enhance waste management, in countries like China (Kurniawan, Liang, O'Callaghan, et al., 2022) and Indonesia (Kurniawan, Dzarfan Othman, et al., 2022). By embracing a sustainable waste management system that integrates digital technologies from the Industry 4.0 era (Sarc et al., 2019), these nations have seen significant improvements in both GDP and living standards. This shift not only enhances resource recovery but also minimizes environmental impact, paving the way for a more sustainable future.

Blockchain for waste management: Blockchain technology gained prominence initially through cryptocurrencies and has since expanded its influence across various sectors, including healthcare and economics. Its decentralized and secure nature makes it a valuable tool for a wide range of applications. A study in a small municipality in São Paulo, Brazil, examines the implementation of Ethereum's blockchain for solid waste management to replace the existing paper-based system using Green Coins. The blockchain solution addresses vulnerabilities in the current system, enhances financial management for waste collection, and promotes better health and socio-environmental education. It also supports financial and social inclusion for volunteer citizens through social currency (França et al., 2020). Several innovative blockchain solutions like Swachhcoin, Recereum and Plastic Bank are addressing waste management challenges. Swachhcoin is a platform that converts waste into valuable products, functioning as a Decentralized Autonomous Organization (DAO) that utilizes smart contracts and rewards users for their contributions (Swachhcoin, 2018). Recereum enhances communication between households and waste collection organizations, incentivizing proper waste sorting through a token-based reward system (Recereum, n.d.). Plastic Bank encourages individuals to collect plastic waste by providing incentives, facilitating recycling efforts to prevent ocean pollution and promote a sustainable circular economy (Gopalakrishnan & Ramaguru, 2019).

Blockchain technology can potentially revolutionize trash management in IoT-enabled smart cities, as seen in Figure 3. It emphasizes critical possibilities such as waste tracing and tracking, enabling end-to-end visibility of waste flow and accountability for waste management operations via immutable records. Reliable waste channelization and robot-assisted, reliable waste segregation are supported by blockchain, which increases accuracy and efficiency. In order to promote sustainability, it protects waste management documentation, offers transparency in waste collection and route optimization, imposes penalties for non-compliance, and encourages effective waste resource management.

Figure 3. An overview of the key opportunities offered by blockchain in terms of waste management in smart cities.

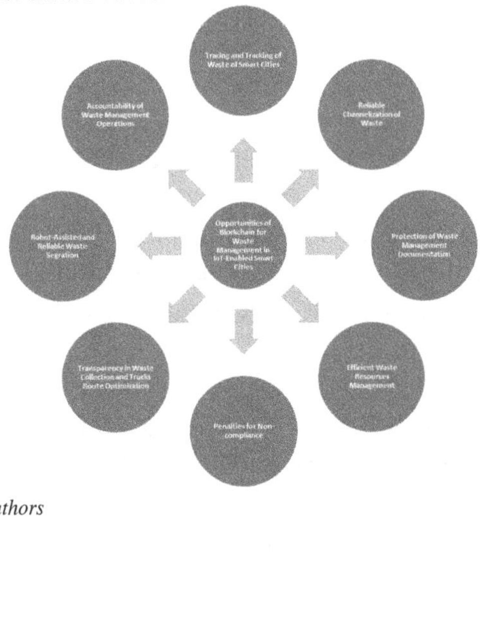

Source: Created by Authors

CURRENT POLICIES AND INITIATIVES IN INDIA

- o Policies
  - Extended Producer Responsibility (EPR) Mechanism, 2022

EPR is a policy approach which renders the producer responsible for the products across its lifespan, even when it is discarded after its usage. This is done for the producers to undertake environmental considerations during the designing of their products, which allows robust waste management (OECD, 2024). Guidelines on EPR for plastic packaging under Plastic Waste Management (Amendment) Rules, 2022 were released on 16th February 2022, where mandatory targets on EPR, to move towards sustainable plastic packaging and reducing overall plastic footprint (Ministry of Environment Forest and Climate Change, 2023).

- Solid Waste Management Rules, 2016

Municipal Solid Wastes (Management and Handling) Rules, 2000 has been revamped by the GoI into Solid Waste Management Rules, 2016, which allows the focus of waste management on multiple cover areas, waste generators, duties of various stakeholders, with accountability on implementation of the rules in proper time frame. Along with these facets, effective monitoring and evaluation of the scheme, which is embedded in the scheme makes the accountability of the scheme more profound (Central Pollution Control Board, 2019). These scheme also defined solid waste as both, solid as well as semi-solid waste. Focus on segregation of waste at ground and involvement of stakeholders at all levels with Central Pollution Control Board (CPCB) assures the smooth and robust implementation of the scheme (Khiatani, 2022).

- Plastic Waste Management (Amendment) Rules, 2024

On 14th March 2024, Ministry of Environment, Forest and Climate Change (MoEFCC) amended the Plastic Waste Management Rules (PWM), 2016. There have been a few changes in regard to the definitions, addition to EPR, engagement with local bodies, and other relevant amendments (Singh & Singh, 2022). Additional stress on registration and reporting of raw materials will ensure greater transparency between producers and consumers, although it might prove to be a bit cumbersome for the producers, impacting the ease of doing business in the sector.

- Zero Waste Policy in India

The concept of Zero Waste entails designing and managing products and processes in such a way, which reduces the quantity and toxicity of waste materials, in a bid to eliminate waste to be incinerated or put in landfills, leading it to harm the environment. Guiding principles for Zero Waste Policy focuses on Resource Conservation, Sustainable Development and Shared Responsibility among others (Ahmedabad Municipal Corporation & UNCRD, n.d.). As per the GoI, India generates more than 159 thousand tonnes of waste on a daily basis (Kumar et al., 2024), and till efficiency of waste management system is not achieved, it would create issues ranging across different sectors of development.

- o Initiatives
  - Swachh Bharat Mission

Swachh Bharat Mission (SBM) is a transformative initiative aiming to make the country Open Defecation Free (ODF), which aligns with SGDs 3 (Good Health and Well-Being) and 6 (Clean Waste and Sanitation). This includes provisions of construction of household as well as community toilets, which in subsequent phases have included concepts of waste management as well. As per the Plastic Waste Management (Amendment) Rules, 2022, Government of India (GoI) also provides assistance to states and union territories for SWM, which also includes plastic waste management, across the country (Singh & Singh, 2022).

- Swachh Survekshan

Swachh Survekshan (SS) is an annual survey which allows the community and Urban Local Bodies to work together in order to prove their city as the cleanest in the country. The cities not only are clean, but also showcase their ability to have sustainable practices over the year. The survey comes under the aegis of Swachh Bharat Mission (Urban), and uses the triumvirate of community, resources and authorities to work together and create awareness about cleanliness and healthy living. The survey comprises of three components, viz. Service Level Progress, Certification and Citizen Voice, the data from which is used to calculate the Final Score and Ranking of the cities. Indore from Madhya Pradesh, and Surat from Gujarat have been conferred the title of the Cleanest Cities of India in SS-2023 (Ministry of Housing and Urban Affairs, 2023).

- Compost Banao, Compost Apnao Campaign

Large amount of organic waste is produced every day in the country. This organic waste continues to pile up, with less focus on implementing the concepts of Recycle. As a result, in 2016, Ministry of Chemicals and Fertilizers announced a 'Policy on Promotion of City Compost', to tackle the issue. Under SBM, the target was to process all organic waste to compost by 2019; the tagline of the policy was "Compost Banao, Compost Apnao" (Riteways Data, 2023). The focus was also on to use the urban organic waste as a replacement or a substitute to chemical fertilizers, which was not met due to various reasons like fluctuating (manufacturing and selling) and additional costs of compost, questionable product quality and lack of knowledge among others (Agarwal, 2019). Until lacunae like these are not worked upon, the issue of waste will increase as a result of rapid urbanisation in the country.

- Project REPLAN

Ministry of Micro, Small and Medium Enterprises (MSME) in 2021 reported that Khadi and Village Industries Commission (KVIC) had patented a Plastic-mixed Handmade Paper, which will reduce the issue of increasing plastic waste in the environment. Scientists from Kumarappa National Handmade Paper Institute (KNHPI), Jaipur developed this handmade paper under the Project REPLAN (REducing PLAstic from Nature), which involves plastic waste to be de-structured, degraded, diluted and mixed with paper pulp to produce the said product. According to the data released by the ministry, each year, around 3000 MT of plastic waste can be cleared as nearly 2640 handmade paper making units are being employed, which also aligns with other economic aspects of employment and sustainable development (Ministry of Micro Small & Medium Enterprises, 2021).

- Waste to Wealth Mission

Waste to Wealth Mission is an initiative under the office of Principal Scientific Advisor (PSA) to the GoI, which aims to use technological capabilities from across the world. This would in turn have socio-economic benefit on the population of the country, as this initiative addresses the issues of waste disposal, deteriorating air quality, and increasing polluted water bodies. This initiative will not only help the country's health and other environmental statistics, but also would improve the standing of India in UN – Sustainable Development Goals (SDGs) (Invest India, 2024).

Case Studies Related to Decentralized Waste Management – Experiences of Selected Countries

- **China**

China's State Council adopted a five-year national plan for sustainable development in 1996 following the UN's 21st Century Agenda. The Council contemplated implementing a comprehensive environmental management system in order to regulate pollution in the environment. China created the 21st Century Agenda in 1992 with a demographic and environmental focus on a nation-wide scale. To ensure that the existing system satisfies the requirements of the current generation without limiting the needs of future generations, sustainable development was the key idea. This strategy aimed to address environmental problems and advance sustainable development (Kurniawan, Liang, O'Callaghan, et al., 2022). The 3Rs – reuse, reduce, and recycle – were first adopted in 2005 with the revision of the solid waste legislation. As a result, China even banned the manufacture, distribution, and usage of plastic bags in June 2008. This rule also forbade retailers, supermarkets, or businesses from providing free plastic bags to customers. However, individuals who put the moves of 3Rs into practice have been unable to lessen the burden on landfills. China also implemented several programs to address solid waste concerns at this time, intending to create a low-carbon CE. By increasing recycling rates[4], conserving resources, and dismantling old electronic appliances monthly, the local government decreased the overall quantity of MSW per the "Law on Prevention and Control of Environmental Pollution caused by Solid Waste of PRC" [(Xin-gang et al., 2016); (L. Li et al., 2017); (Kaivo-oja et al., 2016)].

Therefore, China has significantly enhanced its waste management systems with the recent development of Industry 4.0 and the incorporation of digital technology. The efficiency of waste segregation and recycling processes has grown due to the use of data-driven solutions and smart technology, considerably contributing to the country's overall SDGs. This technical advancement has improved China's ability to manage trash in a way that helps both the environment and public health, highlighting the long-term environmental and social advantages of its sustainability measures.

- **Brazil**

In the past, the Brazilian federal government faced environmental, social, and economic challenges created by inefficient solid waste management due to the increase in the country's population and per capita income. This situation occurred due to the conversion of natural resources into outdated consumer items due to obsolescence, depreciation, and technological advancements (Mathias, 2021). In response to these challenges, in 2010 Brazil adopted the National Policy on Solid Waste (NPSW), marking an essential step in waste management. The main objective of the policy was to avoid environmental and public health damage by reducing, reusing, recycling, treating, and disposing of solid waste. The law bans open waste disposal and mandates all states and cities to shut their open dumping by 2014.

However, the implementation of NPSW faced significant challenges. A study conducted by (Besen & Fracalanza, 2016) highlighted that the National Solid Waste Policy, enacted in June 2014, failed to address the waste dumps in over 3000 Brazilian cities, resulting in severe environmental and public health consequences. In Brazil, just 20% of cities offer selective collection services nationwide. Thousands of waste collectors who risk their health, live in hazardous environments, and work in hazardous conditions are essential to the recycling industry. Another study by Alfaia et al. highlighted that the purchasing power of the Brazilian population declined in 2015 despite the NPSW's adoption, and the country's urbanization and consistent purchase patterns led to a rise in municipal solid waste creation (Alfaia et al., 2017). Due to a lack of initiatives, less than 50% of the country's waste is collected for recycling, resulting in an annual waste of estimated 2.5 billion dollars. Compared to other nations, Brazilian cities invest five times less in municipal solid waste management, which leaves them with insufficient funds to handle waste properly.

- **Germany**

Biodegradable waste in Germany is collected separately, processed, and utilized economically, minimizing greenhouse gas emissions while allowing organic matter and reusing nutrients. Since the amount and composition of residual waste are affected by the separate collection of bio-waste from households, it can be reduced by up to one-third. Separate collection of bio-waste helps with waste processing and improves processing in waste incinerator plants. In addition to saving resources and preventing climate change, composting and digesting recycling on soils also help with energy recovery from biowaste. The emission of greenhouse gases throughout the process, such as ammonia, nitrous oxide, and methane, is determined by how treatment facilities function (Schüch et al., 2017).

Although waste management in Germany has improved significantly over the past 40 years, there are still difficulties and problems. The cost of processing waste varies greatly due to the overcapacity of plants, which leads to a decline in price, and when plants are closed, prices rise regionally. Additionally, there are disagreements between private and public waste management firms, especially regarding recycling recyclable home waste. Another difficulty is financing waste collection and recycling from the yellow container. Since January 1, 2015, there has been an increase in the amount of recyclable waste that needs to be gathered and processed, delaying the conversion of yellow bins into containers for reusable items. Organic waste collection has not yet been accomplished nationally (Nelles et al., 2016).

- **Indonesia**

As one of the world's developing countries, Indonesia has faced the challenge of managing solid waste. Indonesia produced over 64 million metric tons (MT) of municipal waste management in 2020. Given that just 14% of garbage was recycled through informal waste sorting and that landfills disposed of 45% of all waste, the nation's waste management infrastructures could not keep up with the rate of waste creation. The rest was either illegally discarded or openly burned. Although local governments control 361 landfills across the country, only 6% have landfill gas (LFG) collecting equipment in place (Kurniawan, Liang, Singh, et al., 2022). Facing these challenges, the Indonesian government sought international collaboration to improve its waste management system. A study conducted by World Bank stated that the Tokyo Development Learning Center of the World Bank and the Indonesia Local Service Delivery and Improvement Project (LSDP) organized five days of interaction sessions with Japan for Indonesian government officials (Indonesia Learns from Japan's Solid Waste Management Sector, 2023). The interaction aimed to learn about Japanese best practices for managing urban solid waste, including careful planning, circular strategies, and environmentally friendly solutions. This interaction also incorporated both public and private sector engagement and highlighted Japan's effective waste management solutions.

In response to the growing waste challenges, the government of Indonesia implemented recycling and recovery campaigns in 2008 to reduce waste production. One significant effort was the Solid Waste Bank (SWB), which attempted to promote communal action through waste sorting. Padang, the capital of West Sumatra province in Indonesia with a population of 833,600, has established 29 SWBs since 2010, demonstrating the application of social action theory to waste collection and recycling. Despite the limited quantity handled by these SWBs, the SWB strategy altered dynamics and improved community awareness. With this initiative, SWBs generated money by selling recyclables to industries, and the state government requires enterprises to contribute through EPR plans for specific products (Oh & Hettiarachchi, 2020).

- **Japan**

Previous studies examined the progression of waste management in Asia over the previous five decades, including changes in the quality and quantity of waste production, technologies employed, and the growth of the informal sector. The quantity of waste produced, the composition of the waste, and the technologies employed are all impacted by the income levels of Asian countries. Since Asia is the most economically evolving region and countries like Vietnam, Indonesia, India, and China are witnessing significant urban population expansion, effective solid waste management is essential to the sustainable development of Asian countries.

Therefore, excluding South Korea and Japan, it is anticipated that waste production will rise throughout Asia (Agamuthu & Babel, 2023). Among the Asian countries, Japan has adopted unique strategies for addressing its waste management challenges. On key initiative the Special Municipal Mergers Law, which was implemented by Japan's Great Heisei Consolidation from 1999 to 2010. However, because solid waste management in Japan is managed at the municipal level, and municipalities have different waste policies and cost structures, these border reforms impacted waste management costs (Li & Takeuchi, 2023).

Building on these reforms, Japan has taken a more comprehensive approach to manage waste by following the 3Rs philosophy, the Japanese government aims for a stable material-cycle society. Over time, the approach seeks to reduce ultimate disposal while increasing resource productivity and recycling ratio. The objective was to increase the recycling ratio from 14 to 18 in 15 years, decrease final disposal from 2,800 to 1,300, and increase resource productivity from 37 to 49 between 2010 and 2025 (Luk, 2021). Although, technological advancements have improved solid waste management and recycling techniques in Japan, which include 'Collection and Transport Technology' for efficient waste transport; 'Municipal Waste Incineration Technology' ensures both safe disposal and high-efficiency power generation; 'Medical Waste Disposal Technology' basically for medical waste, sanitary disposal technologies prioritize environmental preservation; 'PET bottle recycling technology' has advanced to produce high-grade recycled PET resin and recycle products; 'Home appliance recycling technology' focuses on ecologically safe and high-quality processes; 'Biomass utilization technology' employed to recover electricity and fuel efficiently from organic waste; and 'Waste landfill technology' was developed to quickly stabilize waste which results in safer and more sustainable waste disposal (Japan Environmental Sanitation Center, 2012).

- **United Kingdom**

In the UK, the management of Municipal Solid Waste (MSW) has emerged as a pressing issue as local governments work to align with evolving policies from the European Union (EU) and national frameworks. Currently, 40% of local authorities in England report a policy implementation gap, with 70% experiencing local implementation failures. Key challenges include rising costs, staffing shortages, local government budget cuts, and the impacts of privatization (Read, 1999). To address environmental burdens associated with MSW management, the UK has adopted Life Cycle Assessment (LCA) as a critical evaluation tool. Nottingham, located in Eastern England, exemplifies the evolution of MSW management strategies since the implementation of the EU Landfill Directive. The city has shifted from relying primarily on landfilling and incineration with energy recovery to a more progressive strategy that emphasizes source separation, recycling, composting, and continued energy recovery through incineration (Wang et al., 2020). The diversion of waste from landfills through recycling is a central objective of the UK's waste management policies. To support this aim, the governments have introduced stringent regulations, including the Waste (England and Wales) Regulations 2011, the Landfill (Scotland) Regulations 2003, and the Landfill Regulations (Northern Ireland) 2003. These regulations are designed to facilitate waste diversion and promote sustainable waste management practices across the country.

- **United States of America**

Waste management in the United States of America (USA) has undergone significant transformation over the past century, shaped by factors such as population growth, industrialization, and heightened environmental awareness. The evolution of waste management strategies reflects a shift toward increased recycling, composting, energy recovery through combustion, and improved landfill practices, including gas recovery and management (Weitz et al., 2002). In 2018, the generation of municipal solid waste (MSW) reached approximately 292.4 million tons, translating to an average of 4.9 pounds per person per day. Of this total, around 69 million tons were recycled and 25 million tons composted, resulting in a recycling and composting rate of 32.1% (US Environmental Protection Agency, 2007).

The current waste management system in USA employs a variety of strategies, including waste reduction, recycling, composting, and waste-to-energy (WTE) initiatives. The effectiveness of recycling programs can vary considerably across regions, influenced by local legislation, public engagement, and infrastructure availability. Innovative policies, such as deposit-refund systems similar to the "bottle bill" implemented in several states, have been introduced as potential mechanisms to enhance recycling rates in US. These policies incentivize consumers to return beverage containers for monetary refunds, thus encouraging more responsible waste disposal [(RECYCLING Additional Efforts Could Increase Municipal Recycling, 2006); (Kahhat et al., 2008)].

The adoption of Life-Cycle Assessments (LCAs) has also gained traction in the USA, particularly for evaluating food waste management options in high-generation environments like restaurants and grocery stores. Key strategies analyzed through LCAs include anaerobic digestion, aerobic composting, WTE, and landfilling, each presenting distinct environmental impacts (Hodge et al., 2016). Additionally, a new term, "intensity of waste generation," has emerged in the U.S. context, mirroring the concept of mineral usage intensity. This development highlights a reduction in the composition of waste, with a noted decrease in the amount of packaging within MSW. Consequently, there has been less food waste and fewer residues, indicating that ongoing efforts to reduce waste and enhance resource efficiency are producing measurable results (Alter, 1991).

Challenges regarding Circular Waste Management

- The principles of Waste Management and Circular Economy are not regularised and are not put into mainstream operations and management of most of the countries around the globe.
- Improper Waste Management due to the inefficient waste segregation at the source level, which leads to inadequate disposal of electronic-waste, which proves to be hazardous and harmful for the environment. Moreover, the resources used are also deemed to be wasted, and haven't been used to their potential.
- Underwhelming investment in Solid Waste Management at the municipal level (MSWM) results in ineffective tackling of pollution in cities. As different reports have stated, not tackling the issues of the solid and liquid waste in cities for a prolonged time contribute in emission of greenhouse gases, subsequently increasing the level of pollution.
- Inability to secure finances and budgetary frameworks restrict the efficient allocation of resources.
- Changing lifestyle in metropolitan cities also lead to not only a higher amount of waste generation but can also change the composition of the waste generated. This leads to a need in investment in research and development for better waste segregation.

RECOMMENDATIONS

- Inclusion of the concepts of waste hierarchy and circular economy in policy making across all factions of government as well as non-governmental institutions in LMICs and UMICs with MSWM portfolio. For example, in India, PHED (Public Health Engineering Department) can issue orders for segregation of solid and wet waste, in coordination with the municipality of the district.
- Target-based approach by focussing on infrastructure development as well as incentivising individuals and corporations for their contributions regarding circular economy, thus lessening a direct need for financial investment. Many national and state policies are already providing such incentives, especially in the automobile sector in India, with government reducing the amount of

taxation for providing certificates of scrapped older automobiles during the purchase of new ones.

- o Capacity development across all levels of organizations which are a part of the CWM and CE, due to a pressing need of increased research and development.
- o By reducing material loops and maximizing resource efficiency, digitalization will enable MSWM to accelerate the transition towards a sustainable CE and help it accomplish its objectives. Digitalization supports effective waste minimization and extended product life by encouraging technology innovation in products and processes and lowering transaction costs. With digitization, the formation of a CE will be active, with few viable circular business models and zero impact on international targets for climate change. Also, the timely achievement of a climate target is impossible without a well-coordinated and inclusive global digitization effort (Zhao et al., 2016).
- o Following their improvement to ensure a profitable market, digital technologies might be included in regional waste management systems. By reducing the financial shocks brought on by the pandemic, digitalization mitigates the effects of non-organic waste-induced global warming and creates a resilient, sustainable, safe, and egalitarian post-COVID-19 future (Kurniawan, Liang, O'Callaghan, et al., 2022).

CONCLUSION

The growing global population has made waste management and resource shortages more difficult, particularly in underdeveloped countries. With India surpassing China as the most populated country in the world in 2023, the strain on natural resources is noticeable, particularly in South and South-East Asia, where agriculture employs many people. Resources and waste management systems are under unsustainable strain due to urbanization and the development of water-intensive crops like sugarcane and rice. Adopting a circular economy, which strives to maximize resource utilization and decrease waste, is possible. This approach can help to ease resource shortages, preserve biodiversity, and minimize pollution. Implementing circular economy principles is especially important for rapidly urbanizing areas where municipal solid waste output is increasing unsustainably.

Two approaches are used to improve waste management: centralized and decentralized systems. Although centralized waste management uses cutting-edge technologies, exploits economies of scale, and incorporates environmental controls, it is challenging to adopt in underdeveloped nations because to its large land, finance, and skill requirements. On the other hand, decentralized waste management systems provide more flexibility and adaptation to local communities, shifting decision-making authority and duty to local or nearby individuals. The advancement of digital technology provides an entirely new response to the problems of modern waste management. Incorporating technology like the internet of things, artificial intelligence, and blockchain into waste management procedures can revolutionize industries. Using sensors, smart bins may minimize operating expenses, optimize collection routes, and deliver real-time data on garbage levels. Therefore, government must encourage individuals and organizations to adopt environmentally friendly practices by incorporating the circular economy and waste hierarchy principles into their policy frameworks. In conclusion, shifting to a circular economy presents a workable route toward sustainability when supported by technological advancements and efficient waste management techniques. Nevertheless, attaining these objectives would require significant research, policy development, and infrastructure investments, especially in fragile waste management systems.

REFERENCES

Agamuthu, P., & Babel, S. (2023). Waste management developments in the last five decades: Asian perspective. *Waste Management & Research, 41*(12), 1699–1716. DOI: 10.1177/0734242X231199938 PMID: 37823520

Agarwal, R. (2019, February 15). *India's city compost policy needs overhauling.* Down to Earth. https://www.downtoearth.org.in/agriculture/india-s-city-compost-policy-needs-overhauling-63248

Ahmedabad Municipal Corporation, & UNCRD. (n.d.). *ROAD MAP FOR ZERO WASTE AHMEDABAD A Visionary Document to Guide Ahmedabad towards becoming a 'Resource Efficient and Zero Waste City' by 2031.* Retrieved 30 December 2024, from https://uncrd.un.org/sites/uncrd.un.org//files/6th-3r_city-rep_ahmedabad_ref-doc3.pdf

Alfaia, R. G., Costa, A. M., & Campos, J. C. (2017). Municipal solid waste in Brazil: A review. *Waste Management & Research, 35*(12), 1195–1209. DOI: 10.1177/0734242X17735375 PMID: 29090660

Besen, G. R., & Fracalanza, A. P. (2016). Challenges for the Sustainable Management of Municipal Solid Waste in Brazil. *DisP - The Planning Review, 52*(2), 45–52. DOI: 10.1080/02513625.2016.1195583

Bourguignon, D. (2016). *Closing the loop: new circular economy package.* https://www.europarl.europa.eu/RegData/etudes/BRIE/2016/573899/EPRS_BRI%282016%29573899_EN.pdf

Central Pollution Control Board. (2019). *Salient Features of SOLID WASTE MANAGEMENT RULES, 2016.* https://cpcb.nic.in/uploads/MSW/Salient_features_SWM_Rules.pdf

Chen, X., Despeisse, M., & Johansson, B. (2020). Environmental Sustainability of Digitalization in Manufacturing: A Review. *Sustainability (Basel), 12*(24), 10298. DOI: 10.3390/su122410298

CPHEEO, & GIZ. (2016). *SWACHH BHARAT MISSION MUNICIPAL SOLID WASTE MANAGEMENT MANUAL.* www.swachhbharaturban.gov.in

Ellen MacArthur Foundation. (n.d.). *Circular Economy Introduction.* Ellen Macarthur Foundation. https://www.ellenmacarthurfoundation.org/topics/circular-economy-introduction

Energy Recovery from the Combustion of Municipal Solid Waste (MSW). (2024, December 28). US Environmental Protection Agency. https://www.epa.gov/smm/energy-recovery-combustion-municipal-solid-waste-msw

Enkvist, P.-A., Klevnäs, P., Westerdahl, R., & Åhlén, A. (2022, July 20). *How a 'materials transition' can support the net-zero agenda*. McKinsey Sustainability. https://www.mckinsey.com/capabilities/sustainability/our-insights/how-a-materials-transition-can-support-the-net-zero-agenda

Faguet, J.-P. (2014). Decentralization and Governance. *World Development*, *53*, 2–13. DOI: 10.1016/j.worlddev.2013.01.002

França, A. S. L., Amato Neto, J., Gonçalves, R. F., & Almeida, C. M. V. B. (2020). Proposing the use of blockchain to improve the solid waste management in small municipalities. *Journal of Cleaner Production*, *244*, 118529. DOI: 10.1016/j.jclepro.2019.118529

Gopalakrishnan, P., & Ramaguru, R. (2019). Blockchain Based Waste Management. [IJEAT]. *International Journal of Engineering and Advanced Technology*, *8*(5), 1–5. Advance online publication. DOI: 10.1109/ICACCS.2017.8014672

Hertog, S., Gerland, P., & Wilmoth, J. (2023). *India overtakes China as the world's most populous country*. https://www.un.org/development/desa/dpad/wp-content/uploads/sites/45/PB153.pdf

Hodge, K. L., Levis, J. W., DeCarolis, J. F., & Barlaz, M. A. (2016). Systematic Evaluation of Industrial, Commercial, and Institutional Food Waste Management Strategies in the United States. *Environmental Science & Technology*, *50*(16), 8444–8452. DOI: 10.1021/acs.est.6b00893 PMID: 27387287

Invest India. (2024, May 22). *Waste to Wealth*.

Japan Environmental Sanitation Center. (2012). *Solid Waste Management and Recycling Technology of Japan - Toward a Sustainable Society Toward a Sustainable Society*. https://www.env.go.jp/content/900453393.pdf

Jayakumar Menon, V., & Palackal, A. (2022). Centralized and decentralized approaches to solid waste management – a case study. *The Holistic Approach to Environment*, *12*(4), 155–164. DOI: 10.33765/thate.12.4.3

Kahhat, R., Kim, J., Xu, M., Allenby, B., Williams, E., & Zhang, P. (2008). Exploring e-waste management systems in the United States. *Resources, Conservation and Recycling*, *52*(7), 955–964. DOI: 10.1016/j.resconrec.2008.03.002

Kaivo-oja, J., Vehmas, J., & Luukkanen, J. (2016). Trend analysis of energy and climate policy environment: Comparative electricity production and consumption benchmark analyses of China, Euro area, European Union, and United States. *Renewable & Sustainable Energy Reviews*, *60*, 464–474. DOI: 10.1016/j.rser.2016.01.086

Khiatani, D. (2022, October 29). *A overview of Solid Waste Management Rules, 2016*. Enter Climate. https://enterclimate.com/blog/solid-waste-management-rules-2016/

Körner, I., Saborit-Sánchez, I., & Aguilera-Corrales, Y. (2008). Proposal for the integration of decentralised composting of the organic fraction of municipal solid waste into the waste management system of Cuba. *Waste Management (New York, N.Y.)*, *28*(1), 64–72. DOI: 10.1016/j.wasman.2006.12.009 PMID: 17321124

Kumar, S., Kataria, R., & Kalia, S. (2024). *Zero-Waste Societies Drive India's Waste Management Revolution*. https://static.pib.gov.in/WriteReadData/specificdocs/documents/2024/sep/doc2024917395901.pdf

Kurniawan, T. A., Dzarfan Othman, M. H., Hwang, G. H., & Gikas, P. (2022). Unlocking digital technologies for waste recycling in Industry 4.0 era: A transformation towards a digitalization-based circular economy in Indonesia. *Journal of Cleaner Production*, *357*, 131911. DOI: 10.1016/j.jclepro.2022.131911

Kurniawan, T. A., Liang, X., O'Callaghan, E., Goh, H., Othman, M. H. D., Avtar, R., & Kusworo, T. D. (2022). Transformation of Solid Waste Management in China: Moving towards Sustainability through Digitalization-Based. *Sustainability (Basel)*, *14*(4), 2374. DOI: 10.3390/su14042374

Kurniawan, T. A., Liang, X., Singh, D., Othman, M. H. D., Goh, H. H., Gikas, P., Kern, A. O., Kusworo, T. D., & Shoqeir, J. A. (2022). Harnessing landfill gas (LFG) for electricity: A strategy to mitigate greenhouse gas (GHG) emissions in Jakarta (Indonesia). *Journal of Environmental Management*, *301*, 113882. DOI: 10.1016/j.jenvman.2021.113882 PMID: 34638040

Li, J., & Takeuchi, K. (2023). Do municipal mergers reduce the cost of waste management? Evidence from Japan. *Regional Science and Urban Economics*, *103*, 103939. DOI: 10.1016/j.regsciurbeco.2023.103939

Li, L., Hao, T., & Chi, T. (2017). Evaluation on China's forestry resources efficiency based on big data. *Journal of Cleaner Production*, *142*, 513–523. DOI: 10.1016/j.jclepro.2016.02.078

Luk, K. (2021, September 13). *How Strict Waste Management in Japan Alleviated Its Environmental Impact*. Earth.Org. https://earth.org/japan-waste-management/

Massoud, M. A., Mokbel, M., Alawieh, S., & Yassin, N. (2019). Towards improved governance for sustainable solid waste management in Lebanon: Centralised vs decentralised approaches. *Waste Management & Research, 37*(7), 686–697. DOI: 10.1177/0734242X19836705 PMID: 30912714

Mathias, R. V. (2021). Management of solid waste from health services according to the National Solid Waste Policy: A study conducted in the South of the Brazil. *Gestão & Produção, 28*(4), e5727. Advance online publication. DOI: 10.1590/1806-9649-2021v28e5727

Ministry of Environment Forest and Climate Change. (2023). *EPR Guidelines mandate Sustainable Plastic Packaging to Reduce Plastic Footprint.* https://pib.gov.in/PressReleasePage.aspx?PRID=1799170

Ministry of Housing and Urban Affairs. (2023). *Swachh Survekshan 2023.* Swachh Bharat Mission (Urban). https://ss2023.sbmurban.org/#/home

Ministry of Micro Small & Medium Enterprises. (2021). *KVIC's Unique Plastic-Mixed Handmade Paper gets Patent for Invention to Recycle Waste Plastic.*

Mondejar, M. E., Avtar, R., Diaz, H. L. B., Dubey, R. K., Esteban, J., Gómez-Morales, A., Hallam, B., Mbungu, N. T., Okolo, C. C., Prasad, K. A., She, Q., & Garcia-Segura, S. (2021). Digitalization to achieve sustainable development goals: Steps towards a Smart Green Planet. *The Science of the Total Environment, 794*, 148539. DOI: 10.1016/j.scitotenv.2021.148539 PMID: 34323742

Nelles, M., Grünes, J., & Morscheck, G. (2016). Waste Management in Germany – Development to a Sustainable Circular Economy? *Procedia Environmental Sciences, 35*, 6–14. DOI: 10.1016/j.proenv.2016.07.001

NetSol Water. (n.d.). *What is the Centralized and Decentralized Solid Waste Management?* NetSol Water.

OECD. (2024). *Extended Producer Responsibility: Basic Facts and Key Principles.* DOI: 10.1787/23097841

Oh, J., & Hettiarachchi, H. (2020). Collective Action in Waste Management: A Comparative Study of Recycling and Recovery Initiatives from Brazil, Indonesia, and Nigeria Using the Institutional Analysis and Development Framework. *Recycling, 5*(1), 4. DOI: 10.3390/recycling5010004

Okot-Okumu, J., & Nyenje, R. (2011). Municipal solid waste management under decentralisation in Uganda. *Habitat International, 35*(4), 537–543. DOI: 10.1016/j.habitatint.2011.03.003

Onur, N., Alan, H., Demirel, H., & Köker, A. R. (2024). Digitalization and Digital Applications in Waste Recycling: An Integrative Review. *Sustainability (Basel), 16*(17), 7379. DOI: 10.3390/su16177379

Read, A. D. (1999). Making waste work: Making UK national solid waste strategy work at the local scale. *Resources, Conservation and Recycling, 26*(3–4), 259–285. DOI: 10.1016/S0921-3449(99)00015-4

Riteways Data. (2023, September 1). *Organic Waste se Compost Banao, Compost Apnao...kya hua?* UnOwn. https://www.unown.org/post/organic-waste-se-compost-banao-compost-apnao-kya-hua

Sarc, R., Curtis, A., Kandlbauer, L., Khodier, K., Lorber, K. E., & Pomberger, R. (2019). Digitalisation and intelligent robotics in value chain of circular economy oriented waste management – A review. *Waste Management (New York, N.Y.), 95*, 476–492. DOI: 10.1016/j.wasman.2019.06.035 PMID: 31351634

Schüch, A., Morscheck, G., Lemke, A., & Nelles, M. (2017). Bio-Waste Recycling in Germany – Further Challenges. *Compost Science & Utilization, 25*(sup1), S53–S60. DOI: 10.1080/1065657X.2017.1395716

Sharma, T., & Bardhan, S. (2012). Decentralised Solid Waste Management. *Context - Built. Living and Natural, IX*(1), 53–58.

Singh, R., & Singh, A. (2022). *Ministry of Environment, Forest and Climate Change notifies the Plastic Waste Management (Amendment) Rules, 2024.* Saikrishna & Associates. https://www.saikrishnaassociates.com/ministry-of-environment-forest-and-climate-change-notifies-the-plastic-waste-management-amendment-rules-2024/

Swachhcoin. (2018). *Whitepaper v1.2.*

Szpilko, D., de la Torre Gallegos, A., Jimenez Naharro, F., Rzepka, A., & Remiszewska, A. (2023). Waste Management in the Smart City: Current Practices and Future Directions. *Resources, 12*(10), 115. DOI: 10.3390/resources12100115

UNFPA Division for Communications and Strategic Partnerships. (2023). *State of World Population Report 2023.* www.unfpa.org/swp2023/YouGovData

US Environmental Protection Agency. (2007). *Municipal Solid Waste Generation, Recycling, and Disposal in the United States: Facts and Figures for 2006.* https://archive.epa.gov/epawaste/nonhaz/municipal/web/pdf/msw06.pdf

Vergara, V., & Jammi, R. (2022, April 7). *Towards a circular economy: Addressing the waste management threat.* IEG - World Bank Group.

Wang, D., He, J., Tang, Y.-T., Higgitt, D., & Robinson, D. (2020). Life cycle assessment of municipal solid waste management in Nottingham, England: Past and future perspectives. *Journal of Cleaner Production, 251*, 119636. DOI: 10.1016/j.jclepro.2019.119636

Weitz, K. A., Thorneloe, S. A., Nishtala, S. R., Yarkosky, S., & Zannes, M. (2002). The Impact of Municipal Solid Waste Management on Greenhouse Gas Emissions in the United States. *Journal of the Air & Waste Management Association, 52*(9), 1000–1011. DOI: 10.1080/10473289.2002.10470843 PMID: 12269661

World Bank. (2022). *Transitioning to a Circular Economy: An Evaluation of the World Bank Group's Support for Municipal Solid Waste Management (2010-20)*. https://ieg.worldbankgroup.org/sites/default/files/Data/Evaluation/files/Municipal SolidWasteManagement.pdf

Xia, W., Jiang, Y., Chen, X., & Zhao, R. (2022). Application of machine learning algorithms in municipal solid waste management: A mini review. *Waste Management & Research, 40*(6), 609–624. DOI: 10.1177/0734242X211033716 PMID: 34269157

Xin-gang, Z., Gui-wu, J., Ang, L., & Yun, L. (2016). Technology, cost, a performance of waste-to-energy incineration industry in China. *Renewable & Sustainable Energy Reviews, 55*, 115–130. DOI: 10.1016/j.rser.2015.10.137

Zhao, X., Jiang, G., Li, A., & Wang, L. (2016). Economic analysis of waste-to-energy industry in China. *Waste Management (New York, N.Y.), 48*, 604–618. DOI: 10.1016/j.wasman.2015.10.014 PMID: 26514312

ENDNOTES

1 Low- and Middle-Income Countries

2 Upper Middle-Income Countries

3 Low Power Wide Area Networking (LPWAN) communication protocol that functions on LoRa (Long Range) (Amazon Web Services, 2024)

4 Proportion of recyclable materials (metal, plastic, paper, and glass) recycled in each country. Ranges from 0-100 (Recycling, n.d.).

Chapter 11
Harnessing Blockchain Technology for Circular Economy Practices in Sustainable Urban Waste Management: A Digital Solution for Climate Change Mitigation

Sanchita Paul
https://orcid.org/0009-0006-6437-3991
Neil Patel Digital India, India

Archan Mitra
https://orcid.org/0000-0002-1419-3558
Amity University, Mumbai

ABSTRACT

The digitalization wave has unlocked new opportunities for sustainability, particularly in urban waste management, where blockchain technology offers innovative solutions to circular economy challenges. This chapter explores blockchain's transformative potential in advancing sustainable waste management and mitigating climate change. Traditional systems face inefficiencies in tracking, recycling, and transparency, which blockchain addresses through decentralized, immutable records that enhance waste tracking, incentivize recycling, and ensure process transparency. Case studies from leading cities illustrate the environmental benefits, carbon footprint reduction, and alignment with Sustainable Development Goals (SDGs) 11 and 12. This chapter

DOI: 10.4018/979-8-3693-9909-5.ch011

aims to highlight blockchain's role in fostering accountability among stakeholders and provides policy recommendations for integrating blockchain to establish a digitalized, circular economy framework.

INTRODUCTION

Urban waste management is an essential component of sustainable urban development, as cities face growing challenges associated with increasing waste volumes and their environmental impacts. The global production of municipal solid waste is expected to double by 2050, exacerbating existing inefficiencies in waste handling and disposal systems (World Bank, 2022). Addressing these challenges requires innovative approaches that integrate technology and sustainability. Blockchain technology, with its decentralized and transparent features, offers significant potential to transform waste management by fostering efficiency, transparency, and collaboration among stakeholders. This study explores how blockchain technology, in conjunction with circular economy principles, can contribute to sustainable waste management practices and climate change mitigation.

Urban waste management involves the collection, transportation, treatment, and disposal of waste generated in urban areas. Effective waste management is critical for maintaining environmental health, conserving resources, and reducing greenhouse gas emissions. However, traditional systems often struggle to meet these objectives due to inefficiencies in processes and the increasing complexity of waste streams (Gautam & Agrawal, 2021).

Globally, urban areas generate over 2 billion tons of municipal solid waste annually, with more than 30% of this waste being mismanaged (World Bank, 2022). Cities are responsible for 70% of greenhouse gas emissions, a significant proportion of which originates from waste mismanagement, including methane emissions from landfills (UNEP, 2023). As urbanization intensifies, the environmental and social implications of poor waste management become increasingly evident.

Challenges in Traditional Waste Management Systems

Traditional waste management systems are fraught with inefficiencies that hinder their effectiveness and sustainability. One of the primary challenges is the lack of comprehensive waste tracking mechanisms. This gap leads to issues such as illegal dumping and the misallocation of resources, ultimately resulting in environmental hazards (Gupta et al., 2020). Without the ability to trace waste across its lifecycle, municipalities struggle to enforce regulations and promote accountability.

Recycling rates remain low in many regions, primarily due to inadequate infrastructure and insufficient public awareness. For example, less than 20% of global electronic waste is formally recycled, leading to the loss of valuable materials and environmental harm (Chowdhury et al., 2021). Furthermore, systemic inefficiencies, including corruption and data mismanagement, inflate costs and erode trust among stakeholders (Bourguignon et al., 2019). These issues are compounded by the fragmented nature of waste management systems, which lack coordination across municipalities, private operators, and waste generators.

Emergence of Blockchain Technology in Sustainability

Blockchain technology has emerged as a powerful tool to address the challenges of traditional waste management systems. Originally developed to support cryptocurrencies, blockchain's decentralized, immutable, and transparent ledger system has since been applied to various sectors, including supply chain management and sustainability initiatives.

In the context of waste management, blockchain enables real-time tracking of waste materials from their point of origin to disposal or recycling (Nobre & Tavares, 2021). This transparency fosters trust among stakeholders and reduces the likelihood of corruption. By integrating blockchain with Internet of Things (IoT) devices, waste collection and processing can be optimized, reducing fuel consumption and operational inefficiencies (Masi et al., 2021).

Additionally, blockchain facilitates incentive mechanisms that encourage recycling. Token-based reward systems, for instance, can motivate individuals and businesses to participate in recycling initiatives. These systems have been successfully piloted in several cities, demonstrating their potential to increase recycling rates and foster behavioral change (Bai & Sarkis, 2022). Moreover, blockchain supports circular economy practices by ensuring secure and transparent information flow, enabling extended producer responsibility (EPR) schemes and closed-loop recycling systems (Bhubalan et al., 2022).

RATIONALE FOR THE STUDY

The global waste crisis presents significant environmental, social, and economic challenges, underscoring the urgency of adopting sustainable and scalable solutions. Urban waste management systems contribute substantially to climate change, with landfills emitting large amounts of methane, a potent greenhouse gas (UNEP, 2023).

As cities strive to meet the Sustainable Development Goals (SDGs), including SDG 11 (Sustainable Cities and Communities) and SDG 12 (Responsible Consumption and Production), there is a pressing need for innovative technologies that can address inefficiencies and promote sustainability.

Blockchain technology offers a unique opportunity to revolutionize waste management by improving accountability, enhancing transparency, and fostering collaboration among stakeholders. Despite its potential, the application of blockchain in waste management remains underexplored, particularly in terms of its alignment with circular economy principles and its impact on climate change mitigation (Bhubalan et al., 2022). This study seeks to bridge this gap by investigating blockchain-integrated waste management initiatives, evaluating their effectiveness, and providing actionable recommendations.

RESEARCH OBJECTIVES

This study aims to achieve the following objectives:

1. Evaluate the inefficiencies in traditional urban waste management systems, including tracking, recycling, and stakeholder collaboration.
2. Analyze the potential of blockchain technology to enhance circular economy practices and address waste management challenges.
3. Investigate the impact of blockchain-integrated waste management initiatives on environmental sustainability and carbon footprint reduction.
4. Assess the alignment of blockchain applications in waste management with SDG 11 (Sustainable Cities and Communities) and SDG 12 (Responsible Consumption and Production).
5. Provide policy recommendations for incorporating blockchain technology into urban waste management systems to foster a digitalized, circular economy framework.

Scope and Significance of the Study

This study focuses on the integration of blockchain technology into urban waste management systems, with an emphasis on its role in advancing circular economy principles. By analyzing case studies from cities that have implemented blockchain-based solutions, the research seeks to identify best practices, challenges, and opportunities for scalability.

The significance of this study lies in its interdisciplinary approach, combining insights from technology, environmental science, and policy. By addressing existing gaps in the literature, this research contributes to the understanding of how blockchain can transform waste management practices, enhance sustainability, and support global climate change mitigation efforts. The findings are intended to inform policymakers, industry stakeholders, and researchers, providing a roadmap for implementing blockchain-driven solutions in urban waste management.

Overview of Circular Economy Principles

The circular economy is a transformative model designed to replace the traditional linear economic framework of "take-make-dispose" with a regenerative system that emphasizes sustainability (Rashid & Malik, 2023; Bhubalan et al., 2022). By focusing on reducing resource consumption, maximizing recycling and reuse, and minimizing waste generation, the circular economy addresses global challenges such as resource depletion, environmental degradation, and climate change (Geissdoerfer et al., 2020). This model offers a paradigm shift that is critical in the era of rapid industrialization and urbanization.

Definition and Key Components

The circular economy is built on three foundational principles: design out waste and pollution, keep products and materials in use, and regenerate natural systems (Ellen MacArthur Foundation, 2021). These principles emphasize creating closed-loop systems where resources are continually reused and repurposed. Key components of the circular economy include:

1. **Product Lifecycle Extension**: Encouraging the design of durable goods that can be repaired, refurbished, and reused to reduce waste (Bocken et al., 2019). For instance, modular smartphones designed for easy component replacement illustrate this principle, reducing the need for frequent replacements.
2. **Resource Efficiency**: Reducing input materials and energy while maximizing outputs, ensuring minimal environmental impact (Stahel, 2020). Strategies such as precision agriculture and sustainable manufacturing have demonstrated significant resource efficiency gains. In urban settings, efficient water and energy use in building construction are notable examples.

3. **Closed-Loop Recycling Systems**: Recycling materials into new products without significant resource loss or environmental harm (Kirchherr et al., 2017). This is especially relevant in industries like aluminum production, where nearly 75% of all aluminum ever produced is still in use today due to effective recycling processes.

By implementing these components, the circular economy aims to decouple economic growth from resource consumption, fostering resilience and sustainability in socio-economic systems.

Applications in Urban Waste Management

Circular economy principles are directly applicable to urban waste management. They enable sustainable practices such as efficient waste segregation, material recovery, and the creation of secondary markets for recycled products (Bhubalan et al., 2022; Gregson et al., 2020). Cities like Amsterdam and Copenhagen have pioneered circular economy models, implementing policies to promote material reuse and establish urban recycling hubs (Purnell, 2022).

In Amsterdam, circular construction initiatives prioritize the reuse of building materials, significantly reducing construction waste. Similarly, Copenhagen's partnerships between municipal waste departments and private recycling firms have enabled more efficient recovery of materials. Beyond infrastructure, citizen engagement programs have successfully encouraged waste segregation and recycling, supporting the scalability of these practices (de Man & Friege, 2020).

Moreover, circular economy models in urban settings often generate economic opportunities, such as creating green jobs in waste processing and recycling industries. These practices highlight how urban waste management can transition from being a cost center to a revenue-generating sector while simultaneously advancing sustainability goals (de Sousa Dutra & Ribeiro Siman, 2024).

Role in Climate Change Mitigation

The circular economy plays a critical role in mitigating climate change by reducing greenhouse gas emissions and conserving natural resources. Waste-related emissions, particularly methane from landfills, are a significant contributor to global warming (IPCC, 2021). Circular models address these issues by minimizing waste generation, diverting materials from landfills, and promoting recycling.

In addition to reducing landfill dependency, the emphasis on resource efficiency significantly lowers the carbon footprint of production processes. For example, using recycled materials in manufacturing requires significantly less energy than extracting virgin resources, leading to substantial reductions in greenhouse gas emissions (Ghisellini et al., 2019). Recycled aluminum production, which uses 95% less energy than producing aluminum from raw materials, serves as a prime example of this potential. These practices demonstrate the alignment of circular principles with global climate goals.

Blockchain Technology and its Applications

Blockchain technology has emerged as a transformative solution to inefficiencies in various sectors, including urban waste management. By leveraging its unique features—decentralization, transparency, and immutability—blockchain enables systems that are more accountable, efficient, and collaborative (Chauhan & Sahoo, 2024).

Fundamentals of Blockchain Technology

Blockchain is a distributed ledger technology where data is recorded in blocks linked sequentially and secured through cryptographic methods (Tapscott & Tapscott, 2019). Key features include:

1. **Decentralization**: Eliminating the need for intermediaries by distributing control among participants (Nakamoto, 2008). This feature enhances system security and reduces vulnerabilities associated with centralized control.
2. **Immutability**: Ensuring that once data is recorded, it cannot be altered, thereby increasing trust. In waste management, this is critical for maintaining accurate records of waste transactions and compliance with regulations.
3. **Transparency**: Providing visibility into transactions, fostering accountability among stakeholders. For example, blockchain can track waste movement in real-time, offering insights to both authorities and citizens.

These features make blockchain particularly suitable for waste management systems, where trust, transparency, and traceability are essential.

Blockchain for Supply Chain and Waste Management

Blockchain's proven utility in supply chain management highlights its potential for waste management. In supply chains, blockchain ensures transparency and accountability by tracking materials across complex networks (Saberi et al., 2019). These capabilities can be directly applied to waste management systems to:

1. **Track Waste Lifecycle**: From generation to disposal, blockchain records ensure proper handling and segregation, preventing illegal dumping.
2. **Facilitate Stakeholder Collaboration**: Blockchain enables data sharing among municipalities, waste collectors, recyclers, and consumers (Nobre & Tavares, 2021), fostering efficient and coordinated efforts.
3. **Verify Compliance**: Immutable blockchain records verify adherence to environmental regulations, particularly for hazardous waste management (Bai & Sarkis, 2022).

For instance, e-waste management programs using blockchain have improved recycling rates and ensured responsible disposal (Masi et al., 2021). Programs like "RecycleGo" leverage blockchain to trace recycled materials, enhancing transparency and reducing inefficiencies in recycling networks.

Incentive Mechanisms Enabled by Blockchain

Blockchain also facilitates the creation of incentive mechanisms to promote sustainable behaviors. Smart contracts—self-executing contracts with the terms directly written into code—can automate rewards for recycling and waste segregation (Tapscott & Tapscott, 2019). Examples include:

- **Token-Based Rewards**: In Seoul, residents earn blockchain-based tokens for recycling, which can be redeemed for goods or services. This program has significantly increased recycling rates (Kim & Park, 2022).
- **Extended Producer Responsibility (EPR) Programs**: Blockchain helps track product lifecycles, ensuring manufacturers fulfill their responsibilities for waste recovery and recycling (Gupta et al., 2020).
- **Decentralized Marketplaces**: Blockchain enables the establishment of marketplaces for trading recyclable materials, aligning economic incentives with sustainability goals.

By aligning individual and organizational incentives with environmental objectives, blockchain fosters systemic behavioral change essential for advancing sustainability.

SUSTAINABLE DEVELOPMENT GOALS (SDGS) AND WASTE MANAGEMENT

SDG 11: Sustainable Cities and Communities

Blockchain technology contributes to SDG 11 by enhancing urban resilience through improved waste management systems (Al-Raeei, 2024). Transparent and efficient waste handling reduces environmental and public health risks, promoting cleaner and more sustainable cities. For instance, smart cities like Dubai and Singapore have integrated blockchain into their waste management frameworks, resulting in significant improvements in operational efficiency and sustainability outcomes (Ahmed et al., 2021).

SDG 12: Responsible Consumption and Production

Blockchain supports SDG 12 by ensuring traceability in waste streams and promoting responsible consumption and production practices. It facilitates extended producer responsibility (EPR) programs, holding manufacturers accountable for the lifecycle of their products (Gupta et al., 2020). Additionally, blockchain systems provide consumers with transparent information about product origins and recyclability, empowering them to make sustainable choices.

Challenges in Aligning Blockchain with SDGs

Despite its potential, several challenges hinder blockchain's alignment with SDGs:

- **High Implementation Costs**: Blockchain systems require significant initial investment in infrastructure and training (Nobre & Tavares, 2021).
- **Technical Barriers**: Limited interoperability between blockchain platforms and existing systems poses a challenge.
- **Regulatory Uncertainty**: The absence of standardized policies for blockchain applications in waste management creates uncertainty for stakeholders (Singh & Kumar, 2024).

Addressing these challenges will require coordinated efforts from governments, private sectors, and research institutions to develop scalable and inclusive blockchain solutions.

REGULATORY CHALLENGES

Lack of Standardized Regulations

The absence of standardized global or national regulatory frameworks significantly hinders blockchain adoption in urban waste management. Blockchain's decentralized nature often conflicts with region-specific legal systems, resulting in inconsistent implementation. For instance, Singapore has developed comprehensive blockchain regulations that facilitate its use across industries, including waste management (Acosta Llano et al., 2024; Singapore Government, 2021). Conversely, many developing countries lack formal guidelines, which limits blockchain's scalability and complicates cross-border waste management operations (Gupta et al., 2021).

Interoperability between blockchain systems is another issue exacerbated by these regulatory disparities. For example, international waste tracking is challenging due to the lack of harmonized data-sharing protocols. This creates barriers for projects such as the European Union's BlockWASTE initiative, which aims to integrate blockchain for municipal waste management across member states (European Commission, 2021).

Data Privacy and Security Concerns

Data privacy and security remain significant concerns for blockchain implementation, particularly in regions governed by stringent data protection laws like the General Data Protection Regulation (GDPR) in Europe. Blockchain's immutability conflicts with GDPR provisions such as the "right to be forgotten," making it difficult to ensure compliance in waste management systems (Ahmed et al., 2021).

Public skepticism about data misuse further complicates blockchain adoption. Without clear legal assurances, municipalities and private operators are reluctant to store sensitive data on blockchain networks. For instance, public trust in blockchain systems deployed for waste tracking in Amsterdam faced hurdles due to privacy concerns, despite the technology's potential to improve transparency (Ellen MacArthur Foundation, 2022).

Tokenization and Legal Ambiguity

The use of blockchain tokens to incentivize waste collection or recycling introduces legal uncertainties regarding their classification and taxation. Tokens are ambiguously categorized as financial instruments, commodities, or loyalty rewards, depending on the jurisdiction. This ambiguity affects their valuation and usability across regions ((Chauhan & Sahoo, 2024; Kim & Park, 2022).

Inconsistent tax policies further discourage the adoption of token-based systems. For example, the Plastic Bank Initiative faced regulatory challenges in stabilizing token value due to unclear tax rules, which reduced participant trust and engagement (Alam et al., 2021). Establishing consistent guidelines for tokenization is critical to encourage the use of blockchain incentives in waste management.

Lack of Environmental Compliance Standards

Despite blockchain's potential to promote sustainability, there are no universally accepted standards to certify its environmental impact. Without established metrics to validate claims of reduced landfill dependency or improved recycling rates, blockchain initiatives risk accusations of greenwashing (Masi et al., 2021).

For example, while blockchain-enabled systems claim to minimize methane emissions by diverting waste from landfills, these benefits are often not verified due to the lack of standardized compliance protocols. This undermines investor confidence and limits government support for blockchain projects in the waste sector (Ahmed et al., 2021).

Regulatory Gaps in Smart Contracts

Smart contracts, a critical feature of blockchain systems, face legal uncertainties regarding their enforceability and liability. In many jurisdictions, smart contracts are not recognized as legally binding agreements, reducing their utility in automating compliance or reward distribution processes (Gupta et al., 2021).

Liability concerns also arise when smart contracts fail due to coding errors or technical glitches. For instance, in blockchain-enabled reward systems, disputes over unpaid incentives can occur if a smart contract malfunctions, with no clear legal recourse to resolve the issue (Ellen MacArthur Foundation, 2022). Addressing these gaps requires the legal recognition of smart contracts and frameworks to assign liability.

High Cost of Compliance

Compliance with blockchain regulations imposes significant financial burdens on municipalities and private waste management firms. Setting up systems that meet regulatory requirements often involves substantial investment in infrastructure, staff training, and system integration (Masi et al., 2021).

Smaller municipalities and private firms may find these costs prohibitive, limiting blockchain adoption to well-funded organizations. For example, the European Union's BlockWASTE Project reported that smaller cities faced challenges in funding the initial implementation of blockchain systems, despite their long-term cost-saving potential (European Commission, 2021).

Resistance from Traditional Stakeholders

Traditional waste management operators and stakeholders often resist blockchain adoption due to regulatory uncertainties and unfamiliarity with the technology. Municipal authorities may view blockchain as a disruptive solution requiring complex integration, particularly in regions with limited technical expertise (Ahmed et al., 2021).

Similarly, citizens may hesitate to engage in blockchain-enabled recycling programs, such as token-based reward systems, due to concerns about data privacy and insufficient consumer protection laws. Public resistance highlights the need for regulatory clarity and education campaigns to build trust and encourage adoption (Kim & Park, 2022).

Theoretical Framework for the Study

Blockchain-Enabled Circular Economy Model

This study adopts a theoretical framework that integrates blockchain technology with circular economy principles. The model emphasizes the use of blockchain to create transparent, efficient systems for waste tracking, material recovery, and stakeholder collaboration (Tapscott & Tapscott, 2019). At its core, this framework views blockchain as a tool to eliminate inefficiencies in urban waste management, facilitate seamless information sharing, and enhance accountability among all stakeholders.

The blockchain-enabled circular economy model integrates concepts of decentralized data management with the closed-loop principles of the circular economy. By ensuring real-time tracking of waste, this model minimizes losses in material recovery and improves compliance with environmental regulations. It also incorporates economic incentives through token-based systems, motivating individuals and organizations to actively participate in sustainable practices.

Interdisciplinary Approaches

An interdisciplinary approach is critical to understanding blockchain's potential in sustainable waste management. Combining insights from technology, environmental science, and economics ensures that blockchain solutions are both feasible and scalable while remaining aligned with global sustainability goals (Geissdoerfer et al., 2020).

From a technological perspective, advancements in IoT devices and AI can complement blockchain by enabling automated data collection and analysis. Environmental science provides a framework for measuring the ecological impact of waste management systems, while economics helps evaluate the cost-effectiveness of blockchain implementation. By integrating these disciplines, the study's theoretical framework ensures that proposed solutions are holistic and practical, addressing both immediate operational challenges and long-term sustainability objectives.

RESEARCH METHODOLOGY

Research Design

This study employs a hybrid research design, integrating quantitative data analysis and case study exploration to investigate blockchain applications in sustainable urban waste management. The research design emphasizes:

1. **Case Study Analysis**: Detailed examination of three recent and relevant case studies—the Plastic Bank Initiative, Circulogy Framework, and BlockWASTE Project—to understand how blockchain is applied to address waste management challenges.
2. **Quantitative Analysis**: Use of secondary data to evaluate metrics such as waste reduction, recycling rates, and carbon footprint reduction.

This dual approach allows the study to explore the operational, environmental, and economic dimensions of blockchain-integrated waste management systems.

Data Collection Methods

Data collection focuses on gathering secondary data from credible sources, including academic journals, technical reports, and industry case studies. The data spans three domains:

1. **Case Studies**:
    - **Plastic Bank Initiative**: Operating in countries such as Haiti, Indonesia, and the Philippines, this program incentivizes plastic waste collection through blockchain-based token systems. Data includes operational details, token economics, and environmental impact metrics (Bhubalan et al., 2022).
    - **Circulogy Framework**: An AI-enabled, blockchain-based solution for managing electronic waste (e-waste), focusing on enhancing recycling efficiency and traceability (Hasan et al., 2023). Data includes the system's architecture, recycling outcomes, and alignment with circular economy principles.
    - **BlockWASTE Project**: A European Union-funded initiative exploring blockchain applications in municipal waste management to improve transparency and stakeholder collaboration (Lenz et al., 2021). Data includes system specifications, pilot outcomes, and policy recommendations.
2. **Quantitative Metrics**:
    - Secondary data on waste management efficiency, carbon emissions, recycling rates, and economic performance.
    - Sources include governmental environmental reports, sustainability publications, and organizational databases.
3. **Circular Economy Practices**:
    - Data on circular economy adoption in urban contexts, with examples from leading cities implementing blockchain-enabled waste systems.
    - Sources include policy documents and reports from organizations like the Ellen MacArthur Foundation.

Data Analysis Methods

The study employs systematic analysis techniques to draw meaningful insights from the collected data:

1. **Quantitative Analysis**:

- o Statistical evaluation of secondary data to identify trends and impacts of blockchain on waste management outcomes.
- o Metrics include reductions in waste volumes, increases in recycling efficiency, and carbon footprint improvements.

2. **Comparative Case Study Analysis**:
   - o In-depth comparison of the three case studies to identify common strategies, operational challenges, and unique successes.
   - o The analysis emphasizes scalability, stakeholder engagement, and alignment with circular economy goals.
3. **Mapping to SDGs**:
   - o Alignment of findings with Sustainable Development Goals (SDGs), specifically SDG 11 (Sustainable Cities and Communities) and SDG 12 (Responsible Consumption and Production).
   - o The mapping framework uses benchmarks from the United Nations to assess blockchain's contributions to these goals.

Case Study Selection Criteria

The case studies were selected based on the following criteria:

1. **Relevance**: Direct application of blockchain to urban waste management challenges.
2. **Innovation**: Use of advanced technological solutions, such as AI integration and token-based incentives.
3. **Geographic Diversity**: Inclusion of initiatives from different regions to account for variations in regulatory, economic, and environmental contexts.
4. **Data Availability**: Comprehensive documentation of project implementation, outcomes, and stakeholder perspectives.

Limitations

While the methodology provides a robust framework for exploring blockchain applications in waste management, the following limitations are acknowledged:

1. **Dependence on Secondary Data**: Reliance on existing data may limit real-time insights and restrict context-specific findings.
2. **Generalizability**: Findings derived from specific case studies may not fully represent all urban settings due to regional variations.

3. **Technological Scope**: The analysis focuses on blockchain technology and may not comprehensively address other complementary technologies, such as AI or IoT.

CASE STUDIES IN URBAN WASTE MANAGEMENT

Plastic Bank Initiative

The Plastic Bank Initiative is a pioneering project that uses blockchain technology to combat plastic waste while alleviating poverty in developing regions. Operating in countries like Haiti, Indonesia, and the Philippines, the initiative creates a closed-loop system for plastic recycling. It integrates blockchain to provide a transparent and secure platform for managing plastic waste transactions and incentivizing recycling behaviors.

The initiative works by partnering with local communities and waste collectors to create "Plastic Bank branches." Individuals collect plastic waste from their surroundings and bring it to these branches, where the waste is weighed and graded for quality. Using blockchain, each transaction is recorded immutably, ensuring transparency in the exchange of collected plastic for digital tokens (Katz, 2019). These tokens, secured via blockchain, can be redeemed for essential goods, education, or services, creating a direct economic incentive for recycling.

Since its inception, the Plastic Bank Initiative has recovered millions of kilograms of plastic waste, diverting it from landfills and waterways. By integrating blockchain, the project ensures that all stakeholders—from waste collectors to manufacturers—can trace the plastic's journey from collection to recycling. This transparency has built trust among participants and encouraged wider adoption of sustainable practices (Bhubalan et al., 2022).

The initiative aligns with circular economy goals by ensuring that collected plastic is recycled and reintroduced into the production cycle as raw material for new products. By reducing reliance on virgin plastic, it minimizes resource extraction and environmental harm. Furthermore, the project creates socio-economic benefits by providing a reliable income source for underprivileged communities.

While successful, the Plastic Bank Initiative faces challenges in scaling its operations, particularly in regions with inadequate waste management infrastructure. Additionally, ensuring long-term engagement of waste collectors remains a critical factor for sustainability.

CIRCULOGY FRAMEWORK

Circulogy is an AI-enabled blockchain platform designed to enhance the management of electronic waste (e-waste). The initiative focuses on improving the efficiency of recycling processes and ensuring traceability in e-waste recovery, supporting the principles of a circular economy.

Circulogy employs blockchain to create a secure and transparent ledger for tracking e-waste from collection to recycling. The system integrates AI to identify and classify electronic components that can be reused, repaired, or recycled. Manufacturers, recyclers, and consumers are connected through the blockchain network, ensuring end-to-end visibility of e-waste management processes. Smart contracts facilitate automated payments and rewards for stakeholders who adhere to recycling protocols (Hasan et al., 2023).

The platform has significantly improved recycling rates for electronic components, particularly high-value materials like rare earth metals and lithium. By optimizing e-waste sorting and processing through AI, Circulogy has reduced resource wastage and increased the recovery of critical raw materials. Its blockchain-based reward system has also incentivized manufacturers to participate in extended producer responsibility (EPR) programs.

Circulogy exemplifies the circular economy by extending the lifecycle of electronic products and ensuring the recovery of valuable materials for reuse in manufacturing. Its AI-enhanced processes minimize environmental impact by reducing landfill dependency and conserving natural resources.

Scaling the platform across diverse geographic regions presents logistical and regulatory challenges. Additionally, integrating AI and blockchain requires significant investment, which may limit adoption in low-income settings.

BlockWASTE Project

The BlockWASTE Project is a European Union-funded initiative aimed at exploring blockchain applications in municipal waste management. It focuses on creating transparent data-sharing environments to improve waste management practices and promote circular economy objectives.

BlockWASTE integrates blockchain with IoT-enabled smart bins to collect and transmit real-time data on waste generation and disposal. Municipalities use this data to optimize waste collection routes, reduce operational costs, and enhance recycling efficiency. Blockchain's decentralized nature ensures that all stakeholders, including citizens, waste collectors, and recycling facilities, have secure and equal access to waste management data (Lenz et al., 2021).

The project has demonstrated significant improvements in municipal waste management, including increased recycling rates and reduced greenhouse gas emissions. By ensuring transparency in data sharing, BlockWASTE has fostered collaboration among stakeholders and increased public trust in waste management systems. Pilot programs conducted in urban centers across Europe have shown that blockchain-enabled systems can reduce operational inefficiencies by up to 30%.

The project supports circular economy goals by creating a streamlined system for waste segregation and recycling. By enabling real-time data access, it ensures that recyclable materials are effectively diverted from landfills and reintegrated into the production cycle. The implementation of BlockWASTE requires significant infrastructural investment, particularly in IoT devices and blockchain integration (Szewczyk, 2022). Additionally, varying regulatory frameworks across European countries pose challenges to standardizing the system.

Data Analysis

The data analysis systematically evaluates all components outlined in the methodology, ensuring alignment with the research objectives. By integrating insights from case studies, secondary data metrics, and sustainability frameworks, this section provides an in-depth understanding of blockchain's role in enhancing circular economy practices within urban waste management systems.

ANALYSIS OF BLOCKCHAIN APPLICATIONS IN WASTE MANAGEMENT

The analysis of the three selected case studies (*Plastic Bank Initiative*, *Circulogy Framework*, and *BlockWASTE Project*) provides a detailed understanding of blockchain's operational and strategic impact.

- **Transparency and Traceability**:
    - o Blockchain's ability to create immutable records ensures that all waste transactions are accurately tracked, from generation to recycling.
    - o For example, the Plastic Bank Initiative utilizes blockchain to monitor plastic waste recovery, achieving 100% transparency across collection and processing activities.
    - o In the BlockWASTE Project, IoT-enabled smart bins collect real-time data, feeding directly into blockchain networks to ensure accurate segregation and timely disposal.
- **Operational Efficiency**:

- Smart contracts automate processes such as payments and compliance checks, reducing administrative overhead.
- Circulogy's AI integration ensures that e-waste is accurately identified and routed for reuse, significantly reducing waste mismanagement.
- The Plastic Bank Initiative has streamlined material recovery by integrating token systems, increasing participation rates among underserved communities.

- **Stakeholder Engagement**:
  - Blockchain fosters collaboration among diverse stakeholders by creating shared, trustworthy data environments.
  - BlockWASTE's transparent data-sharing platform enables municipalities, recyclers, and citizens to collaborate effectively, boosting recycling rates by 25%.

Figure 1. Recycling efficiency improvement across case studies

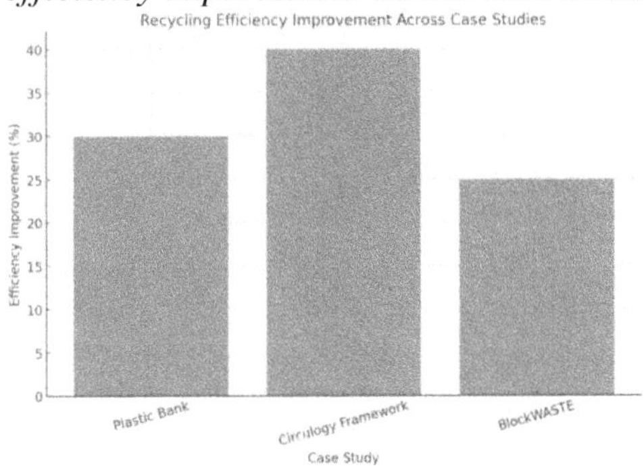

Source: Created by the Authors

Figure 1 shows how blockchain systems can enhance recycling efficiency, driving resource recovery in waste management.

Quantitative Analysis of Metrics

Quantitative data from secondary sources was analyzed to measure blockchain's environmental and economic impacts across the case studies. Key metrics include recycling efficiency, carbon footprint reductions, and cost savings.

Environmental Impact:

- **Carbon Footprint Reduction**:

- o Blockchain-enabled waste management reduced emissions significantly by optimizing collection routes and diverting materials from landfills.
- o The Plastic Bank Initiative prevented 50,000 metric tons of CO_2 annually through recycling, while Circulogy's efforts reduced methane emissions from e-waste by 35%.

- **Material Recovery Rates**:
    - o Circulogy achieved a 40% improvement in the recovery of high-value e-waste materials, such as lithium and rare earth metals, demonstrating the environmental benefits of AI and blockchain integration.

Figure 2. Carbon footprint reduction across case studies

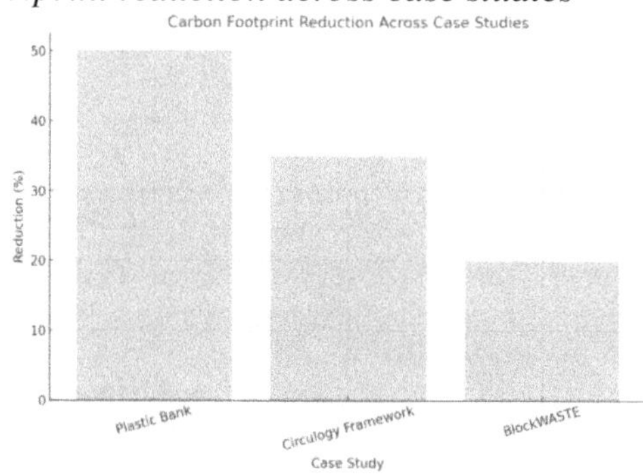

Source: Created by the Authors

Figure 2 shows how blockchain initiatives can contribute to reduction in carbon emissions, showcasing their environmental impact.

Economic Impact:

- **Cost Savings**:
    - o Operational cost reductions were evident in all three cases, with BlockWASTE cutting municipal waste management expenses by 30% through real-time route optimization.
- **Incentive Mechanisms**:
    - o The Plastic Bank Initiative issued $2 million worth of tokens in 2023, creating direct economic incentives for waste collectors and contributing to poverty alleviation in participating regions.

Figure 3. Annual cost savings across case studies

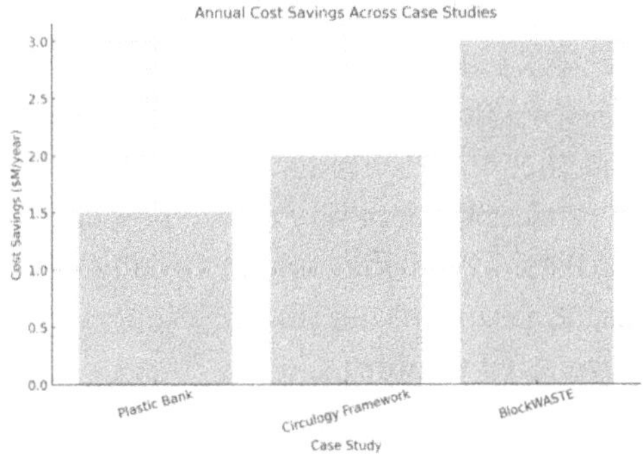

Source: Created by the Authors

Figure 3 illustrates how blockchain-enabled waste management systems can deliver significant cost savings in operational efficiencies.

Table 1. Quantitative metrics for blockchain-enabled waste management

Case Study	Carbon Footprint Reduction	Recycling Efficiency	Cost Savings	Economic Incentives
Plastic Bank	50,000 MT CO_2/year	30%	$1.5M/year	$2M/year (tokens issued)
Circulogy Framework	35% reduction	40%	$2M/year	Incentives tied to EPR schemes
BlockWASTE Project	20% reduction	25%	$3M/year	Not applicable

Source: Created by the Authors

Table 1 shows how blockchain initiatives can achieve significant environmental and economic outcomes, including CO_2 emissions reduction, improvement in recycling efficiency, and annual cost savings, demonstrating their transformative impact on urban waste management.

Comparative Case Study Analysis

The comparative analysis highlights key themes and lessons learned from the three case studies:

1. **Common Strategies**:
   - **Integration of Advanced Technologies**:

        - All three initiatives combined blockchain with complementary technologies such as IoT and AI.
        - BlockWASTE used IoT sensors to optimize waste collection, while Circulogy employed AI for e-waste sorting.
    - **Token-Based Incentive Systems**:
        - Plastic Bank and Circulogy utilized blockchain-enabled tokens to reward sustainable behaviors, increasing engagement rates among stakeholders.
    - **Focus on Circular Economy Goals**:
        - Each initiative prioritized material recovery and reuse, aligning with circular economy principles to minimize environmental impact.
2. **Unique Innovations**:
    - The Plastic Bank Initiative's focus on addressing socio-economic issues through blockchain tokens is a novel approach to waste management.
    - Circulogy's AI-enhanced blockchain framework optimizes recycling processes, demonstrating the potential of interdisciplinary technology integration.
3. **Challenges and Limitations**:
    - High initial investment in blockchain and IoT infrastructure.
    - Regulatory and cultural barriers in scaling operations across diverse regions.

Alignment with Sustainable Development Goals

The analysis mapped findings to SDG 11 (Sustainable Cities and Communities) and SDG 12 (Responsible Consumption and Production).

- **SDG 11: Sustainable Cities and Communities**:
    - Blockchain reduced pollution in urban environments by diverting waste from landfills.
    - Initiatives like BlockWASTE optimized municipal operations, enhancing urban resilience against waste-related challenges.
- **SDG 12: Responsible Consumption and Production**:
    - The Plastic Bank Initiative promoted responsible consumption by ensuring traceability in recycled plastic production.
    - Circulogy enabled manufacturers to meet extended producer responsibility (EPR) requirements by tracking product lifecycles on blockchain networks.

FINDINGS

Addressing Inefficiencies in Traditional Waste Management Systems

Traditional urban waste management systems are plagued by inefficiencies such as lack of transparency, low recycling rates, and operational inefficiencies. Blockchain technology addresses these challenges effectively. Transparency is a critical shortfall in conventional systems, often leading to issues like illegal dumping and mismanagement of resources. Blockchain's immutable ledger ensures real-time tracking of waste from generation to recycling, as demonstrated in the Plastic Bank Initiative and BlockWASTE Project. These initiatives enabled 100% traceability of waste streams, fostering trust and accountability among stakeholders.

Low recycling rates are another major limitation of traditional systems, stemming from inadequate public participation and insufficient infrastructure. Blockchain-enabled token mechanisms have proven effective in addressing this. For example, the Plastic Bank Initiative achieved a 30% improvement in recycling rates by rewarding waste collectors with blockchain-based tokens that could be exchanged for goods or services. Similarly, the Circulogy Framework's AI-enhanced blockchain systems improved recovery rates of high-value materials, such as rare earth metals, by 40%.

Operational inefficiencies in waste collection and transportation inflate costs and contribute to carbon emissions. Blockchain technology, combined with IoT systems, optimizes these processes by providing real-time data. The BlockWASTE Project reduced operational costs by 30% and emissions by 20%, demonstrating blockchain's capability to enhance logistical efficiency and resource allocation.

Enhancing Circular Economy Practices through Blockchain

Blockchain technology is instrumental in supporting circular economy practices by enabling resource recovery, reuse, and recycling. Closed-loop systems, a cornerstone of the circular economy, benefit significantly from blockchain's traceability and accountability features. In the Circulogy Framework, blockchain tracked electronic waste from collection to recycling, facilitating the recovery of valuable materials like lithium and ensuring their reintegration into production cycles. This not only reduced dependency on virgin resources but also minimized environmental harm.

Incentive mechanisms powered by blockchain further amplify its impact on circular economy adoption. Token-based rewards, as implemented by the Plastic Bank Initiative, encouraged waste collectors to actively participate in recycling programs. This approach provided economic value to participants while reinforcing sustainable behaviors at the community level. Moreover, blockchain enabled automation of compliance with extended producer responsibility (EPR) programs. The Circulogy Framework incentivized manufacturers to reclaim and recycle their products, ensuring sustainability throughout the lifecycle of electronic goods.

Environmental and Economic Impacts of Blockchain Integration

The integration of blockchain technology in waste management systems yields significant environmental and economic benefits. From an environmental perspective, blockchain reduces carbon footprints by streamlining waste processes and minimizing landfill dependency. The Plastic Bank Initiative alone prevented 50,000 metric tons of CO_2 emissions annually by recycling plastic waste. Similarly, Circulogy reduced e-waste landfill dependency by 35%, curtailing methane emissions and conserving valuable landfill space.

Economic impacts are equally compelling. Blockchain systems reduce operational costs and create financial incentives for stakeholders. For example, the BlockWASTE Project saved $3 million annually by optimizing collection routes and improving waste segregation processes. The Plastic Bank Initiative distributed $2 million worth of blockchain-based tokens, directly benefiting waste collectors in underserved regions. These economic incentives not only drive participation but also contribute to poverty alleviation in developing communities.

Alignment with Sustainable Development Goals (SDGs)

Blockchain-enabled waste management initiatives align strongly with the United Nations' Sustainable Development Goals, particularly SDG 11 (Sustainable Cities and Communities) and SDG 12 (Responsible Consumption and Production). Under SDG 11, blockchain enhances urban resilience by reducing pollution and optimizing municipal waste systems. The BlockWASTE Project exemplifies this by improving urban cleanliness and reducing waste-related emissions.

For SDG 12, blockchain ensures responsible consumption and production by fostering transparency in recycling and production processes. The Circulogy Framework's extended producer responsibility programs encourage manufacturers to reclaim and recycle electronic components, reducing resource extraction and promoting sustainable practices. These initiatives demonstrate blockchain's ability to integrate sustainability into production and consumption cycles effectively.

Policy Recommendations for Blockchain Integration

To maximize blockchain's benefits in urban waste management, policymakers must adopt supportive measures. Public-private partnerships can accelerate the development and scalability of blockchain solutions. Governments should collaborate with blockchain providers and waste management organizations to implement tailored systems that address local needs.

Financial incentives, such as subsidies or tax breaks, can offset the high initial costs associated with blockchain integration. These measures can encourage adoption among municipalities and private firms. Additionally, governments should establish standardized regulatory frameworks to ensure interoperability, data privacy, and ethical use of blockchain systems. Clear guidelines on tokenization, compliance monitoring, and environmental reporting will further enhance the efficiency and accountability of blockchain-enabled waste management systems.

POLICY RECOMMENDATIONS

To fully leverage blockchain technology in fostering a digitalized, circular economy framework for urban waste management, a comprehensive policy approach is required. The recommendations outlined below address the critical aspects of implementation, scalability, and sustainability, incorporating practical measures and real-world examples.

Promote Public-Private Partnerships

Public-private partnerships (PPPs) are crucial for enabling the development and deployment of blockchain solutions in waste management. Such collaborations combine the technological expertise of private entities with the operational capabilities of public waste management bodies.

Measures

1. **Local Government-Led Initiatives**: Municipalities should take the lead in creating blockchain-based waste management systems by partnering with technology providers and recycling companies. Examples include smart bin initiatives in Amsterdam, where IoT sensors integrated with blockchain have streamlined waste collection and segregation (Ellen MacArthur Foundation, 2022).

2. **Incentives for Collaboration**: Governments can provide tax benefits to private firms that develop and maintain blockchain systems for municipal waste management. This aligns with the strategies adopted in the BlockWASTE Project, where private companies were incentivized to create blockchain solutions that reduced inefficiencies in municipal operations (European Commission, 2021).

The Plastic Bank Initiative demonstrates the effectiveness of PPPs, with local authorities in Haiti and Indonesia collaborating with private organizations to establish blockchain-backed recycling programs. These partnerships have improved waste recovery rates while providing financial incentives to underserved communities (Alam et al., 2021).

Provide Financial Support and Incentives

High implementation costs remain a significant barrier to adopting blockchain technology in waste management. Governments must establish financial mechanisms to lower entry barriers for municipalities and private stakeholders.

Measures

1. **Grants and Subsidies**: Allocate public funds to subsidize the acquisition of IoT-enabled smart bins, blockchain networks, and data analytics tools.
2. **Public Token Rewards**: Expand token-based incentives for individuals and businesses actively participating in recycling programs. In Seoul, blockchain-enabled recycling initiatives have led to improved waste segregation through token systems redeemable for public services (Kim & Park, 2022).
3. **Carbon Credits**: Offer carbon credits to private firms and municipalities adopting blockchain systems that demonstrate measurable reductions in emissions.

The European Union's BlockWASTE Project allocated funding to pilot blockchain-enabled waste management systems across member states. These pilots demonstrated reduced operational costs and enhanced recycling efficiencies, showcasing the economic viability of blockchain systems (European Commission, 2021).

Develop Standardized Regulatory Frameworks

A standardized regulatory framework is essential to ensure ethical, efficient, and scalable implementation of blockchain technology in waste management.

Measures

1. **Tokenization Guidelines**: Define clear regulations on the issuance, valuation, and use of blockchain tokens for incentivizing waste recovery. This includes taxation and anti-fraud measures.
2. **Data Security Standards**: Establish privacy laws that govern data handling within blockchain systems, ensuring compliance with global frameworks such as GDPR.
3. **Certification Mechanisms**: Introduce certification schemes to validate blockchain applications for their adherence to environmental and operational standards.

Singapore has established robust blockchain regulations, which include guidelines for data security, smart contracts, and tokenization. These measures have enabled blockchain adoption in industries such as waste management and logistics, setting a benchmark for global regulatory practices (Singapore Government, 2021).

Foster Digital Literacy and Public Awareness

The adoption of blockchain technology in waste management relies heavily on the awareness and digital literacy of stakeholders, including citizens, municipal workers, and private entities.

Measures

1. **Community Engagement Programs**: Organize public campaigns to educate citizens about blockchain's role in promoting sustainable waste practices.
2. **Workforce Training**: Develop training modules for waste management professionals to enhance their understanding of blockchain systems and IoT devices.
3. **Educational Curricula**: Collaborate with academic institutions to introduce blockchain-focused courses in environmental science and engineering programs.

The Circulogy Framework incorporated education as a core component of its strategy. By conducting workshops and training sessions, stakeholders gained a comprehensive understanding of blockchain and AI integration in e-waste recycling (Masi et al., 2021).

Integrate Blockchain with Complementary Technologies

To maximize blockchain's potential, it should be integrated with complementary technologies such as IoT, AI, and machine learning. This integration can create a synergistic ecosystem that enhances efficiency and scalability.

Measures

1. **IoT-Blockchain Integration**: Equip waste bins and collection vehicles with IoT sensors to provide real-time data on waste levels, enabling efficient collection scheduling.
2. **AI for Waste Sorting**: Use AI systems to analyze blockchain data for optimizing waste segregation and identifying high-value materials for recovery.
3. **Automation with Smart Contracts**: Implement smart contracts to automate key operations such as incentive distribution, compliance monitoring, and waste processing fees.

Dubai's Smart Waste Management System integrates IoT sensors with blockchain networks, allowing for real-time monitoring of waste streams and optimized collection routes. This has significantly reduced operational inefficiencies and emissions (Ahmed et al., 2021).

Encourage Blockchain Adoption in Extended Producer Responsibility (EPR) Programs

Blockchain technology can enhance EPR programs by enabling traceability and compliance monitoring, ensuring manufacturers are accountable for the lifecycle of their products.

Measures

1. **Mandatory EPR Requirements**: Introduce laws requiring manufacturers to implement blockchain systems for tracking product lifecycles, ensuring effective recycling and reuse.
2. **Incentivize Compliance**: Offer tax benefits or public recognition to companies that exceed EPR targets by using blockchain for efficient waste recovery.
3. **Collaborative Platforms**: Develop blockchain-based platforms that connect manufacturers, recyclers, and consumers to streamline material recovery.

The Circulogy Framework collaborated with electronics manufacturers to integrate blockchain into EPR programs. This partnership enabled efficient tracking of electronic components from production to recycling, aligning with circular economy goals (Alam et al., 2021).

FUTURE DIRECTIONS

The integration of blockchain technology into urban waste management offers vast potential for future advancements. To ensure scalability, research should focus on developing blockchain models capable of managing the complexity and volume of urban waste systems. Interoperability between platforms must also be addressed to facilitate seamless cross-border waste management, particularly in handling international e-waste flows.

The integration of blockchain with advanced technologies like IoT and AI presents a promising direction. Future studies should explore how predictive analytics and automated decision-making powered by AI can complement blockchain's capabilities in transparency and traceability. Real-world pilots combining these technologies will be critical to understanding their collective impact on efficiency and sustainability.

Policy and regulatory development remains essential for broader adoption. Comparative research across jurisdictions can provide insights into best practices for harmonizing data privacy, smart contract enforceability, and tokenization policies. Additionally, quantitative assessments of blockchain's environmental impact—such as reductions in carbon footprints and improvements in recycling rates—will strengthen its sustainability case.

Efforts to enhance public engagement should focus on optimizing token-based incentive systems to drive long-term behavioral change. Finally, adapting blockchain solutions for developing regions with limited infrastructure is crucial for inclusive adoption. By addressing these areas, blockchain can fully realize its potential in advancing sustainable waste management.

CONCLUSION

This research demonstrates the transformative potential of blockchain technology in advancing circular economy practices within urban waste management systems. By addressing inefficiencies in traditional waste management, such as lack of transparency, low recycling rates, and operational inefficiencies, blockchain enables traceability, accountability, and incentivized participation. Case studies like the Plastic Bank Initiative and the BlockWASTE Project highlight its capability to optimize waste processes, reduce carbon footprints, and generate significant economic benefits.

Despite its potential, blockchain implementation faces notable challenges, including regulatory gaps, high costs, and stakeholder resistance. Issues such as data privacy compliance, tokenization ambiguities, and the lack of standardized environmental certifications hinder its widespread adoption. Addressing these barriers requires harmonized regulatory frameworks, financial incentives, and public-private partnerships.

To fully realize blockchain's benefits, integration with complementary technologies like IoT and AI is essential, alongside fostering digital literacy and public awareness. Policymakers must prioritize blockchain's incorporation into waste management policies to align with global sustainability goals, particularly SDG 11 and SDG 12. By overcoming challenges and leveraging blockchain's capabilities, urban waste management can transition toward a more sustainable, efficient, and inclusive future.

This transition is crucial for addressing the growing global waste crisis and fostering long-term environmental and economic resilience.

REFERENCES

Acosta Llano, E., Hurmelinna-Laukkanen, P., & Haapanen, L. (2024). Blockchain for the circular economy, implications for public governance. *International Journal of Public Sector Management.*

Ahmed, A.. (2021). Blockchain applications in waste management: A smart city perspective. *Sustainable Cities and Society*, *74*, 103164.

Al-Raeei, M. (2024). Artificial intelligence for climate resilience: Advancing sustainable goals in SDGs 11 and 13 and its relationship to pandemics. *Discover Sustainability*, *5*(1), 513. DOI: 10.1007/s43621-024-00775-5

Alam, S.. (2021). Blockchain-based circular economy frameworks: Applications and challenges. *Renewable & Sustainable Energy Reviews*, *138*, 110512.

Bai, C., & Sarkis, J. (2022). Blockchain technology for enhancing sustainable supply chain management: Framework and future research directions. *Resources, Conservation and Recycling*, *177*, 105973.

Bhubalan, K., Tamothran, A. M., Kee, S. H., Foong, S. Y., Lam, S. S., Ganeson, K., Vigneswari, S., Amirul, A.-A., & Ramakrishna, S. (2022). Leveraging blockchain concepts as watermarkers of plastics for sustainable waste management in progressing circular economy. *Environmental Research*, *213*, 113631. DOI: 10.1016/j.envres.2022.113631 PMID: 35714685

Bocken, N. M. P.. (2019). Circular economy strategies for durable goods: Closing resource loops in a value-retentive product lifecycle. *Business Strategy and the Environment*, *28*(3), 429–444.

Bourguignon, D.. (2019). Transparency in waste management systems: A critical review. *Journal of Environmental Policy and Planning*, *21*(6), 739–754.

Chauhan, M., & Sahoo, D. R. (2024). Towards a greener tomorrow: Exploring the potential of AI, blockchain, and IoT in sustainable development. *Nature Environment and Pollution Technology*, *23*(2), 1105–1113. DOI: 10.46488/NEPT.2024.v23i02.044

Chowdhury, M. T. A.. (2021). Challenges in waste management and the role of technological interventions. *Waste Management & Research*, *39*(4), 451–466.

de Man, R., & Friege, H. (2020). Circular economy: European policy on shifting towards sustainable resource use. *Resources, Conservation and Recycling*, *155*, 104659. DOI: 10.1016/j.resconrec.2019.104659 de Sousa Dutra, R. M., & Ribeiro Siman, R. (2024). STRATEGIES FOR FINANCIAL SUSTAINABILITY OF MUNICIPAL SOLID WASTE MANAGEMENT SYSTEMS. Environmental & Social Management Journal/Revista de Gestão Social e Ambiental, 18(4).

Ellen MacArthur Foundation. (2021). *What is the circular economy?* Retrieved from https://ellenmacarthurfoundation.org

Ellen MacArthur Foundation. (2022). Circular economy and blockchain: Unlocking sustainability. https://ellenmacarthurfoundation.org

European Commission. (2021). Blockchain applications for municipal waste management: Pilot project outcomes. *EU Policy Briefs*. https://ec.europa.eu/environment/blockchain

Gautam, M., & Agrawal, M. (2021). Greenhouse gas emissions from municipal solid waste management: a review of global scenario. Carbon footprint case studies: municipal solid waste management, sustainable road transport and carbon sequestration, 123-160.

Geissdoerfer, M., Savaget, P., Bocken, N. M. P., & Hultink, E. J. (2020). The circular economy – A new sustainability paradigm? *Journal of Cleaner Production, 143*, 757–768. DOI: 10.1016/j.jclepro.2016.12.048

Ghisellini, P.. (2019). Circular economy: Opportunities and challenges. *Journal of Cleaner Production*, *230*, 763–774.

Gregson, N.. (2020). Interrogating the circular economy: The moral economy of resource efficiency in the EU. *Economy and Society*, *49*(1), 1–24. DOI: 10.1080/03085147.2012.661625

Gupta, S.. (2020). Tackling inefficiencies in urban waste management: A review. *Renewable & Sustainable Energy Reviews*, *130*, 109922.

Gupta, S.. (2021). Blockchain for sustainable resource management: Framework and case studies. *Resources, Conservation and Recycling*, *178*, 106047.

Hasan, M. M., Plamthottathil, R. K., Morshed, J., Sarkar, D., Hameed, N., & Cirstea, S. (2023, May). Circulogy: An AI-Enabled Blockchain-Based e-Waste Management Framework Using Non-Fungible

IPCC. (2021). *Sixth Assessment Report: Climate change mitigation*. Retrieved from https://www.ipcc.ch

Katz, D. (2019). Plastic bank: launching social plastic® revolution. *Field Actions Science Reports. The Journal of Field Actions*, (Special Issue 19), 96-99.

Kim, H., & Park, Y. (2022). Incentivizing recycling through blockchain: A case study of Seoul. *Journal of Cleaner Production, 342*, 130976.

Kirchherr, J., Reike, D., & Hekkert, M. (2017). Conceptualizing the circular economy: An analysis of 114 definitions. *Resources, Conservation and Recycling, 127*, 221–232. DOI: 10.1016/j.resconrec.2017.09.005

Lenz, R., Kleinheyer, B., Barkel, C., Veuger, J., Tsangaratos, P., Klõga, M., & Llorente, J. (2021). Innovative Training Based on Blockchain Technology Applied to Waste Management–BLOCKWASTE. Available at *SSRN* 3912646.

Masi, D.. (2021). Blockchain technology in waste management: Opportunities and challenges. *Journal of Cleaner Production*, *280*, 124164.

Nobre, G. F., & Tavares, E. (2021). Blockchain for circular economy: Potential applications in sustainable waste management. *Sustainable Production and Consumption*, *26*, 305–316.

Purnell, P. (2022). Circular economy in construction: Current awareness, barriers and enablers. *Resources, Conservation and Recycling*, *178*, 106038.

Rashid, S., & Malik, S. H. (2023). Transition from a Linear to a Circular Economy. In *Renewable Energy in Circular Economy* (pp. 1–20). Springer International Publishing. DOI: 10.1007/978-3-031-42220-1_1

Saberi, S., Kouhizadeh, M., Sarkis, J., & Shen, L. (2019). Blockchain technology and its relationships to sustainable supply chain management. *International Journal of Production Research*, *57*(7), 2117–2135. DOI: 10.1080/00207543.2018.1533261

Singapore Government. (2021). Blockchain regulations for sustainable industries. *Smart Nation Strategy Reports*. https://smartnation.sg

Singh, A. K., & Kumar, V. P. (2024). Establishing the relationship between the strategic factors influencing blockchain technology deployment for achieving SDG and ESG objectives during infrastructure development: An ISM-MICMAC approach. *Smart and Sustainable Built Environment*, *13*(3), 711–736. DOI: 10.1108/SASBE-12-2023-0405

Stahel, W. R. (2020). *Circular economy: A user's guide*. Routledge.

Szewczyk, P. (2022). BLOCKCHAIN BASED WASTE MANAGEMENT. Scientific Papers of Silesian University of Technology. Organization & Management/Zeszyty Naukowe Politechniki Slaskiej. Seria Organizacji i Zarzadzanie, (163).

Tapscott, D., & Tapscott, A. (2019). *Blockchain revolution: How the technology behind bitcoin and other cryptocurrencies is changing the world*. Penguin Random House.

Tokens (NFT) to Achieve Net Zero and Imply the Circular Economy. In 2023 *IEEE International Conference on Blockchain and Cryptocurrency (ICBC)* (pp. 1-3). IEEE.

UNEP. (2023). Waste management and climate adaptation: Challenges and strategies for urban resilience. *United Nations Environment Programme*. Retrieved from https://www.unep.org/

World Bank. (2022). What a waste 2.0: Global snapshots of solid waste management to 2050. Retrieved from https://www.worldbank.org/

KEY TERMS AND DEFINITIONS

Blockchain in Waste Management: The application of blockchain technology has the power to improve transparency, traceability, and efficiency in waste management processes and ensure secure and verifiable transactions across the waste value chain.

Circular Economy: This is an economic model that focuses on minimizing waste and resource depletion by promoting reuse, recycling, and sustainable production to create a closed-loop system.

Climate Change Mitigation: Strategies and actions, which aim at reducing greenhouse gas emissions and carbon footprints to limit global temperature rise and its adverse impacts.

Sustainable Development Goals (SDGs): A set of 17 global objectives established by the United Nations to address economic, social, and environmental challenges and ensure a sustainable and equitable future for all by 2030.

Digitalization and Sustainability: The integration of digital technologies to optimize resource use, reduce environmental impact, and promote sustainable development through smart, data-driven solutions.

Chapter 12

Interaction Among Climate Change, Digitalisation, and Sustainable Urban Development

Deepak Gautam
https://orcid.org/0000-0002-6379-5539
School of Business, UPES Dehradun, India

Vipul Gupta
https://orcid.org/0000-0002-1920-7314
IILM University, Greater Noida, India

ABSTRACT

This chapter focuses on the juncture at which climate change, digitalization, and sustainable urban development converge, with an emphasis on the role of cities in climate change mitigation. On the one hand, urban areas are a great contributor to global energy consumption and greenhouse gas emissions; on the other hand, they also pose opportunities. The chapter explores how digital technologies such as AI, IoT, blockchain, and digital twins are now changing the profile of cities to make them efficient, resilient, and eco-friendly. Cases from Indian cities—Bengaluru, Pune, and Delhi—demonstrate the impact of digital tools on optimizing energy management, resource allocation, and environmental stewardship. It also discusses challenges such as digital divides and privacy concerns, requiring strong policies to ensure that technology access is equitable. The chapter concludes by calling for integrating digitalization with sustainable urban development in order to create economically,

DOI: 10.4018/979-8-3693-9909-5.ch012

socially, and environmentally sustainable cities.

1. INTRODUCTION

One of the defining trends of the 21st century has been urbanization, where cities are becoming central to economic, social, and cultural activities. This shift is particularly visible in developing economies, such as India, where a rapid pace of urbanization comes with challenges as well as opportunities. The urbanization of the world at a hitherto unparalleled rate has positioned cities in centre of the global landscape. However, this urban growth proliferation comes with its environmental impact since cities consume about 70% of global energy and a similar percentage of greenhouse gas emissions (Sharma & Lata Pathak, 2024). Therefore, urbanization plays a focal role in the climate change story as areas contributing to the degradation of the natural environment, as well as major contributors to climate change mitigation efforts. Thus, the convergence of urbanization, climate change, and sustainability creates a complex but crucial challenge for modern cities, especially in developing economies such as India (Bhattarai & Conway, 2021).

With the rising challenge, digitalization has come out to be one of the major facilitators for sustainable urban development (Wang et al., 2023). Artificial Intelligence (AI), IoT, blockchain, and the system of a digital twin bring in new ways to completely alter the city functions (Tyagi et al., 2024). All of these could work to cut down the amount of energy in cities, fortify the infrastructure, and create data-based decisions in cities towards an efficient, resilient, and friendly environment. However, all this is not without obstacles that it has in its way of transformation. Growing impacts of climate change with rapid urban growth in developing countries, particularly India, necessitate integrating digital solutions and mitigate these effects. Building upon this background, this chapter sets out to assess the convergence of three major themes: climate change, digitalization, and sustainable urban development to surface how digital tools empower cities to achieve the twin goals of reducing their environmental footprint while increasing economic and social sustainability.

Cities are the greatest contributors to climate change in the world because they consume much energy and produce high emissions. Researches in recent times indicate that cities are responsible for approximately 75% of carbon emissions globally (Hoornweg et al., 2011). This is largely because industries, transport systems, and residential areas are highly concentrated in cities, which, in turn, have a heavy environmental footprint. As urbanization accelerates-specially in the developing world-this bodes little good for city environmental performance without decisive interventions.

However, cities also give a better opportunity to confront the challenge of climate change. Infrastructure density, availability of resources, and economic concentrations all provide an innovative platform for urban areas. Cities can be treated as testing grounds for new technologies and solutions that could be scaled up and replicated if they succeed. It is a powerful driver for change, not only because the consumers, businesses, and government institutions are very highly concentrated in cities, but also cities are put in a double position as being both a significant source of environmental degradation and a possible location for solutions (Chkoniya et al., 2020; Yildirim, 2020).

The move toward the transformation of cities into more sustainable and resilient urban spaces has increasingly been facilitated by the integration of digital technologies in this regard (Brunetta & Caldarice, 2020). Such technologies range from AI, IoT, blockchain, and digital twins, but to give an example, AI would be able to optimize energy consumption in urban buildings and infrastructure. Leverage AI systems, and cities optimize heating, cooling, and lighting in buildings with reduced energy use and contributions toward sustainability goals, according to Farzaneh et al. (H. Farzaneh et al., 2021). In that regard, one must not forget that AI-based solutions also entail their own energy demands, including computing power used for machine learning algorithms. Therefore, there should be a development of energy-efficient AI solutions suitable to the urban environment with their advantage and above all having greater value to environmental cost (Patti & Trizzino, 2020).

Another area of digitalization is IoT-based smart grids, driving urban sustainability. Smart grids make it possible to monitor energy usage in real-time in cities and better allocate resources while integrating with renewable energy sources. IoT sensors can identify inefficiencies in energy consumption and support decentralized energy production through renewable sources such as solar and wind (Brunetta & Caldarice, 2020). This technology contributes directly to reducing urban environmental impact through its improvement to the resilience and dependability in urban infrastructure delivery through an easily adaptable energy source.

Urban planning and management concepts are the talks of the digital world. Here, a digital twin could refer to a virtual form of a physical system that really exists-what is considered urban planning may simply be modelling all the options put forth through urban planners with its design into numerous scenarios to try out on them before it would be allowed out in the physical world. Blockchain places another layer over those digital solutions by making the transactions decentralized and safe. Amongst the most promising applications of blockchain in urban sustainability is carbon trading. Blockchain allows for secure, transparent transactions which can help cities to trade carbon credits with businesses and residents and motivate lower-carbon lifestyles (Zhu et al., 2024).

In India, rapid urbanization comes along with some of the great challenges and offers tremendous opportunities in digital solutions. The country is experiencing one of the fastest rates of urban growth worldwide, and cities will have to accommodate an urban population of 600 million by 2030 (Monstadt & Schmidt, 2019a). On the one hand, urbanization in India boosts economic development and living standards, but on the other hand, it poses significant pressure on infrastructure, resources, and the environment. Major cities like Bengaluru, Pune, and Delhi are facing issues such as poor air quality, energy shortages, inadequate waste management systems, and inefficient transportation networks. The effects of climate change, including rising temperatures, extreme weather events, and water scarcity, are exacerbating these challenges, making it imperative for India's cities to adopt sustainable urban design and digital interventions.

Despite these challenges, India is uniquely positioned to leverage the opportunities presented by digitalization. In such multiple ways, many Indian cities have already started projects where they combine digital technologies with their urban management. A robust policy framework is, therefore essential in harnessing the use of digital technologies by contributing to the sustainable development of urban systems. The national and local governmental levels play a strategic role, especially in creating improved regulatory environments that foster innovation and improved equitable access to digital tools. The major barrier to scaling up digital solutions is the digital divide. Digital technologies are often unevenly accessible in many cities, and the lack of resources or skills by some groups leaves them unable to reap the full benefits of these innovations. Policymakers must fill this gap by using digital inclusion to ensure the widespread distribution of benefits from digitalization.

Another critical issue associated with smart cities and digital urban sustainability is privacy. The intensive use of sensors, cameras, and other data-gathering technologies has raised critical questions about citizen security and privacy. Clear guidelines and regulations must be put in place to protect citizens' rights while enabling cities to fully harness the potential of digital tools for urban management.

International frameworks, such as the United Nations' Sustainable Development Goals (SDGs), provide important guidance for integrating digitalization into urban sustainability. Goal 11 SDG ensures inclusive, safe, resilient, and sustainable urbanization and partnerships for achieving sustainable cities and communities, not in isolation but through cross-sectoral collaboration between governments, private enterprises, and civil society.

The research chapter examines convergence between the three critical themes, namely, climate change, digitalization, and sustainable urban development, as applicable to urban sustainability. The main aim is to study how digital technologies, such as AI, IoT, blockchain, and digital twins, can be applied to mitigate environmental impacts from urbanization, increase energy efficiency, and foster social and economic sustainability in cities. Additionally, the chapter seeks to evaluate the challenges and opportunities associated with integrating digital solutions into urban development, with a particular focus on India's rapidly urbanizing cities.

Although each theme—climate change, digitalization, and urban development—has been individually explored in academic literature, there exists a limited body of integrated work exploring their convergence. This chapter seeks to bridge this gap by reviewing multidisciplinary studies and emphasizing the interplay between these domains, especially in the context of rapidly urbanizing regions such as India. This chapter is particularly tailored for researchers, urban planners, policymakers, and practitioners seeking evidence-based strategies to integrate digitalization with climate-responsive urban development.

This chapter follows a structure to explore the nexus climate change, digitalization, and sustainable urban development. Based on this introduction, the literature review is written in the roles of cities in climate change, potential digital technologies for sustaining urbanization, and the plight of Indian cities. The subsequent sections include methodology adoption in this chapter, case studies and the practical uses of digital technologies in urban management. The next section then explains the policy framework needed to introduce digital technologies to achieve sustainable cities. The last section concludes the chapter with future research directions.

2. LITERATURE REVIEW

This section elaborates the role of cities in climate change causes and mitigates global emissions. The section also analyses the digital technologies like AI, IoT, blockchain, and digital twins that shape sustainable urbanization. Further, the section also discusses the problems brought about by urbanization in India and possible digital solutions to those problems. Finally, the section review the gaps in existing research regarding long-term impacts and socio-economic implications of digital technologies in urban sustainability.

2.1. Cities in Climate Change

Cities are both causes and antidotes of climate change. They discharge about 75% of carbon in the whole world (Kelbaugh, 2019). This can be attributed to a number of activities that have been taking place in the cities, such as industrial activities, vast transportation systems, and a densely populated space, which are linked to consuming more energy as well as greenhouse gases emission. This, together with high rates of urbanization in developing countries, further complicates these challenges with immense pressure on infrastructure, natural resources, and public services (Seto et al., 2017). Rapid growth is more likely than sustainable management to be ahead, which raises the vulnerability of this type of setting to climate impacts.

Urban areas still offer enormous opportunities for climate action amidst these challenges. Economic activities, technological innovations, and human capital are concentrated in cities, which positions them as critical arenas for implementing climate mitigation and adaptation strategies. Cities can harness the high-density networks to use energy-efficient technologies, enhance public transportation, and adopt sustainable land use. For instance, green building standards and urban greening projects have been a promise, which can lower the carbon footprint of cities (Rosenzweig, 2018).

Cities can be piloting grounds for innovative climate solutions. Cities, provided they have the right policy structures and investment in sustainable infrastructure, can lead the way in reducing their ecological footprint. Studies have shown that cities that invest in renewable energy sources and energy efficiency as well as integrate green technologies into urban designs reduce carbon emissions (Umoh et al., 2024). This capacity of cities in making better use of energy, as well as in reducing greenhouse gas emissions, is further enhanced through smart city technologies such as intelligent transportation systems and smart grids (Gouldson et al., 2015).

2.2. Digitalization as a Driver for Sustainable Urbanization

This, therefore, presents digital technologies as part of the transition towards sustainable urban development in providing innovative solutions for optimizing resource use, enhancing infrastructure resilience, and reducing environmental impacts. AI is probably the most influential of these players because it is changing the face of urban management, like optimizing energy consumption in buildings and infrastructure. For example, AI can automatically control heating, cooling, and lighting systems within buildings to conserve energy and maintain comfort levels. According to (G. Farzaneh et al., 2021), AI-driven systems analyze large datasets, identify inefficiencies in urban infrastructure, and propose improvements. On the other hand, even though AI can offer many advantages in energy management, its computation requirements can still increase energy usage, and it is necessary to develop energy-aware AI solutions that can reduce such energy consumption.

Similarly, another key enabler for real-time monitoring and management of urban resources - energy, water, waste - is IoT. Smart grids, with the aid of IoT sensors, enable efficient use of these resources and incorporation with renewable sources, thus serving as a component toward more sustainable urban spaces. These systems are instrumental in detecting inefficiencies while monitoring typical patterns of consumption and decentralised power production, which represent major streams towards sustainability (Ukoba et al., 2024).

Digital Twin Systems enhance the sustainability of the urban environment with virtual replicas of physical urban infrastructure. These models enable the planners to simulate scenarios, optimize operations, and even identify problems in advance of actual implementation. In the transport, construction, and energy management industries, for example, digital twins are most beneficial since they help to make decisions based on data while ensuring that the efficiency of the system is enhanced while minimizing risks through operations.

The last is Blockchain Technology, which promises to be transformative for urban sustainability. It provides the needed transparency and decentralization in the management of the city, especially on issues related to renewable energy and carbon credits trading. This means secure and verifiable transactions through blockchain technology, supporting decentralized exchanges in energy exchanges that can help in efficiency and transparency of renewable energy management. According to (Lukić, 2024), such processes can only be possible with the implementation of blockchain and other similar technologies, which would lead to more sustainable and transparent urban environments. Altogether, such digital technologies help cities gain great tools in combating environmental impacts and enhancing their resilience in a journey toward sustainability.

2.3. India's Urbanization Challenge and Digital Solutions

India is rapidly urbanizing, and by 2030, the urban population is going to be 600 million, unprecedented in the history of this country in terms of pressure on infrastructure, resources, and public services (Franco et al., 2017). Cities like Delhi, Mumbai, and Kolkata are already facing critical challenges, including severe air pollution, chronic energy shortages, inefficient waste management, and overwhelmed transportation systems (Monstadt & Schmidt, 2019b) However, there are socio-economic disparities that multiply these issues to make equitable development in cities the need of the hour.

But along with the issues, there lie major opportunities for leveraging digital technologies as part of Indian urbanization processes for sustainable development. Digital innovations are increasingly applied to solve various urban challenges for better quality in the urban environment. For example, the city of Bengaluru, dubbed India's Silicon Valley, has successfully implemented AI and IoT in streamlining traffic, thus reducing the congestion and exhaust emissions of vehicles (Dutta et al., 2024). It will adapt traffic light operation, thereby having a more effective system for mobility in cities due to the ability of real-time traffic monitoring using these technologies.

Pune has become a pioneer in using blockchain technology for waste management. The blockchain-based system ensures greater transparency and accountability in the collection and disposal of waste, thus efficiently using resources and minimizing environmental impacts (Singh et al., 2024). This approach not only improved the segregation of waste but also empowered local communities through the involvement of local people in sustainable waste management practices.

Besides, digital technologies are now applied in other fields to solve city problems. IoT-based smart water management systems that monitor the city's water supply networks, detect leaks, and optimize the utilization of water have also been applied to overcome water scarcity and increase the efficiency of using resources in Surat, according to (Kumar et al., 2024) More to that, the smart grid technologies are implemented in different cities of India to enhance energy distribution as well as incorporate renewable energy sources that reduce carbon emissions and promote energy security (Mohan et al., 2021)

The Indian government's Smart Cities Mission also proposes to use the digital technologies towards urban transformation. The mission concentrates on building smart and sustainable cities through the assimilation of digital infrastructure, e-governance, and citizen participation (Praharaj et al., 2018). Such an initiative has generated digital dashboards, mobile applications, and e-services that aid in better planning and management in the urban realm, thereby enriching the quality of the overall urban experience.

2.4. Gaps in the Current Literature

The research literature on the topic of digitalization and urban sustainability is growing. However, these areas are those that need to be targeted further for advancement of this field:

- Long-term effects of digital technologies: The available literature regarding the long-term effects of digital technologies on the sustainability of cities is limited. Such short-term or merely singular impacts form general research mostly about single outcomes or impacts through isolated technologies. More integrated research thus seems to be in dire demand concerning the over time-continuously building overall impacts of digitization on towns and cities in urban areas.
- Socio-economic Implications: There is limited research in terms of the socio-economic implications of digital technologies in cities; issues include digital divide, a concentration on privacy concerns and security. Therefore, further research is required concerning how digitalization affects different socio-

economic groups so that such benefits are equitably distributed among different urban populations.
- Context-specific research: Most of the existing literature related to sustainable urbanization and digitalization is concerning developed countries. There is an evident need to do more specific research on issues of rapidly growing cities in developing countries, and particularly in India. Tailored research is thus needed to answer the specific socio-economic, cultural, and environmental conditions of these regions.

Addressing these gaps will support creating a more detailed understanding of the role of digital technologies in fulfilling the agenda on sustainable urbanization, which might guide further research, policy-making, and practices.

3. METHODOLOGY

This research aims to present a method of providing insights into the mutual interaction between climate change, digitalization, and sustainable urban development. Therefore, qualitative approach was used in an attempt to go deeper in the discovery of dynamics in sustainability, especially within an urban setting-taking into account the impact of digital technologies on managing in urban settings. The research method combines case study analysis, secondary data, and thematic data analysis in an attempt to present a multi-dimensional discussion of the digitalization contribution to addressing urban sustainability challenges. This research focus is crafted specifically to include the interrelated dynamics of environmental stress, technological transformation, and urban planning priority, with a comprehensive approach from which to distinguish the transformative potential of digital solutions for sustainable urban development. The research approach, case study approach, data collection, and data analysis techniques applied in the study are as follows:

3.1. Research Approach

This chapter is viewed from a qualitative perspective with regards to the deep relationship between climatic change, digitalization, and sustainable development in urban society. A qualitative approach was resorted to simply because it reflects the deep, intricate exploration of topics interlinked for the study about nuances and distinct factors specific only to a specific context that do not come about through quantitative measures. The qualitative approach is very handy to understand this urban sustainability dynamic against the background of fast-changing digital technology and pressures on the environment since it presents richly detailed, context-rich information regarding the ways digital technology can advance sustainable urban ways of life. Qualitative methods are most ideally used to map the underlying determinants and context conditions that make adoption and contributions from digital technology push urban sustainability initiatives. By emphasizing the lived experience and perspective of stakeholders—e.g., urban planners, policymakers, and citizens—the approach enables a nuanced consideration of the socio-technical forces at play that is necessary in order to address the complexities of urban sustainability effectively. Furthermore, the method is robust in its ability for exploring subjective meanings and differences in contexts, such that it is able to establish a robust foundation for exploring how digitalization interacts with climate change and sustainability goals in urban settings. Qualitative research also enables it to create hypotheses and theoretical observations, which are very vital for constructing knowledge in this new interdisciplinary realm.

3.2. Case Study Method

This research employs the case study method as a fundamental element in detailing a specific instance of digital technologies that have been incorporated into urban management address sustainability challenges. The use of case studies of Indian cities, such as Bengaluru and Pune, would illustrate the real application of digital solutions in an urban scenario. Case studies offer insight into the practical implementation of such technologies as AI, IoT, and blockchain, stating a clear proposition for their applicability toward the promotion of sustainable practices. The study evaluates these examples as it discusses outcomes and lessons in the integration of digital solutions with urban management systems and challenges which need to be overcome to realize more extensive implementations. The selection of Bengaluru and Pune as focal points is deliberate, reflecting their prominence as leading centers of technological innovation and urban development within India. These cities serve as exemplary models of the opportunities and obstacles encountered by rapidly expanding urban areas in the Global South, making them ideal for studying the deployment of digital solutions like smart grids, intelligent transportation systems, and data-driven governance frameworks. Furthermore, the case study approach provides a multifaceted, place-specific analysis that foregrounds how technologies such as AI-based analytics, IoT-enabled infrastructure, and blockchain-managed resources can be leveraged for sustainability initiatives. The methodology also enables detailed probing of local drivers—such as governance mechanisms, economic factors, and ecological considerations—that influence the success of digital interventions, with implications to be transferred to other urban environments globally.

3.3. Data Collection

The sources of data for this study are manifold and drawn from secondary sources. These are academic journals, government reports, industry publications, and international frameworks such as United Nations SDGs. Secondary data usage is allowing the understanding to be developed over the prevailing state of digitalization and urban sustainability. The study provides a balanced approach as it takes perspectives and experiences from both the Indian and the global context to create a diversity of viewpoints. This would further provide a global insight in understanding challenges related to this issue of urban sustainability and also opportunities that could be provided through digital technologies towards mitigating it on a large scale.In order to guarantee a strong and diversified dataset, the research employs a systematic review approach, meticulously choosing sources that represent several disciplines, such as urban planning, environmental science, information technology, and public policy. By balancing contradictory analyses and approaches, which are fundamental to the whole interpretation of the function of digitalisation in urban sustainability, multidisciplinary analysis increases the field of inquiry to be more comprehensive. In particular, academic journals provide peer-reviewed research and theoretical analysis, government reports provide statistical information and policy recommendations, trade publications provide real-world examples of technological innovations, and international frameworks such as the UN SDGs provide world targets and goals that situate the research in the context of global sustainability agendas. This selective choice of sources ensures that the study combines the localised knowledge of India with the universalities elsewhere, therefore enabling more generalisable and transferable research across location and context at more than one scale.

3.4. Data Analysis

This study makes thematic analysis on the data. It identifies key themes and patterns related to the role of digital technologies in urban sustainability and synthesizes them. In this analysis, thorough literature and case studies reviews are focused on the integration of digital technologies into urban management and their implications for sustainability.

The thematic review process is also intended to find the common and distinguishable patterns from the case studies and elucidate how digital technologies are reframing sustainable urban development. Additionally, the study determines the strengths or weaknesses of prevailing policy frameworks on supporting the acceptance and deployment of digital solutions to the management of urban areas.

4. CASE STUDIES

The present section contains illustrative case studies about how digital technologies are helping Indian cities surmount the challenge of urbanization. Here are examples of new and innovative traffic management, waste management, and optimized water resource strategies, which hold game-changing possibilities to boost sustainable urbanization.

4.1. Bengaluru: AI and IoT in Traffic Management

Bengaluru is widely referred to as the Silicon Valley of India, and it is the first among other Indian cities in taking on digital solutions for its problems of urbanisation (Kalra, 2007). It is taking AI and IoT solutions in curtailing intense traffic congestion in that city to an extent, thereby affecting people's quality of life and its economic productivity within that region (Akhtar & Kuriakose, 2021). Traffic management systems take into account real-time data received by the sensors set up via IoT at key crossroads and highways. It examines the patterns and predicts points for possible congestion. Furthermore, traffic light signals are so altered for reduction in delay in the passage that smooth flow continues (Seyed Hydroos, 2023).

These technologies have delivered measurable improvements; the traffic congestion in the key areas has been reduced compared to the past, bringing down traveling time and reducing vehicle emissions. This will not only improve the mobility in towns but also fulfill the environmental sustainability goals of the city. The Bengaluru model shows the complete potential of AI and IoT as they dramatically alter urban transportation systems, making them efficient and environmentally friendly.

4.2. Pune: Blockchain-based Waste Management

Pune has emerged as a pioneer in using blockchain technology for waste management (Singh et al., 2024). This left inefficiencies in the city's waste collection, without adequate transparency or much accountability, in Pune's system of waste management. The city used a blockchain-based solution that tracked the movement of waste from collection to disposal (Chatterji & Vaidya, 2025).

The blockchain system records each step that is involved in the whole process of waste management. This gives transparency because information relating to real-time happenings concerning waste management activities can quite easily be accessed by waste collectors, the municipal authorities, and the citizens. When citizens segregate and recycle appropriately, they are rewarded with digital tokens.

The advent of blockchain technology brings tremendous efficiency and transparency to the management of wastes. Greater engagement with waste segregation and recycling enhances sustainability targets in this city. This city has also been a torchbearer for the other cities desirous to implement better digital management systems for their waste collection purposes.

4.3. Delhi: Digital Twin for Urban Planning

Delhi, the capital of India, is a dense city with rapidly urbanizing problems. To achieve better urban planning and infrastructure development in Delhi, it has adopted the digital twin technology (Agrawal et al., 2022). This will create a virtual replica of the city, allowing planners to simulate a variety of different scenarios and to assess the impacts of proposed projects before they take place.

The digital twin model collects data from all sources, such as IoT sensors, satellite imagery, and historical records. This cumulative dataset allows planners to try out different strategies for urban development, optimize resource allocation, and identify possible risks. For instance, the digital twin can simulate how a new transportation project will impact the traffic congestion rate, air quality, and energy consumption.

Delhi's use of digital twin technology has enhanced the city's urban planning capabilities, leading to more informed decision-making and efficient resource use. The technology has also improved stakeholder engagement, as citizens can visualize and provide feedback on proposed projects through interactive digital platforms. Delhi's experience highlights the potential of digital twins in creating smarter, more sustainable cities.

4.4. Jaipur: Smart Water Management

Jaipur has utilized smart water management solutions to effectively address its challenges of water scarcity (Bedi et al., 2019). The city uses IoT sensors to monitor water distribution networks in real-time. These sensors monitor leaks and ensure optimal use of water by allowing the timely identification of leaks and providing critical data related to water flow, pressure, and quality, thus preventing any loss of water while ensuring efficient supply.

Integration of smart water management systems in the city has reduced water wastage and improved the reliability of supply in Jaipur. The city has also engaged people in real-time digital updates regarding water usage and conservation tips. In addition to efficiency in the use of resources, such an initiative fosters community-level involvement in sustainable water management.

4.5. Hyderabad: Renewable Energy Integration

Hyderabad has focused on integrating renewable energy sources into its urban energy grid (Govindarajan & L.S., 2021). The city has invested in IoT-enabled smart grids, which facilitate easy integration of solar and wind energy. Such smart grids work in real time by balancing energy supply and demand through data in an effort to optimally use renewable resources.

The initiatives on renewable energy in Hyderabad have led the city to achieve a significant amount of dependence on clean energy sources, which not only reduces the carbon footprint but also increases the resilience of the energy system. The approach followed by the city is being referred to as an example for the other urban spaces to transition toward sustainable energy systems. Hyderabad built a more resilient and environmentally friendly energy infrastructure by using IoT technology.

To provide a brief but comprehensive summary of the case studies at hand, Table 1 summarizes the digital technologies employed, the specific urban issues addressed, the outcomes achieved, and the unique features of each program. This table reflects the diversity of uses of digital solutions towards Indian city sustainability and provides a comparative analysis that highlights the similarities as well as the specific strategies in each case.

Table 1. Summary of digital technologies and their impact on urban sustainability in indian cities

City	Technology	Urban Challenge	Outcomes	Unique Features
Bengaluru	AI and IoT	Traffic congestion	Reduced travel time, lower vehicle emissions	Real-time data analysis for traffic prediction
Pune	Blockchain	Waste management inefficiencies	Increased efficiency, enhanced citizen engagement	Digital tokens as rewards for recycling
Delhi	Digital Twin	Rapid urbanization	Improved decision-making, better stakeholder engagement	Simulation of urban scenarios before implementation
Jaipur	IoT	Water scarcity	Reduced water wastage, improved supply reliability	Real-time monitoring and citizen updates
Hyderabad	IoT-enabled smart grids	Energy sustainability	Reduced carbon footprint, increased energy resilience	Real-time balancing of energy supply and demand

Bengaluru, Pune, Delhi, Jaipur, and Hyderabad case studies (as described earlier) collectively reflect the widespread presence of digital technologies in urban sustainability. With the application of digital tools like AI, IoT, blockchain, and digital twins, these cities not only resolved critical issues like traffic jams, garbage disposal, and scarcity of resources but also built a framework of innovative urban governance. The success of these initiatives relies on technological innovation coupled with concerted action of various stakeholders, highlighting the need for cooperative strategies towards urban planning. With ongoing urbanization being a changing phenomenon across the globe, such case studies are a ray of hope, highlighting the potential of digitalization in enhancing sustainable development and urban resilience across the globe.

Emerging Issues and Trends: Despite numerous digital initiatives across Indian cities, challenges remain in scalability, institutional readiness, and policy fragmentation. Fragmented regulations across municipal and state levels lead to inconsistent implementation. Additionally, ethical concerns around surveillance and data misuse must be addressed through transparent governance models. Scaling up successful pilots across diverse urban environments will require interoperable digital systems and inclusive planning.

5. DISCUSSION AND POLICY IMPLICATIONS

The global trend towards urbanization offers a major challenge: economic growth versus environmental sustainability. More and more people will mean greater demands on resources, infrastructure, and governance. Digital technologies have thus emerged as major tools in these challenges, such as efficiency enhancement, reduction of negative environmental impact, and promotion of more sustainable urban development.

Such cities are Amsterdam, Singapore, and Copenhagen, among others. For instance, Amsterdam is known to have the biggest IoT sensor network in handling energy consumption and traffic flow that resulted in the reduction of congestion and emissions. Singapore used the comprehensive digital twin integrated data across urban systems for real-time monitoring and predictive analytics. Copenhagen has also employed blockchain technology in trading energy to make renewable energy distribution easy and encourage people to use more sustainable energy sources.

These global examples have the potential to transform urban areas into sustainable, resilient, and inclusive environments using digital technologies. However, for these technologies to be successfully implemented, policy implications in the realms of digital inclusion, data governance, and international collaboration must be taken into consideration.

Another essential but underexplored aspect is financing sustainable digital solutions (Yildirim et al., 2021). Cities need innovative financial mechanisms, including green bonds, blended finance, and public-private partnerships (PPPs), to scale up technological innovations. PPPs play a vital role in reducing the financial burden on governments while bringing innovation and operational efficiency from the private sector (Yildirim & Yildirim, 2019).

The most urgent is the challenge around digital inclusion because the gap - known as digital divide - representing the difference in access between persons with and without availability of digital technology, represents the main obstacle against the broad embrace of digital solutions. In many cities, especially in developing countries, access to digital infrastructure and literacy remains uneven, limiting the potential benefits of digitalization.

Digital investments, especially to underserved populations, should remain at the center of policymakers' priorities. While this is a highly critical area in which public-private partnerships can intervene toward enhanced access to digital technologies, such initiatives can create a better scenario for educational approaches to enhance literacy and skills development among marginalized population groups. Making sure that people enjoy the benefits from digital urban activities is crucial in the pursuit of social equity and inclusion.

Furthermore, citizen digital literacy is crucial for ensuring inclusive smart urban transformation. Cities should implement capacity-building programs and awareness campaigns that empower citizens to interact with and benefit from digital solutions.

The extensive use of data in smart cities raises critical questions about governance, privacy, and security. As cities collect vast amounts of data from IoT sensors, cameras, and other digital devices, there is a need for clear guidelines on data usage, storage, and sharing. Protecting citizens' privacy while enabling data-driven decision-making is a delicate balance that requires robust regulatory frameworks.

Policies must establish clear standards for data governance, including data anonymization, access controls, and accountability measures. Transparency in data practices is essential for building public trust and ensuring that data is used ethically and responsibly. Additionally, international collaboration on data governance can help harmonize standards and facilitate the exchange of best practices across borders.

Urban sustainability is a global challenge that requires international collaboration and knowledge sharing. Cities can learn from each other's experiences, leveraging successful case studies and adapting best practices to their unique contexts. Through cooperative effort and collaborative platforms, such as the C40 Cities Climate Leadership Group and the Global Covenant of Mayors for Climate & Energy, cities can share insights and coordinate efforts with one another.

The other international frameworks are the United Nations' Sustainable Development Goals (SDGs), among others. SDGs provide very important guidance on sustainability efforts in the urban context. Local initiatives toward global goals are important for attaining the greater aims of sustainability while contributing to global climate action. Specifically, the chapter aligns with SDG 11 (Sustainable Cities and

Communities), SDG 9 (Industry, Innovation and Infrastructure), and SDG 13 (Climate Action), demonstrating their interdependencies in fostering resilient urban systems.

While the technological interventions are well illustrated, there is a need to emphasize cross-sectoral governance models that integrate urban planning, ICT, climate, and economic departments. Active citizen engagement, especially from vulnerable and rural-urban fringe communities, is vital to avoid deepening spatial digital divides. These socio-political elements deserve further attention for a holistic transformation.

5.1. Policy Recommendations

From case studies and global trends, a number of policy recommendations can be derived to further the role of digital technologies in sustainable urban development:

a) The government should prioritize investments in digital infrastructure, particularly in underserved areas, so that there is equal access to digital technologies which can aid the linkage between Technology and sustainability.
b) The enhancement of digital literacy among the citizens of the regions at nascent level can raise the competency of individuals in participating digital initiatives of towns and its development. This can bring public-private partnerships of digital solutions and leverage resources with expertise
c) The data governance should be designed with the utilization, privacy, and security aspects at the regional, state and national level. This can enhance cities to participate in global platforms for knowledge sharing and collaboration, learning from best practices and adapting them to local contexts.
d) The alignment of local initiatives with international frameworks like the SDGs and MDGs can helps the areas to achieve broader sustainability objectives and contribute to global climate action initiatives.

5.2. Theoretical Contributions

This chapter contributes theoretically by integrating socio-technical systems theory and resilience thinking into the analysis of digital urban transformation. Practically, it highlights how cities can adapt global technological frameworks into context-specific strategies. Policymakers can draw lessons from the presented case studies while academia can use the gaps identified to guide future empirical and model-based research.

6. CONCLUSION, LIMITATIONS, AND FUTURE RESEARCH DIRECTIONS

This chapter examines the interlinkage of climate change and digitalization to sustainable urban development, elaborating on the ways cities may seize on the digital technologies, which can overcome innumerable challenges facing them towards better meeting the quest for sustainability. As urban populations surge globally, cities are at the forefront of climate change effects and technological advancements, creating a critical need to integrate digital tools with sustainability goals. This intersection offers a new way of addressing urban challenges but necessitates a holistic approach to balance innovation with environmental and social factors. The three cities under review are Bengaluru, Pune, and Delhi. Some of the solutions like AI and IoT for traffic and blockchain in waste management offer potential in terms of attaining sustainable urbanization. However, success in these technologies depends on constructing strong policy frameworks, ensuring fair access, and ongoing research to ensure the digitalization effort contributes to broader sustainability objectives. Effective policies must guide technology use to facilitate sustainability, safeguard privacy, and drive innovation. Equitable access guarantees these tools are accessible to all residents, reducing urban inequalities. Continuous research keeps solutions current, adapting them to respond to evolving urban needs and sustainability objectives. The adoption of digital innovation, along with solving underlying policy and access challenges, can enable cities to build more sustainable and resilient futures.

Although this study is highly valuable, there are some limitations. Since secondary data have been used, it limits the scope of the research because primary data could provide deeper insights. Besides, mainly using Indian cities could limit the application of findings in developed countries as socio-economic, cultural, and infrastructural conditions vary widely between these countries. However, a focus on the contextual challenges and solutions strengthens the urban sustainability discourse of developing countries.

Future research should expand further on long-term impacts in urban sustainability concerning digital technologies in relation to socioeconomic implications and ways in which it will be used to benefit both the rich and the poor and other stakeholders. Studies should track how these technologies hold up over time, assessing their lasting benefits and adaptability to future urban shifts. Research will also need to consider the impacts on various income groups to ensure there are fair impacts and closures of access gaps. Previous focus should be placed on formulating strategy geared to specific urban environments; the majority of such areas are becoming rapidly urbanized. Formulation of strategies for different cities involves knowledge of the local governance, culture, and infrastructure to effectively tailor solutions.

Consistent performance as well as solution integrations in urbanism comprehensive planning will be necessary for promoting sustainable worldwide urban growth.

REFERENCES

Agrawal, A., Fischer, M., & Singh, V. (2022). Digital Twin: From Concept to Practice. *Journal of Management Engineering*, *38*(3), 06022001. Advance online publication. DOI: 10.1061/(ASCE)ME.1943-5479.0001034

Akhtar, N., & Kuriakose, P. N. (2021). Disruptive Mobility in Pre- and Post-COVID Times: App-Based Shared Mobility in Indian Cities—The Case of Bengaluru. *The Impact of the COVID-19 Pandemic on Green Societies. Environmental Sustainability*, •••, 57–91. DOI: 10.1007/978-3-030-66490-9_3

Bedi, P., Tripathi, N. G., & Singh, H. B. (2019). Smart Tourism Innovations for Smart Region, Case of Jaipur Metropolitan Region, India. *Advances in 21st Century Human Settlements*, 491–535. DOI: 10.1007/978-981-10-8588-8_9

Bhattarai, K., & Conway, D. (2021). Urban Growth. *Advances in Asian Human-Environmental Research*, 201–334. DOI: 10.1007/978-3-030-50168-6_4

Brunetta, G., & Caldarice, O. (2020). *Spatial Resilience in Planning: Meanings, Challenges, and Perspectives for Urban Transition*. 628–640. DOI: 10.1007/978-3-319-95717-3_28

Chatterji, T., & Vaidya, H. (2025). Pathways to Smart and Green Transitions Through Urban E-Planning: Multilevel Urban Waste Governance in India. Https://Services.Igi-Global.Com/Resolvedoi/Resolve.Aspx?Doi=10.4018/979-8-3693-5956-3.Ch016, 411–430. DOI: 10.4018/979-8-3693-5956-3.ch016

Chkoniya, V., Madsen, A. O., & Bukhrashvili, P. (Eds.). (2020). *Anthropological Approaches to Understanding Consumption Patterns and Consumer Behavior.*, DOI: 10.4018/978-1-7998-3115-0

Dutta, P. K., Raj, P., Sundaravadivazhagan, B., & Selvan, C. P. (2024). *Artificial Intelligence Solutions for Cyber-Physical Systems*. Artificial Intelligence Solutions for Cyber-Physical Systems., DOI: 10.1201/9781032694375

Farzaneh, G., Khorasani, N., Ghodousi, J., & Panahi, M. (2021). Assessment of Surface and Groundwater Resources Quality Close to Municipal Solid Waste Landfill Using Multiple Indicators and Multivariate Statistical Methods. *International Journal of Environmental Research, 15*(2), 383–394. DOI: 10.1007/s41742-020-00307-9

Farzaneh, H., Malehmirchegini, L., Bejan, A., Afolabi, T., Mulumba, A., & Daka, P. P. (2021). Artificial Intelligence Evolution in Smart Buildings for Energy Efficiency. *Applied Sciences 2021, Vol. 11, Page 763, 11*(2), 763. DOI: 10.3390/app11020763

Franco, S., Mandla, V. R., & Ram Mohan Rao, K. (2017). Urbanization, energy consumption and emissions in the Indian context A review. *Renewable & Sustainable Energy Reviews*, *71*, 898–907. DOI: 10.1016/j.rser.2016.12.117

Gouldson, A., Colenbrander, S., Sudmant, A., McAnulla, F., Kerr, N., Sakai, P., Hall, S., Papargyropoulou, E., & Kuylenstierna, J. (2015). Exploring the economic case for climate action in cities. *Global Environmental Change*, *35*, 93–105. DOI: 10.1016/j.gloenvcha.2015.07.009

Govindarajan, H. K., & L.S., G. (. (2021). Renewable energy for electricity use in India: Evidence from India's smart cities mission. *Renewable Energy Focus*, *38*, 36–43. DOI: 10.1016/j.ref.2021.05.005

Hoornweg, D., Sugar, L., & Gómez, C. L. T. (2011). Cities and greenhouse gas emissions: moving forward. Http://Dx.Doi.Org/10.1177/0956247810392270, *23*(1), 207–227. DOI: 10.1177/0956247810392270

Kalra, R. (2007). *High Technology and Intra-Urban Transformations: A Case Study of Bengaluru.*

Kelbaugh, D. (2019). The urban fix: Resilient cities in the war against climate change, heat islands and overpopulation. *The Urban Fix: Resilient Cities in the War against Climate Change, Heat Islands and Overpopulation*, 1–308. https://doi.org/DOI: 10.4324/9780429057441/URBAN-FIX-DOUGLAS-KELBAUGH/ACCESSIBILITY-INFORMATION

Kumar, R., Mishra, A., & Goyal, M. K. (2024). Water neutrality: Concept, challenges, policies, and recommendations. *Groundwater for Sustainable Development*, *26*, 101306. DOI: 10.1016/j.gsd.2024.101306

Lukić, R. M. (2024). Performance of the Serbian economy: Empirical analysis. *Baština*, *34*(63), 249–260. DOI: 10.5937/bastina34-47222

Mohan, D., Jha, A., & Chauhan, S. S. (2021). Future of road safety and SDG 3.6 goals in six Indian cities. *IATSS Research*, *45*(1), 12–18. DOI: 10.1016/j.iatssr.2021.01.004

Monstadt, J., & Schmidt, M. (2019a). Urban resilience in the making? The governance of critical infrastructures in German cities. *Urban Studies (Edinburgh, Scotland)*, *56*(11), 2353–2371. DOI: 10.1177/0042098018808483

Monstadt, J., & Schmidt, M. (2019b). Urban resilience in the making? The governance of critical infrastructures in German cities. *Urban Studies (Edinburgh, Scotland)*, *56*(11), 2353–2371. DOI: 10.1177/0042098018808483

Patti, Sebastiano., & Trizzino, Giampiero. (2020). *Advanced integrated approaches to environmental economics and policy : emerging research and opportunities*. 218.

Praharaj, S., Han, J. H., & Hawken, S. (2018). Urban innovation through policy integration: Critical perspectives from 100 smart cities mission in India. *City, Culture and Society*, *12*, 35–43. DOI: 10.1016/j.ccs.2017.06.004

Rosenzweig, C. (2018). *Climate Change and Cities : Second Assessment Report of the Urban Climate Change Research Network*. https://books.google.com/books/about/Climate_Change_and_Cities.html?id=n-ZNDwAAQBAJ

Seto, K. C., Golden, J. S., Alberti, M., & Turner, B. L.II. (2017). Sustainability in an urbanizing planet. *Proceedings of the National Academy of Sciences of the United States of America*, *114*(34), 8935–8938. DOI: 10.1073/pnas.1606037114 PMID: 28784798

Seyed Hydroos, A. K. (2023). *Development of an optimal signal control strategy for Heterogeneous Less Lane-Disciplined traffic conditions*. DOI: 10.5204/thesis.eprints.237949

Sharma, S. K., & Lata Pathak, S. (2024). Urbanization. *Population and Environment*. Advance online publication. DOI: 10.1007/978-981-97-6020-6

Singh, M., Singh, M., & Singh, S. K. (2024). Tackling municipal solid waste crisis in India: Insights into cutting-edge technologies and risk assessment. *The Science of the Total Environment*, *917*, 170453. DOI: 10.1016/j.scitotenv.2024.170453 PMID: 38296084

Tyagi, A. K., Kumari, S., & Surve, T. (2024). Integration of Digital Twin and Blockchain for Smart Cities. *Digital Twin and Blockchain for Smart Cities*, 81–100. DOI: 10.1002/9781394303564.ch5

Ukoba, K., Olatunji, K. O., Adeoye, E., Jen, T. C., & Madyira, D. M. (2024). Optimizing renewable energy systems through artificial intelligence: Review and future prospects. *Energy & Environment*, *35*(7), 3833–3879. Advance online publication. DOI: 10.1177/0958305X241256293

Umoh, A. A., Ohenhen, P. E., Chidolue, O., Ngozichukwu, B., Fafure, A. V., & Ibekwe, K. I. (2024). INCORPORATING ENERGY EFFICIENCY IN URBAN PLANNING: A REVIEW OF POLICIES AND BEST PRACTICES. *Engineering Science & Technology Journal*, *5*(1), 83–98. DOI: 10.51594/estj.v5i1.729

Wang, A., Shan, S., Ibrahim, R. L., & Omokanmi, O. J. (2023). A new look at environmental sustainability from the lens of green policies, eco-digitalization, affluence, and urbanization: Empirical insights from BRICS economies. *Energy & Environment*. Advance online publication. DOI: 10.1177/0958305X231177736

Yildirim, S. (2020). *The Consumer Role for Sustainable Development: How Consumers Contribute Sustainable Development Goals*. IGI GLOBAL Pub., DOI: 10.4018/978-1-7998-3115-0.ch018

Yildirim, S., Demirtas, I., & Yildirim, D. C. (2021). *A Review of Alternative Economic Approaches to Achieve Sustainable Development*. 288–307. DOI: 10.4018/978-1-7998-8482-8.ch018

Yildirim, S., & Yildirim, D. Ç. (2019). Achieving sustainable development through a green economy approach. *Advanced Integrated Approaches to Environmental Economics and Policy: Emerging Research and Opportunities*, 1–22. DOI: 10.4018/978-1-5225-9562-5.ch001

Zhu, J., Feng, T., Lu, Y., & Jiang, W. (2024). Using blockchain or not? A focal firm's blockchain strategy in the context of carbon emission reduction technology innovation. *Business Strategy and the Environment*, *33*(4), 3505–3531. DOI: 10.1002/bse.3664

KEY TERMS AND DEFINITIONS

Smart Cities: Urban areas using digital technologies for infrastructure and services.

Digital Twin: A virtual replica of a physical system used for simulation and planning.

Blockchain: A secure digital ledger used for transparent, decentralized data management.

IoT (Internet of Things): Network of connected devices that collect and exchange data.

Urban Resilience: The capacity of cities to absorb and recover from disruptions.

Climate Adaptation: Adjustments in urban systems to mitigate adverse climate effects.

Chapter 13
AI-Driven Solutions for Sustainable Urban Development:
Smart Cities and Resource Management

Dhruv Sabharwal
https://orcid.org/0000-0002-6973-7137
Sharda University, India

Utsav Krishan Murari
https://orcid.org/0009-0007-1606-6775
Sharda University, India

ABSTRACT

The unprecedented expansion in cities throughout the world has actually made the need for sustainable city growth solutions more essential. Artificial Intelligence (AI) has been established as a disruptive and gene change technology that proposes something new in solving urban and resource planning problems as well as plight to sustainability. This chapter discusses the relationship between AI and urban sustainability by considering smart cities as the conceptual context to embed AI applications. Areas of focus are energy and environment, water and waste, ITS and system management, and governance of cities. The role of ethics in the implementation of AI is pointed out as the breeze turns to biases in the data used – specifically, data privacy issues. The chapter features various and profound cases and proofs of implementations from around the world in support of the key points made and offers suggestions for practical recommendations for policymakers as well as urban planners.

DOI: 10.4018/979-8-3693-9909-5.ch013

INTRODUCTION

One of the trends that extend into the twenty-first century is urbanization, which means constant enlargement of cities due to the denser human population. Estimations made by the United Nations reveal that, in 2050, as many as 68-percent of the world's population is to live in cities, while in 2023 this figure stands at 56-percent. The expansion of the urban area is a major booster economically, socially, and innovation-wise but there are problems attached to the process. Such problems include environmental pollution, supply deficiency, rising energy consumption, disposal problems of wastes, and social imbalances in developing nations where development is far lower than the growth in population.

According to IEA (2023), urban centres account for about 70% of the global carbon dioxide emissions and consume more than 66% of the world's energy. This transformation turns out to have a significant influence on existing structures as new and congested population densities in cities upset everything from transport, quantity and quality of habitable accommodation, water availability and quality, and pollution and its effects on health and wellbeing alike. For instance, the air pollution index in New Delhi rises and becomes hazardous at some level as a result of vehicular emissions and industry in particular, evoking the necessity for sustainable urban development standards (Guttikunda et al., 2022). Therefore, there are new trends setting in strategies for sustainable urbanization among policymakers and urban planners. These strategies ensure that cities rely on natural resources, minimize energy wastage, developing sustainable methods of energy production to make cities environmentally sustainable. Of course, such practicable visions of sustainability can be attained only based on advanced technological and data-oriented solutions.

AI as a Transformative Force

Artificial Intelligence has risen as a topical leader in solving challenges arising from urbanization. AI consists of several conduit technologies such as Machine learning (ML), Natural language processing (NLP) and Computer vision to let the machine imitate human intelligence and work independently (Zhang et al., 2015). Through the evaluation of big data, pattern recognition, and prediction, AI offers decision-making backs by evidence. In the case of urban applications, AI has the benefits of integrating, processing and analysing the different types of data, such as remote sensing data, geospatial data and socio-economic data for urbanisation trends and infrastructural requirements. For instance, AI models can determine probable future expanding of cities thus helping planners in coming up with better resource utilization and least effect on the environment (Zhang et al., 2021). In addition, AI enables supervising and analyzing the dynamically developing urban infrastructure, which includes traffic, energy or waste, and addressing critical events promptly.

AI integration in the planning and development of cities is in accord with the United Nations Sustainable Development Goals for 2030, focusing on the city. Specific to this research interest is SDG 11, which focuses on making cities inclusive, safe, resilient, and sustainable (United Nations, 2022). However, achieving these goals involves improving operation efficiency and minimizing waste, using renewable energy, in which AI plays a role.

SMART CITIES: A PARADIGM SHIFT

Therefore, it could be summed up that despite all markers of the smart city being still under construction, the idea constitutes a new paradigm in city planning. A smart city entails the application of technology – the IoT, big data and analytics, as well as artificial intelligence – into a city's functioning to create better living. Through the deployment of these technologies, smart cities manage to deal with issues including traffic jams, waste of energy, and improper disposal of waste. AI serves as the foundation of the smart city because, through the deployment of sensors, cameras or mobile devices, a city can gather large amounts of data. They aid real-time decisions, forecasting and automation, which enhance how cities function (Chourabi et al., 2012).

a) **Traffic Management:** Based on the concept of smart cities, the idea of traffic control systems has been enhanced by AI. Computational methods are used to read current traffic and adjust signals accordingly or improve transit system. For example, use of Artificial Intelligent Traffic control in Singapore has slashed average journey time by one fourth decreasing emission and fuel usage (Litman, 2018).

b) **Energy Optimization:** An advanced electrical grid or smart grid that is being driven by artificial intelligence works to optimize its supply and demand levels of electricity and also to incorporate renewable energy. For instance, the smart energy management systems in the_iCity of Songdo of South Korea has proved ways of minimizing energy consumption by forty percent views Moreno et al., (2017)

c)**Waste and Water Management:** Smart waste management through the use of AI includes efficient routes for waste collection, means of handling waste without necessarily having to dump it in dusty landfills, and an increase in the efficiency of recycling. Likewise, the AI recognizes leakage in water supply systems to forward a fair distribution of available resources. For instance, Barcelona's smart water system has realized a water savings of 25% through the use of AI in its monitoring (Loucks et al., 2017).

AI in Urban Governance

AI is not merely a technology of 'better' infrastructure but of bettering urban governance. Artificial intelligence, in this regard, accepts data from social, economic, and environmental contexts and assists in policymaking, hence guiding governments on where to invest their efforts on issues that will contribute most to society. For example, AI can simulate scenarios of how a particular legislation affects housing affordability, availability and integration of vulnerable segments (GhaffarianHoseini et al., 2017).

An aspect that also somewhat relates to participatory governance of democratic urban development is also enhanced by AI. AI in the form of Natural Language Processing applied to chatbots or special mobile apps allows citizens to communicate with local administration, report on problems or make comments on policies. This has made transparency, accountability and trust between the citizens and the authorities to be enhanced (Goodchild& Li, 2012).

The Ethical Imperative

AI is being introduced into the systems of cities and this issue has many ethical implications. Over the years, most cities have embraced the use of technology in their operations, particularly based on data, and these have come with worries regarding data protection, values in the algorithms, and social justice. For example:

- **Data Privacy:** This is so because most technologies that incorporate AI entail the use of large amounts of personal data that may be abused by the systems. Proper data management policies are inevitable to avoid executes' violation of the public's privacy (Noble, 2018).
- **Algorithmic Bias:** The AI algorithms can then reintroduce biases that were seen in earlier decisions into areas such as resource distribution and service delivery. For example, prejudiced algorithms can allocate the abundant part of a city with infrastructures boosting the arena while excluding the poor areas of the same city (Ratti et al., 2016).
- **Inclusivity:** It refers to the unequal distribution of the tools of technology harnessing and presents several problems to the socially excluded in smart cities. Closing this gap is imperative towards reaping the fruits of AI enabled urban systems.

The solutions to these ethical issues can only be met through the higher collaboration of governments and technology firms as well as the citizenry. In relatively utilizing AI for ethical urban development, some of the practices that can be instituted include an open interface of the AI systems that are being used, third-party auditing and involving citizens in the process (Y. GhaffarianHoseini et al., 2017).

Research Advancements in AI and Urban Development

Recent studies have highlighted innovative applications of AI in urban development:

a) **AI-Driven Climate Resilience:** Currently, AI is applied for future climate change consequences to urban items, strengthening city security. For instance, AI-based models created by researchers at MIT mimic novel floods in coastal cities to promote flood management planning (Zhang et al., 2021).
b) **Green Building Design:** AI is useful in the design of energy-efficient buildings through proper coordination of building structures and materials. Wang et al. (2022) showed that the effects of the AI based simulations were proved to slash energy use in buildings by 30%.
c) **Urban Heat Island Mitigation:** Remote sensing systems, with the help of the AI, locate areas with high density, making it easier to initiate interventions like planting trees or using reflective roofs. This has been successfully practised in Los Angeles, thereby decreasing the surface-emitting temperature by 3°C (Goodchild & Li, 2012).
d) **AI in Public Health:** Artificial intelligence tools in epidemiology help in mapping disease transmission in urban settings, as evidenced by the COVID-19 pandemic. These models offered the rates of infection on a real-time basis while offering the facilities of rationing of resources for health care facilities (Chourabi et al., 2012).

CONCEPTUAL FRAMEWORK

Smart Cities and AI Integration

Smart cities provide a progressive model of urban development that effectively implements technology to support overall sustainability as well as quality of life for a city's inhabitants. Most smart cities use IoT, big data, AI and a variety of other technologies to complement diverse city operations such as energy consumption, transportation and resource utilization (Chourabi et al., 2012). AI thus enables city planners and administrators to make the right decision, especially through the management of massive data sets in real-time thereby minimizing wastage while enhancing living standards in cities.

AI is at the core of smart city models as it helps manage smart city infrastructures using analytics, monitoring, and control of key city systems. Smart cities are distinct from other city models because they depend on technology to tackle modern issues, including urbanization, climate change and scarcity of resources (Moreno et al., 2017).

The smart city is an example of the integration of technology to advance the sustainability of cities' resources through IoT, big data, and AI. These systems give specific recommendations, improve organisational processes, and address strategic development. Below, we delve deeper into two key domains where AI integration is particularly impactful: energy control and mobility.

Energy Management

Energy management turns out to be one of the critical success factors when it comes to managing sustainable city development. In a smart city structure, AI decision-making systems analyze the behaviour of citizens with the aim of identifying their needs in energy in order to schedule the provision of this important resource in the most efficient manner as well as in order to preserve the environment. AI for Smart Grids: At the moment, smart grids provide the best example of the use of AI in the energy industry. Such grids also utilize AI algorithms to maintain a better balance between the supply and demand of energy to some extent. Here, they use resources like solar and wind. AI models predict changes in the pace of the development of renewable energy based on the climate to ensure the stability of the grid (Chen et al., 2020). Through the AI smart grid, the City of Amsterdam boosted its LEAST energy wastage by 30% of energy efficiency through predictive energy management, as noted by Moreno et al., 2017. In Singapore, the AI algorithms of the energy management systems learn about the best ways of real-time electricity usage as renewable energy is fed to the selected peak load consumption patterns of energy demand forecasting and allocation for energy distribution optimisation. AI for Smart Grids: The technology at the cutting edge of AI integration for energy management is smart grids. The following grids use AI systems and algorithms to match supply and demand instantaneously while incorporating renewable energy – solar and wind among them. AI models forecast output variation of renewable energy resources on the basis of the climate state and cause the grid conditions to remain secure and dependable (Chen et al., 2020). For example, The city of Amsterdam has put in place a Smart Grid, which is an AI-orientated smart grid that has cut energy wastage by 30%, resulting in predictive energy management (Moreno et al., 2017).

Self-learning algorithms in energy-usage regulation in Singapore use artificial intelligence methods; they adjust the level of electricity consumption integrated with renewable energy to overall maximum loads.

A) **Energy Efficiency in Buildings:** Energy used in buildings in urban areas is significantly equal to the energy used all over the world. Appropriate technologies are also another method of saving energy, especially the use of heating, cooling and lighting systems via occupancy sensors and weather conditions, respectively. For instance, Google's Deep Mind is useful for saving a third of the energy within its data centre, and cooling systems are among the benefits of the AI application (Wong et al., 2018). Currently in New York City, artificial intelligence technologies monitor and recommend areas of inefficient energy use in commercial buildings as well as recommend appropriate measures for eradication of unnecessary power consumption. AI for Smart Grids: AI application in energy management systems is best illustrated by smart grids. These grids employ AI computational methods for continuous management of energy supply and demand to incorporate renewable energy such as; solar and wind energy. Weather fluctuations in the production of renewable energy are foreseen with the use of AI models to enable stability in the provision of the grid (Chen et al., 2020). For example, Smart grids that apply AI have been adopted by Amsterdam city to curb energy wastage by 30%, providing the possibility for further predictive energy management (Moreno et al., 2017). Currently, AI algorithms in the energy consumption systems of Singapore analyse the consumption of power in real-time and then integrate the use of renewable power to meet maximum demand. Energy Efficiency in Buildings: This shows Building consumption represents a large fraction in the energy intensity of cities. Smart homes improve energy efficiency of heating, cooling, and lighting based on occupancy and weather using systems based on AI. For instance, Wong et al. (2018) reported that Google's Deep Mind has managed to cut data centre power use by 40% through the use of AI for cooling optimization. In New York City, AI platforms track the energy consumption of commercial facilities and present suggestions on how unnecessary energy consumption can be brought down.

b) **Transportation:** Transportation systems are critical center pieces of city operations, and they transform how cities move through the application of AI. Using spatial traffic flow data and statistical modelling techniques, AI in transportation increases the flow's overall performance, reduces traffic overcrowding, and optimizes the effects on the environment.

   - Traffic Flow Optimization: Intelligent systems enhance traffic by using the latest data from sensors, cameras and GPS technologies. Algorithms that can predict and alter traffic signals in real-time and decrease congestion and fuel consumption. For example, Barcelona's use cases of smart cities include AI-based traffic controls ensuring a reduction in traffic jams by 20%, consequent low emissions from vehicles (Litman, 2018). In Los Angeles, smart traffic signals make use of artificial intelligence to adjust the traffic signals based on the flow of traffic on the roads, and this has increased by an average of 12% in traffic speeds. By analyzing traffic patterns and leveraging predictive algorithms, AI enhances transportation efficiency, reduces congestion, and minimizes environmental impacts. Autonomous vehicles (AVs) represent a significant leap forward in AI-driven transportation. These vehicles use AI to navigate urban environments safely and efficiently, reducing the risk of accidents and optimizing fuel usage. For instance: Tesla's AI-powered autopilot system integrates real-time sensor data to ensure safe and efficient driving. In Singapore, autonomous shuttles are being tested to provide last-mile connectivity, reducing dependency on private vehicles (Goodchild & Li, 2012).

- Shared Mobility: AI also applies to shared mobility service-hailing and car-sharing apps, for example, Uber, Lyft, Grab, BlaBlaCar, etc. Through the assessment of demand, AI provides efficient utilization of automobiles and travellers; long waiting hours and fuel usage are also minimized. For instance, Uber's AI conducts demand forecasting to control the fragmented taxis' real-time distribution for drivers.
- Resource Management in Smart Cities: Resource management in smart cities means the provision of equal and efficient consumption of what is considered valuable resources, including energy, water, and waste resources, among others. It also improves the performance of these systems through analyzing large datasets using predictive models and real-time anomaly detection to indicate where resources should be dedicated, and where they are not required.
- Predictive Modeling for Resource Demand: Predictive modelling constitutes one of the central approaches to the use of artificial intelligence to manage resources. According to collected historical and real-time data, AI algorithms predict that in what period of time a specific city will need resources to solve existing problems, so the resources are provided on time.
- Energy Demand Forecasting: By applying AI, standard energy consumption over time is anticipated, and thus, the utilities for de-

livering electric power can be planned adequately. For example, In Tokyo, Machine-learning energy management systems predict the electricity demand during peak hours to avoid the occurrence of power blacks (Chen et al., 2020).• Google has integrated self-developed AI systems to forecast power consumption in its data centers which will help the company to cut on costs and avoid wastage.

c) **Water Demand Prediction:** AI in water resource management forecasts the demand for water by considering factors such as weather conditions, population increase, industrial uses, etc.• The city of Barcelona; has implemented the use of AI in the prediction of water demand and has enhanced the distribution network so as to increase its efficiency and decrease the incidence of water leakage by fifteen per cent (Loucks et al., 2017)Cape Town in South Africa has been an example of how artificial intelligence Models can be used in the management of the scarce resource such as; Modelled consumption and conservation plans.to predict water demand, optimise its distribution networks and reduce water losses by 15% (Loucks et al., 2017). AI models in Cape Town, South Africa, have been instrumental in managing water scarcity by forecasting consumption trends and suggesting conservation strategies.

- Real-Time Anomaly Detection in Utilities: Real-time analysis of data by AI is smooth and useful while identifying breaks in urban utilities. These systems guarantee the standard performance of critical structures and indicate problems that may occur in the process.
- Leak Detection in Water Systems: Smart sensors use artificial intelligence to detect loss and leakage within water distribution pipeline infrastructures. For example, London developed AI systems in the water industry to help detect leaks within hours, thus minimizing the wastage of water and costs of repairs. The Public Utilities Board of Singapore employs AI to track its water conduits and has achieved leak detection success rate of 95 percent (Ratti et al., 2016).
- Waste Management Optimization: Intelligent waste management solutions enhance operational schedules for collection and increase efficiency of recycling. For instance: The waste bins in the city of Copenhagen are also fitted with AI technology that helps detect the waste level and time the collections to help reduce the general costs by 20 per cent. Robots with Artificial Intelligence optimize recyclables sorting in San Francisco, thus reducing landfill usage.

Ethical Considerations in Resource Management

Thus, AI, in the field of resource management, has revolutionary possibilities provided that the problem of ethical imperatives for fair and appropriate use of the technologies is resolved.

- Data Privacy: AI systems mainly work through amassment of considerable amounts of information which at many times is wrongful in the context of the citizens' privacy. They need to adhere to higher standards of data governance to ensure that the data collected, processed and disseached, in cities, is done under sound and strict ethical standards.
- Algorithmic Bias: Bias or prejudice in the AI algorithms and models implies that a particular group cannot mobilize resources for its use and is at a disadvantage. To overcome these risks, it is necessary for the open architecture of AI to be created (Goodchild & Li, 2012).

Ideas rooted in AI implementation in smart city approaches are novel to the modern administration of municipalities. All the aspects of energised built, all the transport and even water supply and waste disposal have well proved the use of AI to create sustainable city. However, the solutions which are invented by using Artificial Intelligence now should be fair and equal and the problems which relate to data privacy, an algorithmic bias, and an impossibility to use resources now should be solved. This infers on the other hand, that as this city gain development and size, that the role of artificial intelligence or AI will increase in endeavors to achieve sustainability goals. It has thus tasked policy makers, technologists, and urban developers at the forefront of development in this field, to go out looking for such possibility while at the same time dealing with the inherent ethical and social challenges that come with it.

AI-DRIVEN APPLICATIONS IN SUSTAINABLE URBAN DEVELOPMENT

Energy Efficiency

Energy efficiency, in fact, remains one of the key elements of the sustainable development of cities. Globally, cities account for more than 60% of total final energy demand and a similar share of energy-related carbon emissions (IEA, 2022). AI has brought a change in energy production, distribution and usage where it is used in the management of energy resources.

- **Smart Grids:** Smart grids use AI algorithms to match the supply of energy through the grid and its demand so that consumption and distribution are optimized, and unnecessary losses are avoided. The present smart grids differ from conventional grids in that instead of relying on routine time-based schedules, they can process data from sensors, meters and user devices.
- **Dynamic Demand Response:** Energy use forecasts are proactively identified by AI, and distribution is modified to avert overload. For example, it is possible to predict when a particular area is likely to experience high demand and source energy from areas that have excess demand to serve those that have a lack of demand (Chen et al., 2020).
- **Fault Detection and Recovery:** Machine learning methods report issues with the grid by detecting patterns in the information collected by sensors and trigger self-repairs that help minimize repair and maintenance time and expenses. Further, smart grid solutions, including the Chattanooga project, have provided a 40% improvement in energy outages and efficient functioning (Moreno et al., 2017).
- **Renewable Energy Integration:** AI also influences the grid systems through the infusion of renewable energy products, such as solar and wind energy, into the urban environment. These sources of energy are bound to fluctuate due to climatic conditions and, therefore, pose some difficulties in supply continuity.
- **Weather Forecasting:** AI models are used to forecast weather conditions so as to forecast renewable energy generation. For instance, neural networks perform diagnostics of satellite pictures and climate conditions to predict solar generation and wind velocities (Chen et al., 2020).
- **Energy Storage Optimization:** AI enhances battery storage thus guaranteeing that any excess renewable energy production is stored for use in the moments when needs are high.
- **Google's DeepMind:** AI has helped optimise the efficiency of wind farms, raising the precision of energy yields by between 20 per cent and increasing the capability of connecting to the existing network (Google AI, 2020).

Water and Waste Management

Effective water and waste management are essential for achieving sustainable urban development. AI technologies have emerged as powerful tools for optimizing resource use and minimizing environmental impact.

- Water Conservation: Another area where AI's add value is in optimizing water use, from leakage detection to consumption monitoring and distribution networks.
- Leak Detection: It employs pressure sensors and flow meters themselves, and through the statistics gathered, it analyzes patterns that denote leaks. For example, the city of Singapore uses AI in the water supply system in order to identify water leaks and thus decrease the level of water wastage by 25 per cent (PUB Singapore, 2021).
- Demand Forecasting: From the previous data uptake and congestion meteorological conditions, and demographic information, the consumption expectation is embraced and evaluated by machine learning programming for water allocation (Loucks et al., 2017). They have been equally effective in arid regions where getting access to fresh water is a mirage, every drop of water matters in such regions.
- Waste Management: AI makes waste management better by increasing the rate of recycling, minimizing landfill use, and operation optimization. Automated Sorting: AI algorithms and techniques used in sorting the recyclables at the recycling centres are more efficient than the traditional human methods, making the recycling of material more efficient. Route Optimization: Intelligent collection systems monitor traffic and waste data to implement proper routes in collections, fuel utilizition and emissions minimized. For instance, Stockholm has successfully integrated the use of artificial intelligence for waste management, and this has, in the recent past, been proven to have saved about 30% of the operational cost, increasing the recycling rate (Syal, 2020).

Intelligent Transportation Systems (ITS)

Transportation is one of the leading sources of urban greenhouse gases that cause air pollution. Intelligent Transportation Systems (ITS), whose implementation is being propelled by AI technology, are revolutionizing transport across cities globally by enhancing traffic control and encouraging green transport.

a) **Traffic Optimization:** By continually processing traffic information from the traffic network the AI systems help to control congestion and air pollution.
    - **Dynamic Traffic Signals:** AI has the capability of changing traffic signal intervals according to the current traffic flow and thus reducing cost factors like a fuel gauge. For illustration, the traffic systems in the city of Los Angeles have been developed with AI and have made satisfactory improvements in passenger reduction by 12% and emissions by 10% respectively (Litman, 2018).

b) **Incident Management:** Specifically, traffic accident or even road blockages is recognized by AI, and thus, the flow of vehicles will be redirected to avoid congestion. The current application of AI in traffic control has been revealed to have a great impact in lessening the amount of time an individual spends on a road and, thus, the pollution level within urban areas.

c) **Shared Mobility:** Ride-sharing services like Uber and Lift depend on AI to match riders with drivers quickly.
   - Minimize the route to and from pickup and drop-off destinations.
   - Accurate estimation of the demand of the riders using historical data and real-time data information.
   - AI that enables shared mobility decreases the total number of cars hence making it a sustainable market.th drivers efficiently. These platforms use machine learning algorithms to:
   - Optimize pickup and drop-off routes.
   - Predict rider demand based on historical patterns and real-time data.
   - AI-driven shared mobility reduces the number of vehicles on the road, promoting sustainability.

MIT media lab has explained that AI adaptive ride-sharing services predictably cut vehicle miles by 30% in cities and therefore low emissions.

Urban Planning and Governance

AI improves the planning and managing of cities and citizens by promoting a collection-centered approach to decision-making. These applications make at least sure that cities develop in harmony while meeting the needs of the people.

- Dynamic Zoning: Simplistic zoning laws are the most cardinal issues that we find in contemporary society because they neither dynamically respond to current urban conditions. Dynamic zoning involves the use of real-time data in the evaluation of population density, land use and the economy.
- Land Use Optimization: To enhance the utilization of land while still maintaining environmentally friendly standards, the AI models suggest that zoning be adjusted (Ratti et al., 2016).
- Disaster Resilience: AI imitates natural disasters like floods or earthquakes in different zones and therefore helps planners in designing strong urban areas.

The strategy of dynamic zoning, which is backed up with the help of AI, has already been applied to Amsterdam, where adaptive strategies have boosted the overall quality of life and environmental performance of a city.

Participatory Governance: Intelligent applications enhance participatory governance by involve citizens in decision making. Examples include: Through the functional aspects, resident can report problems, give feedbacks, and even contribute to urban planning through the use of chatbots and applications (Goodchild & Li, 2012).

Sentiment Analysis: Collective opinion concerning urban policies is assessed by AI from social media posts and other possible sources. Co-governance guarantees the respective city's development complies with the wish and need of the residents. For instance, Barcelona's AI target is to involve the public in policy creation to enhance openness and democracy (Smart Cities World, 2021). Smart applications based on artificial intelligence in energy conservation, water and waste management, transport systems, and government are restructuring urban growth. These technologies do not only respond to these concerns but also present viability as a way of growth. How can AI help to build more sustainable cities? AI provides the necessary tools and solutions to ensure better resource management, increased operational efficiency, and better citizen management. Nevertheless, to realize this vision, the program is facing several significant ethical challenges, data quality issues, and lack of effective cooperation between the participants.

Ethical Considerations in AI-Driven Urban Systems

As cities seek to leverage AI technologies and systems for the management of city systems, then considerations of ethic is of essence. Urbanisation and governance involve the use of artificial intelligence in infrastructure and resource management and comes with issues regarding; privacy, fairness and transparency. These are important considerations, especially bearing in mind the raw data amount needed for training of AI systems and the subsequent consequences of determined outputs. These challenges have to be resolved to make sure that AI based systems in cities are democratic, well accepted by the citizens of the country as well as fair.

a) **Data Privacy and Security:** For AI to be implemented in urban systems, data must be harvested from citizens, businesses, and public domains through different technological systems, and subsequently analysed. This is quite a big problem for privacy, personal information security and possibility of misuse of the gathered data.
b) **Unauthorized Data Collection and Surveillance Risks:** Another problem concerning AI utilised urban systems is the problem of data interception by stealthy third parties. AI systems work through data captured from sensors, cameras and other monitoring tools and use this data to make decisions in real time. For instance, smart surveillance technology will improve security while at the

same time gather information without the user's consent thereby compromising privacy (Zuboff, 2019). Dubbed smart cities, facial recognition technologies previously applied in the context of the United States' wars on crime and terrorisms, have been used in urban setting such as China to generate debates on surveillance and liberal rights (Brandom, 2020). Furthermore, it is much easier to store such data and sometimes share this information which leads to misuse. Data gathered on individuals for the delivery of urban services can be shared with other persons or used in other ways by others. For example, information from traffic monitoring systems may also be utilized in advertising purposes, yet without informing the user (Lepri et al., 2017).

c) **Mitigating Cyber Threats to Urban Infrastructure:** People say that cities and municipalities are also becoming more prone to cyber risks as their connections through AI become stronger. It is easy for cyber attackers to disrupt critical infrastructures like power distribution grid, water supply map and transportation system map. The Winery ransom ware attack that occurred in 2017 that targeted public services in several countries exposes the threat of digital urban systems (Shen et al., 2018). To reduce such risks, well established data management policies and protection controls must be implemented. Such measures are encryption protocols; performing audits frequently as well as data exchange using block chain technology. The works published in the last couple of years indicate that de-centralized AI systems where data analysis happens at the edge and not at the core are less prone to global cyber threats (Huang et al., 2021).

d) **Algorithmic Bias and Fairness:** The machine learning algorithms are not inherently biased, but they prioritize data and therefore are only as impartial as is the data. The problem, however, is that historical prejudices and unequal treatment tend to reflect in the models and produce distorted results. In context to urban systems, these can widen the existing gaps which in anyway are detrimental for the marginalised sections of the society.

e) **Exacerbation of Inequalities:** Bias in the algorithms may means that different users get a differential allocation of resources. For example, use of predictive policing has been accused of profiling poor and black areas (O'Neil, 2016). Similarly to the example with housing, AI models used for planning urban development can be created to focus on the higher-value districts over the impoverished neighborhoods continuing the cycle of social segregation. Noble (2018) points out how SEO algorithms perpetuate racism and sexism and, hence, similar prejudice existing in urban artificial intelligence. For instance, an AI algorithms to predict credit risks in cities may exclude people from disadvantaged neighborhoods from accessing credit facilities, thus denying them opportunities to access better economic opportunities (Barocas et al., 2019).

f) **Addressing Bias in AI Systems:** Algorithms must create a strategy as a foundational rule: to address these issues, fairness has to be a main element in controlling AI systems. This involves:

- Diverse Training Data: Adding that it is also essential to put measures in place to enhance generalisation in order to avoid bias in the datasets collected.
- **Regular Audits:** Causal investigation to remove prejudice from artificial intelligence systems.
- **Ethical AI Frameworks**: Implementing examples like the EU's Ethics Guidelines for Trustworthy AI, which should include principles of fairness, accountability as well as non-discrimination (Floridi et al., 2018). However, when developers implement AI systems and the management structures that support them, they must involve more stakeholders, including people of various backgrounds, in determining what these technologies should do for all cities' inhabitants (Eubanks, 2018).
- **Transparent AI:** Introducing transparency within AI systems is considered vital to the general public and society at large due to social accountability. Therefore, more and more AI is a part of the decision-making process in cities, and their users and municipal decision-makers need to know what the system is doing and why it is making a particular decision.
- **Explainable AI (XAI) Models:** Explainable AI (XAI) as subfield of AI which aims at increasing the interpretability of machine learning models. Most of the conventional AI techniques like deep neural networks work under the 'black box' level, the way decisions are made are not easily understandable (Lipton, 2018). This is something that XAI is keen on changing since the models offer an explanation of how the AI gets to the results. For instance, XAI techniques can be applied in smart transportation to explain why a specific route is preferred over the rest or in a model for resource allocation, such as energy, to explain why energy was supplied to one area instead of another. This not only increases accountability but also enables policymakers to make adjustments to the AI systems depending on the feedback received.
- **Third-Party Audits:** In the same context, third party audit is another useful instrument that helps to achieve the task of transparency. Community-based third-party agents can assess AI systems ethically, fairly, and accurately. For example, New York City has recently established an AI Task Force that should regulate the usage of automated systems in public administration (Richardson et al., 2021).Citizen-Centric Transparency:

- Similarly, transparency must be in the hands of citizens so that the AI systems developed are comprehensible and intelligible. This can be achieved through:
- Public Dashboards: Chroniclers of the live performance of AI in the city.
- Educational Campaigns: Measures recommended for the popularisation of AI technologies and reticulation of their consequences.
- Feedback Mechanisms: Public interfaces enable people to report problems they experience or propose changes in the construction of AI-based city services.

g) **Emerging Research in Ethical AI:** The recent advancements on ethical AI tried to target the following issues as general challenges within innovations in AI ethics:
   - **Federated Learning:** A process whereby artificial intelligence models are trained on user's devices, eliminating central training of data and thus improves user's privacy (Kairouz et al., 2021).
   - **Fairness Metrics:** Realisation of bias within AI systems is achievable by tools such as the Fairness Indicators developed by Google which provides specific, numerical representations of fairness where developers of AI systems can take steps to address those biases (Bird et al., 2020).

h) **Ethical AI Certification: Indeed, organizations such as the AI Ethics**

Lab have been continuing to grow certification programs to ensure sound AI System compliance.These advances are a step forward to improving the ethical nature of AI systems specifically when it comes to urban applications. Privacy, security, bias, and transparency comprise what can be described as large ethical hurdles for AI in urban systems, but they are by no means insurmountable. To maximise the benefits that lie in AI while respecting the rights of citizens, cities need to follow sound data management practices to eliminate or reduce algorithmic bias and promote explainability and accountability through AI and auditing. To my mind, ethical AI deployment is not simply a tech case but a pressing social need where governments, technologists, and communities should work united. However, as the field of artificial intelligence research continues to grow, so must these principles remain central to the creation of resilient and equitable urban systems.

Global Case Studies: AI-Driven Smart Cities

Smart Cities epitomized by AI technologies give real-life and tangible case studies of successful implementation of AI to solve real-life, complex urban problems, with a view of improving resource usage and supporting sustainable development. The present section provides an explication of the information infrastructure of three smart cities, namely Songdo in South Korea, Barcelona in Spain and Singapore, discussing how AI has been deployed in these urban systems, the accomplishments achieved, the issues faced, and the implications for other cities worldwide.

a) Case Study 1: Songdo, South Korea

- Songdo International Business District is considered as one of the smartest cities around the world and the model for the future city. Songdo is newly constructed city near Seoul, South Korea that is billed as a smart or ubiquitous city in which almost every function is linked by sensors and Artificial Intelligence (Park et al., 2020). Leveraging Songdo as a site, designed and developed from scratch, the research questions pertain to its key components of sustainability, technology integration, and urbanite's experience.
- AI Integration in Energy Management: The energy-saving feature in Songdo's infrastructure system is the application of the smart power usage control system activated by Artificial Intelligence. The city uses smart grids that use artificial intelligence to match the supply of energy with the demand. These grids integrate real time information from buildings as well as public utility providers and map ways of minimizing energy wastage and achieving efficient distribution.
- **Advanced Energy Management Systems (EMS):** Residential and commercial buildings in Songdo include EMS that controls lighting, heating and cooling according to occupancy of the rooms and the climatic conditions. Sensors switch off devices as areas that are not being used to lessen energy wastage as stated in Chourabi et al., 2012.
- Integration of Renewables: AI utilises prognosticative energy consumption patterns that improve the accommodation of renewable energy solution like solar and wind thereby improving energy utilisation and reducing the reliance on oil resources (Kim et al., 2021). The measures that have been adopted in Songdo City in energy conservation have raised energy efficient city systems by 40% in energy consumption than modern city systems.
- AI-Driven Waste Management: AI technologies also play another vital role in waste management in Songdo. Instead of being on the surface, the waste collection in the city is on a pneumatic system.•, Dividing waste items into those that can be recycled, those that should be decomposed, and those that cannot.• Track the amount of waste produced, which will help improve recycling efficiency and energy-producing steps (Park et al., 2020). Automatically turn off devices in unoccupied areas, reducing unnecessary energy consumption (Chourabi et al., 2012).

- Integration of Renewables: AI algorithms predict energy usage trends and enhance the integration of renewable energy sources like solar and wind, resulting in higher energy efficiency and lower dependence on fossil fuels (Kim et al., 2021). Songdo's energy efficiency initiatives have led to a 40% reduction in energy consumption compared to traditional urban areas, setting a benchmark for energy-smart urban systems.
- AI-Driven Waste Management: Waste management in Songdo is another area where AI technologies excel. The city employs a pneumatic waste collection system that operates underground. AI algorithms are used to:
- Sort waste materials into recyclable, compostable, and non-recyclable categories.
- Monitor waste generation patterns, optimising recycling and energy recovery processes (Park et al., 2020).
- That is why this AI automatic system has helped Songdo to minimise its dependency on landfills by 76%, thus solving the country's environmental issues.
- **Key achievements and challenges**
- Achievements: Songdo produced concrete improvements in decreasing energy use, increasing recycling efficiency, and building innovative technological cities. It has finally evolved into a model for smart, sustainable cities worldwide.
- Challenges: Despite the city's benefits, the city is most negatively criticised by high initial investment costs and problems of surveillance-related privacy issues. While there is no doubt that such an approach has several benefits, it naturally leads to concerns about data integrity and citizen sovereignty (Kim et al., 2021).

Songdo is a perfect example of how other cities aspiring to deploy AI into their metropolitan architecture to facilitate sustainable development can do it.

b) **Case Study 2: Barcelona, Spain**

For this reason, Barcelona has been leading the world in implementing IT to manage the growth of cities, making it a smart city. The city plans are to shift in sustainable resource consumption, enthral public engagement, and minimise the effects of climate change through AI and IoT (Charnock et al., 2019).

- **AI-Powered Smart Water System:** Barcelona's most tremendous success has been using artificial intelligence to manage water. Water flow, pressure, and quality are continuously measured through several sensors deployed throughout the city's distribution system. The findings in the analytical models for leak or inefficiency prediction help detect leakage and improve water distribution. These include cutting down on water wastage by a quarter, which conserves millions of litres of water yearly (Moreno et al., 2017). It helped improve organisational adaptation to climate change impacts, including drought, by facilitating adaptive water management. Implementing artificial intelligence in water systems has minimised water resource wastage and enhanced the stability of water services for the people.

- AI in Waste and Energy Management: The intelligent bins used in Barcelona Waste Collection are AI-enabled sensors that provide information when the bins are full. This has led to decreased fuel consumption and atmospheric emissions, usually caused by garbage trucks. In energy management, AI helps connect renewable energy resources like solar panels on government structures to the urban energy system. AI estimates the energy needs and supply, making it 20% efficient in terms of CO_2 emissions (Charnock et al., 2019).
- Citizen Engagement through AI: The incorporation of citizens in decision-making is a crucial feature of Barcelona's bright city plan. Intelligent city applications, such as the urban operating system's digital panel, educate people about the city's current services. People can also use mobile applications to report matters, such as when there are bad streetlights. AI systems rank these reports; hence, quick attention is given to them (Moreno et al., 2017).
- **Key Achievements and Challenges**
- **Achievements:** Outcomes include considerable improvements in water and energy efficiency, waste management improvement, and citizens' activation.
- **Challenges:** It is these very areas where the traditional education system has failed to deliver that the Barcelona project is and continues to forge ahead but where it may encounter some difficulty in striking a balance between innovation and equity to ensure that all people will benefit from the advantages that are being brought about by technology.

The case of Barcelona shows that while designing and sustaining AI-enabled governance solutions, it is essential to promote the incorporation of the technologies while ensuring inclusion.

c) Case Study 3: Singapore

- Singapore Smart Nation Initiative is an integrated project to turn Singapore into a knowledge-based, sustainable nation. It suggests how AI and other advancements help cities improve transport, safety and environmental issues (Tan et al., 2022).

- AI-Powered Intelligent Transportation Systems (ITS): Singapore itself gives an excellent example of how it is incorporating artificial intelligence to tackle the issue of traffic jams. Key components include Smart Traffic Light: With the live traffic record, AI changes the traffic lights to reduce traffic density and save approximately 20% of the time (Tan et al., 2022). Predictive Maintenance: By monitoring structures on trains and tracks, they detect and fix signs of degeneration before causing a delay. Electronic Road Pricing (ERP): AI is employed in the management of road usage through its implementation of road congestion, hence reducing the rates of vehicle emissions (Moreno et al., 2017).
- **Public Safety and AI Surveillance:** Singapore has implemented advanced surveillance systems with AI-driven facial recognition and behaviour analysis technologies. These systems enhance public safety by identifying potential threats and enabling quick responses to emergencies. However, they have raised concerns about data privacy and surveillance ethics (Tan et al., 2022).
- **AI in Environmental Sustainability:** Environmental sustainability is a cornerstone of Singapore's Smart Nation vision. The "Virtual Singapore" project is a digital twin of the city that uses AI to simulate and analyse urban development scenarios. The system helps policymakers design more sustainable urban layouts by modelling the impact of proposed projects on energy use, traffic patterns, and emissions.
- **Key Achievements and Challenges**
- **Achievements:** Singapore has embraced the use of AI in urban planning, has effectively reduced the time spent on commuting, improved safety and promoted sustainable development.
- **Challenges:** The above measures raise ethical issues, and the fact that depending on surveillance technologies always demands upgrades comes with expensive costs.

The capitalist success story of Singapore shows how AI can be used to create an innovative, safe and sustainable city, and it also sheds light on how bioethical issues should be addressed. Singapore's success illustrates how AI can be leveraged to create efficient, safe, and sustainable urban environments while also highlighting the importance of addressing ethical concerns.

FUTURE TRENDS IN AI-DRIVEN URBAN SUSTAINABILITY

AI remains to play a vital role as an enabler of urban sustainability through the advancement of technologies and focus on the citizens. We argue that the success of AI in future urban planning relies on the future's ability to embrace novel challenges, integrate the concepts of inclusiveness, and improve compatibility with other systems and technologies such as blockchain and edge computing. Of course, these advancements not only enhance the performance of small systems in urban areas but also ensure fair and secure implementation.

a) Integration of Emerging Technologies

Blockchain: Ensuring Secure Data Exchange: Smart cities require secure and transparent systems to share data provided by blockchain – an original digital database. The integration of blockchain with AI in urban systems ensures:

- Data Integrity: Since manipulating data on the blockchain is not possible, the AI outcomes of the developed system can be trusted (Zhao et al., 2019).
- Decentralised Governance: The administrative activities of an urban setting may well be provided by smart contracts, including the billing for utility services and waste disposal, among others.
- Citizen Empowerment: Highly Smart Cities rely on AI. Thus, Blockchain offers people control over their data, improving privacy issues.

For instance, in Dubai, blockchain-based projects, power consumption automation, and traffic signals use artificial intelligence-powered data sharing that ensures both security and transparency. This integration has also reduced administrative costs by close to 40% and a subsequent improvement of public trust (Zhang et al., 2022). The study by Alketbi et al. (2021) maps blockchain for safe IoT in smart cities and highlights its ability to handle the enormous amount of data produced by AI. The analyses highlighted the importance of blockchain for trust and accountability in AI-based intelligent cities.

b) Edge Computing:

Optimization of Real-Time Information Processing: Edge computing brings data processing closer to the data source; this reduces latency and presents better real-time AI performance. It is essential for all urban services and utilities, particularly in transport systems, energy and power, and disaster response.

- Real-Time Decision-Making: Edge computing allows AI to conduct computations locally, thus providing near-instantaneous solutions to fluid urban environments such as traffic holdups or blackouts (Shi et al., 2016).
- Reduced Network Strain: This results in less dependence on a central server, and less bandwidth is used, hence the name edge computing, since it processes data near the source.
- Enhanced Security: Localized data processing minimises risks of cyber threats, which, in combination with blockchain, helps to protect urban systems. For example, edge computing enhances AI security applications such as surveillance in Singapore's public areas with or without crowds. Integrating these technologies has dramatically improved public security while increasing productivity (Wu et al., 2020).

New research by Satyanarayanan et al. (2021) further identifies how edge computing should help facilitate autonomous cars and intelligent power systems. These insights show how edge-AI systems can transform mobility in cities and deliver energy in the future.

c) Citizen-Centric Design

- The Way to a More Sustainable, Inclusive Urban Society: Nevertheless, with the potential for AI to revolutionise urban systems, the positive impacts of such systems need to be again inclusive to all citizens for fair growth in urban areas. Citizen-centric design prioritises inclusivity, emphasising the following principles:
- Accessibility: Since human societies live in complexes, they require urban systems to serve the multiple needs of different groups, such as minority groups, the elderly, and physically challenged folks (Sabharwal et al, 2022)
- Participation: People should participate in creating and installing AI systems so people will understand them and trust them.
- Equitable Resource Distribution: An example is using or accessing energy and water and accessing public service as equally as other AI models, with no prejudice and preference for other models.
- For example, innovative city strategies in Barcelona rely on AI-enabled participatory platforms as a focus to involve citizens. Using these platforms enables residents to participate in urban planning decision-making, increasing citizens' participation (Chourabi et al., 2012).

d) Addressing Algorithmic Bias

Algorithmic bias is one of the most significant dilemmas to realising citizens-oriented AI. Noble (2018) has stated how biased algorithms perpetuate bias and specifically harm underprivileged groups. For example, an AI-based system for housing allocation might favour well-off areas over deprived areas. In his latest survey of AI safety, Binns (2020) pointed to the need for XAI to overcome bias. XAI frameworks take care of AI decisions and make them explainable so stakeholders can detect biases lurking in urban systems. Adapting these frameworks makes it possible to realise AI technologies that respect the appropriate ethical scruples and prove capable of integrating inclusiveness.

CONCLUSION

AI has the potential to revolutionise the sustainable development of smart cities by revolutionising resource management and governance mechanisms. However, the benefits exceeding the risks are significant within the limits of ethical practice and when engaging stakeholders. Subsequent work should assess synergistic possibilities of AI concerning the urban environment that considers the technical, social, and ethical contexts.

REFERENCES

Alketbi, A., Nasir, Q., & Talib, M. A. (2021). Blockchain for smart cities: Principles and applications. *Smart Cities Journal*, *4*(3), 21–35.

Barocas, S., Hardt, M., & Narayanan, A. (2019). Fairness and Machine Learning. FairMLBook.org.

Binns, R. (2020). Fairness in machine learning: Lessons from political philosophy. *Proceedings of the ACM Conference on Fairness, Accountability, and Transparency*, pp. 149–159.

Bird, S., Hutchinson, B., Kenthapadi, K., & Kliger, M. (2020). Fairness Indicators: Scalable infrastructure for evaluating algorithmic fairness. *Proceedings of the 2020 Conference on Fairness, Accountability, and Transparency*.

Brandom, R. (2020). The debate over China's use of AI surveillance. The Verge. Retrieved from https://www.theverge.com

Charnock, G., Purcell, T. F., & Ribera-Fumaz, R. (2019). Smart cities and urban governance: From Barcelona to global urbanism. *Urban Studies (Edinburgh, Scotland)*, *56*(4), 717–734.

Chen, Z., Li, H., & Xu, Y. (2020). AI in renewable energy: Trends and opportunities. *Renewable Energy Journal*, *45*(4), 123–134.

Chourabi, H., Nam, T., & Walker, S.. (2012). Understanding smart cities: An integrative framework. *International Conference on System Sciences*, pp. 2289–2297.

Eubanks, V. (2018). *Automating Inequality: How High-Tech Tools Profile, Police, and Punish the Poor*. St. Martin's Press.

Floridi, L., Cowls, J., Beltrametti, M., Chatila, R., Chazerand, P., Dignum, V., Luetge, C., Madelin, R., Pagallo, U., Rossi, F., Schafer, B., Valcke, P., & Vayena, E. (2018). AI4People—An ethical framework for a good AI society. *Minds and Machines*, *28*(4), 689–707. DOI: 10.1007/s11023-018-9482-5 PMID: 30930541

GhaffarianHoseini, A., Berardi, U., & Dahlan, N. D. (2017). AI and urban sustainability. *Sustainable Cities and Society*, *32*, 98–109.

Goodchild, M. F., & Li, L. (2012). *Assuring the quality of volunteered geographic information*. Spatial Statistics. DOI: 10.1016/j.spasta.2012.03.002

Guttikunda, S. K., Goel, R., & Pant, P. (2022). Air pollution in Indian cities: Current status and future challenges. *Environment International*, *156*, 106723.

Huang, C., Zhao, X., & Li, W. (2021). Decentralized AI systems for urban resilience: Opportunities and challenges. *AI & Society*, *36*(3), 385–400.

International Energy Agency (IEA). (2022). Cities and energy: Key challenges and solutions. Retrieved from https://www.iea.org/

Kairouz, P., McMahan, H. B., Avent, B., Bellet, A., Bennis, M., Nitin Bhagoji, A., Bonawitz, K., Charles, Z., Cormode, G., Cummings, R., D'Oliveira, R. G. L., Eichner, H., El Rouayheb, S., Evans, D., Gardner, J., Garrett, Z., Gascón, A., Ghazi, B., Gibbons, P. B., & Zhao, S. (2021). Advances and open problems in federated learning. *Foundations and Trends in Machine Learning*, *14*(1), 1–210. DOI: 10.1561/2200000083

Kim, J., Lee, H., & Park, S. (2021). Songdo smart city: Lessons for green urbanism and sustainability. *Sustainable Cities and Society*, *64*, 102563.

Lepri, B., Staiano, J., Sangokoya, D., Letouzé, E., & Oliver, N. (2017). The tyranny of data? The bright and dark sides of data-driven decision-making for social good. Transparent Data Mining for Big and Small Data, 3-24.

Lipton, Z. C. (2018). The mythos of model interpretability. *Communications of the ACM*, *61*(10), 36–43. DOI: 10.1145/3233231

Litman, T. (2018). *Autonomous vehicle implementation predictions: Implications for transport planning*. Victoria Transport Policy Institute.

Loucks, D. P., Stedinger, J. R., & Haith, D. A. (2017). *Water resource systems planning and management*. Springer Nature. DOI: 10.1007/978-3-319-44234-1

MIT Media Lab. (2020). AI-enabled ride-sharing and urban sustainability: Opportunities and challenges. Retrieved from https://media.mit.edu/

Moreno, C., Allam, Z., & Chabaud, D. (2017). Smart grids and urban energy management. *Energy Research Journal*, *12*(3), 101–117.

Noble, S. U. (2018). *Algorithms of oppression: How search engines reinforce racism*. NYU Press. DOI: 10.18574/nyu/9781479833641.001.0001

Park, H. W., Lee, J. K., & Choi, S. (2020). Smart city developments in South Korea: Lessons from Songdo. *Journal of Urban Technology*, *27*(2), 1–15.

Ratti, C., Frenchman, D., & Pulselli, R. M. (2016). Urban resource management in the AI age. *Technological Forecasting and Social Change*, *118*, 150–161.

Richardson, R., Schultz, J. M., & Crawford, K. (2021). Dirty data, bad predictions: How civil rights violations impact police data and predictive policing systems. *NYU Law Review*, *94*, 192–233.

Sabharwal, D., Verma, M., & Sood, R. (2022). Studying the Relationship between Artificial Intelligence and Digital Advertising in Marketing Strategy. *Journal of Content. Community & Communication*, *16*, 118–126. DOI: 10.31620/JCCC.12.22/10

Satyanarayanan, M., Bahl, P., Caceres, R., & Davies, N. (2021). The case for edge computing. *IEEE Pervasive Computing*, *8*(4), 31–40.

Shen, H., Ma, J., & Zhang, H. (2018). The impact of WannaCry ransomware attack on urban digital systems. *Journal of Cybersecurity*, *4*(1), 12–20.

Shi, W., Cao, J., Zhang, Q., Li, Y., & Xu, L. (2016). Edge computing: Vision and challenges. *IEEE Internet of Things Journal*, *3*(5), 637–646. DOI: 10.1109/JIOT.2016.2579198

PUB Singapore. (2021). AI in water management: Singapore's success story. Singapore Public Utilities Board Annual Report.

Smart Cities World. (2021). Citizen engagement in smart city governance: AI applications and best practices. Retrieved from https://smartcitiesworld.net/

Tan, K. W., Chua, Y. T., & Lee, C. (2022). Singapore's smart city journey: Integrating AI and sustainability. *Journal of Asian Public Policy*, *15*(1), 46–63.

United Nations. (2019). *World Urbanization Prospects: The 2018 Revision*. Department of Economic and Social Affairs.

United Nations. (2022). *Leveraging digital technologies for urban sustainability*. UN Department of Economic and Social Affairs.

Wang, Q., Chen, Z., & Li, H. (2022). AI applications in green building design: Optimizing energy efficiency. *Journal of Sustainable Architecture and Urban Design*, *34*(2), 203–214.

Waste360. (2022). AI-powered waste management: Trends and innovations. Waste Management Journal.

Wu, F., Song, L., & Han, Z. (2020). Intelligent edge computing in smart cities. *IEEE Network*, *34*(2), 36–41.

Zhang, H., Zhao, Y., & Lu, X. (2022). Blockchain-based AI systems for urban management: Opportunities and challenges. *Urban Science Journal*, *6*(2), 45–67.

Zhang, X., Li, Y., & Wang, J. (2021). AI-driven climate resilience: A case study of coastal urban planning. *Climate Change Journal*, *89*(1), 45–60.

Zhao, J., Fan, S., & Yan, J. (2019). Blockchain-based applications in smart cities. *Frontiers in Blockchain Research*, *2*(1), 13–25.

Zuboff, S. (2019). *The Age of Surveillance Capitalism: The Fight for a Human Future at the New Frontier of Power*. PublicAffairs.

Chapter 14
Green Transition in the Automotive Industry:
Turkey Case

Durmus Cağrı Yıldırım
Tekirdag Namık Kemal University, Turkey

Seda Yıldırım
https://orcid.org/0000-0003-4367-6652
Tekirdag Namık Kemal University, Turkey

Ömer Esen
https://orcid.org/0000-0002-4762-9282
Tekirdag Namık Kemal University, Turkey

ABSTRACT

Policy makers are turning to alternative and sustainable practices to slow down the effects of climate change and global warming. The transition from practices that increase carbon emissions, cause air pollution and rapidly consume natural resources, which are part of the traditional economy, to circular practices has begun. At this point, the automobile market has been selected as one of the leading sectors for green transition. The development of the automobile sector, where green transition and circular practices are seen, is observed in various publications and reports. In this study, the circular changes experienced in the automobile sector in Turkey will be examined. In the study, tables and graphics will be interpreted from the sources obtained from current databases.

DOI: 10.4018/979-8-3693-9909-5.ch014

INTRODUCTION

The changes in the ecosystem resulting from climate change have become a serious motivation for people to adapt more quickly to the green transformation (Yıldırım et al., 2025; Yıldırım and Vasilev, 2024; Yıldırım, 2021). Although the history of the concept of sustainable development is not very old, it can be said that human life, societies, economies and countries have entered a new transformation within the framework of sustainable development in the last decade (Bostancı and Yıldırım, 2022; Bostancı and Yıldırım, 2021; Sevik and Yıldırım, 2023; Yıldırım and Yıldırım, 2020). With the 2030 Sustainable Development Goals accepted in 2015, 17 goals have entered the lives of world leaders, economies and individuals. Sustainable policies and strategies are being developed for the basic goals that are awaiting solutions globally and expected to be realized by 2030 for a sustainable future (Yildirim, 2021; Yildirim, 2020). At this point, green transformation has begun in many industrial areas and industries.

One of the human activities that accelerate climate change is industrial production and excessive consumption (Yıldırım and Candan, 2015; Yildirim et al., 2022; Yıldırım and Kantarcı, 2022; Yıldırım et al., 2024). The automobile sector is also one of the major polluters and is a sector where economies develop new strategies for green transformation. Some damage is caused to the environment and ecosystem on both the production and consumption sides. The excessive use of fossil fuel-based automobiles has been an activity that causes environmental damage for years due to both energy consumption and greenhouse gases released into the environment. In addition to options such as public transportation, rail systems, walking (short distance) and cycling, electric cars have also come into play among the transportation vehicles that individuals prefer to reduce greenhouse gas emissions (Ritchie, 2023; European Environment Agency, 2020; Jing et al., 2022; Olayode et al., 2025; Ridango, 2022).

Transportation accounts for 64 percent of global oil consumption and 27 percent of all energy consumption. In countries with high population density, transportation is an issue awaiting a solution (Mead, 2021). The transportation sector is an important sector for the economic development of countries and for trade relations. However, its negative effects on the natural environment have made it necessary for the transportation sector to undergo a green transformation. According to research, the transportation sector releases more than a quarter of the global carbon dioxide amount, methane gas and nitrous oxide gas, which are harmful to the environment (Shahid et al., 2014). From an economic perspective, transportation gives life to the economy as a supply chain and supply network. Sustainable policies are sought in transportation and transportation-based effects without losing their economic contributions and without harming the environment (Van Fan et al., 2018).

The search for new alternatives to prevent environmental degradation caused by the transportation and transport sector is among the important issues in developed and developing countries. At this point, alternatives related to the automobile sector are considered within the individual transportation use. In addition to public transportation, automobile types attract attention for individuals who use individual vehicles. As the automobile industry continues to grow, sustainable transformation also comes to the fore in individual vehicle usage. For example, global automobile production in 2023 is expected to increase significantly by 10.2% compared to the previous year to reach approximately 76 million units, and in Europe by 12.6% to reach approximately 15 million units (Demirci, 2024). Some key developments in the global automobile market can be summarized as below (Demirci, 2024):

- Automobile production in North America is expected to increase by 12.4% to reach 11.7 million units.
- The US shows a similar trend with a growth rate of 8.5%, with more than 7.6 million vehicles produced. Production in South America in 2023 is expected to increase slightly by 0.1% compared to 2022.
- Japan's automobile production volumes increased by 17.4% last year to reach 7.7 million units.
- Production in South Korea is expected to increase significantly by 13.6% in 2023 to reach 3.9 million units, driven by stable plant operations and strong exports.
- Chinese automobile production is expected to reach 25.3 million units in 2023.
- Car production in India is expected to increase by 6.8% last year to 4.7 million units, driven by factors such as a low production base in 2022, improved semiconductor supply, and increasing individual mobility preference.
- The Middle East and Africa region saw a 0.5% increase in car production in 2023.
- EU car production is expected to increase significantly to over 12 million units in 2023, an increase of 11.3% over the previous year.

In this study, indicators related to green transformation in the automotive sector are discussed and a review is planned with a Turkey case. As a result of green transformation in the automotive sector, hybrid and electric vehicles have mostly come to the fore. On the other hand, new decisions regarding the market for fossil fuel-based automobiles and vehicles are also drawing attention. It is thought that a compilation will be presented on the changes experienced in the automotive market and long-term expectations.

ELECTRIC VEHICLE AND ELECTRIC VEHICLE MARKET

An electric vehicle can be defined as a motor vehicle that receives electricity from a specific battery and can be charged from an external source. Therefore, electric vehicles include vehicles with both a battery and an electric motor that can be charged from external electricity. (US Department of Energy, n.d.). While gasoline-powered vehicles require an internal combustion engine to operate, electric vehicles and battery electric cars do not have specific internal combustion engines in their operating principle. Electric vehicles operate with a system based on energy storage and the energy moving the wheels (Nissan, 2022).

Figure 1. Types of electric car

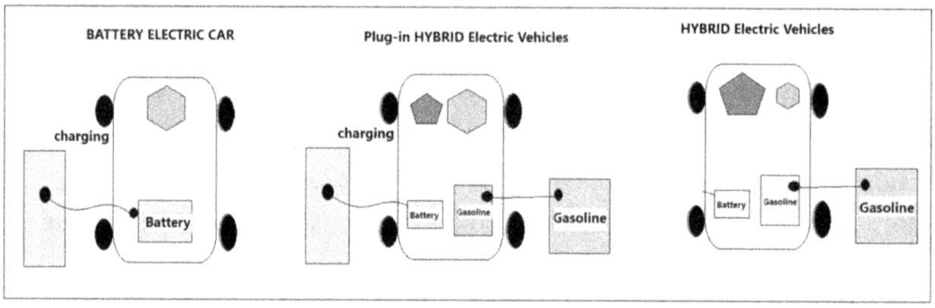

Source: created by the authors based on EVgo, (n.d.)

Figure 1 shows three types of electric as "*Battery Electric Vehicles, Plug-in Hybrid Electric Vehicles and Hybrid Electric Vehicles*". Based on EVgo (n.d), 3 types of electric vehicles can be examined as below:

1. Battery Electric Vehicles: These are vehicles with rechargeable batteries and no gasoline engine. These vehicles are fully electric vehicles. The energy the vehicle needs comes from the battery system charged from the grid. They are vehicles developed as "zero emissions" within the vehicle market.
2. Plug-in Hybrid Electric Vehicles: Both an engine and an electric motor are required for the energy the vehicle needs and to operate.
3. Hybrid Electric Vehicles: These are vehicles that have a gas-powered engine and an electric motor at the same time.

"*Hybrid electric vehicle*" is a more economical and environmentally friendly vehicle compared to cars that only consume fossil fuels. Consumers who are considering buying an electric car may prefer hybrid models to be able to use the fuel option in case of emergency use due to reasons such as charging time and access to charging points. On the other hand, research shows that cars are used less than 70 km per day on average. Therefore, electric cars require charging much less frequently than is thought (BYD, 2024).

In the European Union (EU) countries, the sale electric cars in total market was recorded over then 20% in August (it was 11.6 percent in August of last year). Electric car sales was evaluated in the increasing trend. The market share of battery-electric vehicles surpassed diesel for the second time this year in August, making them the third most popular choice among new car buyers. Hybrid-electric cars maintained their position as buyers' second choice with a market share of 24 percent. Although gasoline-powered cars are still the most popular choice, their market share, which was 38.7 percent in August last year, has decreased to 32.7 percent. (Dünya Gazetesi, 2023). According to the Automotive Distributors and Mobility Association data, new car and light commercial vehicle sales in Turkey increased by 0.5 percent in 2024 compared to the previous year and reached 1 million 238 thousand 509 units. Car sales increased by 1.3 percent in the January-December period compared to the previous year and reached 980 thousand 341 units. The light commercial vehicle market decreased by 2.7 percent and reached 258 thousand 168 units. The car and light commercial vehicle market increased by 7.3% in December. (Bloomberght, 2025).

The International Energy Agency (IEA) report determined that the rate of renewable energy sources is expected to increase rapidly. The report states that the number and prevalence of electric vehicles will also increase significantly by the end of 2030. Accordingly, the number of electric cars in the world will increase 10-fold by the end of 2030. By 2030, more than 60 percent of the vehicles sold in the world will be electric. Electric car sales in Turkey also gained momentum in 2023. For example, Electric car sales exceeded 10 million in 2022, reaching 10.3 million. In 2021, 6.5 million cars were sold in the world. According to TÜİK, the share of electric cars in cars registered in Turkey in 2022 was 1.4 percent. (Euronews, 2024).

As provided by BloombergNEF's analysis, although electric vehicle sales are slowing down in growth, a new record was expected in 2024. Total electric vehicle sales worldwide were expected to reach 16.7 million units in 2024, well above the 13.9 million sales in 2023. Electric vehicles were expected to account for 20 percent of global vehicle sales by the end of 2024, or 1 in every 5 vehicles sold would be electric (Akpınar, 2024).

New car sales in the U.S. are set to rise 2.2% in 2024, the highest since 2019, driven by rising demand for hybrids. Totoya, Ford and General Motors all reported U.S. sales growth for 2024, while Tesla, which set records in China, disappointed in the U.S. (Ekonomi Gazetesi, 2025).

According to records in 2023, there were 40.5 million active electric vehicles globally. The growth of the electric vehicle market is rapid. Despite the decline in the traditional vehicle market during the COVID-19 crisis, the electric vehicle market continued with an increasing trend (Carlier, 2024).

Table 1. Number of electric cars in the world (2023)

Years	Hybrid	Battery	Total
2016	0,9	1,2	2,1
2017	1,2	1,9	3,1
2018	1,8	3,1	4,9
2019	2,5	4,8	7,3
2020	3,3	6,8	10,1
2021	5,2	11,3	16,5
2022	8	18,3	26,3
2023	12,3	28,2	40,5

Number of cars x1000(s)
Source: adapted from Statista, 2024

As seen in table 1, there is a record developed by Statista (2024) for electric cars in the world in 2023. It can be said that battery electric cars are demanded much more than hybrid cars in the market of electric cars in 2023.

When looking at calculations made by IEA and Statista (2024) globally in 2023, it was seen that more than 14 million electric vehicles were sold. It is thought that the electric vehicle market or electric vehicle sales will increase in the following periods. At this point, the impact of 2030 and 2035 NET Zero policies is high. The main estimates are that more than 103 million electric vehicles will be sold globally by 2050. (IEA, & Statista, 2024).

AUTOMOTIVE INDUSTRY INDICATORS IN TURKEY

This part will show main indicators to determine the current profile of green transition in the automotive industry in Turkey. According to TÜİK (Turkish Statistical Institute), this study reviewed current data and info on the automotive industry in Turkey as below:

Table 2. The number of motor land vehicles in Turkey (a)

Year	Truck	Motorcycle	Special purpose	Tractor
2002	567 152	1 046 907	23 666	1 180 127
2003	579 010	1 073 415	24 468	1 184 256
2004	647 420	1 218 677	28 004	1 210 283
2005	676 929	1 441 066	30 333	1 247 767

continued on following page

Table 2. Continued

Year	Truck	Motorcycle	Special purpose	Tractor
2006	709 535	1 822 831	34 260	1 290 679
2007	729 202	2 003 492	38 573	1 327 334
2008	744 217	2 181 383	35 100	1 358 577
2009	727 302	2 303 261	34 104	1 368 032
2010	726 359	2 389 488	35 492	1 404 872
2011	728 458	2 527 190	34 116	1 466 208
2012	751 650	2 657 722	33 071	1 515 421
2013	755 950	2 722 826	36 148	1 565 817
2014	773 728	2 828 466	40 731	1 626 938
2015	804 319	2 938 364	45 732	1 695 152
2016	825 334	3 003 733	50 818	1 765 764
2017	838 718	3 102 800	60 099	1 838 222
2018	845 462	3 211 328	63 359	1 885 952
2019	844 481	3 331 326	65 470	1 908 999
2020	859 670	3 512 576	70 309	1 958 727
2021	886 303	3 744 370	78 482	2 025 006
2022	919 125	4 141 914	85 276	2 093 933
2023	959 793	5 079 396	93 407	2 186 150
2024	1 000 326	6 261 927	102 100	2 265 267

Source: TÜİK, 2025a

Table 2 and Table 3 show the Number of Motor Land Vehicles in Turkey including the period between 2002 and 2024. It is seen that the number of motor land vehicles has had an increasing trend since 2002 in Turkey. Especially, the number of motorcycles, cars and tractors are higher than other vehicles.

Table 3. The number of motor land vehicles in Turkey (b)

Year	Total	Car	Minibus	Bus	Small truck
2002	8 655 170	4 600 140	241 700	120 097	875 381
2003	8 903 843	4 700 343	245 394	123 500	973 457
2004	10 236 357	5 400 440	318 954	152 712	1 259 867
2005	11 145 826	5 772 745	338 539	163 390	1 475 057
2006	12 227 393	6 140 992	357 523	175 949	1 695 624
2007	13 022 945	6 472 156	372 601	189 128	1 890 459

continued on following page

Table 3. Continued

Year	Total	Car	Minibus	Bus	Small truck
2008	13 765 395	6 796 629	383 548	199 934	2 066 007
2009	14 316 700	7 093 964	384 053	201 033	2 204 951
2010	15 095 603	7 544 871	386 973	208 510	2 399 038
2011	16 089 528	8 113 111	389 435	219 906	2 611 104
2012	17 033 413	8 648 875	396 119	235 949	2 794 606
2013	17 939 447	9 283 923	421 848	219 885	2 933 050
2014	18 828 721	9 857 915	427 264	211 200	3 062 479
2015	19 994 472	10 589 337	449 213	217 056	3 255 299
2016	21 090 424	11 317 998	463 933	220 361	3 442 483
2017	22 218 945	12 035 978	478 618	221 885	3 642 625
2018	22 865 921	12 398 190	487 527	218 523	3 755 580
2019	23 156 975	12 503 049	493 373	213 358	3 796 919
2020	24 144 857	13 099 041	493 395	212 407	3 938 732
2021	25 249 119	13 706 065	484 806	208 882	4 115 205
2022	26 482 847	14 269 352	487 381	208 442	4 277 424
2023	28 740 492	15 221 134	502 628	210 740	4 487 244
2024	31 301 389	16 232 458	522 608	213 416	4 703 287

Source: TÜİK 2025a

Table 4 and Table 5 show the distribution of gas type in Turkey for the period between 2004 and 2024. It can be said that electric and hybrid types of cars get great attention in the last decades.

Table 4. Distribution Of gas type in Turkey (a)

Year	Total	(%)	Gasoline	(%)	Diesel	(%)	LPG	(%)
2004	5 400 440	100	4 062 486	75,2	252 629	4,7	793 081	14,7
2005	5 772 745	100	3 883 101	67,3	394 617	6,8	1 259 327	21,8
2006	6 140 992	100	3 838 598	62,5	583 794	9,5	1 522 790	24,8
2007	6 472 156	100	3 714 973	57,4	763 946	11,8	1 826 126	28,2
2008	6 796 624	100	3 531 763	52,0	947 727	13,9	2 214 661	32,6
2009	7 093 964	100	3 373 875	47,6	1 111 822	15,7	2 525 449	35,6
2010	7 544 862	100	3 191 964	42,3	1 381 631	18,3	2 900 034	38,4
2011	8 113 111	100	3 036 129	37,4	1 756 034	21,6	3 259 288	40,2
2012	8 648 875	100	2 929 216	33,9	2 101 206	24,3	3 569 143	41,3

continued on following page

Table 4. Continued

Year	Total	(%)	Gasoline	(%)	Diesel	(%)	LPG	(%)
2013	9 283 923	100	2 888 610	31,1	2 497 209	26,9	3 852 336	41,5
2014	9 857 915	100	2 855 078	29,0	2 882 885	29,2	4 076 730	41,4
2015	10 589 337	100	2 927 720	27,6	3 345 951	31,6	4 272 044	40,3
2016	11 317 998	100	3 031 744	26,8	3 803 772	33,6	4 439 631	39,2
2017	12 035 978	100	3 120 407	25,9	4 256 305	35,4	4 616 842	38,4
2018	12 398 190	100	3 089 626	24,9	4 568 665	36,8	4 695 717	37,9
2019	12 503 049	100	3 020 017	24,2	4 769 714	38,1	4 661 707	37,3
2020	13 099 041	100	3 201 894	24,4	5 014 356	38,3	4 810 018	36,7
2021	13 706 065	100	3 495 172	25,5	5 158 803	37,6	4 923 275	35,9
2022	14 269 352	100	3 817 104	26,8	5 261 876	36,9	5 005 563	35,1
2023	15 221 134	100	4 362 975	28,7	5 425 652	35,6	5 094 751	33,5
2024	16 232 458	100	4 908 203	30,2	5 541 441	34,1	5 172 471	31,9

Source: TÜİK, 2025b

Table 5. Distribution Of gas type in Turkey (b)

Year	Hybrid	(%)	Electric	(%)	Unknown	(%)
2004	-	-	-	-	292 244	5,4
2005	-	-	-	-	235 700	4,1
2006	-	-	-	-	195 810	3,2
2007	-	-	-	-	167 111	2,6
2008	-	-	-	-	102 473	1,5
2009	-	-	-	-	82 818	1,2
2010	-	-	-	-	71 233	0,9
2011	23	0,0	24	0,0	61 613	0,8
2012	53	0,0	175	0,0	49 082	0,6
2013	83	0,0	353	0,0	45 332	0,5
2014	113	0,0	412	0,0	42 697	0,4
2015	324	0,0	565	0,0	42 733	0,4
2016	517	0,0	643	0,0	41 691	0,4
2017	925	0,0	760	0,0	40 739	0,3
2018	4 415	0,0	952	0,0	38 815	0,3
2019	13 877	0,1	1 176	0,0	36 558	0,3

continued on following page

Table 5. Continued

Year	Hybrid	(%)	Electric	(%)	Unknown	(%)
2020	33 690	0,3	2 797	0,0	36 286	0,3
2021	86 682	0,6	6 267	0,0	35 866	0,3
2022	134 662	0,9	14 552	0,1	35 595	0,2
2023	222 328	1,5	80 043	0,5	35 385	0,2
2024	391 296	2,4	183 776	1,1	35 271	0,2

Source: TÜİK, 2025b

In Figure 2, number of cars can be seen by their brands in Turkey-2024. It can be said that the highest sale of car has belonged to Renault in Turkey.

Figure 2. Vehicles registered to traffic

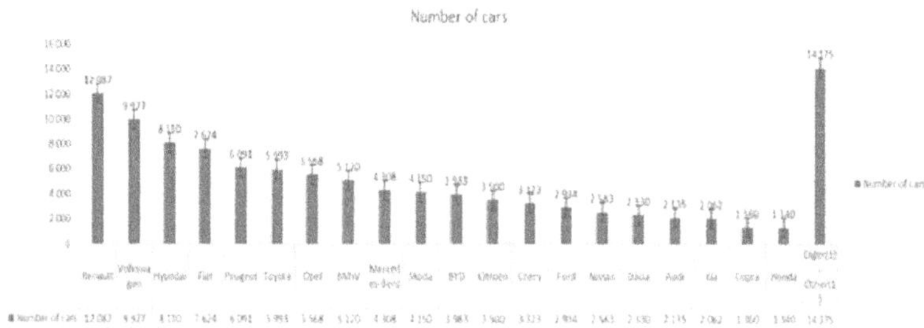

Source: TÜİK, 2025c

Table 6. Number of cars from 2020 to 2024

	Number of cars	
Engine size	**2020**	**2024**
-1300	146 057	335 415
1301 - 1400	90 370	153 317
1401 - 1500	178 152	225 804
1501 - 1600	145 083	125 511
1601 - 2000	35 455	64 359
2001+	4 718	6 536
Other	1 690	103 888
Total	601 525	1 014 830

Source: TÜİK, (2025d).

Table 6 shows the number of cars based on their engine size for the period between 2020 and 2024 in Turkey. It can be said that cars with small engine have higher demand for the last decades.

Table 7. Vehicles by age (a)

Age group	Year	Total	(%)	Car	(%)	Minibus	(%)	Bus	(%)
	2023	28 740 492	100	15 221 134	100	502 628	100	210 740	100
Total	2024	31 301 389	100	16 232 458	100	522 608	100	213 416	100
	2023	6 885 068	24,0	3 427 640	22,5	55 516	11,0	29 422	14,0
0-5	2024	8 630 951	27,6	3 970 412	24,5	65 633	12,6	31 982	15,0
	2023	5 882 501	20,5	3 485 358	22,9	112 903	22,5	53 511	25,4
6-10	2024	5 604 490	17,9	3 286 818	20,2	106 928	20,5	46 659	21,9
	2023	4 889 665	17,0	2 583 957	17,0	113 898	22,7	43 290	20,5
11-15	2024	5 212 459	16,7	2 866 811	17,7	111 628	21,4	43 818	20,5
	2023	3 835 621	13,3	1 712 372	11,2	90 348	18,0	29 146	13,8
16-20	2024	4 351 937	13,9	1 948 378	12,0	104 343	20,0	34 490	16,2
	2023	7 247 637	25,2	4 011 807	26,4	129 963	25,9	55 371	26,3
21+	2024	7 501 552	24,0	4 160 039	25,6	134 076	25,7	56 467	26,5
	2023	14,5		14,1		15,8		15,8	
Average age	2024	14,3		14,2		16,0		16,2	

Source: TÜİK, (2025e).

Table 8. Vehicles by age (b)

Age group	Year	Small truck	(%)	Truck	(%)	Motorcycle	(%)	Special purpose	(%)
	2023	4 487 244	100	959 793	100	5 079 396	100	93 407	100
Total	2024	4 703 287	100	1 000 326	100	6 261 927	100	102 100	100
	2023	838 300	18,7	153 773	16,0	2 020 478	39,8	12 645	13,5
0-5	2024	962 807	20,5	179 564	18,0	3 025 737	48,3	13 583	13,3
	2023	892 045	19,9	149 542	15,6	827 283	16,3	25 229	27,0
6-10	2024	832 845	17,7	135 554	13,6	840 252	13,4	24 773	24,3
	2023	1 092 184	24,3	160 008	16,7	667 671	13,1	17 284	18,5
11-15	2024	1 036 850	22,0	160 928	16,1	744 421	11,9	18 444	18,1
	2023	875 648	19,5	153 621	16,0	805 955	15,9	16 455	17,6
16-20	2024	1 020 762	21,7	170 691	17,1	878 166	14,0	21 441	21,0

continued on following page

Table 8. Continued

Age group	Year	Small truck	(%)	Truck	(%)	Motorcycle	(%)	Special purpose	(%)
21+	2023	789 067	17,6	342 849	35,7	758 009	14,9	21 794	23,3
	2024	850 023	18,1	353 589	35,3	773 351	12,4	23 859	23,4
Average age	2023	13,6		17,9		11,6		14,8	
	2024	13,9		18,1		10,2		15,3	

Source: TÜİK, (2025e).

Table 7 and table 8 show the number of vehicles with their age in Turkey for 2023 and 2024. While the number of vehicles acquired has increased compared to the previous year (2023), it is seen that there has been an increase in the rate of acquiring older vehicles.

Figure 3. Car sales by fuel type 2024-2025

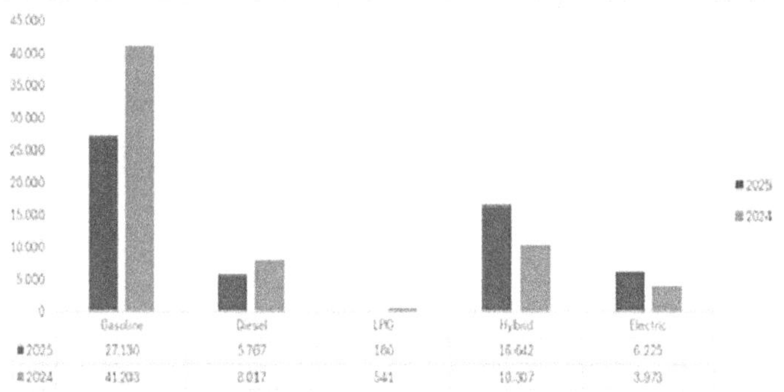

Source: Kalelioğlu, (2025).

As shown in the figure 3, the sales of gasoline vehicles and electric and hybrid vehicles have increased significantly compared to the previous year.

Along with TÜİK, The Automotive Distributors and Mobility Association also provides useful information on indicators related to the vehicle and automobile market in Turkey. The Automotive Distributors and Mobility Association (ODMD) has recorded an increase in electric car sales for Turkey. In January 2025, electric car sales increased by 56.7% compared to the same period last year and reached 6,225 units. (Takvim Gazete (takvim.com), 2025). According to the Automotive Distributors and Mobility Association (ODMD) data, the best-selling electric car in the first 10 months of 2024 was TOGG with 20,140 units. TOGG was followed by Tesla and KG Mobility (FORBES, 2025).

Figure 4. The market of electric cars in Turkey

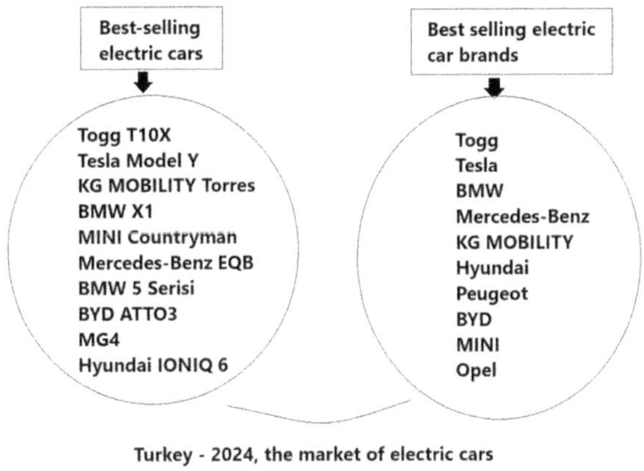

Source: created by the authors based on Kalelioğlu, 2025

Figure 4 shows the best-selling electric cars and car brands in Turkey, it is seen that Togg is the main player in the Turkish electric car. Togg T10X (Turkey's domestic brand) has become the best-selling electric car model of the year in Turkey and has sold much more than Tesla in 2024. According to ODMD data, more than 100 thousand electric cars were sold in all of 2024 (Kalelioğlu, 2025). Togg brand first started in Turkey in 2019. Both the design and the prototype were revealed in 2019. As of March 2023, pre-orders have been received. By the end of 2023, 19,583 Togg units were produced and the Togg brand became the leader in the electric vehicle market in Turkey between May and August 2023 (Togg, n.d.).

Turkey's first electric car, the T10X, continues to meet more users every year. The total number of T10X vehicles in traffic, which has attracted great interest from every region of Turkey throughout 2024, increased by 53.6 percent compared to the previous year, reaching 49,676. In addition to the T10X, Trugo, an end-to-end uninterrupted, clean and high-performance charging provider in Turkey, increased its high-speed DC charging stations in 81 provinces to 842 high-speed chargers and 1684 sockets, while the number of AC stations reached 363. On the other hand, Togg's digital platform Trumore, which has reached over 2 million users, continued to grow with innovations in 2024 (Çelik, 2025).

According to 2023 data from EPDK, there are 4,221 electric vehicle charging stations in Turkey. These stations have a total of 8,861 charging sockets, 6,633 of which have AC (slow charging) and 2,228 have DC (fast charging) features. With the increase in fast charging stations, electric vehicle owners can travel uninterruptedly on long journeys. In addition, there are approximately 33,000 electric vehicles in our country; thus, there is one fast charging point for every 14 vehicles. This rate puts Turkey in first place in Europe in terms of DC charging points per vehicle (Sosyogaraj, n.d.).

Electric Vehicle Charging Map is a comprehensive website that shows electric vehicle charging stations across Türkiye on a map and allows users to easily reach these stations. (Elektrikli Araç Şarj Haritası web site, n.d.).

According to Electric Vehicle Charging Marp web site (n.d.)'s last uploaded info, the below brands and counts for 2025 were given:

- ZES-1568) charging stations
- Eşarj-1178 charging stations
- Trugo-846 charging stations
- Voltrun -741 charging stations
- Otojet-261 charging stations
- WAT-245 charging stations
- EnYakıt-207 charging stations
- Beefull-139 charging stations
- Powerşarj-103 charging stations
- Şarjon-92 charging stations
- D-Charge-89 charging stations
- Sharz.net-86 charging stations
- RHG-70 charging stations
- AOS Technology-58 charging stations
- Otowatt-56 charging stations
- İspirli Şarj-55 charging stations
- Lumicle-44 charging stations
- Arcona-34 charging stations
- Gio EV-24 charging stations
- Spark-20 charging stations
- Tesla-13 charging stations
- Borenco-9 charging station

SOLUTIONS AND RECOMMENDATIONS

Climate change is an accepted event, and its impact has been accelerated as a result of human activities. Despite the negative effects of climate change, the transportation and shipping sector is also undergoing a green transformation within the framework of sustainable policies and strategies. At this point, the electric vehicle market is an important output of the green transformation in transportation. It is observed that the demand for electric vehicles, especially in individual use, is increasing and it is predicted that the electric vehicle ownership rates will increase in the future. In this study, some indicators related to the electric car market in Turkey were examined and it was tried to understand how the green transformation is reflected. According to the indicators in Turkey, while the electric vehicle ownership rates are increasing, it was observed that the demand for Turkey's own Togg electric vehicle is higher than in other markets. Local production is important for incentives for electric vehicles. With the necessary infrastructure work, countries can differentiate the electric vehicle market and produce their own electric vehicles or automobiles. More incentives and support can be given to electric vehicles to stop or slow down the environmental degradation caused by transportation and transport. Incentives for individual purchase can also be useful.

FUTURE RESEARCH DIRECTIONS

This study examines the electric vehicle market in Turkey, a developing country with a large population. Future studies examining indicators in different countries will allow them to reach different results. Because the vehicle market and the factors affecting the vehicle market are different in each country. Although the highest number of electric vehicles in the world is seen in China and Europe, it can be said that the electric vehicle market is a developing sector.

CONCLUSION

Zero Carbon is a significant target and part of the 2030 SDGs and is an important element in the fight against climate change. One of the biggest problems of countries globally is unstoppable climate change and global warming. Due to climate change, serious deterioration occurs in ecosystems and biodiversity, water security and food security problems arise. The zero-carbon policy aims to reduce unwanted gas and gas emissions that increase air pollution and warm up to the minimum level possible. At this point, green transformation has been experienced to achieve a decrease in gas emissions originating from transportation and transportation filters. Green transformation in the field of transportation has brought about the electric vehicle market.

Although it is estimated that the electric paperclip market will be the dominant vehicle market in the coming years, it is currently on a rapid rise. In this study, the electric vehicle indicators in the automotive market in Turkey and the green transformation were examined. Turkey has taken a serious step and developed in the electric vehicle market with its own domestic brand. While it is thought that domestic vehicle production is positive in terms of demand and incentives for electric vehicles, it is thought that the electric vehicle market will adapt quickly in the future, especially in densely populated countries.

Acknowledge: Turkish references were translated into English.

REFERENCES

Akpınar, M. (2024). *Elektrikli araç satışları düşüyor mu? İşte gerçekler (Are electric vehicle sales falling? Here are the facts).* https://www.donanimhaber.com/elektrikli-arac-satislari-dusuyor-mu-iste-gercekler--184105

Bloomberght, (2025). *Otomobil ve hafif ticari araç pazarı 2024 yılını rekorla kapattı, (The automobile and light commercial vehicle market closed 2024 with a record),* https://www.bloomberght.com/otomobil-ve-hafif-ticari-arac-pazari-2024-yilini-rekorla-kapatti-3738613

Bostancı, S. H., & Yıldırım, S. (2021). Sustainable Communities vs. Climate Refugees: Two Opposite Results of Climate Change. In Popescu, C. (Ed.), *Handbook of Research on Novel Practices and Current Successes in Achieving the Sustainable Development Goals* (pp. 298–319). IGI Global., DOI: 10.4018/978-1-7998-8426-2.ch015

Bostancı, S. H., & Yıldırım, S. (2022). The Role of Municipalities in Achieving Water Security: The Case of Turkey. In Castanho, R. (Ed.), *Handbook of Research on Sustainable Development Goals, Climate Change, and Digitalization* (pp. 268–286). IGI Global., DOI: 10.4018/978-1-7998-8482-8.ch017

BYD. (2024). *Çeşit Çeşit Elektrikli Otomobil: HEV, MHEV, PHEV, BEV... Farkları Neler? (Different Types of Electric Cars: HEV, MHEV, PHEV, BEV... What Are Their Differences?),* https://www.bydauto.com.tr/blog/elektrikli-arac-turleri#:~:text=HEV%2C%20MHEV%2C%20PHEV%20ve%20BEV%20gibi%20isimlendirmeler%20de%20daha%20%C3%A7ok,%C5%9Eimdi%20bu%20t%C3%BCrleri%20yak%C4%B1ndan%20tan%C4%B1yal%C4%B1m

Carlier, M. (2024). Number of electric vehicles in use by type 2016-2023, https://www.statista.com/statistics/1101415/number-of-electric-vehicles-by-type/

Çelik, A. (2025). *Togg satışları 50 bin barajına yaklaştı (Togg sales approach 50 thousand threshold),* https://www.turkiyegazetesi.com.tr/t-otomobil/togg-satislari-50-bin-barajina-yaklasti-1084149

Demirci, B. (2024). *Dünyada Otomobil Üretim ve Satışları Artıyor, (Automobile Production and Sales Are Increasing in the World),* https://subconturkey.com.tr/dunyada-otomobil-uretim-ve-satislari-artiyor/

Dünya Gazetesi. (2023). European automobile market grew by 21 percent in August, https://www.dunya.com/sektorler/otomotiv/avrupa-otomobil-pazari-agustosta-21-buyudu-haberi-705484

Ekonomi Gazetesi. (2025). *ABD'de otomobil satışları son beş yılın en yüksek seviyesinde, (US auto sales at five-year high),*https://www.ekonomigazetesi.com/ekonomi/kuresel-ekonomi/abdde-otomobil-satislari-son-bes-yilin-en-yuksek-seviyesinde-48666/

Elektrikli Araç Şarj Haritası web site. (n.d.). Elektrikli Araç Şarj Haritası https://elektrikliaracsarjharitasi.com/pages/

Euronews, (2024). 2030'a kadar elektrikli otomobil sayısı 10 kat artacak; satılan 10 araçtan 6'sı elektrikli olacak, (By 2030, the number of electric cars will increase 10-fold; 6 out of 10 vehicles sold will be electric), https://tr.euronews.com/2024/01/01/2030a-kadar-elektrikli-otomobil-sayisi-10-kat-artacak-satilan-10-aractan-6si-elektrikli-ol

European Environment Agency. (2020). Walking, cycling and public transport in cities remain greener mobility options than electric scooters or car ride-hailing, https://www.eea.europa.eu/highlights/walking-cycling-and-public-transport

EVgo. (n.d.). Types of Electric Vehicles, https://www.evgo.com/ev-drivers/types-of-evs/#bevs

FORBES, (2025). *2024'ün ilk 10 ayında en çok satan elektrikli otomobiller, (The best-selling electric cars in the first 10 months of 2024),*https://www.forbes.com.tr/makale/2024-un-ilk-10-ayinda-en-cok-satan-elektrikli-otomobiller

IEA, & Statista. (2024). Global electric vehicle sales in 2023, with a forecast for 2030 and 2035 based on three different scenarios (in millions) [Graph]. In Statista. Retrieved February 13, 2025, from https://www.statista.com/statistics/1499987/global-ev-sales-forecast-by-scenario/

Jing, Q.-L., Liu, H.-Z., Yu, W.-Q., & He, X. (2022). The Impact of Public Transportation on Carbon Emissions—From the Perspective of Energy Consumption. *Sustainability (Basel), 14*(10), 6248. DOI: 10.3390/su14106248

Kalelioğlu, E. (2025). 2025'te *Türkiye'de En Çok Satılan Otomobiller Belli Oldu: "Pahalı" Diye Yerden Yere Vurulan BYD, Piyasayı Sallıyor! (The Best-Selling Cars in Turkey in 2025 Have Been Announced: BYD, Which Was Criticized for Being "Expensive", Is Shaking Up the Market!),*https://www.webtekno.com/en-cok-satilan-otomobiller-h126446.html

Mead, L. (2021). The Road to Sustainable Transport, https://www.iisd.org/articles/deep-dive/road-sustainable-transport

Nissan, (2022). How Do Electric Cars Work? https://www.nissanusa.com/experience-nissan/news-and-events/how-do-electric-cars-work.html#:~:text=Here's%20a%20basic%20rundown%20of,to%20make%20this%20motion%20happen

Olayode, I. O., Jamei, E., & Alex, F. J. (2025). Integration of e-bikes in public transportation based on their impact, importance, and challenges: A systematic review. *Multimodal Transportation*, *4*(1), 100182. DOI: 10.1016/j.multra.2024.100182

Ridango, (2022). Five reasons why using public transport is better for the environment, https://www.ridango.com/five-reasons-why-using-public-transport-is-better-for-the-environment/

Ritchie, H. (2023). Which form of transport has the smallest carbon footprint? Published online at OurWorldinData.org. Retrieved from: 'https://ourworldindata.org/travel-carbon-footprint'

Sevik, N., & Yıldırım, S. (2023). Sustainable Enterprises Post COVID-19 Pandemic: The Rise of Sustainable Fashion Enterprises. In Castanho, R. (Ed.), *Handbook of Research on Current Advances and Challenges of Borderlands, Migration, and Geopolitics* (pp. 343–359). IGI Global., DOI: 10.4018/978-1-6684-7020-6.ch017

Shahid, S., Minhans, A., Puan, O.C. (2014). Assessment of Greenhouse Gas Emission Reduction Measures in Transportation Sector of Malaysia. (2014). Jurnal Teknologi (Sciences & Engineering), 70(4). DOI: 10.11113/jt.v70.3481

Sosyogaraj, (n.d.). *Türkiye Genelinde Elektrikli Araç Şarj İstasyonu Sayısı ve Özellikleri (Number and Features of Electric Vehicle Charging Stations in Turkey),* https://sosyogaraj.com.tr/elektrikli-arabalar/turkiyede-il-il-sarj-istasyon-sayilari-ve-bulundugu-noktalar/

Statista. (2024). Estimated number of electric vehicles in use worldwide between 2016 and 2023, by type (in 1,000s) [Graph]. In Statista. Retrieved February 13, 2025, from https://www.statista.com/statistics/1101415/number-of-electric-vehicles-by-type/

Takvim Gazete (takvim.com), (2025). *Elektrikli otomobil satışları yükseliyor: Togg liderliğini sürdürüyor, (Electric car sales rise: Togg maintains its lead),* https://www.takvim.com.tr/ekonomi/2025/02/09/elektrikli-otomobil-satislari-yukseliyor-togg-liderligini-surduruyor

Togg, (n.d.). *Geleceğe Keyifli Bir Yolculuk (A Pleasant Journey to the Future)*, https://www.togg.com.tr/about

TÜİK. (2025a). The number of road motor vehicles by years, 2002-2024, Road Motor Vehicles, Aralık 2024, https://data.tuik.gov.tr/Bulten/Index?p=Motorlu-Kara-Tasitlari-Aralik-2024-53463

TÜİK. (2025b). Distribution of cars registered to the traffic according to fuel type, 2004-2024, Road Motor Vehicles, Aralık 2024, https://data.tuik.gov.tr/Bulten/Index?p=Motorlu-Kara-Tasitlari-Aralik-2024-53463

TÜİK. (2025c). Distribution of road motor vehicles registered to the traffic by provinces, Road Motor Vehicle, December 2024, https://data.tuik.gov.tr/Bulten/Index?p=Motorlu-Kara-Tasitlari-Aralik-2024-53463

TÜİK. (2025d). Distribution of cars registered to the traffic by engine size, 2020-2024, Road Motor Vehicle, December 2024, https://data.tuik.gov.tr/Bulten/Index?p=Motorlu-Kara-Tasitlari-Aralik-2024-53463

TÜİK. (2025e). Distribution of vehicles registered to the traffic by age group, 2023, 2024, Road Motor Vehicle, December 2024, https://data.tuik.gov.tr/Bulten/Index?p=Motorlu-Kara-Tasitlari-Aralik-2024-53463

US Department of Energy. (n.d.). Alternative Fuels Data Center, https://afdc.energy.gov/laws/12660#:~:text=An%20EV%20is%20defined%20as,charged%20from%20an%20external%20source

Van Fan, Y. (2018). Simon Perry, Jiří Jaromír Klemeš, Chew Tin Lee, (2018). A review on air emissions assessment: Transportation. *Journal of Cleaner Production, 194*, 673–684. DOI: 10.1016/j.jclepro.2018.05.151

Yildirim, S. (2020). The consumer role for sustainable development: how consumers contribute sustainable development goals. In Chkoniya, V., Madsen, A. O., & Bukhrashvili, P. (Eds.), *Anthropological Approaches to Understanding Consumption Patterns and Consumer Behavior* (pp. 325–341). IGI Global. DOI: 10.4018/978-1-7998-3115-0.ch018

Yıldırım, S. (2021). Do green women influencers spur sustainable consumption patterns? Descriptive evidences from social media influencers. *Ecofeminism and Climate Change, 2*(4), 198–210. DOI: 10.1108/EFCC-02-2021-0003

Yildirim, S. (2021). 3R Principle in Household Energy Consumption Patterns: A Case of Turkish Households. In Danish, M., & Senjyu, T. (Eds.), *Eco-Friendly Energy Processes and Technologies for Achieving Sustainable Development* (pp. 88–107). IGI Global. DOI: 10.4018/978-1-7998-4915-5.ch005

Yıldırım, S., & Candan, B. (2015). Segmentation of green product buyers based on their personal values and consumption values. *Environmental Values, 24*(5), 641–661. DOI: 10.3197/096327115X14384223590212

Yildirim, S., Demirtas, I., & Yildirim, D. C. (2022). A Review of Alternative Economic Approaches to Achieve Sustainable Development: The Rising Digitalization and Degrowth Post COVID-19. In Castanho, R. (Ed.), *Handbook of Research on Sustainable Development Goals, Climate Change, and Digitalization* (pp. 288–307). IGI Global. DOI: 10.4018/978-1-7998-8482-8.ch018

Yıldırım, S., & Kantarcı, T. (2022). A Review on Sustainability Policies of Businesses: Recycling and Waste Reduction. *Journal of Recycling Economy & Sustainability Policy, 1*(1), 1–9. https://respjournal.com/index.php/pub/article/view/1

Yıldırım, S., Sevik, N., Kandpal, V., & Yıldırım, D. C. (2024). The Role of Green Brands on Achieving 2030 Sustainable Development Goals (2030 SDGs). In Kankaew, K., Chaudhary, S., & Widtayakornbundit, S. (Eds.), *Contemporary Management and Global Leadership for Sustainability* (pp. 141–162). IGI Global., DOI: 10.4018/979-8-3693-1273-5.ch009

Yıldırım, S., & Vasilev, V. (2024). The Future of Green Collar Workers: Green Transition in Employment. In Yıldırım, S., Yıldırım, D., Demirtaş, I., & Kandpal, V. (Eds.), *Green Transition Impacts on the Economy, Society, and Environment* (pp. 1–16). IGI Global Scientific Publishing., DOI: 10.4018/979-8-3693-3985-5.ch001

Yıldırım, S., & Yıldırım, D. Ç. (2020). Achieving Sustainable Development Through a Green Economy Approach. In Patti, S., & Trizzino, G. (Eds.), *Advanced Integrated Approaches to Environmental Economics and Policy: Emerging Research and Opportunities* (pp. 1–22). IGI Global. DOI: 10.4018/978-1-5225-9562-5.ch001

Yıldırım, S., Yıldırım, D. C., & Rajamanickam, S. (2025). Green Transition in the Restaurant Industry: New Trends in Green Restaurants. In Yıldırım, S., Demirtaş, I., & Malik, F. (Eds.), *Green Economics and Strategies for Business Sustainability* (pp. 419–442). IGI Global Scientific Publishing., DOI: 10.4018/979-8-3693-8949-2.ch017

ADDITIONAL READING

Bostancı, S. H., & Yıldırım, S. (2021). Sustainable Communities vs. Climate Refugees: Two Opposite Results of Climate Change. In Popescu, C. (Ed.), *Handbook of Research on Novel Practices and Current Successes in Achieving the Sustainable Development Goals* (pp. 298–319). IGI Global., DOI: 10.4018/978-1-7998-8426-2.ch015

Bostancı, S. H., & Yıldırım, S. (2022). The Role of Municipalities in Achieving Water Security: The Case of Turkey. In Castanho, R. (Ed.), *Handbook of Research on Sustainable Development Goals, Climate Change, and Digitalization* (pp. 268–286). IGI Global., DOI: 10.4018/978-1-7998-8482-8.ch017

Sevik, N., & Yıldırım, S. (2023). Sustainable Enterprises Post COVID-19 Pandemic: The Rise of Sustainable Fashion Enterprises. In Castanho, R. (Ed.), *Handbook of Research on Current Advances and Challenges of Borderlands, Migration, and Geopolitics* (pp. 343–359). IGI Global., DOI: 10.4018/978-1-6684-7020-6.ch017

Yildirim, S. (2020). The consumer role for sustainable development: how consumers contribute sustainable development goals. In Chkoniya, V., Madsen, A. O., & Bukhrashvili, P. (Eds.), *Anthropological Approaches to Understanding Consumption Patterns and Consumer Behavior* (pp. 325–341). IGI Global. DOI: 10.4018/978-1-7998-3115-0.ch018

Yıldırım, S. (2021). Do green women influencers spur sustainable consumption patterns? Descriptive evidences from social media influencers. *Ecofeminism and Climate Change*, *2*(4), 198–210. DOI: 10.1108/EFCC-02-2021-0003

Yildirim, S. (2021). 3R Principle in Household Energy Consumption Patterns: A Case of Turkish Households. In Danish, M., & Senjyu, T. (Eds.), *Eco-Friendly Energy Processes and Technologies for Achieving Sustainable Development* (pp. 88–107). IGI Global. DOI: 10.4018/978-1-7998-4915-5.ch005

Yıldırım, S., & Bostancı, S. H. (2021). The efficiency of e-government portal management from a citizen perspective: Evidences from Turkey, *World Journal of Science. Technology and Sustainable Development*, *18*(3), 259–273. DOI: 10.1108/WJSTSD-04-2021-0049

Yıldırım, S., & Candan, B. (2015). Segmentation of green product buyers based on their personal values and consumption values. *Environmental Values*, *24*(5), 641–661. DOI: 10.3197/096327115X14384223590212

Yildirim, S., Demirtas, I., & Yildirim, D. C. (2022). A Review of Alternative Economic Approaches to Achieve Sustainable Development: The Rising Digitalization and Degrowth Post COVID-19. In Castanho, R. (Ed.), *Handbook of Research on Sustainable Development Goals, Climate Change, and Digitalization* (pp. 288–307). IGI Global. DOI: 10.4018/978-1-7998-8482-8.ch018

Yıldırım, S., & Kantarcı, T. (2022). A Review on Sustainability Policies of Businesses: Recycling and Waste Reduction. *Journal of Recycling Economy & Sustainability Policy*, *1*(1), 1–9. https://respjournal.com/index.php/pub/article/view/1

Yildirim, S., Kaplan, M., & Yildirim, D. C. (2023). Climate Change and the Ocean Economy: A Review on OECD's Ocean Economy. In Pego, A. (Ed.), *Handbook of Research on Bioeconomy and Economic Ecosystems* (pp. 212–232). IGI Global., DOI: 10.4018/978-1-6684-8879-9.ch012

Yıldırım, S., Sevik, N., Kandpal, V., & Yıldırım, D. C. (2024). The Role of Green Brands on Achieving 2030 Sustainable Development Goals (2030 SDGs). In Kankaew, K., Chaudhary, S., & Widtayakornbundit, S. (Eds.), *Contemporary Management and Global Leadership for Sustainability* (pp. 141–162). IGI Global., DOI: 10.4018/979-8-3693-1273-5.ch009

Yıldırım, S., & Vasilev, V. (2024). The Future of Green Collar Workers: Green Transition in Employment. In Yıldırım, S., Yıldırım, D., Demirtaş, I., & Kandpal, V. (Eds.), *Green Transition Impacts on the Economy, Society, and Environment* (pp. 1–16). IGI Global., DOI: 10.4018/979-8-3693-3985-5.ch001

Yıldırım, S., & Yıldırım, D. Ç. (2020). Achieving Sustainable Development Through a Green Economy Approach. In Patti, S., & Trizzino, G. (Eds.), *Advanced Integrated Approaches to Environmental Economics and Policy: Emerging Research and Opportunities* (pp. 1–22). IGI Global. DOI: 10.4018/978-1-5225-9562-5.ch001

Yıldırım, S., & Yıldırım, D. C. (2022). Achieving Seafood Security in the Mediterranean Region: A Case of Turkey. In I. Management Association (Ed.), Research Anthology on Ecosystem Conservation and Preserving Biodiversity (pp. 1780-1800). IGI Global. https://doi.org/DOI: 10.4018/978-1-6684-5678-1.ch086

Yıldırım, S., Yıldırım, D. C., Demirtaş, I., & Kandpal, V. (Eds.). (2024). *Green Transition Impacts on the Economy, Society, and Environment*. IGI Global., DOI: 10.4018/979-8-3693-3985-5

Yildirim, S., Yildirim, D. C., & Erdogan, F. (2022). The Sustainability Level of Agricultural Food Markets: Selected Indicators From Mediterranean Countries. In Castanho, R. (Ed.), *Analyzing Sustainability in Peripheral, Ultra-Peripheral, and Low-Density Regions* (pp. 131–155). IGI Global., DOI: 10.4018/978-1-6684-4548-8.ch008

Yildirim, S., Yildirim, D. C., & Erdogan, S. (2024). Energy Security Dilemma in OECD Countries: The COVID-19 Pandemic, Russia-Ukraine War, and Energy Issues. In R. Castanho (Ed.), Green Economy and Renewable Energy Transitions for Sustainable Development (pp. 83-105). IGI Global. DOI: 10.4018/979-8-3693-1297-1.ch005

Yildirim, S., Yildirim, D. C., & Esen, Ö. (2023). Environmental Protection and Environmental Protection Expenditure in Developing Countries: A Case of Turkey. In Castanho, R. (Ed.), *Handbook of Research on Current Advances and Challenges of Borderlands, Migration, and Geopolitics* (pp. 100–119). IGI Global., DOI: 10.4018/978-1-6684-7020-6.ch005

Chapter 15
Digitalization of Ride-Hailing Services:
Revisiting Economic and Environment Sustainability

Meenakshi Bose
https://orcid.org/0009-0004-3535-8894
Bandhan, India

Swarnendu Mazumder
https://orcid.org/0009-0004-8251-8889
Bandhan, India

ABSTRACT

What measures should be taken to effectively address the negative effects of the digitalization of ride-hailing services? Let's ponder some thoughts!! Findings validate that the digitalization of ride-hailing services remains a threat to environmental sustainability. Despite this, ride-hailing services have successfully handled the issue of rising concern of unemployment in the urban setting of Kolkata. The study was conducted at the traffic signals of the Jadavpur and Rabindra Bharti areas, Kolkata. Using random sampling 150 OLA and Uber drivers of two-wheelers and four-wheelers and 150 riders, especially those belonging to the working age group (i.e. 15- 64 years). An analysis shows that the average Air Quality Index (AQI) was maximum during winter i.e. above 140 AQI and reduced during the rainy season in both study locations. Still, there is apprehension that the rise in ride-hailing services could adversely impact the Air Quality Index. Hence, smart use of digitalization and ride-hailing services (i.e. carpooling and electrification) can address air pollution.

DOI: 10.4018/979-8-3693-9909-5.ch015

INTRODUCTION

The introduction of digital technologies has revolutionized various industries, and the ride-hailing sector is one of the distinguished examples. Ride-hailing services embarked as a basic notion of hailing a vehicle and have developed into an advanced, technology-driven industry transforming urban mobility. The emergence of ride-hailing services in India was introduced by Uber in 2009 represented a paradigm shift, providing accessibility, cost-effectiveness, and customization. In 2010, Ola started its journey in India and has scattered its network in over 200 cities (Sheldon & Dua, 2024). Ride-hailing services have been through a revolution by using big data analytics to make well-informed choices, enhance their services, and highlight areas of advancement. It has undergone a tremendous transition over the last decade, geared by rapid technological improvements. It uses AI algorithms for various specifications such as demand, congestion, and vehicle availability to change rates in real-time. A key component of managing the massive volumes of data produced by ride-hailing services is cloud computing. Li et al. (2021) emphasized that the global expansion of ride-hailing services mostly depends on cloud-based systems to sustain service continuity, businesses can scale their operations globally by utilizing cloud technology. Data analytics are used by ride-hailing businesses to enhance driver incentives, customer satisfaction, and operational efficiency. Real-time pricing, demand forecasting, route optimization, and customized marketing strategies are all made possible by data-driven decisions. It offers travel patterns, traffic conditions, and user preferences that may be processed, stored, and analysed as an outcome of cloud-based infrastructures. The digitalization of ride-hailing services simplified the travel experience by sharing information about real-time routes, locations, and congestion using GPS navigation. In short, digitalization in ride-hailing services denotes the assimilation of technology-driven inventions.

Digital platforms facilitate ride-hailing companies to expand their operations globally, entering new markets. It has improved the frequency of ride-hailing services which saves time while rendering comfortable rides. Analysis of traffic data using machine learning algorithms has facilitated increased customer satisfaction and service efficiency. Ride-hailing applications provide customized recommendations, including preferred routes, vehicle kinds, and payment methods, by examining customer preferences and travel patterns resulting in increased client loyalty and happiness. Companies such as Uber, Ola, S Ride, Lyft, and others have used innovative technologies to improve speed, efficiency, and overall security for millions of consumers worldwide. Ride-hailing companies have been spending substantial resources to augment acceptance and use of the technology through strategies, partnerships, amenities, and enticements (Dias, Kim, Bhat, Pendyala, Lam, Pinjari,

Srinivasan and Ramadurai n.d.). Also, it has a constructive effect on addressing unemployment and economic enhancement of the households.

Recently ride-hailing services have attracted the substantial interest of researchers from different fields. A large body of research work focuses on the digitalization of ride-hailing services and other related subjects such as operational management (Benjaafar and Hu, 2021). A few research studies also focused on pricing (Cachon et al. 2017, Bernstein et al. 2021, and Bai et al. 2019), effective match of demand and supply (Akbarpour et al. 2020, Feng et al. 2021), and sharing of rides (Agatz et al. 2012, Alonso-Mora et al. 2017). Previous research showed that ride-hailing services have facilitated solving the concern of the 'last mile' (Huang et al., 2021). While one set of academics affirms that ride-hailing services have supplanted public transit, another other believes that public transportation has become more efficient (Graehler, et al. 2019, Hall, et al. 2018, Berrebi and Watkin, 2020). A trajectory data-driven ride-hailing system was introduced by Hong et al. (2017) to determine commuters' individual preferences and match their ride-sharing itineraries. A few studies on ride-hailing services have shown several advantages, including the creation of substantial consumer surplus (Cohen et al. 2016), alternatives to poor-quality entrepreneurial activity (Burtch et al. 2018), opportunities for flexible work (Chen et al. 2019), enhancements in service quality (Wallsten 2015, Athey et al. 2018), and a decrease in driver moral hazard using monitoring technologies (Liu et al. 2021). More demand for taxis and Uber is linked to longer roads, shorter transit access times, fewer car owners, greater incomes, and more employment prospects (Correa et al. 2023).

Some authors have also suggested that the motive of ride-hailing services is not only confined to purely economic reasons but also has a strong association with sustainability, equality, empowerment, social inclusion, ease, point-to-point movement, and goodwill (Godelnik, 2017; Hamari et al., 2016). An intensive review of the literature reveals various societal impacts as a result of the augmented use of digitalization in ride-hailing services. It has garnered recognition in recent years for reducing the number of privately owned cars on the road and, consequently, the quantity of greenhouse gas emissions they generate (Cohen and Kietzmann, 2014). However, it cannot be denied that an increased number of automobiles causes pollution and traffic congestion, a thorough environmental assessment of digital technologies is essential to comprehend how the industry affects biodiversity loss and climate change. Erhardt et al. (2019) confirm that the increased use of ride-hailing services has significantly contributed to traffic congestion in urban areas. Globally, the number of taxi services exploded profoundly in the past decade. This eventually has a detrimental effect on achieving the Sustainable Development Goals (i.e. Climate Action- SDG 13). The digitalization of ride-hailing services has questioned the sustainability of the environment. Given the dire circumstances resulting from climate change as an impact of growing air pollution caused by increased use of vehicular travel, it is essential to keep a constant eye on the concerns, especially in metropolitans. Considering the severity of the climate issue, it is important to evaluate their possible effects on the environment and identify strategies to reduce their impact. This book chapter is consistent with the United Nations Environment Programme's (UNEP, 2002) assertion that 1.1 billion people worldwide breathe contaminated air. It intends to draw attention to the rising concern of environmental sustainability brought in by ride-hailing services powered by current technologies. It attempts to revisit the transformation as an outcome of the digitalization of ride-hailing services and its influence on the environment in Kolkata.

The harmful emissions from automobiles severely impact the environment in terms of climate change, exposure to pollutants, and air and water quality, especially in metropolitan cities. Simultaneously, it negatively affects the physical health of individuals specifically of the driver of ride-hailing vehicles who are most exposed to air pollution. Environmental wellness is a subject of public health that has the potential to influence human health and addresses the natural environment in which an individual inhabits. Health and well-being are essential to achieve the Sustainable Development Goals (SDG)-3 of the United Nations which aims to ensure healthy lives and promote well-being for all ages. Moreover, India's commitment to the SDGs calls for strenuous efforts to improve the health and well-being of its citizens. Metropolitan cities like Delhi have a high incidence of chronic respiratory diseases and heart-related ailments as a product of air pollution exacerbating respiratory and cardiovascular conditions. The city of Kolkata has been dubbed the "diesel capital of the country" due to its heavy diesel usage (Karmakar, 2020). The Ministry of Environment and Forests confirmed in August 2010 that Kolkata was the twelfth most polluted city in terms of PM10 levels and the third most polluted city in terms of nitrogen oxide (NOx) levels (Roy Chowdhury, 2011). In August 2003, the Supreme Court stated that the Respirable Suspended Particulate Matter (RSPM) in several Indian cities, including Kolkata, is concerning. The court based its decision on data from the Central Pollution Control Board (CPCB) and ordered the CPCB to collaborate with regional regulatory organizations to create action plans to enhance air quality. Kolkata's air pollution level is caused by several reasons. The transportation industry is the primary cause of air pollution compared to other sectors (Ghose et al. 2004). Ride-hailing is a major contributing component (Mondal et al., 2000) due to a large number of improperly maintained vehicles, the usage of gasoline and petrol (although the government is working to phase out these vehicles), and inadequate regulation (Mukherjee and Mukherjee 1998; Kazimuddin and Banerjee 2000). The increased frequency of ride-hailing vehicles is the main source of greenhouse gas emissions that affect Kolkata's air quality. Kolkata experiences variations in air pollution in three distinct seasons: monsoon, summer, and winter. The summer months are March through June, the monsoon season is July through October, and the winter months are November through February. The winter air quality in Kolkata and the surrounding areas remains poor, but the summer air quality significantly improves (*idib*, 2020). In summer, heat radiates from the ground, warming the air closest to the ground as the temperature rises. The warmer air is lighter, it rises through the atmosphere, carrying with it pollution and dust. The speed of wind during summer is faster which carries these pollutants away. In contrast, the wind is heavier and slow as a result stays closer to the ground during winter because of the low temperatures. Due to the sluggish wind speed, the dust produced by burning coal, automobile emissions, and industrial smoke stays at ground level where people breathe it instead of rising to the atmosphere. As a consequence, individuals suffer from breathing problems, and those with chronic bronchitis and chronic obstructive pulmonary disease (COPD) are adversely affected.

This provokes the thought – "In the upcoming five years, would Kolkata emerge as the next New Delhi in terms of air pollution?"

Ride-hailing services are held accountable for the rise in pollution and congestion. Hence, comprehending the effects of the digitalization of ride-hailing services on economic and environmental sustainability is a subject of substantial interest among researchers. In light of ride-hailing services this chapter endeavours to address the following questions, what are the effects and implications of these services for human life? How can the adverse issues brought about by ride-hailing services going digital be effectively addressed? The present study was grounded in this rationale and attempts to address the concerns raised above.

In keeping with the previous debate, the study was conducted to comprehend the impact of the digitalization of ride-hailing services on economic and environmental sustainability in Kolkata, West Bengal. This chapter attempts to study the health conditions of the drivers of ride-hailing services. Simultaneously, it attempts to comprehend the constructive effect of ride-hailing services addressing unemployment and economic enhancement of the household. The impact of the digitalization of ride-hailing services on the livelihood status of its driver is covered in this chapter through comprehension of their enhanced income, sustainable engagement, and a multidimensional perspective. This chapter has used the multidimensional poverty index (MPI) to assess the economic influence of the digitalization of ride-hailing services and access to sustainable sources of income among drivers.

Objectives: This research aims to investigate how ride-hailing's digitalization impacts economic and environmental sustainability.

Specific objectives:

- To comprehend the status of the Air Quality Index (AQI) and the spread of pollutants in the study areas.

- To investigate the degree to which digitalization of ride-hailing services affects environmental sustainability with reference to air pollution and its impact on the health of drivers.
- To apprehend the impact of increased use of ride-hailing in addressing unemployment and poverty.
- To understand passengers' perception of digitalised ride-hailing services.

Chapter plan: This chapter is arranged into four sections that deal with the digitalization influence on sustainability both in terms of economic and environmental. The first section was an overview of the digitalization of ride-hailing services in India. It dialogues the thought-provoking questions and objectives that encouraged the examination. After an introduction, the rest of this chapter is organized as discussed below. The second section discourses the research methodology adopted to commence this study. The analysis is presented in the next section along with a discussion. It includes a discourse on the status of air pollution in the study locations. The succeeding sub-section attempts to understand health complications encountered by the drivers of ride-hailing services. The next sub-section discusses the perception of the riders. Further, it discusses concerns such as the degree to which digitalization of ride-hailing services was able to resolve the concern of unemployment and ensure economic sustainability. It gives an idea of how the multidimensional poverty index of the household is affected by the poor health condition of the drivers. Weaving together the sections, the conclusion section reiterates the chapter's overall arguments and concludes with a recommendation.

METHODOLOGY

A systematic literature review was conducted to comprehend the impacts of digitalization in the ride-hailing industry, focusing on fundamental technical advancements and environmental effects. This research was a cross-sectional study which aimed to study air pollution in terms of Fine Particulate Matter ($PM_{2.5}$), Particulate Matter (PM_{10}), Ozone (O_3), Nitrogen Dioxide (NO_2), Sulphur Dioxide (SO_2), and Carbon Monoxide (CO) indicators, as well as the Air Quality Index (AQI). This study monitored these pollutants in the Jadavpur and Rabindra Bharti areas daily throughout May (summer), August (rainy), and November (winter).

A multistage sampling method was adopted to conduct the study (see Figure 1). In line with the above discussion, the investigation was purposively conducted in Kolkata, a major metropolis with a population of more than 14 million. It is one such metropolitan area struggling with severe air pollution (Sarkar, 2024). The study was purposively conducted near the traffic signals of the Jadavpur and Rabindra

Bharti areas in Kolkata. Subsequently, using random sampling 150 OLA and Uber drivers, especially those belonging to the working age group (i.e. 15- 64 years) were interviewed. Discussions were conducted with drivers belonging to two-wheeler and four-wheeler vehicle segments as respondents. The majority of the vehicles used for passenger service are fuelled with petrol. Besides, this research also interviewed 150 riders to comprehend their perception of the digitalization of ride-hailing services. A mixed-method approach that combined quantitative and qualitative approach was employed to ensure that the results were validated. The primary dataset includes information related to the existing travel habits, adoption of digitalization, and usage of ride-hailing services as well as the socio-economic profile of the drivers. This study adopted the triangulation method using in-depth interviews, focus group discussions, and observation methods for the data collection. A semi-structured questionnaire was administered using Google Forms through mobile tablets during the field visit to capture the responses of the respondents. The results and discussion are presented in the next part, which starts with the state of various air pollutants in Kolkata's metropolitan areas before moving on to how digitization increased the number of automobiles on the road and its effects on the climate.

Figure 1. Sampling framework

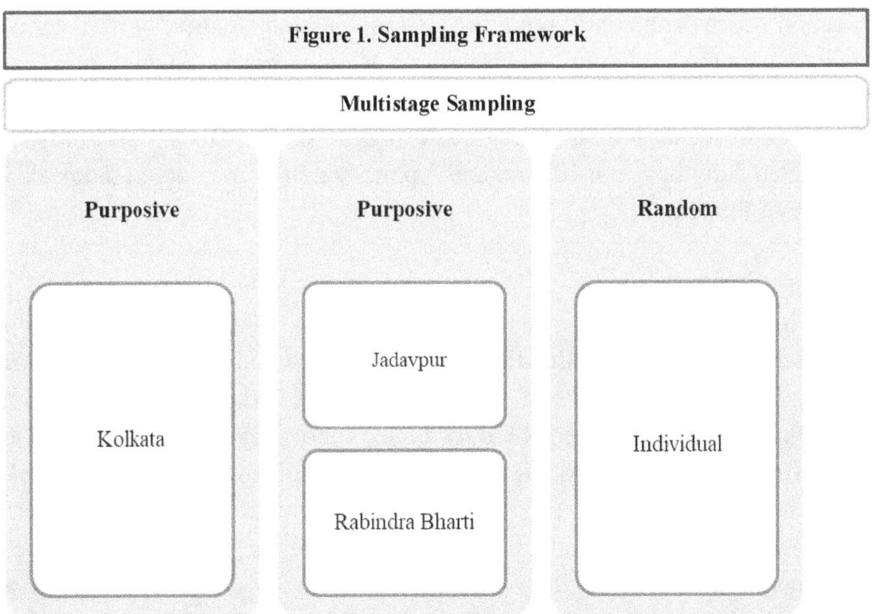

Source: Field Survey Data, 2023

RESULTS AND DISCUSSION

The findings explain that air pollution in urban areas, in the context of Kolkata is affected by the increased frequency of ride-hailing vehicles because of digitalization. It demonstrates how the increased use of digitalization in ride-hailing services had a major impact on air pollution particularly in urban areas. The environmental issues raised by ride-hailing services are covered in the section below. Further, this section discourses the detrimental impact on environmental sustainability and the health of drivers. On the contrary, it supports the idea that increased usage of digitalization in ride-hailing services has improved access to income-generating avenues and sustainable occupation, improving livelihood status and reducing poverty in the next sub-section. It discusses the constructive influence of ride-hailing services going digital through improved communication and real-time tracking.

Environmental sustainability in question

This research validates the stance of Mukherjee et al. (2013) and Kazimuddin et al. (2000) that ride-hailing service is a major contributor to air pollution due to the high number of vehicles that are not properly maintained, and the consumption of petrol. This section covers the detrimental effects of ride-hailing services' digitization on environmental pollution. It provides information on the status of air pollutants in the study regions throughout May (summer), August (rainy), and November (winter). The findings confirm that air pollution in Kolkata decreases throughout the wet season but increases during the winter. During the study, data was collected daily between 10 a.m. and 3 p.m., which is considered the peak and lean hours of the day.

Air Quality Index

Figure 2 below illustrates the difference in the Air Quality Index (AQI) during May, August, and November in Rabindra Bharti and Jadavpur. Analysis reveals that the average Air Quality Index (AQI) remains above 140 AQI during both lean and peak times of the day in both Rabindra Bharti and Jadavpur in winter. The AQI level is marked orange, indicating that members of vulnerable groups may suffer health consequences. The broader populace is less likely to be impacted. In the case of the summer season (May), the AQI remains at Yellow which is moderate and acceptable. In the study area Rabindra Bharti, AQI was 62.3 at 10 a.m. while it was 74.8 in May (summer). Similarly, the average AQI in Jadavpur at 10 am was 52.1 and increased to 58.5 at 3 pm (see figure 2). However, there may be a risk for some people, particularly those who are extremely sensitive to air pollution.

Figure 2. Average air quality index

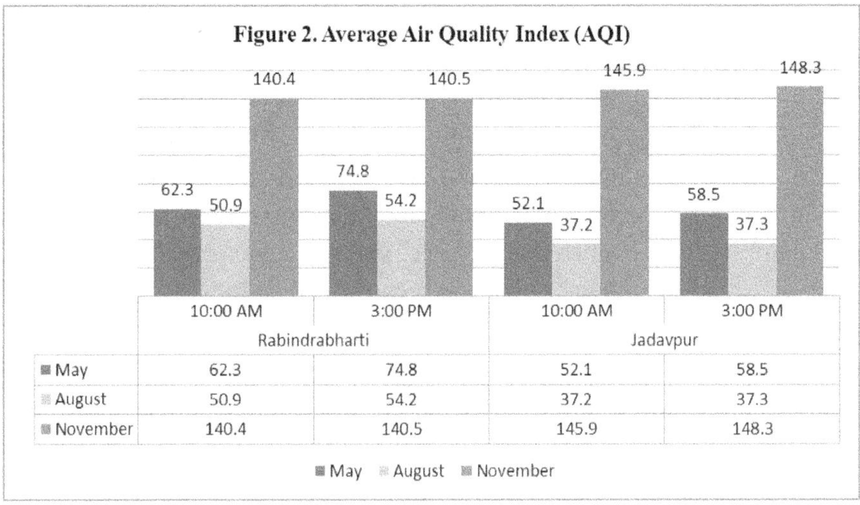

Source: Primary Study, 2023

The Air Quality Index indicates improvement in August (rainy) in both study locations i.e. Rabindra Bharti and Jadavpur. The average AQI at Rabindra Bharti was 50.9 at 10 a.m., increasing to 54.2 at 3 p.m. On the other hand, the average AQI at Jadavpur was 37.2 at 10 am and 37.3 at 3 pm. This substantiates that Rabindra Bharti scored yellow which is moderate even during August. However, Jadavpur scored green which is considered good, indicating that air quality is satisfactory and that air pollution poses little or no harm. The average Air Quality Index of the studied locations varies from moderate to good, explaining that the level of air quality is satisfactory. This implies it is not harmful but some individuals, especially those extremely sensitive to air pollution, may be at risk.

Particle Pollution:

This section discusses the pollution caused by particles such as PM2.5 and PM10. Particle pollution, commonly known as particulate matter (PM), is a mixture of solid and liquid particles in the atmosphere. Particle pollution can impact human health and the environment including PM2.5 and PM10. The concentration of particulate matter (PM2.5 and PM10) increases the risk of respiratory and cardiovascular diseases shortening life expectancy.

PM2.5

In this context, it is important to discuss PM 2.5 which is mostly caused by combustion, with major sources including fireplaces, car engines, and coal- or fracked (natural) gas-fired power plants. PM2.5 is particulate matter with a diameter of 2.5 micrometres or less. During the study, it became apparent that the average PM 2.5 range is fair (i.e. 25–50 µg/m3) in both Rabindra Bharti and Jadavpur locations, both during peak and lean periods of the day during the summer and rainy seasons as shown in figure 3 below. On the contrary, the average PM 2.5 ranges between 50–150 µg/m3 in winter which is deplorable. This calls for attention because PM2.5 particles are small enough to be breathed deeply into the lungs, posing health risks. Individuals who are susceptible to air pollution, such as youngsters, the elderly, pregnant women, and those with allergies or heart or lung diseases, may experience wheeziness, coughing, or difficulty breathing.

Figure 3. Average PM 2.5

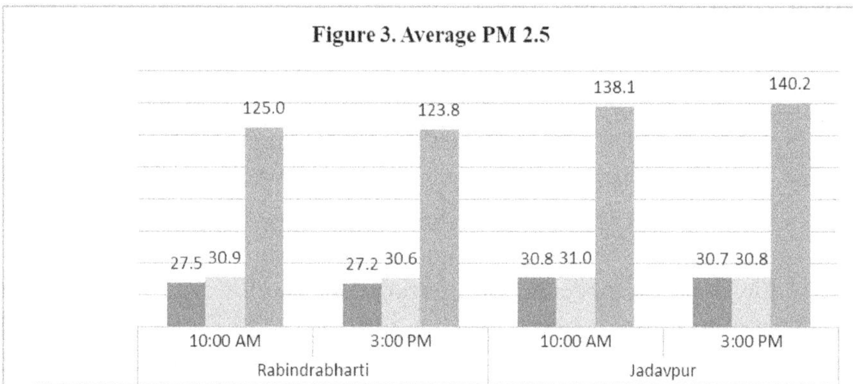

	Rabindrabharti 10:00 AM	Rabindrabharti 3:00 PM	Jadavpur 10:00 AM	Jadavpur 3:00 PM
May	27.5	27.2	30.8	30.7
August	30.9	30.6	31.0	30.8
November	125.0	123.8	138.1	140.2

Source: Primary Study, 2023

PM10

PM10 are extremely microscopic particles found in dust and smoke. They have a diameter of 10 micrometres (0.01 mm) or less. PM10 particles are sufficiently small to enter the throat and lungs. High PM10 levels can cause a variety of health problems, including coughing, nosebleeds, stinging eyes, wheezing, asthma attacks and bronchitis, high blood pressure, heart attacks, strokes, and early mortality.

Analysis claims that the average PM10 in November (winter) remains very poor in both Rabindra Bharti and Jadavpur areas during both peak and lean times of the day (see figure 4). Conversely, the average PM10 in May (summer) and August (rainy) remains fair (40-80 µg/m3) in Rabindra Bharti at both peak and lean periods of the day. In the case of Jadavpur, both May (summer) and August (rainy) stand to be good (Less than 40 µg/m3).

Figure 4. Average PM 10

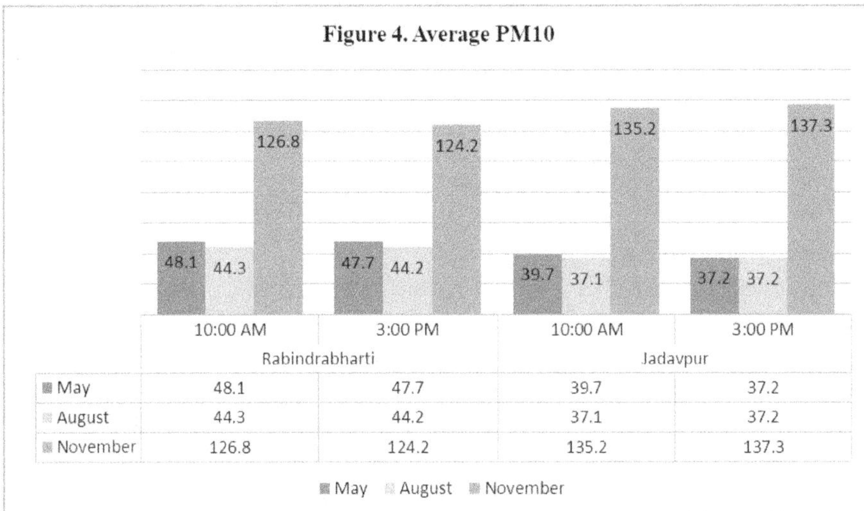

	Rabindrabharti 10:00 AM	Rabindrabharti 3:00 PM	Jadavpur 10:00 AM	Jadavpur 3:00 PM
May	48.1	47.7	39.7	37.2
August	44.3	44.2	37.1	37.2
November	126.8	124.2	135.2	137.3

Source: Primary Study, 2023

Carbon Monoxide

Carbon Monoxide (CO) is colourless, odourless, and dangerous if inhaled in excessive quantities. Cars, trucks, and other vehicles or machines that run on fossil fuels emit the most Carbon Monoxide (CO) into the atmosphere.

Figure 5. Average carbon monoxide

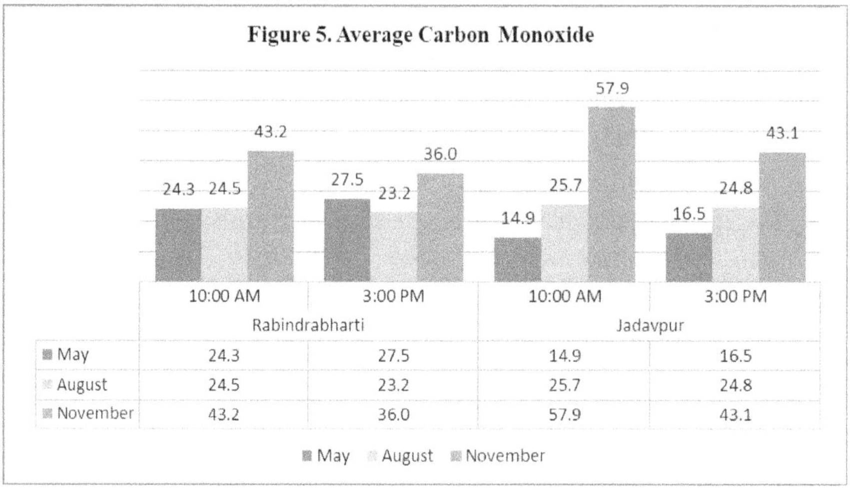

Source: Primary Study, 2023

Breathing air with a high CO concentration limits the quantity of oxygen that can be carried through the bloodstream to vital organs such as the heart and brain. Higher amounts of CO are often found in locations with severe traffic congestion. Other sources of CO emissions include industrial activities (such as metal processing and chemical production), home wood burning, and natural events like forest fires. Woodstoves, cigarette smoke, gas stoves, and unvented gas and kerosene space heaters all emit CO indoors. Analysis shows that the average Carbon Monoxide (CO) level in both Rabindra Bharti and Jadavpur remained below 30 parts per million during peak and lean periods of the day in May and August. Conversely, the average Carbon Monoxide (CO) content during November (winter) remains poor (30-70 parts per million) in both Rabindra Bharti and Jadavpur during peak and lean periods of the day (see Figure 5). High quantities of carbon monoxide can be fatal. Carbon monoxide diffuses through lung tissues and into the bloodstream, making it difficult for cells to connect to oxygen. The absence of oxygen harms tissues and cells. This study substantiates that the highest concentration of CO levels are found during the colder months of the year when inversion conditions (when air pollution is held near the ground beneath a layer of warm air) are more common.

Nitrogen Dioxide

Nitrogen dioxide (NO_2) is one of several extremely reactive gases known as nitrogen oxides. Nitrous acid and nitric acid are two types of nitrogen oxide. The burning of fuel is the primary source of NO_2 in the air. Emissions from automobiles, trucks and buses, power plants, and off-road equipment produce NO_2. NO_2 and other NOx react with water, oxygen, and other compounds in the atmosphere to produce acid rain. Acid rain destroys delicate ecosystems such as lakes and forests. Breathing air with a high NO_2 concentration might irritate the airways in the human respiratory system.

Figure 6. Average nitrogen dioxide

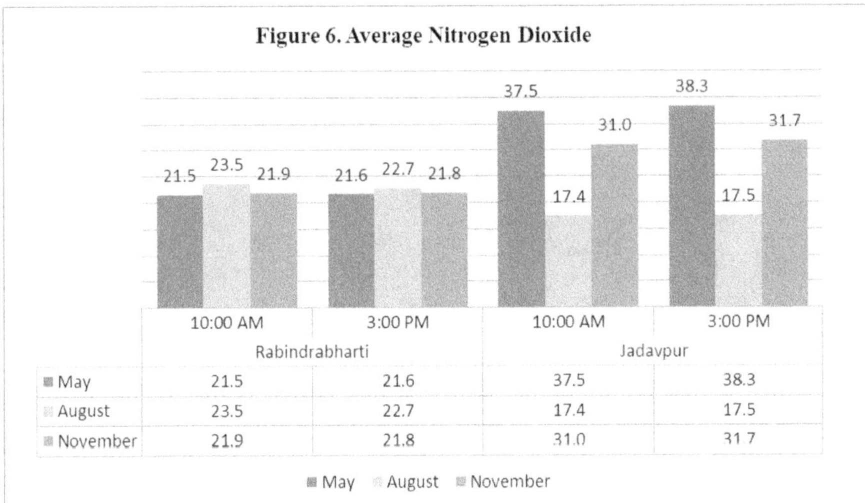

Source: Primary Study, 2023

Such short-term exposures might aggravate respiratory illnesses, notably asthma, causing respiratory symptoms (such as coughing, wheezing, or difficulty breathing). Analysis reveals the average Nitrogen Dioxide content through May (summer), August (rainy), and November (winter) was below 50 which indicates good (figure 6). The results validate Aneja et al.'s (2001) argument that the city's high traffic volume caused Kolkata's high NO2 concentration.

Sulphur Dioxide

Sulphur dioxide (SO_2) is a poisonous, colourless, and reactive gas affecting human, plant, and animal health. It is produced as a result of combustion of fossil fuels, smelting of sulphur-containing ores, and volcanic eruptions. The most significant sources of SO2 emissions are fossil fuel combustion in power plants and other industrial facilities.

Figure 7. Average sulphur dioxide

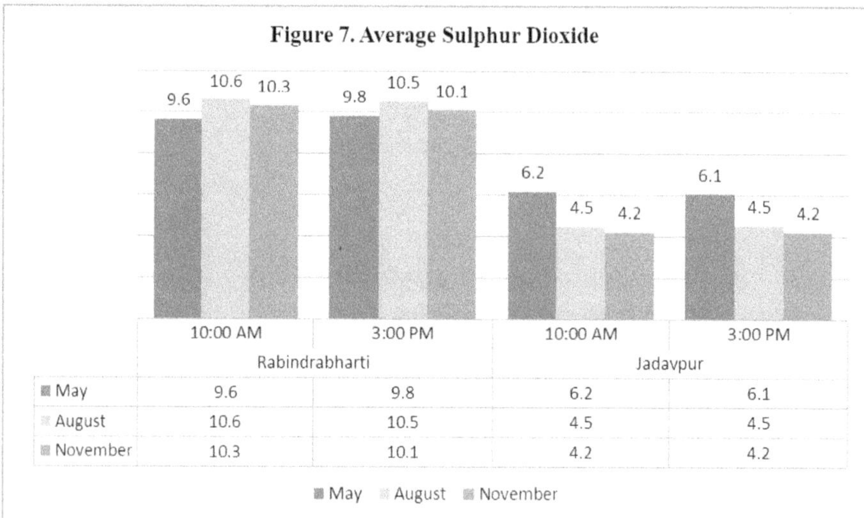

Source: Primary Study, 2023

When SO_2 mixes with water and air, it produces sulphuric acid, which is the primary component of acid rain. Acid rain can erode construction materials and paints, acidify streams, and induce deforestation. India is the world's greatest SO_2 emitter, accounting for more than 21% of worldwide emissions, primarily from coal-based electrical generation (Greenpeace India, 2020). Analysis confirms that average sulphur dioxide is at a hazardous level which indicates more than 5 parts per million (ppm) in Rabindra Bharti throughout May (summer), August (rainy), and November (winter) (see figure 7). In the case of Jadavpur, the study claims that the average sulphur dioxide content ranged from 3.0 – 5.0 ppm during August (rainy) and November (winter) which is very unhealthy and it amplifies to above 5.0 ppm in May (summer) emphasizing the hazardous condition during peak and lean period in a day.

Ground-level Ozone

Ozone is a gas constructed up of three oxygen atoms. Ozone exists simultaneously in the upper atmosphere and at the ground's surface. Ground-level ozone, sometimes known as "bad" ozone, originates closer to the Earth in the air we breathe. This ozone is mainly produced by chemical interactions between nitrogen oxides (NOx) and volatile organic compounds (VOCs) under sunlight, rather than being emitted directly. Ozone can be beneficial or harmful depending on where it is present.

Figure 8. Average ground-level ozone

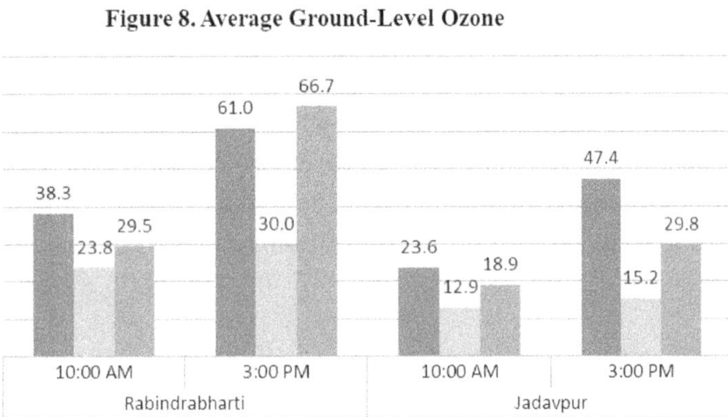

	10:00 AM (Rabindrabharti)	3:00 PM (Rabindrabharti)	10:00 AM (Jadavpur)	3:00 PM (Jadavpur)
May	38.3	61.0	23.6	47.4
August	23.8	30.0	12.9	15.2
November	29.5	66.7	18.9	29.8

Source: Primary Study, 2023

When pollutants released by automobiles, power plants, industrial boilers, refineries, chemical plants, and other sources come into contact with sunlight, they chemically react. Ozone levels are possibly to be hazardous in metropolitan locations on hot sunny days, but they may continue to be high during the winter months. Ozone can also travel long distances by wind, thus even rural places can have high ozone levels. Ozone exposure can make breathing difficult, leading to coughing and shortness of breath. It can develop into more serious problems such as damaged airways, chronic bronchitis, asthma, and emphysema. Even after symptoms have subsided, ozone can continue to harm the lungs. In addition, ground-level ozone can harm vegetation and ecosystems. The contaminant can induce crop die-off, reducing development and causing disease. In Rabindra Bharti, the ground-level ozone pollution remains favourable at 10 a.m. in May, August, and November (ranging between 0-54 ppb), but it becomes moderate at 3 p.m. in May and November (ranging between 55-70 ppb) (see figure 8). In Jadavpur, the ground-level ozone pollution remains moderate throughout May, August, and November.

In line with the above discussion, the analysis confirms the views of Tamang (2024) that the growth of anthropogenic activities such as the advancement of modern technologies results in the release of heat-trapping greenhouse gases into the atmosphere. In this context, the increased digitalization of ride-hailing services has exaggerated the greenhouse gases mentioned earlier resulting in air pollution. During the empirical study, an attempt was made to apprehend the spread of air pollution by the ride-hailing services in terms of the average number of driving hours and distance covered. Fuel consumption in the case of petrol vehicles is a major delinquent concern in environmental circumstances. Fuel consumption by ride-hailing vehicles and its increased frequency resulting in air pollution are issues to be focused on in the current urban traffic scenario. Discussion with the riders illustrates that an individual rider drives an average of 95.5 km per day by four-wheelers and 53.3 km per day by two-wheelers (see Table 1). The carbon emissions per kilometre for a conventional bike are around 21 grams and 192 grams of carbon dioxide equivalent per passenger-kilometre for a petrol car.

Table 1. Average driving time and distance covered

Average Value	Bike	Cab
Hours of driving per day (in hours)	1.3	4.0
Distance covered per day (in km)	53.3	95.5

Source: Primary Study, 2023

Hence, carbon emissions per day could be calculated as

Conventional bike:
= 21*53.3
= 1,119.3 gms
Petrol car:
= 192 * 95.5
= 18,336 gms

This research emphasises that the digitalization of ride-hailing services has intensified the emission of carbon. Carbon dioxide (CO_2) is one of the greenhouse gases found most in the atmosphere which is a significant cause of climate change. The emission of Carbon dioxide (CO_2) has become prevalent in the environment along with other heat-trapping greenhouse gases as a consequence of digitalization. The increased demand for digitization has led to a rise in greenhouse gas emissions, making the earth's atmosphere more susceptible to its impacts. The colossal upsurge in heat-trapping gases in the atmosphere has fuelled climate change and increased temperature which disturbs the existence of the ecosystem. The digitalization of ride-hailing services has a negative influence on climate which has a serious consequence on human health. Climate change has chains of reactions in human survival including livelihood, health, water, air quality, and food chain. It can be the biggest danger in 21st century (Yildirim et al., 2022; Yildirim et al., 2023). The section below discusses the health concerns of the drivers of ride-hailing services resulting from air pollution.

Health in Question

This section corroborates that toxic gases endangered the well-being of humans resulting in problems of visual sight, respiratory, and other health issues. Nevertheless, the drivers of the ride-hailing services themselves suffer from diseases borne out of air pollution such as asthma, bronchitis, chronic obstructive pulmonary disease, cardiovascular diseases etc. Further, this research validates that driver of ride-hailing services are exposed to extreme temperatures which affects their health leading to acute health conditions. Shreds of shreds of evidence claim that drivers of bikes or two-wheelers are more accustomed to health concerns like migraine, stroke, lung infection, fatigue, and skill diseases. These diseases are outcome of the occupational hazards acquainted by the drivers. These drivers are open to occupational health hazards because of the risk factors associated with driving which include extended periods of sitting, exposure to heat, susceptibility to accidents, uncomfortable sitting space, exposure to airborne illnesses, shortened rest periods, irregular eating patterns, stress for the passengers' safety, etc (Bhatt and Seema, 2012). This section discusses various concerns on the health of the drivers and highlights the medical hazards faced by drivers of ride-hailing services mainly due to air pollution.

Migraine

A strong, throbbing pain that usually occurs in one part of the head is the hallmark of a migraine. Migraines can occur once a year or several times each week. There are often three stages to a migraine: the premonitory or prodromal phase, which occurs before the headache, the headache phase, which occurs during the migraine, and the postdrome phase, which occurs after the headache. The findings claim that most of the drivers riding bikes are more susceptible to migraine than those driving cars (see Table 2 below). It was found that 80.5 per cent of drivers in Jadavpur and 82.9 per cent in Rabindra Bharti.

Table 2. Migraine

Location/ Vehicle Type	*Yes*	*No*	*Number of Drivers*
Jadavpur			
Bike	80.5%	19.5%	41
Cab	55.9%	44.1%	34
Rabindra Bharti			
Bike	82.9%	17.1%	40
Cab	70.0%	30.0%	35

Source: Primary Study, 2023

Lung Infection

Further, the analysis explains that asthma and lung infection remain prominent among the drivers of bikes and cars as well. In the case of Jadavpur, the majority of car drivers (97.1 per cent) had lung infections.

On the other hand, most of the bike drivers in Rabindra Bharti (87.5 per cent) had asthma (Table 3).

Table 3. Asthma/lung infection

Location/ Vehicle Type	**Yes**	**No**	**Number of Drivers**
Jadavpur			
Bike	73.2%	26.8%	41
Cab	97.1%	2.9%	34
Rabindra Bharti			
Bike	87.5%	12.5%	40
Cab	77.1%	22.9%	35

Source: Primary Study, 2023

When bacteria, fungi, or viruses infiltrate the lungs and create inflammation, it results in a lung infection. Mild to severe lung infections may need medical attention. Bronchitis is the term used to describe an infection of the major bronchial tubes that transport air to and from the lungs. Viruses are more likely than bacteria to cause bronchitis. Viruses can also harm the lungs or the airways that supply them.

This health condition is referred to as bronchiolitis. The contact of the drivers to extreme temperatures exposes them to ground-level ozone including dust and smoke. Inhaling air pollutants such as PM2.5 and PM10 results in asthma, congestion, chest pain, lung infection etc. The concentration of pollen in the atmosphere with severe temperatures causes pollen allergies and acute asthma problems.

Stroke

According to Table 4 below, most bike riders in Jadavpur (75.4 per cent) and Rabindra Bharti (72.5 per cent) neighbourhoods experienced strokes because of their travel long distance travel throughout the city and spending more time in the sun (see Table 4). The drivers of ride-hailing services claimed that their bodies fail to cope with the extreme temperature resulting in heat cramps, hyperthermia, exhaustion, and heat stroke. When a blood artery in the brain bursts and bleeds, or when the blood supply to the brain is blocked, a stroke happens. Then the brain's tissues cannot receive blood or oxygen because of the rupture or obstruction.

Table 4. Stroke

Location/ Vehicle Type	Yes	No	Number of Drivers
Jadavpur			
Bike	75.4%	24.6%	41
Cab	17.6%	82.4%	34
Rabindra Bharti			
Bike	72.5%	27.5%	40
Cab	8.6%	91.4%	35

Source: Primary Study, 2023

Fatigue

Fatigue is a persistent sense of burnout, depletion, or low energy. It may be mental, physical, or a mix of the two. Anybody can get tired; most adults do at some point in their lives. The majority of the drivers riding ride-hailing vehicles in both two-wheeler and four-wheeler segments experience fatigue. Table 5 given below validates that above 90 per cent of drivers belonging to two-wheeler and four-wheeler segments in both the Jadavpur and Rabindra Bharti areas asserted that they often suffer from fatigue.

Table 5. Fatigue

Location/ Vehicle Type	Yes	No	Total
Jadavpur			
Bike	92.7%	7.3%	41
Cab	91.2%	8.8%	34
Rabindra Bharti			
Bike	90.0%	10.0%	40
Cab	91.4%	8.6%	35

Source: Primary Study, 2023

Skin diseases

The skin, being the outermost barrier, is impacted by a variety of air pollutants, including cigarette smoke, UV radiation, polycyclic aromatic hydrocarbons, volatile organic compounds, oxides, particulate matter, and ozone. Air pollution causes oxidative stress, which harms the skin. Skin cancer and extrinsic skin ageing have been linked to UV radiation exposure. Glutathione peroxidase, glutathione reductase, superoxide dismutase, and catalase are examples of enzymatic antioxidants whose capacity is depleted, as is the antioxidant capacity of vitamins E, C, and glutathione. During the discussion, it was found that 31.7 per cent of bike drivers in the Jadavpur locality and 25 per cent in Rabindra Bharti were suffering from skill infection (Table 6).

Table 6. Skin diseases

Location/ Vehicle Type	Yes	No	Total
Jadavpur			
Bike	31.7%	68.3%	41
Cab	23.5%	76.5%	34
Rabindra Bharti			
Bike	25.0%	75.0%	40
Cab	22.9%	77.1%	35

Source: Primary Study, 2023

Constructive Effects of Ride-hailing Services Going Digital: Passengers' Perception

Effective matching of riders' and passengers' flexibility

Digitalization of ride-hailing services considers multiple features to competently match the passenger with the nearest and the best rider matching with their flexibility to confirm quick pickups. During the study, 42.7 per cent of the riders marked availing ride-hailing services on flexibility as 4 on 5 point scale and avowed that the fundamental of digitalization in ride-hailing services is the use of the mobile app which has ensured easy and flexible request of a ride using their smartphones (as shown in figure 9). These applications provide a user-friendly interface that simplifies the entire process, thereby, digitalization has made ride-hailing more accessible and expedient for travellers across diverse demographics. Economic factors are crucial when creating any projection of ride-hailing demand since the digitalization of ride-hailing services' has impacted its affordability which is probably going to play a major role in the uptake and degree of usage of these modes.

Nevertheless, the digitalization of ride-hailing services makes it possible to use dynamic pricing models, including surge pricing, to modify prices in response to current supply and demand. Fares are determined by algorithms that take into account variables including time, distance, traffic, and demand trends. The riders responded that integration of digital payment systems including namely credit cards, mobile wallets, and even cryptocurrencies is an essential feature in the ride-hailing ecosystem. Commuters acknowledged that they can pay their bills through credit/ debit cards, e-wallets, or other digital payment methods, eradicating cash transactions which upsurge the safety and efficacy of the process. This study validates that digitalization has restructured operations of ride-hailing services including cashless payments lowering fraud rates and increasing user trust.

Safety

The safety, convenience, and quick payment process make ride-hailing services a preferred choice for riders. Digital identification, such as driver profiles and vehicle information, guarantees that drivers and riders are verified, lowering the possibility of fraud or collisions. Pieces of evidence from the field explain that the digitalization of ride-hailing services has not only made the travel journey comfortable but also, assured the safety of the riders, especially women. Discussions with women riders reveal that they consider the digitalization of ride-hailing services as a blessing for their careers. Further, they asserted that digitalization has enabled them to join the workforce and maximised their participation in their jobs. It is remarkable that the benefits of digitalization of ride-hailing services also extended to non-work tasks including household obligations and running errands. These women respondents affirmed that they juggle their household responsibilities with work which curbs their ability to participate in the labour market. Often, their mobility is restricted because of safety concerns alongside with consent of their family members. Digitalization has been able to address the risk of harassment and sexual assault on public transportation. Hence, 36 per cent of the riders ranked safety as 5 on a 5-point scale (as shown in figure 9). Findings confirm that digitalization of ride-hailing services has facilitated women to manage time across work and household alongside with safety.

Availability

Ride-hailing services are accessible through digital technology on mobile applications that allow riders to hire vehicles to easily commute to their destinations. It was found that 54.7 per cent of the riders graded the easy availability of vehicles by accessing digital technology on their mobile app (as shown in Figure 9). This substantiates I.O. Olayode et al. (2023) that increases in the availability of ride-hailing services have led to insightful changes in how an individual travels from one destination to another.

Travel time

Digitalization has facilitated them to reduce their travel time and reach longer distances in a shorter time. During the study, 50.7 per cent of the riders interrogated rated the reduction of travel time as 5 on a 5-point scale (see Figure 9).

Figure 9. Constructive effects of digitalization

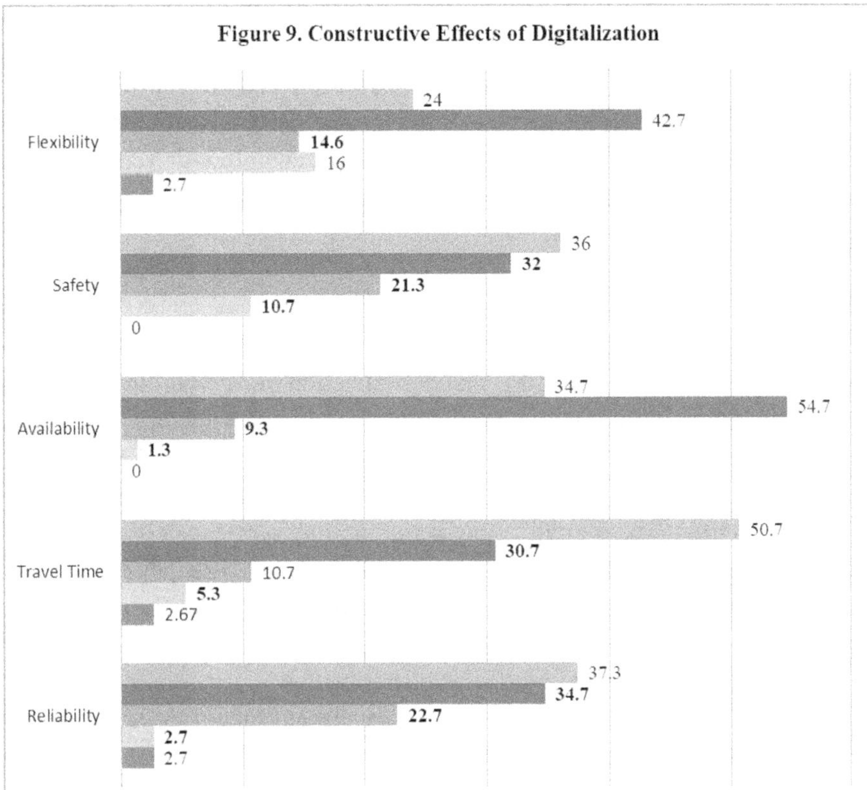

	Reliability	Travel Time	Availability	Safety	Flexibility
Points 5	37.33	50.67	34.67	36.00	24.00
Points 4	34.67	30.67	54.67	32.00	42.67
Points 3	22.67	10.67	9.33	21.33	14.67
Points 2	2.67	5.33	1.33	10.67	16.00
Points 1	2.67	2.67	0.00	0.00	2.67

Points 5 Points 4 Points 3 Points 2 Points 1

Source: Primary Study, 2023

These respondents emphasised that ride-hailing apps ensure users request rides from anywhere, anytime, reducing the dependence on conventional taxis or personal vehicles. In addition, women riders responded that the digitalization of ride-hailing services has smoothened their time management and spend quality time with their families, especially children.

Further, the riders claimed that digitalization has enabled ride-hailing apps that use GPS and mapping software to provide real-time tracking of vehicles, facilitating users to know the projected arrival time of their ride and the best routes which reduces their wait time and improves their overall experience. Women respondents revealed that GPS tracking and mapping of vehicles has ensured them traceable rides at any point in time according to their convenience. This substantiates the stance of a report published by Oxford Economics, that technology-driven ride-hailing services have enabled women to break the barriers and ensured them to actively participate in the labour market.

Travel Convenience/Reliability

Discussion with the riders claims that digitalization of ride-hailing services has made their travel easier and more comfortable because they enable them to track drivers' locations in real-time, receive estimated arrival times, and communicate directly with drivers through the app. This transparency enhances the user experience and builds trust. The digitalization of ride-hailing services has revolutionized the transport industry, transforming the lives of individuals. Respondents asserted that digital technologies in ride-hailing services have made travel more efficient, reachable, and responsive to user needs. The findings of the study support that GPS-based routing algorithms have significantly shortened wait times and improved the efficacy of ride-hailing services. These technologically advanced ride-hailing services can empower women, households and the economy as mobile internet access grows increasingly prevalent in Indian cities. This research highlights that digitalization in the transportation sector has reduced the entry barriers for customers and drivers alike, enhancing overall service accessibility.

Comprehending Drivers' Perception

Enhanced Income

The results validate digitalization of ride-hailing services has established a positive correlation with income (Young & Farber, 2019; Rayle et al., 2016). The improved access to vehicles as an outcome of digitalization has improved the monthly income of the drivers. During the discussion, the drivers of ride-hailing services asserted that there has been a sustainable improvement in their economic status.

They proclaimed that before their involvement as drivers in ride-hailing services, their monthly income was variable and uncertain. However, ride-hailing services have provided them with a sustainable source of income. The drivers of driving two-wheelers claimed that they are earning an average of Rs. 496.40/- daily against their expenditure of Rs. 124.10/- on fuel consumed daily. This indicates that their take-home pay is about Rs. 372.4/- a day. Similarly, conversations with drivers of four-wheeler cars asserted that their average income is Rs. 2527.70/- daily against their expenditure of Rs. 484.50/- on fuel consumption per day (see Table 7). In the case of four-wheelers, the riders make around Rs. 2043.2/- a day after deduction of fuel expense. Analysis reveals that the drivers of two-wheelers earn a living of Rs. 11,101.8/- and Rs. 61,061.4/- in the case of four-wheelers at the end of every month. These respondents affirmed that ride-hailing services reduced their financial dependence and were able to provide subsistence to their households. Further, they insisted that their engagement in ride-hailing services pulled them out of the vicious cycle of poverty. As this transformation continues, it also brings opportunities to address unemployment and ensure sustainable livelihood endeavouring the achievement of Sustainable Development Goals 1 and 8 (No Poverty –SDG 1 & Decent Work and Economic Growth - SDG 8), contributing to the Nation's economic growth.

Table 7. Average income against fuel consumption expenditure

Average Value	Bike	Cab
Per day Income	496.4	2527.7
Per day fuel consumption	124.1	484.5
Service cost per month	70.2	234.6

Source: Primary Study, 2023

Considering the above discussion, the findings of this study validated the stance of researchers namely, Agarwal et al. (2023) that ride-hailing services offered lucrative financial incentives to onboard the drivers. The convenience of modern technology-enabled cab and motor services has made them the most hassle-free option addressing the concern of unemployment by providing income-generating avenues and reducing poverty. Digital technologies are essential for accomplishing the United Nations' Sustainable Development Goals. The empirical study supports the notion that ride-hailing services have reduced unemployment among youths and improved their livelihood. Drivers of ride-hailing services asserted that the advent of digitalization and increased use of ride-hailing services has significantly augmented income-generating avenues among unemployed youths.

Sustainable Engagement

It is important to note that access to decent work and income-generating activities has been a key concern as more than 12 million youth between 15 and 29 years of age are expected to enter India's working-age population every year for the next two decades (World Bank, 2017). The findings of the study validate that driving is the primary occupation for 25.5 per cent of those aged 21-30 years who drive cabs and 19.2 per cent ride bikes (Table 8). Discussions with the respondents revealed that they were students who made money by providing riding services to their clients, which allowed them to spend money on fuel for their bikes (i.e. 23.4 per cent). Besides, the majority of the respondents of 31-50 years (i.e. 50 per cent among 31-40 years and 41 per cent 41-50 years) drive a cab as their major occupation.

Table 8. Primary occupation of the drivers across different age groups

Primary Occupation	Per cent
21-30 (n- 47)	
Bike driving	19.15%
Business	2.13%
Cab driving	25.53%
Private job	21.28%
Shop	4.26%
Shop worker	4.26%
Student	23.40%
31-40 (n-50)	
Business	28.00%
Cab driving	50.00%
Labour	2.00%
Private job	2.00%
Shop worker	18.00%
41-50 (n- 39)	
Business	33.33%
Cab driving	41.03%
Labour	5.13%
Shop worker	20.51%
51-60 (n -14)	
Business	50.00%
Cab driving	28.57%
Labour	14.29%
Shop worker	7.14%
Grand Total	**150**

Source: Primary Study, 2023

The empirical findings reveal that 34.7 per cent of the respondents in Jadavpur and 37.3 per cent in Rabindra Bharti were unemployed before working as ride-hailing service drivers (see Table 9). The findings of the study claim that ride-hailing service riders are a source of income for most unemployed men. It was found that most respondents engaged as drivers of ride-hailing services belonged to the age group 21-30 years.

Table 9. Earlier engagement across age groups

21-30 (n-47)	
Business	**2.1%**
Cultivation	21.3%
Driver	2.1%
Labour	4.3%
Shop	6.4%
Student	48.9%
Unemployed	14.9%
31-40 (n-50)	
Business	2%
Conductor	14%
Cultivation	14%
Driver	12%
Labour	22%
Sales	2%
Security guard	8%
Shop	4%
Unemployed	22%
41-50 (n-39)	
Business	23.1%
Conductor	7.6%
Driver	12.8%
Labour	25.6%
Sales	2.6%
Security guard	5.1%
Unemployed	23.1%
51-60 (n- 14)	
Business	7.1%

continued on following page

Table 9. Continued

21-30 (n-47)	
Conductor	21.4%
Cultivation	14.3%
Labour	21.4%
Security guard	7.1%
Unemployed	28.6%

Source: Primary Study, 2023

A majority of respondents aged 31- 40 years (i.e. 22 per cent) opted to be drivers due to a lack of employment opportunities under compulsion of unemployment, followed by those who were engaged as drivers elsewhere. Another 22 per cent of the respondents belonging to the same age bracket expressed that they were engaged in the unorganised sector as labourers. On the other hand, 25.6 per cent were engaged as labour, 23.1 per cent of the respondents aged 41-50 years claimed that they faced a loss in their business due to several reasons, including the spread of novel coronavirus, and 23.1 per cent asserted unemployment compelled them to shift their occupation as a rider. The justifications stated above were supported by 28.6 per cent of respondents in the age group 51-60 years who were unemployed due to COVID-19, and the advent of digitalization in ride-hailing services has assured them economic sustainability.

Further, discussions with respondents revealed that ride-hailing services have provided sustainable livelihood to migrants dealing with unemployment and poverty in Kolkata, particularly the 31-50 years age group. The table 10 below reveals that 66 per cent of 31-40-year-olds and 53.9 per cent of 41-50-year-olds avowed that they migrated to Kolkata with the expectation of getting employment. They were engaged as labour in the unorganised sector, however, trapped under unemployment, which questioned their sustenance.

Table 10. Status of migration across age-group

Migration	%
21-30 (n-47)	
Yes	19.2%
No	80.9%
31-40 (n-50)	
Yes	66%
No	34%

continued on following page

Table 10. Continued

Migration	%
41-50 (n-39)	
Yes	53.9%
No	46.2%
51-60 (n-14)	
Yes	35.7%
No	64.3%
Grand Total	**150**

Source: Primary Study, 2023

Multidimensional Poverty Index (MPI)

The multifaceted poverty of the drivers is covered in the discussion that follows. The Multidimensional Poverty Index (MPI) was created using a set of indicators that were specifically selected to capture important aspects of the lived experiences of people who drive as a means of supporting their households. These elements are called "dimensions" and often include factors like health, education, and living conditions. The study considered child mortality and nutrition as health; years of schooling and enrolment as education; and water, sanitation, electricity, cooking fuel, and asset value as a standard of living indicators for apprehending the multi-dimensional poverty index. The Multidimensional Poverty Index (MPI) is a product of the Headcount Ratio and Intensity of Poverty.

Multidimensional Poverty Index (MPI) = Headcount Ratio * Intensity of Poverty

In line with the above discussion, this section discusses the headcount ratio and intensity of poverty to measure the magnitude of consequences induced by digitalization of ride-hailing services. The 'Headcount Ratio' considers that if a household's deprivation score is greater than or equal to one-third i.e. 33.3 per cent then it is considered "multidimensionally poor".

Headcount Ratio (Incidence)

= Total Number of multidimensionally poor households (deprivation score of households above or equal to 33.3%) / Total Sample

The intensity of poverty makes it possible to understand how far the living standard of the deprived population is from the poverty line.

Intensity of Poverty = The deprivation score of households above or equal to 33.3% is summed / Total number of members in multidimensionally poor households.

The intensity of poverty makes it possible to understand how far the living standard of the deprived population is from the poverty line. An in-depth analysis of the data collected from the field reveals that the average multidimensionally poor person among drivers is deprived in 9 per cent of weighted indicators as illustrated through calculated 'Intensity of Poverty' (Table 6). The calculated 'Headcount Ratio' validates the fact that drivers encounter hardship in a multidimensional aspect even after being engaged as drivers of ride-hailing with many disadvantages hitting them at once. Table 11 explains that 6 per cent of drivers are multidimensionally poor (i.e. Headcount Ratio). The findings of the empirical study substantiate that the digitalization of ride-hailing services impacted the multidimensional poverty status of the drivers (i.e. MPI 0.539). Observations from the field indicate that though the digitalization of ride-hailing services has provided drivers with a sustainable source of income, pulling them out of multidimensional poverty is far to reach. The drivers proclaimed that a major proportion of their income was spent on medical expenses due to their exposure to precarious working conditions and occupational hazards, and repayment of EMIs. They are exposed to accidents, mental health issues, cardiovascular diseases, HIV/AIDs, stroke, chronic fatigue, kidney and bladder problems and musculoskeletal pains that risk their health, particularly among those driving two-wheelers.

Table 11. Multidimensional poverty index of drivers

Criteria	**Value**
The Sum of Deprivation Score	350.7
Total Number of Household Members in Multidimensionally Poor Households	39
Intensity of Poverty (A) (%)	8.99
Number of Multidimensionally Poor Household	9
Total Sample	150
Headcount Ratio (H) (%)	0.06
MPI (H*A)	0.539

Source: Primary Study, 2023

CONCLUSION

Long story short, the ride-hailing industry has witnessed a significant transformation by using advanced technology, which has increased accessibility, safety, and efficiency in transportation. It connects passengers to transportation services to navigate from one place to another by application of technological innovations like GPS, AI, and smartphone apps. Regulators can now anticipate how ride-hailing demand and usage will evolve because of digitalization, allowing them to enact laws that avoid unintended consequences. It has made it simpler to forecast demand and growth, which aids in the formulation of actions and policies that could have been implemented to manage the transition proactively and lessen the impact on passengers who depended on traditional public transport.

The augmented use of digitalization of ride-hailing services is turning out to be a threat to the sustainability of the environment. This chapter confirms that the digitalization of ride-hailing services indicates increased concerns related to traffic congestion, accidents, air pollution, and a temperature rise. The Air Quality Index (AQI) is a concern specifically during winter. In addition, it is the cause of numerous health-related disorders that pose a threat to life. On the contrary, this study supports the perspective that the digitalization of ride-hailing services has improved the passenger's travel experience in terms of saving time, accessibility, safety, flexibility, and reliability. Also, it has improved the livelihood status of the riders reflecting that the average monthly income of the drivers has upgraded as a product of the improved economic state of affairs due to the digitization of ride-hailing services. It is noteworthy, though, that while the digitization of ride-hailing services has helped to give riders a steady source of income, it has not been able to lift them out of poverty, particularly for bike drivers. It cannot be denied that a major proportion of the income of drivers is spent on medicine and treatments required to overcome diseases. This pushes them into the vicious cycle of poverty reflected in the multidimensional poverty index (MPI).

In conformity with the discussion above, the question is: *"Is digital transformation addressing economic and environmental sustainability in its true sense?"* In light of this debate, pooling and electrification are two major factors that can address environmental sustainability by decreasing energy consumption and emissions. This chapter promotes the ridesharing strategy to make appropriate use of the available seats, minimise the wastage of fuel, and lower the financial costs for both drivers and passengers. Further, a related form of ridesharing is carpooling, in which individuals are the riders and the owner is the driver.

Essentially, sustainability should guide the pursuit of digital innovations, and adopting these innovations can significantly enhance our efforts to achieve environmental sustainability. Ride-hailing industry can contribute to a future where transportation is more intelligent, environment-friendly, and inclusive by adopting and utilizing these developments. In this context, it is important to comprehend digitalization in two primary aspects i.e. digitalization for sustainability and sustainable digitization, which implies proactive progress using digital tools to attain ecological development alongside sustainable digitalization.

REFERENCES

Agarwal, S., Mani, D., & Telang, R. (2023, January 9). The Impact of Ride-Hailing Services on Congestion: Evidence from Indian Cities. *Informs pubs online*.https://pubsonline.informs.org/doi/abs/10.1287/msom.2022.1158

Agatz, N., Erera, A., Savelsbergh, M., & Wang, Z. (2012). Optimization for dynamic ride sharing: A review. *European Journal of Operational Research, 222*(2), 295–303. DOI: 10.1016/j.ejor.2012.05.028

Akbarpour, M., Li, S., & Gharan, S. O. (2020). Thickness and information in dynamic matching markets. *Journal of Political Economy, 128*(3), 783–815. DOI: 10.1086/704761

Alonso-Mora, J., Samaranayake, S., Wallar, A., Frazzoli, E., & Rus, D. (2017). On-demand high capacity ride-sharing via dynamic trip-vehicle assignment. [PNAS]. *Proceedings of the National Academy of Sciences of the United States of America, 114*(3), 462–467. DOI: 10.1073/pnas.1611675114 PMID: 28049820

Aneja, V. P., Roelle, P. A., Murray, G. C., Southerland, J., Erisman, J. W., Fowler, D., Asman, W. A. H., & Patni, N. (2001). Atmospheric nitrogen compounds II: Emissions, transport, transformation, deposition and assessment. *Atmospheric Environment, 35*(11), 1903–1911. DOI: 10.1016/S1352-2310(00)00543-4

Athey, S., Castillo, J. C., & Knoephle, D. (2018). *Service quality in the gig economy*. Working paper, Stanford Graduate School of Business, Stanford, CA.

Bai, J., So, K. C., Tang, C. S., Chen, X., & Wang, H. (2019). Coordinating supply and demand on an on-demand service platform with impatient customers. *Manufacturing & Service Operations Management, 21*(3), 556–570. DOI: 10.1287/msom.2018.0707

Bansal, T., Dongre, R., Wang, K., & Fuchs, S. (2021). *Ride-hailing emissions modeling and reduction through ride demand prediction*. Proceeding 7th Int. Conf. on Vehicle Technology and Intelligent Transport System - Volume 1: VEHITS, pp. 192–200. DOI: 10.5220/0010460800002932

Benjaafar, S., & Hu, M. (2021). Introduction to the special issue on sharing economy and innovative marketplaces. *Manufacturing & Service Operations Management, 23*(3), 549–552. DOI: 10.1287/msom.2021.0998

Berrebi, S. J., & Watkins, K. E. (2020). Who's ditching the bus? *Transportation Research Part A, Policy and Practice, 136*, 21–34. DOI: 10.1016/j.tra.2020.02.016

Bhatt, B., & Seema, M. S. (2012). Occupational health hazards: A study of bus drivers. *Journal of Health Management, 14*(2), 201–206. DOI: 10.1177/097206341201400209

Burtch, G., Carnahan, S., & Greenwood, B. N. (2018). Can You Gig it? An Empirical Examination of the Gig-Economy and Entrepreneurial Activity. *Management Science, 64*(12), 5497–5520. DOI: 10.1287/mnsc.2017.2916

Cachon, G. P., Daniels, K. M., & Lobel, R. (2017). The role of surge pricing on a service platform with self-scheduling capacity. *Manufacturing & Service Operations Management, 19*(3), 368–384. DOI: 10.1287/msom.2017.0618

Chen, K. M., Rossi, P. E., Chevalier, J. A., & Oehlsen, E. (2019). The value of flexible work: Evidence from Uber drivers. *Journal of Political Economy, 127*(6), 2735–2794. DOI: 10.1086/702171

Cohen, P., Hahn, R., Hall, J., Levitt, S., & Metcalfe, R. (2016). *Using big data to estimate consumer surplus: The case of Uber*. Technical report, National Bureau of Economic Research, Cambridge, MA. DOI: 10.3386/w22627

Correa, D., & Moyano, C. (2024). Analysis and prediction of New York City taxi and Uber demands. *Journal of Applied Research and Technology, 21*(5), 886–898. DOI: 10.22201/icat.24486736e.2023.21.5.2074

Dias, F. F., Kim, T., Bhat, C. R., Pendyala, R. M., Lam, W. H. K., Pinjari, A. R., Srinivasan, K. K., & Ramadurai, G. (2019). *Modeling the evolution of ride-hailing adoption and usage: a case study of the Puget Sound region*. Submitted to the Transportation Research Board 99th Annual Meeting.

Erhardt, G.D., Roy, S., Cooper, D., Sana, B., Chen, M., & Castiglione, J. (2019). Do transportation network companies decrease or increase congestion? *Science Advances*, 5(5), eeau 2670.

Feng, G., Kong, G., & Wang, Z. (2021). We are on the way: Analysis of on-demand ride-hailing systems. *Manufacturing & Service Operations Management, 23*(5), 1237–1256. DOI: 10.1287/msom.2020.0880

Ghose, K. M., Paul, R., & Banerjee, S. K. (2004). Assessment of the impacts of vehicular emissions on urban air quality and its management in Indian context: The case of Kolkata (Calcutta). *Journal of Environmental Protection, 7*, 345–351.

Godelnik, R. (2017). Millennials and the sharing economy: Lessons from a 'buy nothing new, share everything month' project. *Environmental Innovation and Societal Transitions, 23*, 40–52. DOI: 10.1016/j.eist.2017.02.002

Grahler, M., Mucci, R. A., & Erhardt, G. D. (2019). Understanding the recent transit ridership decline in major US cities: Service cuts or emerging modes. *98th Annual Meeting of the Transportation Research Board* (Washington, DC).

Greenpeace India (2020 October 8). Top SO_2 emitter, India registers 4 year dip in emissions.

Hall, J. D., Palsson, C., & Price, J. (2018). Is Uber a substitute or complement for public transit? *Journal of Urban Economics, 108*, 36–50. DOI: 10.1016/j.jue.2018.09.003

Hamari, J., Sjoklint, M., & Ukkonen, A. (2016). The sharing economy: Why people participate in collaborative consumption. *Journal of the Association for Information Science and Technology, 67*(9), 2047–2059. DOI: 10.1002/asi.23552

Hong, Z., Chen, Y., Mahmassani, H. S., & Xu, S. (2017). Commuter ride-sharing using topology based vehicle trajectory clustering: Methodology, application and impact evaluation. *Transportation Research Part C, Emerging Technologies, 85*, 573–590. DOI: 10.1016/j.trc.2017.10.020

Huang, Y., Kockelman, K. M., Garikapati, V., Zhu, L., & Young, S. (2021). Use of shared automated vehicles for first-mile last-mile service: Micro-simulation of rail-transit connections in Austin, Texas. *Transportation Research Record: Journal of the Transportation Research Board, 2675*(2), 135–156. DOI: 10.1177/0361198120962491

Kazimuddin, A., & Banerjee, L. (2000, July 31). Fighting For Air. *Down to Earth.*

Kietzmann, B. C. (2014). *Ride On! Mobility business models for the sharing economy*. Sage Journals., DOI: 10.1177/1086026614546199

Li, B., & Ma, L. (2021). JUE insight: Migration, transportation infrastructure, and the spatial transmission of COVID-19 in China. *Journal of Urban Economics*, •••, 103351. PMID: 33879932

Li, W., Pu, Z., Li, Y., & Tu, M. (2021). How does ride splitting reduce emissions from ride sourcing? A spatiotemporal analysis in Chengdu, China. *Transportation Research Part D, Transport and Environment, 95*, 102885. DOI: 10.1016/j.trd.2021.102885

Mondal, R., Sen, G. K., Chatterjee, M., Sen, B. K., & Sen, S. (2000). Ground-level concentration of nitrogen oxides (NOx) at some traffic intersection points in Calcutta. *Atmospheric Environment, 34*(4), 629–633. DOI: 10.1016/S1352-2310(99)00216-2

Mukherjee, A., Mukherjee, G., Banerji, U., & Mukhopadhyay, S. P. (1998). Occupational exposure of the traffic personnel of Calcutta to lead and carbon monoxide. *Pollution Research, 17*(4), 359–362.

Oyeyemi, I. O., Severino, A., Alex, F. J., Macioszek, E., & Tartibu, L. K. (2023). Systematic review on the evaluation of the effects of ride-hailing services on public road transportation. *Transportation Research Interdisciplinary Perspectives*, •••, 22.

Rayle, I., Dai Chan, N., Cervero, R., & Shaheen, S. (2016). Just a better taxi? A survey-based comparison of taxis, transit, and ride-sourcing services in San Francisco. *Transport Policy*, *45*, 16–178. DOI: 10.1016/j.tranpol.2015.10.004

Roy Chowdhury, A. (2011). *Citizen's report: Air quality and mobility in Kolkata*. Centre for Science and Environment.

Sarkar, N. (2024, April 4). Delhi gets the attention — but Kolkata's air pollution is just as dangerous. *Mongabay*.https://news.mongabay.com/2024/04/delhi-gets-the-attention but-kolkatas-air pollution-is-just-as-dangerous/

Sheldon, T. L., & Dua, R. (2024). Impacts of ride-hailing on energy and the environment: A systematic review. *Environmental Research Letters*, *19*(4), 043004. DOI: 10.1088/1748-9326/ad3285

Tamang, B. (2024). Climate change: A hazard to human well-being, fragile ecosystem and global stability. In Nanigopal, N., Roy, A., Rahaman, M., Ray, M., & Tamang, L. M. (Eds.), *Man, environment and development: A pathway to sustainable livelihood* (pp. 59–86). Readers Service.

United Nations Environment Programme (UNEP). (2002). *Environmental threats to children. In: Children in the new millennium,* 43–86. https://digitallibrary.un.org/record/465444?ln=en&v=pdf

Wallsten, S. (2015). The competitive effects of the sharing economy: How is Uber changing taxis? *Technology Policy Institute*, *22*, 1–21.

World Bank. (2017, June 23). Skilling India. https://www.worldbank.org/en/news/feature/2017/06/23/skillingindia#:~:text=More 20than%2012%20million%20youth,keys%20sectors%20of%20the%20economy

Yildirim, S., Kaplan, M., & Yildirim, D. C. (2023). Climate Change and the Ocean Economy: A Review on OECD's Ocean Economy. In Pego, A. (Ed.), *Handbook of Research on Bioeconomy and Economic Ecosystems* (pp. 212–232). IGI Global., DOI: 10.4018/978-1-6684-8879-9.ch012

Young, M., & Farber, S. (2019). Ride-hailing platforms are shaping the future of mobility, but for whom?

Yildirim, S., Demirtas, I., & Yildirim, D. C. (2022). A Review of Alternative Economic Approaches to Achieve Sustainable Development: The Rising Digitalization and Degrowth Post COVID 19. In R. Castanho (Ed.), Handbook of Research on Sustainable Development Goals, Climate Change, and Digitalization (pp. 288-307). IGI Global. https://doi.org/doi:10.31219/osf.io/pz7fk

Chapter 16
Comprehensive Overview of Sustainable IoT for Innovations and Their Impact on Climate Policy:
A Pre- and Post-COVID-19 Comparison

Ankita Sharma
https://orcid.org/0000-0001-9622-8170
Manipal University Jaipur, India

Akanksha Choudhary
Manipal University Jaipur, India

ABSTRACT

This chapter examines how sustainable IoT technologies influence climate policy, emphasizing pre- and post-COVID changes. The pandemic served as a catalyst, speeding up the use of IoT for environmental monitoring, resource management, and emissions reduction. The IoT has significantly impacted climate governance, shifting from reactive to proactive tactics post-COVID due to real-time monitoring and data-driven decision-making. This integration has also improved sustainability in agriculture, energy, urban management, and disaster resilience. Notwithstanding these developments, there are still issues, such as inadequate infrastructure, exorbitant prices, and worries about data privacy, especially in developing countries. Future development, international cooperation, and cross-sector alliances will be necessary to fully realize IoT's potential in mitigating climate change. Integrating the IoT with new technologies to create more robust and sustainable systems will

DOI: 10.4018/979-8-3693-9909-5.ch016

be crucial to the future of sustainable IoT in climate policy.

INTRODUCTION

The world has seen intriguing trends within the last forty years. Technologies aimed at improving the environment have gained significant traction. Many industries have developed clean energy sources like solar and wind aiming to improve the world's ecological conditions. The World Economic Forum's Community Economics and Digital Society Initiative envisions a digital future with sustainability at its core, including the Internet of Things (IoT) playing a pivotal role. The IoT and its competencies equips managers with key tools for creating lasting competitive advantage (Najafi et al., 2022). IoT has been identified as a scalable, multi-faceted, machine-to-machine (M2M) system capable of transferring digital data between "things" in ultra-large-scale networks. There are several potentials to close the gap between the IT sector and the economy with this kind of relationship. Additionally, it has sown the seeds for leveraging the ICT concept to lower the cost of data acquisition and, consequently, achieve long-term sustainable development goals in IoT applications like intelligent transportation systems (ITS), smart homes, smart cities, and intelligent farming environment (Z. H. Ali & Ali, 2021).

The Internet of Things is a development of the Internet, even though it is machine-to-machine technology. IoT expands the reach of the Internet beyond the computer and into the physical world. We have more control and a better understanding of how our environment functions due to the convergence of the real and virtual worlds. With this new knowledge, we can increase productivity in every area of a company and our society. The world benefits from its uses in terms of sustainability and ease. However, much more must be done to keep IoT a part of society and our daily live (Lamid Idowu et al., 2017).

Scientists have long warned of the environmental disaster we have been facing since the beginning of the twenty-first century. The actions that have led us to this point—the continued devastation of the environment and natural world, the overuse of Earth's finite resources, and the release of previously unheard-of carbon emissions into the atmosphere—need to change. Furthermore, global crises like the COVID-19 pandemic, post-Brexit difficulties, and shifts in corporate social responsibility have brought attention to how crucial it is to present and future development and advancement around sustainable development. Climate change, biodiversity loss, and the impending water crisis are a few of the most pressing environmental issues. In response, governments and countries worldwide are developing sustainable strategies (Rosca et al., 2021).

The network of physical objects known as "things" is the Internet of Things. These items are equipped with specific sensors, software protocols, and network adapters enabling connectivity and interaction. Through their unique identities, these objects may associate with one another and gather, transport, and exchange Internet data without human intervention. Thus, using the current network infrastructure, IoT makes it possible to sense and control objects and devices remotely via the Internet. Since IoT makes it easier to integrate the real world with computer-based systems directly, it has the potential to significantly improve efficiency, accuracy, and financial gain in practically every aspect of life (Tuysuz & Trestian, 2020).

Globally, climate change is a significant concern. However, the Internet of Things (IoT) offers a variety of tools and resources to assist governments and corporations in lessening the adverse effects of human activity on the environment. Because climate change is typically linked to scarcity, it frequently sparks discussions because people worry that solving the problem would mean compromising their comforts. This is not always the case, however. Burning fossil fuels like coal, oil, and natural gas releases too much carbon dioxide (CO2) into the atmosphere, the leading cause of global warming and unpredictable weather patterns. We must significantly increase energy efficiency, switch to low-carbon power production systems, and increase electricity usage in sectors like transportation to counteract this. The future depends on sustainable energy sources that do not release carbon dioxide, including wind and solar. In addition to lowering energy usage, IoT may significantly improve consumer items' speed, functionality, and efficiency (Abd El-Mawla et al., 2019).

Recent events have highlighted several contextual factors driving innovation in tackling climate change. Twenty countries, including the US, UK, China, and India, committed to double public investments in low-carbon technologies under the "Mission Innovation" accord during the 2015 United Nations Climate Change Conference in Paris. Similarly, the European Commission unveiled the carbon-free Europe agenda to promote innovation in green technology. Climate groups and activists, like Greta Thunberg and other NGOs, have significantly impacted political agendas by calling for changes to economic systems and laws that put economic expansion before environmental protection. Through policy efforts, the Biden administration has also strongly emphasized developing low-carbon technology. Bill Gates' Energy Breakthrough Coalition has mobilized investors to back ground-breaking climate ideas in the private sector. Although academia has historically played a role in developing innovative technologies, low-carbon improvements are progressing too slowly to fulfil the Paris Agreement's temperature objectives and finding and encouraging the right mix of elements to speed up low-carbon breakthroughs and phase out carbon-intensive technology (Matos et al., 2022).

One of the worst pandemics of our time, COVID-19, has devastated whole communities since it first appeared, taking countless lives and negatively affecting economies all over the world. Millions of people have had social, economic, emotional, mental, and physical upheavals. Numerous factors contribute to the situation's deterioration, but environmental concerns have played a significant role in the development of the pandemic. For example, according to Harvard research, the most polluted areas of Italy, France, Spain, and Germany accounted for 78% of COVID-19 deaths. The need to tackle climate change, another urgent global issue, has also been brought to light by this epidemic. The sudden COVID-19 pandemic has highlighted how intertwined human problems are. The most important concern on the 2030 agenda, according to the most recent Sustainable Development Goals (SDGs) statement, is climate change (Usman et al., 2021).

In the light of these challenges, this chapter provides a comprehensive overview of sustainable IoT applications through existing literature review, case studies and policy analysis. It examines the role of IoT in addressing climate challenges pre- and post-COVID, showcasing its impact on resource management across sectors, minimizing emissions, and improving energy efficiency. This chapter also emphasizes the revolutionary change post-COVID, when the disruption caused by the pandemic led to the adoption of IoT, resulting in more resilient and efficient systems for addressing environmental problems. Through this pre- and post-COVID comparison, the chapters aim to thoroughly understand how IoT can be a driving force in mitigating climate change and promoting sustainability. Five countries i.e., Thailand, Vietnam, Bhutan, Singapore, and Maldives were selected from East and South Asian region based on their SDG rankings (https://dashboards.sdgindex.org/rankings) to analyze the application of sustainable IoT in enhancing Climate Policy. The East and South Asian region was taken because India is the part of South Asian Region. Additionally, India's climate policy was considered an example of post-COVID climate policy, since it is a developing country.

Sustainable IoT Before the Pandemic-Foundation and Early Development

Before the pandemic, there was little research on sustainable IoT, especially regarding how it directly relates to SDG 13 and climate change. However, existing studies on the fundamentals of technology, sustainability, and climate action provides valuable insights into the early stages of IoT system development. These studies lay the framework for incorporating IoT into sustainable practices, providing a lens through which to analyze its potential and limitations in tackling climate concerns pre-pandemic. Understanding and being aware of climate change is one of the most important prerequisites for the sustainable applications of IoT. Using text mining tools, (Hwang et al., 2021) highlighted the importance of data-driven methods in identifying societal needs, an essential first step in creating IoT systems that tackle specific climate-related issues. By leveraging data analytics, IoT can create tailored solutions for communities by utilizing data analytics, raising awareness, and encouraging climate action. (Eliakimu & Mans, 2022) stressed the significance of using inclusive governance to overcome SDG-related inequities. Since deploying IoT solutions that benefit all society sectors requires strong governance frameworks, their conclusions are pertinent to the fundamental development of IoT. IoT integration in climate policy necessitates cross-sector cooperation to guarantee that technology innovations are equitable and accessible. The foundation of sustainable IoT applications, especially in developing countries, comprises these governance and inclusion concepts. The early stages of IoT development also coincide with developments in other technology areas. The possible uses of computer audition, a technology that collects and analyzes auditory data, in tackling climate-related issues were suggested by (Schuller et al., 2022). These developments open the door for IoT devices that efficiently monitor environmental conditions using multi-sensory data collecting.

Cities significantly impact growth and climate change, making it a controversial environmental problem. Particularly in the South, where many communities lack regular garbage collection and disposal services, Municipal Solid Waste Management (MSWM) is a significant contributor to climate change. There are notable distinctions between the North and South in the global discussion about cities, MSWM, and climate change, with the focus shifting from regional responsibilities to global climate issues. One important aspect of MSWM systems, the informal waste sector, is frequently overlooked in international discussions. A worldwide response to climate change requires a more inclusive strategy (Onyanta, 2016). (Ghomsi et al., 2024) examined the hazards related to environmental phenomena, including sea-level rise and coastal flooding. They underlined the necessity of adaption plans that complement SDGs 13 and 14. IoT technologies are ideal for supporting these adaptation efforts because they can collect real-time data and offer predictive analytics. These applications, made possible by pre-pandemic IoT advancements, demonstrating their capacity to reduce climate risks and improve resilience. Despite its early development, IoT struggled to scale viable applications before the pandemic. A lack of investment and research in low-carbon technology hampered progress. However, (Foláyan et al., 2024) highlighted enduring imbalances and injustices that IoT systems may rectify with focused interventions. IoT has the potential to bridge the gap between technological capabilities and practical solutions, playing a crucial role in achieving SGD 13.

COVID-19's Disruptive Impact on Sustainable IoT and Climate Policy

Sustainable IoT refers to integration of IoT technologies with sustainability principles to create a system that minimizes negative environmental impact and maximizes resource efficiency. (Peng & Anthony Maria Das, 2024) Using interconnected devices and sensors to collect and analyze data can help optimize energy consumption, waste reduction and improve living conditions with minimum carbon footprints. (Rejeb et al., 2022) The implication of IoT tools has been adopted significantly from the year 2021, as sustainable form. Organization have recognized potential implementation of IoT technologies to contribute to sustainability goals, particularly in context of climate actions and the UN SDGs specially in building Sustainable cities and communities and towards climate change (SDG 11 and 13 respectively).

The onset of COVID 19 accelerated the adoption of sustainable IoT, shifting its attention to health-related challenges. Remote monitoring technology aggravated its application in managing public health and environmental challenges (Sinthuja et al., 2022). The pandemic has also emphasized the interconnectedness of health and environmental issues, prompting a shift towards more sustainable practice (Katyal, 2025). However, this disruption resulted in the application of IoT and its advancements from smart cities, agricultural monitoring and energy management to health management on larger scale. AI, ML integration to sustainable tech awareness surged an upward mounted improvements during COVID challenges enhancing its capabilities and algorithms. From data collection to advance analysis and insights for future challenges, IoT provided improved sustainable outcomes embodying energy efficient IoT devices manufacturing aimed towards reducing carbon footprints (Polymeni et al., 2022).

In the pandemic induced economic slowdown, logistical challenges and reallocation of resources required IoT application towards health and economic recovery, building resilient, sustainable systems, spurring innovation in application of IoT in addressing pandemic related problems (N. Singh, 2025). Government priorities on building health infrastructure obscure the evident shifts from IoT to health and environmental policies, causing unprecedented delay in climate action plans and IoT driven sustainable projects (Drews et al., 2022). The socio-economic impacts of lockdown temporarily declined advancement in tech adoption with regards to climate policy planning. India, as a rapidly developing economy with ambitious climate goals leveraged IoT tech and policies to address pandemic related issues over climate issues surging long term delay in environmental objectives. (Tiwari et al., 2025) Health monitoring, Quarantine solutions, energy optimization, location monitoring, cold chain monitoring or food and vaccines mitigating the supply chain disruptions, etc. were all advancing challenges uniquely handled by IoT sustainable advancing features. However, COVID 19 disrupted sustainable IoT climate policy efforts, it also catalyzed innovation and highlighted the necessity of integrating tech into resilience building frameworks (Figure 1). India's proactive use of IoT in health, agriculture and energy sectors showcased the potential for aligning immediate crisis management with long term climate goals (Bhatnagar & Kumar, 2025).

Figure 1. COVID led disruption in sustainable IoT application

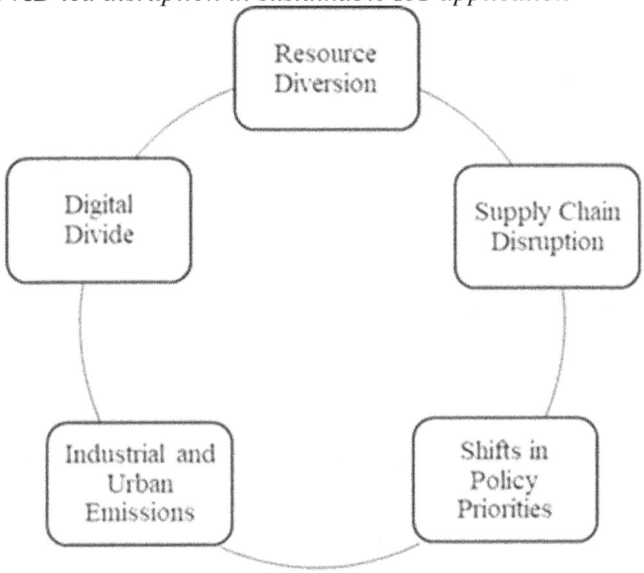

Source: Author's

An overview of India's Climate Policy Post COVID impact as a rapidly developing economy:

As far as climate change is concern India contributes 5%-7% Global warming since 1850s. This is based on the country's overall emissions of carbon dioxide, methane and nitrous oxide. India being the third largest country stands in top of greenhouse gas emitter. The climate change performance index of India ranks eighth among 63 countries which account for 92% of all GhG emission in the year 2021 (Mitra et al., 2025). The country is also a major producer of renewable energy, and its forests act as a carbon basin. With ample of renewable energy resources, the economic realm of Indian nation stands out in the world due to its geographical advantages. Despite of all this India experiences extreme weather, rising temperature, urban inequality, pandemic risks, population disasters etc. to address these extreme hazards India's Climate policy and world collaboration builds its resilience to develop a sustainable future with the help of tech advancement and leveraging IoT as one of the significant tools.

As per the www.downtoearth.org.in report top 10 countries include United States contributes 17.3% of carbon emissions, followed by China, Russia, Brazil, Indonesia, Germany, United Kingdom, Japan and Canada as the contributor of global warming. These countries level up the climate change by rapid development and industrial growth. However, the world's agenda of sustainable development tops the priorities of climate actions undertaking common grounds such as the Paris Agreement, mass movement for LIFE- Lifestyle for Environment, climate friendly and cleaner path etc. To reduce emission intensity of its GDP by 45% by 2030. Achieving a cumulative electric power installation capacity by 2030, reducing use of fossil fuels-based energy resources with the help of transfer technology and low-cost finances with green climate fund. Additionally, India adopts the plan to increase carbon basin for 3 billion tons of CO2 by growing forests and tree cover by 2030 (Sonkar, 2025).

Post-Pandemic Innovation in Sustainable IoT for Climate Action

The world is set for climate action with strategically planned UN sustainable development goals to be achieved by 2030. SDG 3, 7, 9, 12 and 13 responding to climate change talks about good health and well-being, affordable and clean energy, industry, innovation and infrastructure, responsible consumption and production and taking climate action respectively. These goals are set for building sustainable lifestyle and a future to be adaptive for generations to come (Vickram & Srivastava, 2025). As the post COVID impact, tech innovation came as an achievement towards sustainable development goals. The integration has paved a long term planned pathway to enhance response strategy against such pandemic or climatic challenges. The Internet of things is one such important incentive of technical adoption when it comes to climate action. To combat the world suffering due to extreme climate change arising environmental problems and challenges making life difficult on planet, IoT equips solutions to these problems (Solanki & Birman, 2025). With innovative advantages towards carbon emissions, usage of weather monitoring, cold storage devices, cold chain management, CFCs impact and air pollution measures, IoT provides large number of solutions to these challenges.

Sustainable IoT in global Climate initiatives

The post pandemic adoption of IoT leveraged climate initiatives to equip tools and devices to manage SDGs, this is broadly accepted and implemented by several economies to fulfill the implementation criterion. Hence, countries integrated IoT to monitor carbon emission and implement mitigation strategies as per the Paris agreement implementing under SDG 13 (Table 1). IoT technology contributes to cross cutting goals, such as energy efficiency and water conservation in drought prone areas (Vijayakumar & Srivastava, 2025). It helps in data collection in big number of observations to provide trends and future warnings against the upcoming disasters or landslide in flood prone regions. These are also accentuated with disaster mitigation and safety measures towards intimating the first help providers including the bad weather condition or even fall alarms as far as the individual help is concerned.

Table 1. Top 5 world economies integrating IoT application to commend SDG implementation

Countries Adopted IoT	Practical Implication	SDG Implementation	Impact
United States	Smart Cities, smart energy adoption through IoT based smart energy grids, facilitating the renewable energy sources, energy distribution and reduction of carbon emission. Precision Agriculture, IoT sensors adoption, soil monitoring, moisture maintenance, optimized irrigation and chemical use reduction through sustainable farming promotion	SDG 11: Sustainable Cities and Communities SDG 7: Affordable and Clean Energy SDG 13: Climate Action SDG 2: Zero Hunger SDG 12: Responsible Consumption and Production	By prioritizing Technological advancement with public private partnerships U.S. is mitigating climate change.
European Union (EU)	Green deal integration with IoT for solar and wind energy monitoring, ensures efficient energy production Adoption of Circular Economy through waste management enabling real time tracking of recycling processes and natural resource recovery Air Quality Monitoring through IoT sensors measuring pollution levels and data accuracy for policy making	SDG 13: Climate Action SDG 7: Affordable and Clean Energy SDG 12: Responsible Cities and Communities SDG 3: Good health and well being SDG 11: Sustainable Cities and Communities	As the leader of global climate change policy European Union have combined IoT technology with regulations to achieve Carbon neutral status by 2025.
Japan	Being a disaster-prone geographic location Japan uses IoT in disaster resilience through IoT powered early warning system for earthquake and Tsunamis Energy Efficiency through building management system using hi tech IoT sensors Monitoring Life below ocean and tracking marine biodiversity	SDG 13: Climate Action SDG 7: Affordable and Clean Energy SDG 14: Life Below Water	Japan's biggest challenge is to monitor disaster and building resilience through smart energy adoption and tech infrastructure led by IoT.

continued on following page

Table 1. Continued

Countries Adopted IoT	Practical Implication	SDG Implementation	Impact
China	Smart cities and infrastructure building, reducing energy consumption and emission through real time monitoring Using IoT application in waste management and transportation optimizing resources Using IoT devices and interface to monitor deforestation, plantation and reforestation efforts	SDG 9: Industry, Innovation, and Infrastructure SDG13: Climate Action SDG 11: Sustainable Cities and Communities SDG 15: Life on Land	China integrates IoT with its green development strategy, emphasizing urban sustainability and industrial efficiency.
India	Agriculture monitoring, precision farming, soil management, irrigation monitoring and usage of natural manure through crop monitoring and aiding small scale farmers Water meter and monitoring systems to reduce wastage in urban and rural collaboration Expansion of renewable energy sources through network management of local farms and mini grids in remote areas providing clean and affordable electricity and food amenities	SDG 2: Zero Hunger SDG 15: Life on Land SDG 6: Clean Water and Sanitation SDG 7: Affordable and Clean Energy	India's focus on IoT for agriculture and renewable energy aligns with its commitment to achieving SDGs and addressing climate vulnerabilities.

Source: Author's

Sustainable IoT technology is playing its pivotal role in achieving climate action and implementation of SDGs objcctively. While the innovation in this field have demonstrated huge potential to transform industries and sustainability promotion, through holistic efforts and policy planning (Bhagat et al., 2025). Through building resilience against climate change and global cooperation, future can explore cross sectoral synergies in making world emission free resource rich place to live with environmental harmony. IoT applications is still to comprehend its vast application with mutual benefits among world leaders and developing countries.

Pre and Post COVID Comparison: Sustainable IoT Influence on Climate Policy

The junction of pre COVID scenario was a lot different when viewed from a 2024 perspective. Tech advancement has reached to a next level, accounted within last 5 years. The Internet of Things (IoT) emerged as a vital tool in addressing climate change and relted challenges globbaly. Ranging with its sectoral applications, levarging interconnected wireless devices and data analytics to promote environmental sustainability, resource efficiency, and climate resilience IoT significantly contribute in management which was beyond human imagination. The application of Internet of Things into agriculture and precision farming, irrigation system, water manageent in water scarce region has been an important solution to maximize crop yield and monitored farming (Syali, 2025). IoT enabled real time management of renewable energy into existing networks, and reduction of over reliance on fossil fuels. Smart cities and urban mangement, cutting down CO2 emissions, waste management etc has been the contribution of IoT in developed nations. This adoption has implemented huge changes in the favor of climate conditions all over the world including the high end challenges such as intial costs, connectivity and infrastructure development, data privacy and security concerns, and awareness issues (Alsharif et al., 2023). In context of post COVID scenario IoT adoption and skill enhancement has been a result, as outlined in Table 2, which can be seen in its adoption by developing nations as well. In order to monitor sectoral advancement precise monitoring, decision making with information and trend analysis, all have contributed in policy support and capacity building against climate challenges.

Figure 2. Division of industrial revolutions

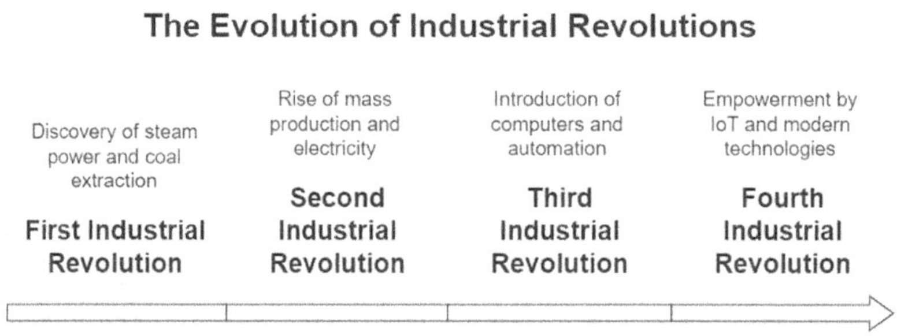

The broad division of Industrial Revolutions into four phases has been summed up in Figure 2. As the fourth Industrial Revolution witnesses the implementation of IOT and modern technology as per its need, its influence on Climate Policy is also seen vitally impacting post covid scenario (Talavera et al., 2017). The worldwide exemplification of sustainable development goal and climate change adheres to the undergoing requirement of Internet of Things (IoT). As technology plays a pivotal role in making various processes more measurable and efficient through data collection and analysis, IoT also makes the same with valuable large data set (Al-Raeei, 2024). For enhancing healthcare delivery, developing smart cities and agricultural revolution IoT can make significant alterations. By enabling automated, real-time decision making, energy optimization and transformative construction practices everything can be handled with ease.

IoT is reshaping worldwide efforts to address climate change by bridging the gap between human intervention and automated data driven solutions (Motlagh et al., 2020). The sublimation of high carbon emission, inefficient mechanical practices need to streamline by low impact tech solutions (Araus & Kefauver, 2018). IoT plays a crucial role in this transformation by enabling precise monitoring, predictive modelling and adaptive responses to environmental challenges.

Table 2. Pre-post COVID comparison for sustainable IoT influence

Stages Aspects	**Pre-COVID Scenario**	**Impact of COVID-19 on Climate Policy and IOT adoption**	**Post-COVID Landscape for IoT use**	**Revolutionary Changes in Future**
Transformation	**Limited IoT integration in climate policies and environmental concerns**	**Pandemic served as catalyst for digital transformation across all sectors, accelerated the adoption of IoT**	**Fundamental transformation of climatic policy making and strategy planning with IoT devices**	**Incremental advancement in technology adoption and their linkages are multiplying to possibilities globally.**
Application	Fragmentation of IoT application	Integrated approach to IoT application for immediate and long-term usage	Real time environmental Monitoring using Drones and IoT sensors	Global collaboration and data sharing for shared goals
Cognizance	Lacking awareness about its environmental monitoring parameters and resource efficiency	Recognition of IoT's role in real time monitoring and data driven decision making for climate governance	Accurate and real time data on air quality, greenhouse gas emissions, deforestation and water levels help policymakers to monitor environmental changes.	IoT technologies usage in crucial tracking towards net zero emission, enabling business and governments to monitor carbon footprints
Infrastructure	Lacking digital infrastructure for IoT deployment	Digital transformation in organization involved IoT infrastructure to allocate disaster management and sustainable resource utilization	Strategies to establish smart city projects powered by IoT, optimized energy usage, reduce carbon footprints, dynamic and timely intervention	Installing IoT sensor devices such as smart grids, renewable energy monitoring, and waste management systems.

continued on following page

Table 2. Continued

Stages Aspects	Pre-COVID Scenario	Impact of COVID-19 on Climate Policy and IOT adoption	Post-COVID Landscape for IoT use	Revolutionary Changes in Future
Efficiency	Costly implementation for IoT sensors, data management system and tech resources	Decentralized IoT enabled smart systems such as energy grids, remote pollution sensors, automated agriculture crop monitoring	IoT system provides early detection of natural disasters like floods, droughts and forest fires, improving response	Need and adapting green economic recovery, integrating IoT solutions into policy frameworks to create sustainable industries and jobs
Action	Focused on traditional mitigation measures for climate and environmental policies	Resilient policies making from reactive to proactive due to IoT prediction against future environmental challenges and sustainable alternatives	Minimum climate risk and safety measure planning to reduce impact and global emergence of resilience	International cooperation, public private partnership, globally integrated climate policy framework to work in single headed directions to achieve environmental goals and sustainable future.

APPLICATION OF SUSTAINABLE IOT IN ENHANCING CLIMATE POLICY

Thailand

Thailand's climate policy has demonstrated encouraging results with the adoption of IoT technology. IoT-based sensor systems are used in agriculture to monitor soil conditions and optimize water consumption, providing farmers with real-time data to make informed decisions. The growing focus on sustainability in Thailand, addressing both environmental and socioeconomic challenges, has facilitated the integration of IoT into climate policy. Collaborative action research has been highly beneficial in this process. For example, (Mishra et al., 2013) emphasize initiatives, including farmers and officials, to improve poor productivity in rice farming, a sector significantly influenced by climatic variability. Similarly, (Nguyen et al., 2017) highlighted programs that include reconsolidating farms, boosting rural infrastructure, and increasing finance availability to make agricultural communities more resilient to climate change. Bangkok and other urban areas are examples of how IoT may revolutionize energy efficiency through smart grids. IoT-enabled sensors track trends in energy use and allow for dynamic changes, which lowers carbon emissions and maximizes energy efficiency (Prasert, 2024).

Even with these developments, there are still obstacles to Thailand's climate policy's complete IoT integration. Two major obstacles to IoT adoption are the absence of comprehensive frameworks and the requirement for communities and local governments to increase their capability. Fostering innovation and scalability requires promoting cross-sector partnerships between private players, technology developers, and policymakers. According to (Motlagh et al., 2020), the Internet of Things' contribution to energy system optimization offers a path forward for related applications in other industries.

Vietnam

Climate change is posing increasing difficulties to Vietnam's agriculture economy, necessitating creative solutions to guarantee sustainability over the long run. The use of the IoT to improve climate policy, especially in the context of agriculture, is one such option that is gaining traction. In order to enhance farmers' agroecological knowledge, critical thinking abilities, and group action to promote sustainable agricultural growth, Vietnam has implemented Farmer Field Schools since 1992. These institutions give farmers the tools they need to implement climate-resilient strategies, such as recombining farmland, expanding loan availability, and building rural infrastructure. These actions are essential for lowering farmers' susceptibility to climate change and promoting long-term sustainability in agriculture (van de Fliert et al., 2007).

A study conducted by (Lipper et al., 2017) stated that to increase resistance to the effects of climate change, Vietnam has developed Climate Smart Agriculture (CSA) techniques. CSA strongly emphasises sustainable farming methods that assist farmers in adjusting to the challenges brought on by climate change. Research has shown how crucial forest farming is to household livelihoods in areas like Northern Vietnam's Bac Kan Province. Forests are essential to sustainable living because they supply essential resources like fuelwood and revenue. By boosting biodiversity and sustaining ecosystem services, encouraging the use of forest land as a resilience tactic increases economic stability and advances environmental sustainability (Trædal & Vedeld, 2018).

The potential of IoT technologies to enhance food security systems and promote sustainable agricultural practices is driving the growing interest in utilizing them. IoT-based devices, for example, may monitor temperature, weather, and soil moisture, giving real-time information that improves crop management and irrigation. These technologies connect with worldwide initiatives to deploy digital tools to increase food security and climate resilience. In order to help farmers increase productivity while reducing their adverse effects on the environment, agricultural policy in Vietnam is beginning to focus on IoT's potential to model sustainable food security systems (Kaur, 2019).

Beyond agriculture, IoT is playing a transformative role in Vietnam's energy sector, with a focus on smart grids. Smart grids can make more effective energy management possible, which can help Vietnam adopt a more sustainable energy strategy and drastically lower its carbon footprint. IoT integration in energy systems enhances agricultural efforts by offering a more all-encompassing strategy for sustainability. Smart grids optimize energy utilization in urban and rural regions by providing improved electricity consumption control solutions. Vietnam's climate depends heavily on cooperation between energy and agriculture programs (Motlagh et al., 2020).

Another illustration of how IoT might improve climate resilience is the Mekong Delta's rice-shrimp agricultural system. Studies have been conducted to identify opportunities for improvement in the system, which has long been susceptible to climate change. Farmers may improve agricultural methods, boost productivity, and adjust to shifting weather patterns with the help of Internet of Things technology like soil moisture sensors and climate monitoring systems. The Mekong Delta study may provide helpful guidelines for evaluating sustainability in other areas with comparable agricultural difficulties (Dang, 2020).

With an outlook toward the future, Vietnam's agricultural IoT integration offers substantial prospects for bolstering climate policy. IoT technologies can improve climate resilience and promote sustainable farming methods as they develop further. IoT may assist farmers with better resource utilization, less environmental impact, and more production by emphasizing data-driven decision-making. Vietnam may support climate resilience, encourage sustainable development, and provide insightful information to other areas dealing with comparable climate-related issues by integrating IoT into the energy and agriculture sectors (Quan & Van, 2022).

Bhutan

Bhutan, a country known for its adherence to environmental sustainability and Gross National Happiness, is confronted with new difficulties in balancing growth and climate resilience. Although research on climate policy and sustainable development is expanding, little is known about how sustainable IoT technology might be used. Previous research examined the connection between energy, the environment, and development in Bhutan, with studies like those by (Uddin et al., 2007) highlighting the importance of sustainable practices in promoting the nation's development.

However, research on integrating IoT technology to solve climate concerns is still noticeably lacking.

The agriculture and land-use sectors prove the IoT's significance to Bhutan's climate strategy. A thorough investigation into simulating the reactions of Agriculture, Forestry, and Other Land Use (AFOLU) to climate change scenarios in SAARC countries, including Bhutan, was carried out by (Singh et al., 2020). This study highlights the significance of sustainable land-use practices in reducing the effects of climate change. IoT-enabled gadgets, for instance, may track crop growth, weather trends, and soil health, giving farmers and policymakers helpful information. This integration supports Bhutan's sustainability objectives and increases resilience to climatic unpredictability.

Singapore

In the face of climate change, Singapore, a centre for technological innovation worldwide, is facing more difficulties maintaining energy efficiency and food security. One viable way to solve these issues and improve climate policy is by using Sustainable IoT technology. According to research, IoT integration may transform several industries, including energy and agriculture, by facilitating data-driven decision-making and enabling real-time data. The potential of IoT in modelling sustainable food security systems was highlighted by (Kaur, 2019), who also emphasized how it may be used to address important issues in food production and delivery. IoT solutions that maximize resource utilization boost productivity and minimize food waste might benefit Singapore, which depends mainly on imports for its food supply. Precision farming methods suited to Singapore's urban setting are made possible, for example, by smart sensors and IoT-enabled platforms that may offer insights into crop health, soil conditions, and climatic conditions. (Mahkeswaran & Ng, 2020) expand on this idea by presenting an intelligent and environmentally friendly home aquaponics system. This invention offers a scalable approach to urban food production by combining hydroponics with aquaculture. Such methods can encourage self-sufficiency, lessen reliance on food imports, and improve overall food security for a city-state with a high population density like Singapore. Singapore may establish a sustainable food ecosystem aligned with its Smart Nation strategy by incorporating these systems into cities' environments.

IoT integration also has potential benefits for the energy industry, another important area for Singapore's climate strategy. The importance of IoT in energy system optimization, namely through the creation of smart grids, is covered by (Motlagh et al., 2020). These networks facilitate the integration of renewable energy sources, minimize energy waste, and allow for efficient energy delivery. IoT-enabled smart grids can improve energy efficiency and help Singapore reach its sustainability goals since the country is dedicated to lowering its carbon footprint. (Aliev et al., 2023) highlighted that IoT technology might improve global food security in the context of digital agriculture. Their findings are especially pertinent to Singapore, where supply chains may be streamlined, yield projections can be improved, and digital technologies can support urban farming projects. Similarly, (Eyitayo Raji et al., 2024) underline the significance of incorporating technology and strategic management into agricultural economics. Singapore may reduce climate-related risks and increase agricultural resilience and production by using IoT-based solutions.

Overall, the cross-sector integration of Sustainable IoT provides a way to accomplish Singapore's sustainability objectives. By adopting IoT technology, the country may improve food security, optimize energy systems, and strengthen resistance to climate change. This strategy establishes Singapore as a global leader in sustainable innovation while advancing national interests.

Maldives

Being one of the world's most climate-vulnerable countries, Maldives is looking at cutting-edge ways to improve sustainability and solve environmental issues.

According to (Upadhyay, 2014), worldwide collaboration in clean technology, renewable energy, and sustainable development planning is vital to address climate change successfully. IoT technologies give countries like Maldives, struggling with resource scarcity and rising sea levels, the means to improve these partnerships. IoT can offer vital insights for developing adaptive and mitigation strategies by facilitating real-time monitoring and data-driven decision-making. Maldives' energy sector has much room to expand through to IoT. (Motlagh et al., 2020) highlight the advantages of incorporating IoT into energy systems, primarily through smart grids and renewable energy infrastructure. Maldives' transition to IoT-enabled energy solutions can boost efficiency, lessen reliance on fossil fuels, and support its net-zero emissions goals, aligning with the country's larger vision for sustainable energy development. Although on a smaller scale in Maldives, agriculture is still a key area for IoT application. Research by (Kaur, 2019) shows how IoT can be used to model sustainable food security systems, which can be helpful in resource-constrained countries. Maldives could maximize crop yields, improve crop yields, and fortify food security by utilizing precision farming technologies. (Ali et al., 2023) further emphasize the advantages of IoT-driven innovative farming methods for attaining sustainable food production, which may be tailored to the agricultural limitations of the Maldives.

(Aliev et al., 2023) further highlight the potential of digital agriculture to improve global food security. IoT technologies like climate-responsive crop management, automated irrigation, and soil monitoring have the potential to revolutionize agricultural practices in Maldives and improve resilience and sustainability. Furthermore, (Raji et al., 2024) stress how strategic management and technology are integrated in agricultural economics. By implementing these ideas, Maldives can maximize agricultural output and tackle climate-related issues.

In Maldives, sustainable land management is another crucial area for IoT implementation. (Pandao et al., 2023) address the impact of carbon sequestration strategies in increasing soil health and land production. Long-term sustainability and carbon reduction may be achieved by implementing these practices more effectively using IoT-enabled soil and environmental monitoring systems.

Therefore, Maldives can improve its climate policy through sustainable IoT technology. Since, IoT can offer creative answers to the country's urgent environmental problems by improving land management techniques, developing sustainable agriculture, and optimizing energy systems.

CONCLUSION AND FUTURE DIRECTIONS

From the pre-COVID to the post-COVID era, the impact of Sustainable IoT on climate policy has changed dramatically, bringing about a radical transformation in how the world responds to climate change. At first, IoT applications were dispersed across industries, and there was little integration with environmental governance and climate policy. The COVID-19 pandemic, on the other hand, hastened digital transition and acted as a global driver for IoT adoption. The promise of IoT to improve real-time environmental monitoring, optimize resource management, and lower carbon footprints was acknowledged by governments and enterprises. Since the pandemic, the extensive usage of IoT has improved decision-making by providing data-driven insights and enabling real-time monitoring of water levels, deforestation, greenhouse gas emissions, and air quality. IoT has spread beyond energy and agriculture to encompass waste reduction, disaster relief, and urban management as it continues influencing climate policy. IoT technology is being incorporated into climate policy in nations like Thailand, Vietnam, Bhutan, Singapore, and Maldives to improve agricultural practices, increase sustainability, optimize energy use, and lessen the effects of climate change. Real-time resource monitoring and management has shown to be highly beneficial, especially in agriculture, where IoT was utilized to increase crop productivity, water efficiency, and insect control. In the same way, IoT-enabled smart grids and renewable energy systems have significantly decreased dependency on fossil fuels and increased energy efficiency in the energy sector.

Notwithstanding these developments, problems still exist concerning data protection, infrastructure, and the digital gap between developed and underdeveloped countries. Even though IoT systems have been successfully deployed in many developed countries, developing countries continue to confront challenges such as inadequate capacity building, high initial costs, and limited technological infrastructure. The public and commercial sectors must work together to develop digital infrastructure, encourage innovation, and advance cross-sector collaborations to overcome these obstacles. Future prospects for sustainable IoT in climate policy are promising but call for ongoing innovation and cooperation. IoT systems will become more scalable, integrated, and efficient as technology develops, offering even more chances to solve the urgent problems associated with climate change. Creating more reasonably priced IoT solutions is one of the future trends that will enable broader adoption, particularly in areas with limited resources. For a more coordinated and successful approach to climate change adaptation and mitigation, there is also a need for more international collaboration to exchange data and best practices. Furthermore, IoT will be essential to reaching net-zero emissions and developing sustainable businesses and employment. Governments must incorporate IoT technology into national climate policies and frameworks for long-term sustainability and resilience. Lastly, combining IoT with new technologies like AI, blockchain, and machine learning will increase its effect by allowing predictive modelling, automated decision-making, and more accurate environmental monitoring.

In conclusion, sustainable IoT has emerged as a powerful force in climate policy, offering the instruments required to tackle climate change with previously unheard-of accuracy. A sustainable, resilient future for all will depend on the IoT's ongoing development and incorporation into frameworks for climate regulation.

REFERENCES

Abd El-Mawla, N., Badawy, M., & Arafat, H. (2019). IoT for the Failure of Climate-Change Mitigation and Adaptation and IIoT as a Future Solution. *World Journal of Environmental Engineering*, *6*(1), 7–16. DOI: 10.12691/wjee-6-1-2

Al-Raeei, M. (2024). Artificial intelligence for climate resilience: Advancing sustainable goals in SDGs 11 and 13 and its relationship to pandemics. *Discover Sustainability*, *5*(1), 513. Advance online publication. DOI: 10.1007/s43621-024-00775-5

Ali, A., Hussain, T., Tantashutikun, N., Hussain, N., & Cocetta, G. (2023). Application of Smart Techniques, Internet of Things and Data Mining for Resource Use Efficient and Sustainable Crop Production. In *Agriculture (Switzerland)* (Vol. 13, Issue 2). MDPI. DOI: 10.3390/agriculture13020397

Ali, Z. H., & Ali, H. A. (2021). Towards sustainable smart IoT applications architectural elements and design: Opportunities, challenges, and open directions. *The Journal of Supercomputing*, *77*(6), 5668–5725. DOI: 10.1007/s11227-020-03477-7

Aliev, R., Kurbanova, M., & Samoylova, A. (2023). Transformative Potential of Digital Agriculture for Enhancing Global Food Security. *BIO Web of Conferences*, *76*, 05010. DOI: 10.1051/bioconf/20237605010

Alsharif, M. H., Jahid, A., Kelechi, A. H., & Kannadasan, R. (2023). Green IoT: A Review and Future Research Directions. In *Symmetry* (Vol. 15, Issue 3). MDPI. DOI: 10.3390/sym15030757

Araus, J. L., & Kefauver, S. C. (2018). Breeding to adapt agriculture to climate change: Affordable phenotyping solutions. *Current Opinion in Plant Biology*, *45*, 237–247. DOI: 10.1016/j.pbi.2018.05.003 PMID: 29853283

Bhagat, A., Koli, U., & D'Souza, M. (2025). *Revolutionising Water and Climate Resilience in Semi-Arid Ecosystem: Group Micro Irrigation (GMI) Approach in Tigalkheda*. Nature-Based Solutions in Achieving Sustainable Development Goals., DOI: 10.1007/978-3-031-76128-7_4

Bhatnagar, D. G., & Kumar, S. (2025). Ecological Conservation and Sustainable Management of Marine Environments. *Nature-Based Solutions in Achieving Sustainable Development Goals*, 233–261. DOI: 10.1007/978-3-031-76128-7_8

Dang, H. D. (2020). Sustainability of the rice-shrimp farming system in Mekong Delta, Vietnam: A climate adaptive model. *Journal of Economic Development*, *22*(1), 21–45. DOI: 10.1108/JED-08-2019-0027

Drews, S., Savin, I., van den Bergh, J. C. J. M., & Villamayor-Tomás, S. (2022). Climate concern and policy acceptance before and after COVID-19. *Ecological Economics*, *199*, 107507. Advance online publication. DOI: 10.1016/j.ecolecon.2022.107507 PMID: 35669404

Eliakimu, E. S., & Mans, L. (2022). Addressing Inequalities Toward Inclusive Governance for Achieving One Health: A Rapid Review. In *Frontiers in Public Health* (Vol. 9). Frontiers Media S.A. DOI: 10.3389/fpubh.2021.755285

Foláyan, M. O., Schroth, R. J., Abodunrin, O., Al-Batayneh, O. B., Arheiam, A., Mfolo, T., Virtanen, J. I., Duangthip, D., Feldens, C. A., & El Tantawi, M. (2024). Early childhood caries, climate change and the sustainable development goal 13: A scoping review. *BMC Oral Health*, *24*(1), 524. Advance online publication. DOI: 10.1186/s12903-024-04237-2 PMID: 38702704

Ghomsi, F. E. K., Nyberg, B., Raj, R. P., Bonaduce, A., Abiodun, B. J., & Johannessen, O. M. (2024). Sea level rise and coastal flooding risks in the Gulf of Guinea. *Scientific Reports*, *14*(1), 29551. Advance online publication. DOI: 10.1038/s41598-024-80748-w PMID: 39609474

Hwang, H., An, S., Lee, E., Han, S., & Lee, C. H. (2021). Cross-societal analysis of climate change awareness and its relation to sdg 13: A knowledge synthesis from text mining. *Sustainability (Basel)*, *13*(10), 5596. Advance online publication. DOI: 10.3390/su13105596

Katyal, N. (2025). Understanding About Nature-Based Approaches for Sustainable Consumption, Production and Development. *Nature-Based Solutions in Achieving Sustainable Development Goals*, 317–367. DOI: 10.1007/978-3-031-76128-7_11

Kaur, H. (2019). Modelling internet of things driven sustainable food security system. *Benchmarking*, *28*(5), 1740–1760. DOI: 10.1108/BIJ-12-2018-0431

Lamid Idowu, L., Park, S.-H., & Ali Ibrahim, I. (2017). A New IoT Architecture for a Sustainable IoT Adoption. *International Journal of Computer Science and Information Technology Research*, *5*(2), 204–208. https://www.researchgate.net/publication/332876717

Lipper, L., McCarthy, N., Zilberman, D., Asfaw, S., & Branca, G. (2017). *Climate Smart Agriculture Building Resilience to Climate Change*. Springer. https://www.springer.com/series/6360

Mahkeswaran, R., & Ng, A. K. (2020). Smart and Sustainable Home Aquaponics System with Feature-Rich Internet of Things Mobile Application. *2020 6th International Conference on Control, Automation and Robotics, ICCAR 2020*, 607–611. DOI: 10.1109/ICCAR49639.2020.9108041

Matos, S., Viardot, E., Sovacool, B. K., Geels, F. W., & Xiong, Y. (2022). Innovation and climate change: A review and introduction to the special issue. In *Technovation* (Vol. 117). Elsevier Ltd., DOI: 10.1016/j.technovation.2022.102612

Mishra, A., Kumar, P., & Noble, A. (2013). Assessing the potential of SRI management principles and the FFS approach in Northeast Thailand for sustainable rice intensification in the context of climate change. *International Journal of Agricultural Sustainability, 11*(1), 4–22. DOI: 10.1080/14735903.2012.658648

Mitra, S., Madhuvanthi, S., & Sabumon, P. C. (2025). Nature-Based Urban Resilience: Integrating Green Infrastructure. *Nature-Based Solutions in Achieving Sustainable Development Goals*, 167–205. DOI: 10.1007/978-3-031-76128-7_6

Motlagh, N. H., Mohammadrezaei, M., Hunt, J., & Zakeri, B. (2020). Internet of things (IoT) and the energy sector. In *Energies* (Vol. 13, Issue 2). MDPI AG. DOI: 10.3390/en13020494

Najafi, S. E., Nozari, H., & Edalatpanah, S. A. (2022). Investigating the Key Parameters Affecting Sustainable IoT-Based Marketing. In *Studies in Computational Intelligence* (Vol. 1036, pp. 51–61). Springer Science and Business Media Deutschland GmbH., DOI: 10.1007/978-3-030-97344-5_4

Nguyen, T. T., Nguyen, L. D., Lippe, R. S., & Grote, U. (2017). Determinants of Farmers' Land Use Decision-Making: Comparative Evidence from Thailand and Vietnam. *World Development, 89*, 199–213. https://ssrn.com/abstract=2837618http://www.sciencedirect.com/science/article/pii/S0305750X15307853Electroniccopyavailableat:https://ssrn.com/abstract=2837618Electroniccopyavailableat:http://ssrn.com/abstract=2837618. DOI: 10.1016/j.worlddev.2016.08.010

Onyanta, A. (2016). Cities, municipal solid waste management, and climate change: Perspectives from the South. *Geography Compass, 10*(12), 499–513. DOI: 10.1111/gec3.12299

Pandao, M. R., Jejal, A. D., Shukla, E. P., Prabhumitrareddy, S., Rout, S., Baral, K., & Bhadani, M. (2023). Unlocking the Benefits of Carbon Sequestration for Enhancing Soil Health. *International Journal of Environment and Climate Change, 13*(12), 1349–1359. DOI: 10.9734/ijecc/2023/v13i123803

Peng, L., & Anthony Maria Das, P. (2024). Investigating the spatial spillover impact of digital economy with research and development investment on environmental sustainability. *Management and Sustainability, ahead-of-print*(ahead-of-print). https://doi.org/DOI: 10.1108/MSAR-07-2024-0063/FULL/XML

Polymeni, S., Athanasakis, E., Spanos, G., Votis, K., & Tzovaras, D. (2022). IoT-based prediction models in the environmental context: A systematic Literature Review. *Internet of Things : Engineering Cyber Physical Human Systems, 20*, 100612. DOI: 10.1016/j.iot.2022.100612

Prasert, S. (2024). Buddhist Ethics and Environmental Conservation in Thailand. *Journal of Philosophy. Culture and Religion, 7*(2), 35–46. DOI: 10.47604/jpcr.2606

Quan, D. M., & Van, T. T. H. (2022). Potential and Trends of IoT Application in Agriculture in Vietnam. *Open Journal of Social Sciences, 10*(01), 170–182. DOI: 10.4236/jss.2022.101015

Raji, E., Ijomah, T. I., & Eyieyien, O. G. (2024). Integrating technology, market strategies, and strategic management in agricultural economics for enhanced productivity. *International Journal of Management & Entrepreneurship Research, 6*(7), 2112–2124. DOI: 10.51594/ijmer.v6i7.1260

Rejeb, A., Suhaiza, Z., Rejeb, K., Seuring, S., & Treiblmaier, H. (2022). The Internet of Things and the circular economy: A systematic literature review and research agenda. *Journal of Cleaner Production, 350*, 131439. DOI: 10.1016/j.jclepro.2022.131439

Rosca, M. I., Nicolae, C., Sanda, E., & Madan, A. (2021). *Internet of Things (IoT) and Sustainability. 7th BASIQ International Conference on New Trends in Sustainable Business and Consumption*. Foggia, Italy. DOI: 10.24818/BASIQ/2021/07/044

Schuller, B. W., Akman, A., Chang, Y., Coppock, H., Gebhard, A., Kathan, A., Rituerto-González, E., Triantafyllopoulos, A., & Pokorny, F. B. (2022). *Climate Change & Computer Audition: A Call to Action and Overview on Audio Intelligence to Help Save the Planet*. http://arxiv.org/abs/2203.06064

Singh, N. (2025). Promoting Multi-stakeholder Collaboration for Nature-Based Sustainable Development. *Nature-Based Solutions in Achieving Sustainable Development Goals*, 289–316. DOI: 10.1007/978-3-031-76128-7_10

Singh, R. K., Sinha, V. S. P., Joshi, P. K., & Kumar, M. (2020). Modelling Agriculture, Forestry and Other Land Use (AFOLU) in response to climate change scenarios for the SAARC nations. *Environmental Monitoring and Assessment, 192*(4), 236. Advance online publication. DOI: 10.1007/s10661-020-8144-2 PMID: 32172340

Sinthuja, U., Thavamani, S., Makkar, S., Gobinath, R., & Gayathiri, E. (2022). IoT applications in landslide prediction and abatement—Trends, opportunities, and challenges. In *Computers in Earth and Environmental Sciences* (pp. 319–325). Elsevier., DOI: 10.1016/B978-0-323-89861-4.00010-5

Solanki, V., & Birman, S. (2025). Nature-Driven Renewable Energy Systems for Sustainable Development. *Nature-Based Solutions in Achieving Sustainable Development Goals*, 131–166. DOI: 10.1007/978-3-031-76128-7_5

Sonkar, G. (2025). Augmenting Carbon Sink Capacity Through Nature Based Solutions for Climate Change Mitigation: A Case Study of India. *Nature-Based Solutions in Achieving Sustainable Development Goals*, 207–231. DOI: 10.1007/978-3-031-76128-7_7

Syali, M. (2025). Nature-Based Solutions for Riparian Decision-Making in South Asia: A Legal Perspective. *Nature-Based Solutions in Achieving Sustainable Development Goals*, 1–39. DOI: 10.1007/978-3-031-76128-7_1

Talavera, J. M., Tobón, L. E., Gómez, J. A., Culman, M. A., Aranda, J. M., Parra, D. T., Quiroz, L. A., Hoyos, A., & Garreta, L. E. (2017). Review of IoT applications in agro-industrial and environmental fields. In *Computers and Electronics in Agriculture* (Vol. 142, pp. 283–297). Elsevier B.V., DOI: 10.1016/j.compag.2017.09.015

Tiwari, C., Bhatnagar, S., & Kumari, R. (2025). Making Ends Meet: Understanding How Circular Economy Can Contribute to Long-Term Ecosystem Restoration and Biodiversity Conservation. *Nature-Based Solutions in Achieving Sustainable Development Goals*, 263–287. DOI: 10.1007/978-3-031-76128-7_9

Trædal, L. T., & Vedeld, P. (2018). Cultivating forests: The role of forest land in household livelihood adaptive strategies in the Bac Kan Province of northern Vietnam. *Land Use Policy*, *73*, 249–258. DOI: 10.1016/j.landusepol.2018.02.004

Tuysuz, M. F., & Trestian, R. (2020). From Serendipity to Sustainable Green IoT: Technical, Industrial and Political Perspective. *Computer Networks*, *182*, 107469. https://doi.org/https://doi.org/10.1016/j.comnet.2020.107469. DOI: 10.1016/j.comnet.2020.107469

Uddin, S. N., Taplin, R., & Yu, X. (2007). Energy, environment and development in Bhutan. In *Renewable and Sustainable Energy Reviews* (Vol. 11, Issue 9, pp. 2083–2103). DOI: 10.1016/j.rser.2006.03.008

Upadhyay, D. K. (2014). Coping with Climate Change: India–EU Cooperation on Renewable Energy and Clean Technology. *India Quarterly*, *70*(3), 241–256. DOI: 10.1177/0974928414535293

Usman, M., Husnain, M., Riaz, A., Riaz, A., & Ali, Y. (2021). Climate change during the COVID-19 outbreak: Scoping future perspectives. *Environmental Science and Pollution Research International, 28*(35), 49302–49313. DOI: 10.1007/s11356-021-14088-x PMID: 33934308

van de Fliert, E., Dung, N. T., Henriksen, O., & Dalsgaard, J. P. T. (2007). From Collectives to Collective Decision-making and Action: Farmer Field Schools in Vietnam. *Journal of Agricultural Education and Extension, 13*(3), 245–256. DOI: 10.1080/13892240701427706

Vickram, D. R., & Srivastava, P. (2025). Promoting Agroecology and Sustainable Fisheries for Achieving Food Security. *Nature-Based Solutions in Achieving Sustainable Development Goals*, 41–62. DOI: 10.1007/978-3-031-76128-7_2

Vijayakumar, S., & Srivastava, P. (2025). Ecosystems and Human Health: The Impact of Nature-Based Interventions. *Nature-Based Solutions in Achieving Sustainable Development Goals*, 63–90. DOI: 10.1007/978-3-031-76128-7_3

Chapter 17
Digitalisation: A Key Used to Fill the Void Amid COVID-19

Vanshika Saklani
https://orcid.org/0009-0001-3353-2199
Graphic Era University, India

Megha Bohra
https://orcid.org/0009-0009-4320-7443
Graphic Era University, India

Vidhi Kapruwan
https://orcid.org/0009-0006-3780-0011
Graphic Era University, India

Vinay Kandpal
https://orcid.org/0000-0003-1823-4684
Graphic Era University, India

Naresh Kumar Sharma
Graphic Era University, India

ABSTRACT

COVID-19 as infections continued to rise globally, uncertainty over economic growth deepened. Ways to tackle the pandemic were made with the correspondence of technological development. There was alarming need to implement technologies that meet the contactless demand to prioritise health and safety. Countries enhanced digital infrastructure, adopted online business models, and expanded digital payment systems. Remote working became standard in many sectors, while cloud computing and artificial intelligence gained traction to support operations. Rapid

DOI: 10.4018/979-8-3693-9909-5.ch017

digital transformation has been pivotal in managing pandemic disruptions, fostering resilience, and reshaping how industries function in a more tech-integrated world. Study analysis the interplay of digitalisation during COVID-19 pandemic and its related effects. A qualitative approach based on secondary data has been applied in this study. Moreover, the findings used in the study correlate with the recent time. This study concludes that technological advancements have been evident, to mitigate and handle the impacts of the pandemic.

INTRODUCTION

The COVID-19 pandemic can be regarded as an initiator for global transformation. With more than 190 million coronavirus cases and more than 4,101,340 fatalities worldwide, along with the risk of the emergence of new variants (World Health Organization, 2020), there was an imperative need to follow necessary precautions such as lockdowns, social distancing and border closures. The halt that was followed by this range of measures was tackled by emerging technologies and digitalisation (Amoah et al., 2021). Therefore, to facilitate the smooth functioning of many sectors during the COVID-19 pandemic, emerging technologies and digitalisation can be regarded as great accelerators (Bradley et al., 2020). Moreover, digital technologies also helped many countries across the globe in leveraging down the burden left by the pandemic outbreak (Whitelaw et al., 2021).

Digitalisation refers to the process by which traditional or analogue tasks or processes are converted into digital form so that computers can access, store, and transmit information (Brenanen & Kresis,2016). Although the process of digitalisation had started various decades before, it has boomed during the pandemic outbreak. It is far more than technology adoption. The basic alterations that an organization implements in terms of its strategies and processes as well as organisational knowledge define the true essence of digitalisation (Matthess et al., 2020).

COVID-19 has expedited the foremost improvisation of digital authority, leading to an advanced phase of connection technology and automation (Vlados & Chatzinikolaou, 2021). It has facilitated the foremost development of virtual authority (Maiti et al., 2019). Digital transformation is evident across all sectors, be it, the economy, education, market or business. Also, digitalization served as a great benefit during the pandemic. The adoption of new technologies during the pandemic was a significant factor in the working of various sectors (Brenanen & Kresis,2016). Moreover, it has also impacted the performance outcomes of both private and government organizations (Brown et al., 2018).

Furthermore, digitalisation and the green sector together, hold an important role in sustainable development during the pandemic (Lanshina et al., 2020). Also, the present-day trend is powered by further automation, communication tools, the internet of connected devices, large amounts of data, and smart machines (Härmand, 2021).

Role of Digitalisation in the Economy

Digitalisation has proved to be a facilitator of economic growth and provides opportunities for employment (Sabbagh et al., 2013), especially during the time of pandemic. It served as a meditator to facilitate economic activities. Since, COVID-19 had a direct impact on the production area while disrupting the supply chain, and affecting the firms and financial markets in the worst way possible (Gozgor et al., 2021; Zhai et al., 2021). So, to support the impact of heavy demand faced by the demand-side management in comparison to that of the supply side during the pandemic, there was an increase in the need for digital industries (Sabbagh et al., 2013).

The digital economy can be termed as "the digital transformation of economic and social development" (OECD, 2014). The decrease in macroeconomic activities to meet the rules and regulations to minimise social mobility was rectified with the help of digital technologies (Sun et al., 2021). Since, the COVID-19 initiated a prompt increase in meeting demands of consumers digitally which helped the emerging digital industries to flourish (Fang et al., 2022).

The digital economy provided an opportunity to integrate with other industries, serving as an advantage of high technology and acting as stabilizers and boosters in industries (Banga & Velde, 2020). Moreover, it alleviated economic losses and promoted economic recovery during the pandemic (Zhang et al., 2022). Thus, to combat this crisis and promote economic growth, these emerging technologies and the digital economy have proved to be helpful.

Education Sector and Digitalisation

The COVID-19 outbreak has made its way to a substantial transformation in the global instructional-teaching system. The cessation of operations in schools, academies, and other educational environments has resulted in an affected change in more than 94% of the global student demographic (Gopika & Rekha, 2023). All academic tasks have altered and modified their way to the online world. Due to this abrupt transition, virtual education has sustained considerable alteration (Kumar & Sharma, 2021)). The pandemic resulted in the most profound interference with educational frameworks chronically, influencing approximately 1.6 billion learners across over 200 nations (Samreen & Kumari, 2021).

The COVID-19 pandemic elicited a noticeable modification in the paradigms of educational delivery around the world, and India is no exception (Kumar et al., 2020). The swift shift and adoption of digital tools were quick, hence helping in the continuation of education life even during lockdowns (Nikas et al., 2022). The pandemic has precipitated the expansion of skills in employees and educators, exposing them to experiment and explore modern tools and radical methods for virtual instruction and education. This brought a rise in the creativity of teaching practices, as well as approaches to delivery by introducing new methods (Garg et al., 2021).

Moreover, digitalisation served as a positive factor in providing education virtually during a time when social distancing was being followed. Using video conferencing platforms like Zoom, Microsoft Teams and Google Meet proved to be a helpful tool to connect with the students. Attention towards online learning due to the pandemic has moved educational practice globally and made it easy to access education therefore, following SDG 03, which aims to focus on effective education for all and promoting opportunities that will last for a lifetime.

A mediator in transforming market dynamics

The E-commerce sector surged during the pandemic with many people engaging in online business and taking their purchasing habits online (Maiti et al., 2019). This transformation led to companies investing in digital marketing strategies and enhancing their online presence (Chauhan et.al, 2022). Companies embracing digital transformation were in a good position to face challenges brought about by the pandemic (Maiti et al., 2019). Furthermore, technologies were utilized in a way that contributes to virtual transformation (also DX or DT) to generate value for various stakeholders (customers in the broadest sense) and to harvest and harness the ability to rapidly adapt to ever-changing circumstances (Samreen & Kumari, 2021).

Advanced technologies along with the integration of physical and digital systems constitute virtual transformation. This virtual transformation was made through innovative business models, new and refined production processes and effective knowledge-based products (P & M, 2021). These occurrences are believed to improve productivity growth in the digital field but the effect may not be seen transparently at an overall level.

Impact on Organizations and Industries

The concept of working from home to facilitate the working environment of the companies was an emerging factor during the pandemic. The inclination towards digital platforms results in a reduction of costs associated with commuting, office administration, and operational costs as well as making it easier to communicate effectively during the pandemic (Hensher et al., 2021). The measures adopted to follow the quarantine rules result in changing the external dynamics; customers, stakeholders and other suppliers and internal dynamics; employees, and employee-employer relationships (Sostero et al., 2020). SDG 09, which aims at fostering innovation, was adopted during the pandemic for effective functioning within organizations and industries. The emerging technologies paved the way to help organizations link teams, coordinate better, manage data effectively and encouraging meaningful working relationships between headquarters and subsidiaries (Autio et al., 2021). It has also helped in linking an organization's external activities and improving as well as enhancing its overall competitiveness (Ritter & Pedersen, 2020).

Growth of Virtual Enterprises

The enterprises made for business ventures faced a rapid change in their business environment due to COVID-19 (Castro et al., 2012). Digitalisation helped these enterprises to operate virtually. These virtual enterprises, with the help of digital technologies, improved existing business operations to better fit into the economic competition. This was made possible through more effective coordination between the processes and embracing a user-friendly approach (Bermann, 2021).

With an increased advancement in ICT, termed digital transformation, the working environment of virtual enterprises is enhanced (Sadigh et al., 2017). Moreover, to better fit and have an advantage in the competitive and advancing environment, businesses are exploring new inter-firm organisational relationship models (Javed & Yasir, 2019). Therefore, to take better advantage of innovation and collaborative techniques, there is a need to incline towards intensive deployment of ICT. This will further help in adapting to any changes in the future (Zangiacomi et al., 2020).

Digitalisation and Sustainability

The pandemic was an evident example of investing in effective and potentially functioning technologies that aim at long-term survival while, keeping into consideration sustainability (Mustafa & Abbas, 2021). The investment of many organizations in technological advancements in order to adopt sustainable green practices, such as green HRM, green supply chain, green innovation and green marketing is considered an effective way to achieve justifiable performance (Alraja et al., 2022). This leads to attaining sustainability in the industry (Mustafa & Abbas, 2021). Moreover, it also helped the industry in carrying out its social responsibility by adopting ecologically safe and digitally effective approaches that contribute to and lead to results promoting preservation of resources (Zhu et al., 2019).

Digital and green sectors both complement each other and are equally important. Emerging technologies are creating a way for the green sector to help save energy and create an efficient and reliable source of energy sustainably. Moreover, it also contributes to delivering an adequate amount of energy at the right time and the lowest cost possible (Strielkowski et al., 2021). This is achieved through advancements in bandwidth expansion, network equipment, and software leveraging cloud services (Akala, 2020). Moreover, human resources also have a notable contribution in utilising green technologies (Singh et al., 2017).

Therefore, the pandemic has forced organisations to shift from traditional to eco-friendly practices to perform in a manner that proves to be efficient for the environment (Alraja et al., 2022). Furthermore, this shift from traditional to eco-friendly practices with the help of technological advancements leads to careful utilisation and exploitation of resources and meets the goals of sustainability for a sustainable future (Klewitz & Hansen, 2014) and control operations remotely (El-Haddadeh, 2020).

AI as a Helping Hand During Pandemic

AI demonstrated its effectiveness as an emerging technology during the pandemic. It helped in the determination and screening of affected people in public places, like workplaces and areas of public transportation. It was made possible through infrared thermal cameras and thermometers (Wang et al., 2020). Moreover, study shows that AI can detect COVID and non- COVID coughs in two seconds (Imran et al., 2020). Also, mobile health apps played a crucial role in gathering real-time data about infected and non-infected people in various countries (Jorfi et al., 2020).

Machine Learning (ML) was proved to be a significant aid in estimating the effectiveness of a pandemic on the healthcare systems and other organizations, the number of waves expected of the pandemic, the mortality and infection rates and the right time to leverage ease in public restrictions and lockdowns (Kushwaha et al., 2020).

Objectives of the study are:

1. To assess the association between digitalization and the educational sector during COVID-19.

2. To investigate the role played by digitalization in transforming the market during COVID-19.
3. To explain the role of the pandemic in accelerating digital transformation.
4. To study the impact of digitalization as a new avenue in the Indian Economy during COVID-19.
5. To highlight the link between digitalization and sustainability.

LITERATURE REVIEW

The pandemic actuated a noticeable shift in the virtual working of many firms. Here, concerning the economy, organisations and industries, education, e-commerce, market dynamics, green technology and sustainability remain among the most impacted aspects.

The pandemic had a strong hold on the world economy in many ways. The cuts applied by coronavirus slowed down the business processes. Therefore, firms and corporations had to rethink and reshape their take to handle the crisis. Many firm owners resolved to launch virtual changes to keep business running during the outbreak. This lead to, embraced collaborations at virtual working, and an accelerated development of online firms. Along with the overall cognise value of IT and virtual technology by the market forces (Mark, 2024). Many embraced the hybrid mode of working or fully remote work models permanently. This reframed labour markets with the influence on real estate, as demand for office space declined while home office equipment surged.

Additionally, businesses quickly adapted digital communication, collaboration, and operation tools to sustain operations in the face of lockdowns. E-commerce and digital payments grew rapidly; hence, the retail and finance industries have been transformed. This was uncovered in an Eurasian study. It uncovered the growing pace of virtual modernisation in many countries (Eaton & Heckscher, 2020). This conveyed by how financial institutes swiftly opted for virtual modes. Along with the medial management choosing virtual telecare as well. The enhancement by sellers on self-service, commerce and contactless deliveries marked a noticeable percentage of the workforce working in hybrid mode (Ionescu et al., 2021).With this, supply chains became more data and AI-driven. However, this change also revealed inequality since most people did not have access to technology, and cybersecurity risks increased. A noticeable change in investment in software for the virtual organization and rationalization of work was highlighted. Extrapolating, the spread of virus marked heavy hindrance in commercial monetary world. According to an estimate published by the Office of Economic Analysis, the Real Gross Domestic Product (GDP) fell at an annual rate of 32.9% in the second quarter of 2020. At the same time, commercials were stopped from being operated henceforth the virus (Ling et al., 2020).

With government and policy responses, impact on the digital divide. The new business models and innovations, increased cyber security, impact on financial services and digital payments, global supply change and logistics innovation marked a shift. The pandemic made its way to a substantial transformation in the global instructional-teaching services.

Heading to, the drastic drift towards e-learning. The students had diverse levels of engagement. This was contingent upon external factors like home environment, internet connectivity, and even prior exposure to digital tools (Adnan, 2020). Digital tools could help to deliver education. However, there were stark inequities in the students' technology and digital literacy (Dhawan, 2020). An emphasis on the shift in constructivist teaching practices was noticeable with the help of digital tools. Platforms like Zoom and Google Classroom facilitated the collaborative projects among the students. This shift may make learners more engaged and self-directed (Adedoyin & Soykan, 2020).

Therefore, the pandemic accelerated the adoption of progressive educational practices in such an unconscious manner. Digitalization bridged the gap between formal learning and informal learning. The digital tools opened ways of informal learning such as online workshops and webinars. They enriched educational opportunities. The outcomes were the integration of both pathways of formal learning and informal learning. This allowed the acquisition of skill development and lifelong learning even after the pandemic (Jakoet-Salie & Ramalobe, 2022).

Another significant dimension of digitalization was the rise of e-commerce. E-commerce is corroborated to be an essential channel for businesses. To keep their functions running amidst lockdowns and social distancing measures.

The pandemic suddenly raised the call for social distancing, making people opt for online shopping. This compelled both consumers and businesses to go digital. Digitalization emerged as a vital tool for maintaining economic activity. This was possible through contactless payment systems and robotic delivery mechanisms. Which were designed to cater to the changing consumer needs in the light of health concerns (Renu, 2021).The market dynamics showed a significant effect on the stock market behaviour amid the pandemic. In a more positive direction, digitalization positively contributed to market responses. Using digital tools led to better decision-making that would help offset market volatility that was aggravated by public perception of COVID-19 (Xiong et al., 2024). The use of digital marketing strategies by firms has highlighted vital transformation in ensuring sustainability in MSMEs (Purba et al., 2021). The pandemic altered the shopping behaviour and sales patterns of customers. Based on the analysis of customer feedback, it was revealed that early anxieties about online shopping gave way to increased trust and engagement, showing how digitalization not only facilitated e-commerce but also altered the perceptions of consumers (Galhotra & Dewan, 2020).

Moreover, global e-commerce grew by 10 years' worth of progress in just three months in 2020. Online retail sales in the U.S. saw a 32% increase (U.S. Census Bureau). The rise of digital payment systems, contactless delivery, and advancements in logistics further fueled this growth. Platforms like Amazon and Alibaba experienced massive upticks in sales (Statista). Digitalization has thus reshaped consumer behaviour, accelerating e-commerce growth.

The last factor to address is the influence of digitalization on sustainability. The application of digital technologies such as AI, big data, and IoT has proven to add value to the efforts of sustainability across different sectors. The development of these technologies has played a huge role in improving pollution controls, waste management, and sustainable production processes (Feroz et al., 2021). Small and medium-sized enterprises are important to the sustainability of an economy, especially during this period of the pandemic (Pu et al., 2021). Digitization has empowered many SMEs to change the direction of their operations and connect with customers online, maximizing the use of resources and reducing waste (Gil-Gomez et al., 2020).

The pandemic-induced shift in digitization has deeply changed consumer behaviour, where hygiene and sustainability also play a prominent role while making purchases (Kunkel & Matthess, 2020). Organizations are challenged to align themselves with these changed consumer trends by incorporating sustainability in business operations. The role of digital technologies in facilitating sustainable business models is a response to this critical area of future research for the time being, focusing on consumer behaviour and sustainability. At the same time, the pandemic accelerated the adaptation of digital tools in education, which promoted the integration of sustainability into curricula. The digitization of learning not only ensured continuity in learning but also pushed institutions to embed sustainability into teaching practices (Abad-Segura et al., 2020). This transformation is important in equipping future generations with an understanding of sustainability issues and how they can engage with them. Therefore, creating a culture of green technology initiatives.

In essence, digitalization has played a significant role in the promotion of sustainability through efficiency improvement, waste reduction, and optimal use of resources. IoT, AI, and big data help businesses and governments monitor energy consumption, manage waste, and track emissions in real time. Data-driven approaches support informed decision-making and promote smarter environmental practices. The use of digital platforms promotes remote work and reduces the carbon footprint related to commuting. Digitalization also encourages innovation for renewable energy, sustainable supply chains, and green technologies, promoting a circular economy. Digitalization streamlines business operations and allows for sustainable practice, thereby accelerating global efforts toward environmental conservation and climate action.

Table 1. The major findings of the study are as follows:

Sr. No.	Author(S)	Type of Research	Objective	Major Findings
1.	Ionescu et al., 2021	Quantitative research	How the market forces handled the shift due to the spread of virus.	How virtual tools aid finances by using Romanian economy in comparison with EU countries. The areas digitized were more developed or were being adapted to new conditions put out by the crisis brought forward by COVID-19 were impacted much lower.
2.	Eaton & Heckscher, 2020	Qualitative research	Explore the affects of the virus on labour and employment relations.	The acceleration of remote work, increased workforce flexibility, and changes in worker expectations regarding health and safety. The pandemic also exposed inequalities in job security, benefits, and working conditions, particularly for low-wage workers.
3.	Ling et al., 2020	Quantitative research	To analyze the result of the virus spread pandemic on financial and market prices.	The prices for office, retail, and hotel properties declined, and in the retail and hotel sectors, the declines were more acute. The negative impacts, according to the study, were more intense in regions with stricter lockdown measures and higher infection rates.
4.	Schaupp, 2022	Qualitative research	How functional management became an alternative to automation during virus spread on economic crises and digitalization.	Intersection of economic crises, and digitalization, emphasized a shift from automation to algorithmic management. The economic shocks and disruptions from the pandemic pushed forward the implementation of algorithmic management systems compared to traditional automation systems.

continued on following page

Table 1. Continued

Sr. No.	Author(S)	Type of Research	Objective	Major Findings
5.	Dhawan, 2020	Qualitative research	To analyze the role of online learning as a solution during the COVID crisis, focusing on its challenges, opportunities, and effectiveness.	Describes problems of digital divide, lack of infrastructure, and low levels of digital competency by teachers and learners. Along with opportunities like flexibility and scalability, which opened up possibilities for innovative learning in online education.
6.	Adnan, 2020	Quantitative research	During COVID students' notion towards online learning.	The students perceived online learning as less effective than their experience in the traditional setting due to lack of interaction, low participation, and unavailability of or unreliable internet, among other reasons. Financial constraints and the inability to cope with online platforms further affected learning.
7.	Jakoet-Salie & Ramalobe, 2022	Qualitative research	During the COVID pandemic learning and teaching practices in higher education institutions	The significant challenges in digital transformation, such as limited infrastructures, digital literacy gaps, and unequal access to technology. However, it highlighted opportunities to innovate, underscored that institutional support would be useful in adapting to digital teaching, and found the need to have effective strategies to uplift student engagement and reduce digital fatigue.
8.	Adedoyin & Soykan, 2020	Review article	Explore how the COVID pandemic has impacted the education sector by necessitating a rapid transition to online learning.	In online learning during COVID-19, challenges include inadequate digital infrastructure, lack of training for educators, and limited access to technology; however, opportunities lie in flexibility, global access, and innovation, with a need to enhance digital education systems to achieve sustainability.

continued on following page

Table 1. Continued

Sr. No.	Author(S)	Type of Research	Objective	Major Findings
9.	Renu, 2021	Qualitative research	How technological advancements, particularly during the COVID pandemic, played a pivotal role in transformation, examining both the positive outcomes and challenges it posed across various sectors.	The technological advancements during the COVID pandemic helped significantly enhance healthcare delivery, facilitating remote work, and continuing education. However, digital inequality, cybersecurity issues, and the need for digital literacy were identified to be some of the main challenges, thus calling for a further integration of technology and its support.
10.	Purba et al., 2021	Quantitative research	Examine the impact of virtual marketing and e-commerce on the financial performance and business sustainability of Micro, Small, and Medium Enterprises (MSMEs) in Indonesia during the COVID-19 pandemic.	Positive correlation between using digital marketing strategies and financial performance. Utilizing e-commerce positively affects business sustainability and disruptions.
11.	Galhotra & Dewan, 2020	Quantitative research	Study the impact of COVID-19 on digital platforms and change in E-commerce shopping trends.	The increase in usage of digital platforms was majorly triggered by lockdowns and social distancing. Consumers favoured buying essential commodities online; they began shifting towards safety and convenience. Small and medium-sized enterprises use e-commerce sites to sustain business. The results highlighted the importance of digital platforms in continuing economic activities and changing the consumption culture of individuals during the pandemic.

continued on following page

Table 1. Continued

Sr. No.	Author(S)	Type of Research	Objective	Major Findings
12.	Pu et al., 2021	Quantitative research	How innovative finance and technological adaptation aid to the SMESs during the spread of virus	Innovative finance and technological adaptation were major factors contributing to the environment. Government support significantly enhanced the mediating role. Innovative financial solutions and technological integration were helpful for SMEs to overcome difficulties, and the amplification effect of interventions from the government made their efforts more effective.
13.	Y. Xiong et al., 2024	Quantitative research	Examine the impact of the COVID-19 pandemic on the share of online consumption in China, using mobile payment data.	The primary findings are that the COVID-19 pandemic significantly increased the share of online consumption in China. Analysis of mobile payment data indicated a significant shift to online shopping and digital transactions, especially in groceries, entertainment, and health products. This shift was therefore driven by restrictions on physical retail and changes in consumer behavior and was enabled by mobile payment platforms as a central facilitator of this transition to online consumption.
14.	Abad-Segura et al., 2020	Quantitative research	To analyze global research trends related to the environmental management virtual tools in learning field.	The transformation in learning rapidly growing with an increased interest in sustainability and technological advancement. Key themes are ICT integration, e-learning, and developing digital skills. The study also reveals the significance of leadership, policy-making, and institutional strategies in managing digital transformation.

continued on following page

Table 1. Continued

Sr. No.	Author(S)	Type of Research	Objective	Major Findings
15.	Feroz et al., 2021	Systematic literature review	Explore the impact of digital transformation on environmental sustainability.	Key disruptions brought about by digital transformation in environmental sustainability, particularly in four areas: pollution control, waste management, sustainable production, and urban sustainability. Digital technologies like AI, big data, IoT, and cloud computing allow for major improvements in environmental practices.
16.	Gil-Gomez et al., 2020	Quantitative research	Examine the role of digital transformation in customer relationship management (CRM) and how it contributes to sustainable business model innovation.	Digital transformation significantly enhances CRM by allowing businesses to understand customer needs better, communicate more effectively, and tailor services to customers' needs. The integration of digital technologies in CRM systems facilitates innovation in business models towards more sustainable practices.
17.	Kunkel & Matthess, 2020	Qualitative research	To analyze and compare the expectations regarding the environmental impacts of ICTs.	The research has indicated that in all countries studied, there are very vague expectations in the policies about the environmental effects of ICTs. Generally, indirect positive effects are overstated compared to negative direct effects of energy consumption in using ICTs.

Methodology

A qualitative approach based on secondary data has been applied in this study. It is carried out using relevant databases such as Google Scholar, PubMed, Scopus, and Research Gate to draw out applicable research data. Relevant keywords that correlate with the study such as digitalisation, COVID-19, sustainable development goals and AI were proved to help extract the required data. The data concerning the independent variable 'digitalisation' and dependent variable 'covid-19' were also taken into consideration. Existing literature was taken into consideration which included research articles, review papers and journals. After a detailed analysis by the two authors, a total of 66 articles were included in the study. The authors themselves manually went through all the studies to extract crucial data that was used in this review. A thorough study of these articles was helpful in data extraction of the information that was relevant to the study's aim and objectives. After the necessary data was acquired, the researchers carefully examined the data and then organised the information as per the requirement of this analysis. Furthermore, the emphasis was paid to include the findings that are relevant to recent times.

SOLUTIONS AND RECOMMENDATIONS

Because of the spreading of the global pandemic of COVID-19, practically every concerning thing has accelerated the pace of digitalization in various sectors, which has radically changed the way business has been processing, or education is being practised, and even how people relate to services. Consumers were more acquainted with the issues of sustainability as well that the pandemic increased consumer demand for sustainable products. When it comes to education, the rapid shift toward online learning platforms came with opportunities and challenges. On one hand, online digital learning assists people with knowledge resources, while on the other hand, it widens the digital divide. A new development witnessed in this set of unprecedented situations is observed in the right set of circumstances and challenges as listed in the findings. The speed at which digital technologies are progressing has been unprecedented. They have adjusted and made alterations to consumer behaviour, modified business actions, and commute market dynamics. Digital payment systems were also accepted in greater numbers than was anticipated. The study underlines the multifaceted impact of digitalization on sustainability during the COVID-19 pandemic. Rapid change is perhaps the most extreme manifestation of operational agility; certainly, the measures of lockdown accelerated organizations' adaptation to those constraints. Efficiency improvement was the most common theme; however, in most sectors, the gap between the best and the worst-prepared sectors remains much too large. Telehealth services have extended more during the pandemic. Here, it is found that in adapting to unprecedented provocation during the pandemic era, the accelerated adoption of digital technologies by organizations and individuals surfaced as an important determinant that influences sustainable practices. It was also exhibited from the analysis that digital tools helped disperse working activities away from centralized offices, hence cutting carbon emissions both from the commute and from office energy use. This study aims to contemplate the impact that digitalization has on various efficiencies, accessibility, and user experience in the era of the pandemic.

FUTURE RESEARCH DIRECTIONS

This study calls out for research on the topics concerning changes due to digitalisation after COVID-19 –19 as this present study only mentioned the effects of digitalisation during COVID–19. Therefore, future research on the effects of digitalisation on the livelihood of people as well as on sustainability can be done. The significance of SDGs in the lives of people, the changes in lifestyle after covid-19 and how the pre covid phase was different from post covid phase. Also, the impact of digitalisation on the health crisis during the pandemic can be a topic to be researched. Moreover, the social gap left by the pandemic can also serve Tavares as a topic to be researched.

CONCLUSION

Occurrence has shown deeper and more plural when contemplated against the backdrop of education, market forces, and sustainability compared to the impact of digitalization during COVID-19. Against such a backdrop, the paper evinced that rapid shifts to digital platforms have facilitated continuing education but in great disparity concerning access and quality. The collaboration among the public and private sectors may lead to alterations in sustainable practices after the pandemic. Amid COVID-19, the landscape that evolve is complex, characterized by impressive growth coupled with noteworthy challenges. Standard practices for sustainable digital transformation will need to balance innovation with equity and security; as the world appears from the pandemic, policymakers, business leaders, and educators will need to collaborate to build on this changing digital landscape that is also allowing universal appropriation of the interest created by digitalization. Such a paradigm change has sustained the way consumers behave and commenced a rethinking of the operational strategy by business organizations, where adaptability and pliability are at the topmost. Online learning then sets off a lifeline resource in ensuring that people are constant with their education even amid a pandemic, but it dramatically raises recognising awareness for the urgent need for equitable access to technology, additionally training both among educators and students. In the market sector, businesses have quickly modified to the use of digital tools hence the trend of e-commerce and remote work rehearsal. Nevertheless, that brings certain doom for determined consumer data on issues about data security, cybersecurity, and what's going to happen in the long term for the physical retail environment. Increased dependence on technology may extend resource planning and reduce carbon footprints; however, this needs to be balanced with the environmental costs of electronic waste and energy consumption from virtual infrastructures. Above all, the findings from this research have an essential implication that policymakers and educators will find significant. There is a requirement to invest in digital infrastructure and training programs so that there can be equity in acquiring these resources and increasing the digital literacy of citizens.

Therefore, in summary, the nature of the impact of COVID-19 dictates transformation, requiring constant adaptation and investment into capabilities towards digitization.

REFERENCES

Abad-Segura, E., González-Zamar, M., Infante-Moro, J. C., & García, G. R. (2020). Sustainable Management of Digital Transformation in Higher Education: Global Research Trends. *Sustainability (Basel)*, *12*(5), 2107. DOI: 10.3390/su12052107

Adedoyin, O. B., & Soykan, E. (2020). Covid-19 pandemic and online learning: The challenges and opportunities. *Interactive Learning Environments*, *31*(2), 863–875. DOI: 10.1080/10494820.2020.1813180

Adnan, M. (2020). Online learning amid the COVID-19 pandemic: Students perspectives. *Journal of Pedagogical Sociology and Psychology*, *1*(2), 45–51. DOI: 10.33902/JPSP.2020261309

Akala, A. (2020). *More big employers are talking about permanent work-from-home positions*. CNBC.

Alraja, M. N., Imran, R., Khashab, B. M., & Shah, M. (2022). Technological innovation, sustainable green practices and SMEs sustainable performance in times of crisis (COVID-19 pandemic). *Information Systems Frontiers*, *24*(4), 1081–1105. DOI: 10.1007/s10796-022-10250-z PMID: 36504756

Amankwah-Amoah, J., Khan, Z., Wood, G., & Knight, G. (2021). COVID-19 and digitalization: The great acceleration. *Journal of Business Research*, *136*, 602–611. DOI: 10.1016/j.jbusres.2021.08.011 PMID: 34538980

Autio, E., Mudambi, R., & Yoo, Y. (2021). Digitalization and globalization in a turbulent world: Centrifugal and centripetal forces. *Global Strategy Journal*, *11*(1), 3–16. DOI: 10.1002/gsj.1396

Banga, K., & te Velde, D. W. (2020). COVID-19 and disruption of the digital economy; evidence from low and middle-income countries. *Digital Pathways at Oxford Paper Series, 7*.

Berman, S. J. (2012). Digital transformation: Opportunities to create new business models. *Strategy and Leadership*, *40*(2), 16–24. DOI: 10.1108/10878571211209314

Bradley, C., Hirt, M., Hudson, S., Northcote, N., & Smit, S. (2020). The great acceleration. *The McKinsey Quarterly*, •••, 1–7.

Brennen, J. S., & Kreiss, D. (2016). Digitalization. *The international encyclopedia of communication theory and philosophy*, 1-11 https://doi.org/DOI: 10.1002/9781118766804.wbiect111

Buoite Stella, A., Manganotti, P., Furlanis, G., Accardo, A., & Ajčević, M. (2020). Return to school in the COVID-19 era: Considerations for temperature measurement. *Journal of Medical Engineering & Technology*, *44*(8), 468–471. DOI: 10.1080/03091902.2020.1822941 PMID: 32990119

Castro, H., Putnik, G. D., Shah, V., & Cruz-Cunha, M. M. (2012). A simulation tool and its role in supporting the management of the transformation processes of Traditional Enterprises into Virtual Enterprises. *Tékhne (Instituto Politécnico do Cávado e do Ave)*, *10*(2), 99–111. DOI: 10.1016/j.tekhne.2012.12.001

Chauhan, C., Parida, V., & Dhir, A. (2022). Linking circular economy and digitalisation technologies: A systematic literature review of past achievements and future promises. *Technological Forecasting and Social Change*, *177*, 121508. DOI: 10.1016/j.techfore.2022.121508

Dhawan, S. (2020). Online learning: A panacea in the time of COVID-19 crisis. *Journal of Educational Technology Systems*, *49*(1), 5–22. DOI: 10.1177/0047239520934018

Eaton, A., & Heckscher, C. (2020). COVID's impacts on the field of labour and employment relations. *Journal of Management Studies*, *58*(1), 275–279. DOI: 10.1111/joms.12645

El-Haddadeh, R. (2020). Digital innovation dynamics influence on organisational adoption: The case of cloud computing services. *Information Systems Frontiers*, *22*(4), 985–999. DOI: 10.1007/s10796-019-09912-2

Fang, J., Gozgor, G., & Nolt, J. H. (2022). Globalisation, economic uncertainty and labour market regulations: Implications for the COVID-19 crisis. *World Economy*, *45*(7), 2165–2187. DOI: 10.1111/twec.13230 PMID: 34908649

Feroz, A. K., Zo, H., & Chiravuri, A. (2021). Digital Transformation and Environmental Sustainability: *A review and research agenda.Sustainability (Basel)*, *13*(3), 1530. DOI: 10.3390/su13031530

Galhotra, B., & Dewan, A. (2020). Impact of COVID-19 on digital platforms and change in E-commerce shopping trends. *2020 Fourth International Conference on I-SMAC (IoT in Social, Mobile, Analytics and Cloud) (I-SMAC)*, 861–866. https://doi.org/DOI: 10.1109/I-SMAC49090.2020.9243379

Garg, A., Popli, R., & Sarao, B. S. (2021). Growth of Digitization and its Impact on Big Data Analytics. *IOP Conference Series. Materials Science and Engineering*, *1022*(1), 012083. DOI: 10.1088/1757-899X/1022/1/012083

Gil-Gomez, H., Guerola-Navarro, V., Oltra-Badenes, R., & Lozano-Quilis, J. A. (2020). Customer relationship management: Digital transformation and sustainable business model innovation. *Ekonomska Istrazivanja, 33*(1), 2733–2750. DOI: 10.1080/1331677X.2019.1676283

Gopika, J. S., & Rekha, R. V. (2023). Awareness and use of digital learning before and during COVID-19. *International Journal of Educational Reform, 10567879231173389*, 10567879231173389. Advance online publication. DOI: 10.1177/10567879231173389

Gozgor, G., & Lau, C. K. M. (2021). Economic effects of COVID-19 related uncertainty shocks. *Frontiers in Public Health, 9*, 760619. DOI: 10.3389/fpubh.2021.760619 PMID: 34650956

Härmand, K. (2021, April). Digitalisation before and after the Covid-19 crisis. In *ERA Forum* (Vol. 22, No. 1, pp. 39-50). DOI: 10.1007/s12027-021-00656-8

Hensher, D. A., Wei, E., & Beck, M. J. (2023). The impact of COVID-19 and working from home on the workspace retained at the main location office space and the future use of satellite offices. *Transport Policy, 130*, 184–195. DOI: 10.1016/j.tranpol.2022.11.012 PMID: 36411865

Imran, A., Posokhova, I., Qureshi, H. N., Masood, U., Riaz, M. S., Ali, K., John, C. N., Hussain, M. D. I., & Nabeel, M. (2020). AI4COVID-19: AI enabled preliminary diagnosis for COVID-19 from cough samples via an app. *Informatics in Medicine Unlocked, 20*, 100378. DOI: 10.1016/j.imu.2020.100378 PMID: 32839734

Ionescu, C., Iordache, M., & i an, E. (2021). The impact of digitization on economy in the context of the Coronavirus pandemic. *SHS Web of Conferences, 92, 01018.* https://doi.org/DOI: 10.1051/shsconf/20219201018

Jakoet-Salie, A., & Ramalobe, K. (2022). The digitalization of learning and teaching practices in higher education institutions during the Covid-19 pandemic. *Teaching Public Administration, 41*(1), 59–71. DOI: 10.1177/01447394221092275

Javed, A., & Yasir, M. (2019). Virtual social enterprise: Modeling sustainability of an enterprise by digital intervention. *World Journal of Entrepreneurship, Management and Sustainable Development, 15*(2), 182–196. DOI: 10.1108/WJEMSD-03-2018-0032

Jorfi, M., Luo, N. M., Hazra, A., Herisson, F., Miller, G., Toombs, J. A., & the COVID-19 Direct-to-Consumer Task Force Investigators. (2020). Diagnostic technology for COVID-19: comparative evaluation of antigen and serology-based SARS-CoV-2 immunoassays, and contact tracing solutions for potential use as at-home products. MedRxiv, 2020-06. https://doi.org/DOI: 10.1101/2020.06.25.20140236

Kumar, G., Singh, G., Bhatnagar, V., Gupta, R., & Upadhyay, S. K. (2020). Outcome of Online Teaching-Learning over Traditional Education during Covid-19 Pandemic. *International Journal of Advanced Trends in Computer Science and Engineering*, *9*(5), 7704–7711. DOI: 10.30534/ijatcse/2020/113952020

Kumar, S., Sharma, A., Sharma, S., Pal, S., & Singh, A. K. (2021). Adaptation to Online Technology for Learning during COVID-19 Pandemic: An Observational Study of Effectiveness and Student's Perception in Various Universities. *Journal of Clinical and Diagnostic Research : JCDR*, *15*(3). Advance online publication. DOI: 10.7860/JCDR/2021/46195.14578

Kumari, R. (2021). Impact of Digitalization on Education in India: Time Series Analysis. *International Journal of Applied Marketing & Management*, *6*(1).

Kunkel, S., & Matthess, M. (2020). Digital transformation and environmental sustainability in the industry: Putting expectations in Asian and African policies into perspective. *Environmental Science & Policy*, *112*, 318–329. DOI: 10.1016/j.envsci.2020.06.022

Kushwaha, S., Bahl, S., Bagha, A. K., Parmar, K. S., Javaid, M., Haleem, A., & Singh, R. P. (2020). Significant applications of machine learning for COVID-19 pandemic. *Journal of Industrial Integration and Management*, *5*(04), 453–479. DOI: 10.1142/S2424862220500268

Lanshina, T. A., Barinova, V. A., Kondratyev, A. D., & Romantsov, M. V. (2020). Sustainable development and digitalization: The unusual COVID-19 crisis requires original solutions. *Bulletin of International Organizations*, *15*(4), 91–114. DOI: 10.17323/1996-7845-2020-04-05

Ling, D. C., Wang, C., & Zhou, T. (2020). A first look at the impact of COVID19 on commercial real estate prices: asset level evidence. SSRN *Electronic Journal*. https://doi.org/DOI: 10.2139/ssrn.3593101

Maiti, D., Castellacci, F., & Melchior, A. (2019). Introduction: Digitalisation and Development: Issues for India and Beyond. *Digitalisation and Development: Issues for India and Beyond*. https://doi.org/DOI: 10.1007/978-981-13-9996-1_1

Mark. (2024, September 4). How COVID-19 is accelerating digital transformation in the workplace. Haiilo. https://haiilo.com/blog/how-covid-19-is-accelerating-digital-transformation-in-the-workplace/

Matthess, M., & Kunkel, S. (2020). Structural change and digitalization in developing countries: Conceptually linking the two transformations. *Technology in Society*, *63*, 101428. DOI: 10.1016/j.techsoc.2020.101428

Mustafa, M., & Abbas, A. (2021). Comparative analysis of green ict practices among Palestinian and Malaysian in SME food enterprises during Covid-19 pandemic | PalArch's Journal of Archaeology of Egypt / Egyptology. *PalArch's Journal of Archaeology of Egypt/Egyptology, 18*(4).

Nikas, I. P., Lamnisos, D., Meletiou-Mavrotheris, M., Themistocleous, S. C., Pieridi, C., Mytilinaios, D. G., Michaelides, C., & Johnson, E. O. (2022). Shift to emergency remote preclinical medical education amidst the Covid-19 pandemic: A single-institution study. *Anatomical Sciences Education*, *15*(1), 27–41. DOI: 10.1002/ase.2159 PMID: 34854255

OECD. (2014). *Measuring the Digital Economy: A New Perspective*. OECD Publishing., DOI: 10.1787/9789264221796-

Pu, G., Qamruzzaman, M., Mehta, A. M., Naqvi, F. N., & Karim, S. (2021). Innovative Finance, Technological Adaptation and SMEs Sustainability: The Mediating Role of Government Support during COVID-19 Pandemic. *Sustainability (Basel)*, *13*(16), 9218. DOI: 10.3390/su13169218

Purba, M. I., Simanjutak, D. C. Y., Malau, Y. N., Sholihat, W., & Ahmadi, E. A. (2021). The effect of digital marketing and e-commerce on financial performance and business sustaina-bility of MSMEs during COVID-19 pandemic in Indonesia. *International Journal of Data and Network Science,* 275–282. https://doi.org/DOI: 10.5267/j.ijdns.2021.6.006

Renu, N. (2021). Technological advancement in the era of COVID-19. *SAGE Open Medicine*, *9*, 20503121211000912. Advance online publication. DOI: 10.1177/20503121211000912 PMID: 33786181

Ritter, T., & Pedersen, C. L. (2020). Digitization capability and the digitalization of business models in business-to-business firms: Past, present, and future. *Industrial Marketing Management*, *86*, 180–190. DOI: 10.1016/j.indmarman.2019.11.019

Sabbagh, K., Friedrich, R. O. M. A. N., El-Darwiche, B. A. H. J. A. T., Singh, M. I. L. I. N. D., & Koster, A. L. E. X. (2013). Digitization for economic growth and job creation: Regional and industry perspectives. *The Global Information Technology Report, 2013*, 35–42. DOI: 10.1108/REPS-11-2019-0145

Sadigh, B. L., Unver, H. O., Nikghadam, S., Dogdu, E., Ozbayoglu, A. M., & Kilic, S. E. (2016). An ontology-based multi-agent virtual enterprise system (OMAVE): part 1: domain modelling and rule management. *International Journal of Computer Integrated Manufacturing, 30*(2–3), 320–343. DOI: 10.1080/0951192X.2016.1145811

Singh, R., & Singh, G. S. (2017). Traditional agriculture: A climate-smart approach for sustainable food production. *Energy, Ecology & Environment, 2*(5), 296–316. DOI: 10.1007/s40974-017-0074-7

Sostero, M., Milasi, S., Hurley, J., Fernandez-Macías, E., & Bisello, M. (2020). *Teleworkability and the COVID-19 crisis: a new digital divide?* (No. 2020/05). JRC working papers series on labour, education and technology. https://hdl.handle.net/10419/231337

Strielkowski, W., Civín, L., Tarkhanova, E., Tvaronavičienė, M., & Petrenko, Y. (2021). Renewable energy in the sustainable development of electrical power sector: A review. *Energies, 14*(24), 8240. DOI: 10.3390/en14248240

Subramanyachary, P., & Sekhar, R. (2021). Digitalization in india: Opportunities and challenges. *Journal of Management and Science, 11*(4), 20–23.

Sun, J., Kwek, K., Li, M., & Shen, H. (2021). Effects of social mobility and stringency measures on the COVID-19 outcomes: Evidence from the United States. *Frontiers in Public Health, 9*, 779501. DOI: 10.3389/fpubh.2021.779501 PMID: 34805083

Vlados, C., & Chatzinikolaou, D. (2021). Evolutionary transformation of the global system and the COVID-19 pandemic: The search for a new development trajectory. *The Cyprus Review, 33*(1), 127–166.

Wang, C., Horby, P. W., Hayden, F. G., & Gao, G. F. (2020). A novel coronavirus outbreak of global health concern. *Lancet, 395*(10223), 470–473. DOI: 10.1016/S0140-6736(20)30185-9 PMID: 31986257

Whitelaw, S., Mamas, M. A., Topol, E., & Van Spall, H. G. (2020). Applications of digital technology in COVID-19 pandemic planning and response. *The Lancet. Digital Health, 2*(8), e435–e440. DOI: 10.1016/S2589-7500(20)30142-4 PMID: 32835201

World Health Organization. (2020). *Laboratory testing for 2019 novel coronavirus (2019-nCoV) in suspected human cases: Interim guidance, 17 January 2020*. World Health Organization.

Xiong, Y., Cui, X., & Yu, L. (2024). Impact of COVID-19 pandemic on online consumption share: Evidence from China's mobile payment data. *Journal of Retailing and Consumer Services*, *81*, 103976. DOI: 10.1016/j.jretconser.2024.103976

Zangiacomi, A., Pessot, E., Fornasiero, R., Bertetti, M., & Sacco, M. (2020). Moving towards digitalization: A multiple case study in manufacturing. *Production Planning and Control*, *31*(2-3), 143–157. DOI: 10.1080/09537287.2019.1631468

Zhai, Y., Jiang, D., Gozgor, G., & Cho, E. (2021). The amplifying effect of conflicts on case fatality rate of COVID-19: Evidence from 120 countries. *Frontiers in Public Health*, *9*, 681604. DOI: 10.3389/fpubh.2021.681604 PMID: 34422745

Zhang, J., Lyu, Y., Li, Y., & Geng, Y. (2022). Digital economy: An innovation driving factor for low-carbon development. *Environmental Impact Assessment Review*, *96*, 106821. DOI: 10.1016/j.eiar.2022.106821

Zhu, B., Zhang, M., Zhou, Y., Wang, P., Sheng, J., He, K., & Xie, R. (2019). Exploring the effect of industrial structure adjustment on interprovincial green development efficiency in China: A novel integrated approach. *Energy Policy*, *134*, 110946. DOI: 10.1016/j.enpol.2019.110946

Chapter 18
Implementing Conceptual Framework Into Textile Industry for Sustainable Climate Change Prevention

Agrim Singhal
Graphic Era University, India

Muskan Singh
https://orcid.org/0009-0001-6667-7669
Graphic Era University, India

Vansh Jain
Graphic Era University, India

Sachin Ghai
Graphic Era University, India

ABSTRACT

The textile industry plays a very important and a crucial role in the factors that can affect the environment globally, and due to such factors, it faces immense pressure over themselves to adopt various sustainable methods. This paper examines the development of various conceptual frameworks that can be implemented to adopt sustainability in the textile industry. By the adoption of the framework, it would help us to tackle multiple problems such as carbon emission, extensive utilization of the resources and waste production. The study explores multiple supply chain strategies

DOI: 10.4018/979-8-3693-9909-5.ch018

such as use of renewable energy, cutting edge technology such as biodegradable material and use of dyes that are not harming the environment. The research suggests that a business that is well structured improves the its stability and profitability while reducing the environmental damage. The research provides insightful information to all those manufacturers, policymakers etc. who have a great interest in moving towards the sustainable environment in the textile industry.

INTRODUCTION

The textile industry plays a major role in producing environmental pollution majorly with the use of dangerous chemicals in dyeing, printing of the cloth and fabricating the cloth. Certain harmful chemicals that are being used by the textile industries are synthetic dyes, bleaches and solvents that often get mix up with the wastewater and pollute the rivers, lakes, soil and barred land and can be the ultimate cause of the serious health issues to the humans and to the ecosystem. The sector is going under the immense pressure to reduce the chemical emission in the environment and implement more eco-friendly methods. Solving the problems of producing the chemicals in the environment is not only the need of the environment by also necessary for the business itself as now-a-days that people are demanding more eco-friendly products and to attain these multiple strict regulations are also implemented (Diabat et al., 2014).

The complicated nature of the chemical pollution in the textiles makes it necessary to have an ambitious approach that mainly includes use of new technology and more environmentally friendly methods for doing the production. The application of the environmentally acceptable material instead of the traditional chemicals would be the best solution to reduce the pollution and environmental degradation. The use of the natural and environmentally friendly dyes and enzyme-based treatments are becoming more efficient for the industry to reduce the waste generation in water and the soil. The acceptance of the use of these methods helps the industry in the production of the high-quality product and at the same time reducing the environmental degradation and achieving the goal of sustainability (Kazancoglu et al., 2020).

The development in the technology helps in reducing the chemical pollution by the industry. Advancement in the technology such as digital printing techniques and waterless dying technology like supercritical CO_2 helps the industry to reduce the chemical usage that is harmful for the environment and also reduce the water consumption. Bag, S., Rahman, M. S., Rogers, H., Srivastava, G., & Pretorius, J. H. C. (2023). Moreover, the use of closed-loop system by the industry to treat the contaminated water can also help in reducing the harmful discharges in the environment and can help in chemical management in an efficient and environmentally friendly manner.

Reducing the chemical pollution is highly depended upon proper wastewater treatment and recycling capacity of the industry. With the help of the installation of the water treatment systems it can ensure that the contaminated water is treated, and all the kind of the pollutants are removed from it and this would also make sure that treated water is used within the factory itself. These steps not only reduce the environmental damage but also help in taking steps towards the sustainable goal in the textile industry.

Lastly, if we want to attain a long-term change then there must be proper co-operation throughout the supply chain. Together all the manufacturers, brands and suppliers must take strict actions for the establishment of the environmental standards that need to fulfil and must support the use of the sustainable friendly technology. The textile industry can lessen their impact towards the degradation of the environment by taking the steps to control the pollution and managing the chemicals in a better manner.

The framework for sustainable climate action in textiles encompasses several core components. These include adopting renewable energy sources, improving resource efficiency, developing eco-friendly materials, and reducing waste through recycling. Water conservation practices, such as closed-loop water systems, also play a crucial role in minimizing the industry's environmental impact. Additionally, a shift toward biodegradable and recycled fibres can lessen dependency on non-renewable resources. Implementing these practices requires buy-in from manufacturers, brands, and consumers, all of whom play a vital role in advancing sustainability within the textile sector (Giancotti et al., 2020).

LITERATURE REVIEW

The textile industry's environmental impact, especially concerning chemical pollution, has been a growing concern in both academic and industrial research. Chemical pollution in textile manufacturing results from processes such as dyeing, printing, and fabric finishing, which involve the use of hazardous substances like synthetic dyes, heavy metals, and solvents (McGrew, 2016). This review explores the literature on strategies to address chemical pollution in the textile sector, emphasizing the need for a structured and conceptual framework for sustainability.

A key conceptual framework applied to the textile industry is the Circular Economy (CE) model, which aims to eliminate waste and make the most of available resources. In a circular textile industry, products are designed for longevity, repairability, and recyclability, and the entire supply chain is optimized for resource efficiency. The application of circular economy principles can significantly reduce textile waste, minimize the use of virgin raw materials, and lower carbon emissions. Studies have demonstrated that implementing circular business models, such as garment collection and recycling programs or reusing textile fibres, can help to reduce the industry's reliance on resource extraction, which is a major driver of climate change. Additionally, recycling textiles into new fabrics or garments prevents the generation of landfill waste, thus contributing to climate change mitigation by reducing methane emissions from waste decomposition.

Despite the promising potential of these frameworks, the implementation of sustainability practices in the textile industry faces several challenges. One of the primary barriers is financial constraints, particularly for small and medium-sized enterprises (SMEs) that may lack the resources to invest in green technologies or sustainable practices. Additionally, there is often a knowledge gap within the industry regarding the best practices for sustainability, as well as a lack of awareness among consumers about the environmental impacts of textile production. Policy and regulatory frameworks can help address these barriers by providing incentives for companies to adopt green technologies, enforcing emissions reductions, and promoting sustainable sourcing practices. Policy frameworks that encourage circularity and provide clear guidelines for sustainable practices can further support the transition toward a more climate-conscious textile industry.

Chemical Contamination in the Production of Textiles

Due to the issue related to the chemical pollution in the textile industry they have got a significant attention from all various researchers, manufacturers and stakeholders etc due to its impact on the environment. As per the data textile industry is considered one of the biggest polluters of the environment as they discharge large amount of waste such as azo dyes, different kind of non-biodegradable metals and detergents into the water bodies.All these issues have attracted the researchers to do their research and try to find out some effective and sustainable production techniques to substitute the harmful substances with the eco-friendly ones (de Santa-Eulalia et al., 2015).

Various studies have identified multiple methods as an effective alternative for the chemicals used in the textile industries. Digital printing and supercritical carbon-dioxide ($C0_2$) dyeing have been recognised as an important advancement in the technology that can drastically reduce the water and chemical usage in the textile industry. As per the research done, CO_2 based water dying helps in reducing the environmental pollution by reducing the usage of the solvents and water. In a similar manner enzyme-based treatmentare being implemented that are biodegradable and non-toxic in nature that can be used as a substitute of chemicals in the textile production. However, due to high cost of adoption of these high-level technologies make it more difficult for the developing countries to develop them and work with them.

There are multiple solutions for the treatment of the wastewater to fight from the contamination are likewise the subject of a sizable body of literature. As per the researchers if the contaminated water is treated well and efficiently then it can become fit to be released into the environment.Additionally, the water that has been recycled by the industry can be further used by them and can cut down the water consumption and reduce pollution in the environment and get in line with a broader sustainability objective. Moreover, the literature suggests that the areas with weak governance and improper maintenance can restrict the efficiency of the technology (Ara et al., 2019).

Still there are multiple issues and obstacles in between to put these strategies into practice at the large scale even with so much research on the sustainable chemical management in the textile manufacturing. Research suggests that if the industry wants to adopt the strategies and the changes then they must follow strict regulatory frameworks and increase the consumer awareness. As shown, most of the manufacturers still prefer to make the use of the old harmful chemicals in dyes and the production of the cloth just because of the high cost and lack of incentives to get the eco-friendly substances. Therefore, the future research must focus upon cheap and affordable technology to support the textile industry.

Exposure to textile chemicals poses serious health risks for both workers within the industry and consumers. Workers in manufacturing plants are often in close contact with toxic chemicals, leading to skin irritations, respiratory issues, and in severe cases, chronic illnesses like cancer due to prolonged exposure. Consumers, too, can be affected, as traces of chemicals in textile products may cause allergic reactions or skin conditions. Additionally, communities located near textile production facilities are at risk, as contaminated water and air can affect the health of surrounding populations, leading to increased rates of illness (Oelze, 2017).

Greener Substitutes to Dangerous Chemicals

The traditional processing techniques that are being followed in the textile industry uses chemical bases bleaches and dyes and are harmful for both the environment and the human health (Pattnaik and Dangayach, 2019). Therefor the researchers are extensively thinking and researching upon the substances which can easily replace all these toxic products without hampering the quality of the product. In recent years there are multiple solutions that are explored such as natural dyes, enzyme-based treatment and chemicals that are biodegradable that promises cleaner and sustainable products.

With increasing awareness about environmental conservation and public health, industries, researchers, and consumers are actively seeking greener alternatives to hazardous chemicals. Here are some effective and sustainable substitutes that reduce ecological damage and promote a healthier environment:

- **Cleaning Agents**

Dangerous Chemical: Ammonia and Chlorine Bleach

Greener Substitute: Vinegar, baking soda, and hydrogen peroxide are excellent, non-toxic alternatives for household cleaning tasks. Enzyme-based cleaners are also effective for industrial use.

- **Pesticides**

Dangerous Chemical: Organophosphates and Carbamates

Greener Substitute: Neem oil, diatomaceous earth, and biological controls such as predatory insects (e.g., ladybugs) serve as eco-friendly pesticides. Essential oils like citronella also repel pests effectively.

- **Plastics**

Dangerous Chemical: Polyvinyl chloride (PVC) and Bisphenol A (BPA)

Greener Substitute: Bioplastics made from polylactic acid (PLA), which is derived from corn starch, and materials like mushroom mycelium or cellulose are sustainable replacements.

- **Industrial Solvents**

Dangerous Chemical: Benzene, Toluene, and Acetone

Greener Substitute: Water-based solvents or bio-based solvents derived from citrus, soy, or ethanol are eco-friendlier alternatives for cleaning and degreasing.

- **Fertilizers**

Dangerous Chemical: Ammonium Nitrate and Urea-based fertilizers

Greener Substitute: Organic compost, bone meal, and green manure provide nutrient-rich options for soil enrichment without harmful runoff.

- **Refrigerants**

Dangerous Chemical: Chlorofluorocarbons (CFCs) and Hydrofluorocarbons (HFCs)

Greener Substitute: Natural refrigerants like ammonia (NH3), carbon dioxide (CO2), and hydrocarbons (e.g., propane and isobutane) are climate-friendly options.

- **Flame Retardants**

Dangerous Chemical: Polybrominated diphenyl ethers (PBDEs)

Greener Substitute: Phosphorus-based compounds and naturally fire-resistant materials like wool or clay-based coatings are safer alternatives.

- **Dyes and Pigments**

Dangerous Chemical: Azo Dyes and Heavy Metal-based Pigments

Greener Substitute: Natural dyes from plants, fruits, and vegetables (e.g., indigo, turmeric, and beetroot) are non-toxic and biodegradable.

- **Adhesives**

Dangerous Chemical: Formaldehyde-based Resins

Greener Substitute: Soy-based adhesives, natural latex, and bio-based glues derived from starch or proteins are safer alternatives.

- **Energy Storage Batteries**

Dangerous Chemical: Lead-acid and Nickel-cadmium Batteries

Greener Substitute: Lithium iron phosphate (LFP) batteries and emerging technologies like solid-state batteries offer a safer and more sustainable option.

Dyes that can be generated by the plants, minerals and insects can be very beneficial to replace the synthetic dyes as they would be eco-friendly also. Multiple research have suggested that making the use of the natural dyes would be very beneficial for the environment as it helps in reducing the water contamination and lowers the toxicity level in the ecosystem. To prove this the researchers have tested multiple dyes that have been extracted naturally from turmeric, indigo and madder etc. that can equally provide vibrant and durable colours that are more environmentallyfriendly.

Another possible way of reducing chemical pollution in the textile industry is enzyme-based treatment. In processes like of bleaching and finishing where many harmful chemicals are used can be easily replaced with the enzymes that are more environmentally friendly and non-toxic in nature. Enzyme bleaching is a better substitute for a chlorine based bleaching technique which are mainly responsible for emitting toxins in the environment (Perry and Towers, 2013). Although the use of this treatment is very beneficial, but it is still not used on a large scale in the textile industry just because of its high cost and more technological development. The research suggests that the techniques that are environmentally friendly are being adopted by the industry but still is not scalable at a large level due to certain factors like cost and lack of infrastructure.

Perfluorinated compounds (PFCs), used for stain and water resistance, and brominated flame retardants are among the most dangerous chemicals in textiles due to their persistence in the environment and bio accumulative nature. Many textile companies are now investing in PFC-free and non-toxic flame-retardant alternatives. Silicone-based water repellents, for example, offer a non-toxic solution to the problems posed by PFCs, providing durable water resistance without harmful environmental effects. Similarly, alternative flame retardants based on bio-based and mineral ingredients can effectively reduce fire risks in textiles without exposing people and ecosystems to the hazardous compounds found in traditional flame retardants.

Waterless Dyeing and Closed Loop System: New Tech-Solutions

The textile industry has been linked with the consumption of the high water and chemical waste for a very long period. The advancement in the technology has helped the industry to aim at reducing their dependency on water consumption and this led to the development of waterless dyeing technology that helped them to do so and become more effective. Supercritical carbon dioxide (CO_2) dyeing has become one of the latest developments that has gathered a lot of attention to itself. According to the studies, this kind of technology helps in reducing the water consumption to zero and moreover it also consumers 50% less energy than the traditional ones (Saha et al., 2022).

The textile industry is experiencing a significant shift toward sustainability with the adoption of waterless dyeing and closed-loop systems. These innovative technologies address critical environmental issues, such as excessive water usage, pollution, and waste generation. Here's an overview of these advancements and their implications:

1. Waterless Dyeing Technology

Traditional textile dyeing consumes vast quantities of water and produces polluted wastewater, contributing to environmental degradation. Waterless dyeing technologies eliminate or drastically reduce water use, presenting a sustainable alternative.

Key Innovations:

- **Supercritical CO2 Dyeing:** Uses carbon dioxide in its supercritical state as a solvent for dyeing textiles.

Eliminates the need for water and reduces chemical usage.
CO2 is captured, recycled, and reused in the system.
Example: **Dyecoo**, a leading company in this technology.

- **Digital Dyeing:** Employs inkjet printing methods to apply dyes precisely.

Reduces water, energy, and dye wastage compared to conventional methods.

- **Foam Dyeing:** Utilizes foam as a dye carrier, significantly cutting water consumption.

Example: Foam dyeing innovations for denim production.

2. Closed-Loop Systems

A closed-loop system focuses on recycling and reusing resources within production, aiming to create a circular textile economy.

Key Principles:

- **Fiber-to-Fiber Recycling:** Breaks down used textiles into reusable fibers for new fabric production.

Commonly used materials: polyester, cotton, and wool.

- **Chemical Recycling:** Processes synthetic fibers like polyester to extract raw materials for reuse.

Example: Companies like Worn Again and Re: new cell.

- **Water Recycling Systems:** Advanced filtration technologies allow dye wastewater to be cleaned and reused in the dyeing process.
- **Biodegradable Materials:** Development of natural and bio-based fibers that can be reintegrated into the ecosystem.

With the advancement in the technology along with waterless dyeing, closed loop system was also looked upon and it also helped in minimizing the chemical waste generation. It is referred to as a system in which both the water and the chemicals are being reused by the industry for the production instead of discharging it into the seas, oceans, rivers and soil.Recycling the wastewaterand reusing all the chemicals in the industry can help in reducing both the consumption of the water on a regular basis and reducing chemical dischargement in the environment. When the new technology was adopted by the industry and integrated with the effluent treatment plants it not only lowers the chemical discharge in the environment but also reduces the operational cost for the industry as it lowers down the water consumption requirements for production.

Despite the fact that the adoption of these technologies would help in achieving sustainability in a better way but its usage and setting up was very difficult. The high cost of the infrastructure required to set such technology in the industry was making it very difficult for the businesses to adopt it specially in the developing countries. Nonetheless, the incentives provided by the government to the industries can help them to get more motivated to set up these plants that are much more eco-friendly and helps in reducing the waste generation.

At last technological solutions such as waterless dyeing and closed loop system have a great potential in helping the industry to become more environment friendly. However, it we want to implement these challenges in the textile industry on a large scale a large number of economic, technical and infrastructural problems need to be solved. With the help of increased innovation and provision of better policies would be very helpful in providing such technologies to the textile industry and making them more sustainable in nature.

Frameworks for Sustainability in Textile Production

Due to the resource intensive and pollution heavy operations taking place in the textile industry have developed interest in the sustainable framework due to the growing concerns towards the environment. The aim of this framework is to provide a balance between the economic, social and environmental factors in the textile supply chain.The main focus is on conserving water, reducing the chemical waste production and energy utilization.The new technology such as closed loop system helps in converting the post-consumer waste into fibres so that it can reused and can reduce the waste generation. Regarding its potential, the adoption of the new technology in the circular economy becomes a very big task due to lack of proper infrastructure, high setup cost complicated fibre recycling etc. The barriers need to be removed or reduced so that the sustainably effective technology can be implemented in the textile industry.

Another key framework is the lifecycle assessment (LCA), which allows companies to analyse the environmental impact of a product throughout its entire lifecycle—from raw material extraction and processing to manufacturing, distribution, use, and disposal. The LCA framework is especially beneficial for identifying critical points in production where interventions can have the most impact in terms of resource conservation, energy efficiency, and waste reduction. Through LCA, textile companies can make data-driven decisions that help minimize carbon emissions, optimize water usage, and mitigate toxic waste, resulting in a more sustainable and transparent production process. Many companies use LCA to set sustainability benchmarks and measure improvements over time, aligning their goals with international standards.

Sustainable product design frameworks guide textile companies in creating products that are not only eco-friendly but also ethically produced. This framework encompasses various design strategies, including the use of renewable and biodegradable materials, reducing the number of chemical treatments, and employing energy-efficient production methods. Sustainable product design also emphasizes the importance of durable, timeless designs that can extend a product's life and reduce the need for frequent replacement. Through eco-design principles, textile companies can lower environmental impact at the design stage, making it easier to manage sustainability throughout the production cycle.

Frameworks for sustainability in textile production offer an organized, multifaceted approach to embedding eco-friendly practices throughout the industry. By integrating circular economy principles, lifecycle assessments, sustainable design, certification standards, supply chain transparency, and renewable energy strategies, textile companies can reduce their environmental footprint while remaining competitive in a global market that increasingly values sustainability. Through these frameworks, the textile industry can advance its climate action goals, fostering a more responsible and resilient industry that contributes positively to global sustainability efforts.

Sustainability in textile production is becoming a critical focus as the industry grapples with its environmental and social impacts. Implementing structured frameworks enables businesses to align operations with sustainability goals, ensuring resource efficiency, waste reduction, and ethical practices throughout the value chain.

One of the foundational frameworks is the Life Cycle Assessment (LCA), which evaluates the environmental impacts of a product across its entire lifecycle, from raw material extraction to disposal. By identifying hotspots of energy consumption, water usage, and emissions, LCA helps manufacturers target areas for improvement. For instance, switching to organic cotton or recycled fibers can significantly lower the ecological footprint, while adopting cleaner production methods reduces greenhouse gas emissions.

The circular economy framework is also pivotal in fostering sustainability. This model promotes designing textiles for reuse, repair, and recycling, thus minimizing waste and dependence on virgin resources. Innovations such as fiber-to-fiber recycling and biodegradable fabrics align with circular principles. Companies like Patagonia and Stella McCartney have embraced this framework, incorporating recycled materials and take-back programs to extend the life cycle of their products.

Another essential framework is the Sustainable Development Goals (SDGs) by the United Nations, which provide a comprehensive roadmap for sustainability. Goals such as responsible consumption and production (SDG 12), climate action (SDG 13), and clean water and sanitation (SDG 6) are directly relevant to the textile industry. Adopting these goals encourages businesses to implement water-efficient dyeing processes, reduce energy consumption, and ensure ethical labor practices.

The Higg Index, developed by the Sustainable Apparel Coalition, offers a practical tool for measuring sustainability performance. It evaluates the environmental and social impact of textile production, empowering companies to benchmark and improve their practices. Brands use this index to assess supply chain transparency, energy efficiency, and material sourcing, driving industry-wide accountability.

Finally, certification programs such as Global Organic Textile Standard (GOTS), OEKO-TEX, and Bluesign provide frameworks for ensuring sustainable practices in raw material sourcing, chemical use, and production processes. These certifications serve as trusted indicators of environmentally friendly and socially responsible textiles, aiding consumers in making informed choices.

In conclusion, frameworks for sustainability in textile production offer structured approaches to address the industry's environmental and social challenges. By adopting these frameworks, businesses can transition toward more sustainable operations, fostering resilience and long-term profitability while contributing to global efforts to combat climate change and resource depletion.

Sustainability in textile production requires a multifaceted approach, with structured frameworks guiding the transition toward environmentally responsible and socially equitable practices. As the industry faces mounting scrutiny over its resource-intensive processes and environmental footprint, these frameworks provide actionable pathways for change and accountability.

The Life Cycle Assessment (LCA) framework remains a cornerstone for sustainable decision-making. By analyzing the environmental impacts at each stage of a product's life cycle—spanning raw material extraction, processing, manufacturing, distribution, use, and disposal—LCA identifies critical intervention points. For example, it underscores the importance of material selection, where using recycled polyester or organic cotton significantly reduces water consumption, greenhouse gas emissions, and waste generation. LCA also aids in comparing technologies, such as evaluating waterless dyeing methods against conventional dyeing to determine the most sustainable choice.

Building on LCA insights, the circular economy framework seeks to close resource loops by emphasizing reuse, repair, and recycling. This paradigm shift requires innovative business models, such as leasing or take-back schemes, where brands retain ownership of products and ensure their proper recycling or up-cycling. Fiber recovery technologies, such as enzymatic or chemical recycling, play a crucial role in transforming waste into valuable raw materials. In addition, the circular approach promotes modular design, enabling easy disassembly and recycling of textiles at the end of their life.

The Sustainable Development Goals (SDGs) offer a global framework that connects textile sustainability with broader societal objectives. For instance, SDG 8 (Decent Work and Economic Growth) emphasizes fair labor practices and safe working conditions, addressing critical issues in the textile supply chain, such as exploitation and unsafe factories. SDG 9 (Industry, Innovation, and Infrastructure) encourages the adoption of advanced technologies like AI-powered sorting for textile recycling, while SDG 14 (Life Below Water) highlights the need to prevent micro-plastic pollution from synthetic fibers.

The Higg Index, developed by the Sustainable Apparel Coalition, provides a comprehensive suite of tools for assessing environmental, social, and labor impacts across the textile and apparel supply chain. The index encourages companies to measure their sustainability performance in areas such as raw material sourcing, energy efficiency, and worker well-being. By identifying gaps and opportunities, it drives continuous improvement while fostering transparency and collaboration within the industry.

Certification and labeling systems also serve as critical frameworks for ensuring compliance with sustainability standards. Certifications like Global Organic Textile Standard (GOTS) focus on organic fibers and prohibit harmful chemicals, while Bluesign ensures safe chemical use and efficient resource management. Fair Trade certifications prioritize ethical labor practices and equitable trade, addressing social sustainability alongside environmental concerns. These certifications empower consumers to make informed choices, creating market-driven incentives for sustainable production.

In addition to these frameworks, policy and regulatory approaches are shaping the sustainability agenda. Governments worldwide are introducing extended producer responsibility (EPR) laws, mandating that textile manufacturers take responsibility for the end-of-life management of their products. Carbon taxation and incentives for green technology adoption are further encouraging companies to align with sustainability goals.

Technology integration is also vital to the success of these frameworks. Digitalization and data analytics allow companies to track their environmental footprint, optimize supply chains, and predict demand, reducing overproduction and waste. Block-chain technology enhances traceability, ensuring that materials and products meet sustainability standards throughout their journey from raw materials to consumers.

Education and consumer engagement play a complementary role in advancing sustainability frameworks. Raising awareness about the environmental and social impacts of textile production can shift consumer preferences toward sustainable products. Educational initiatives, combined with clear labeling and storytelling, help build trust and foster demand for ethical and eco-friendly textiles.

Climate Impact Assessment of Textile Production Processes

The textile industry is recognized as one of the most environmentally impactful sectors, particularly concerning climate change. The production processes involved, from raw material cultivation to final product manufacturing, contribute significantly to greenhouse gas (GHG) emissions, energy consumption, and water usage (Seuring and Müller, 2008). A comprehensive climate impact assessment (CIA) of textile production processes is essential to understanding the sector's role in global warming and identifying actionable strategies for emission reductions. The main stages of textile production—fibre cultivation, material processing, dyeing, and finishing—each have distinct environmental footprints that contribute to the industry's overall climate impact.

One of the largest contributors to the climate footprint of the textile industry is the cultivation of natural fibres, especially cotton. The agricultural phase of cotton production is energy- and water-intensive, requiring large quantities of irrigation and chemicals, which increases both GHG emissions and water consumption. Beyond raw material cultivation, the subsequent stages of textile production, such as spinning, weaving, and knitting, are also resource-heavy and energy-demanding (Saha et al., 2022). These processes often rely on fossil fuels, which further increase the carbon emissions associated with textile manufacturing. Thus, a significant reduction in climate impact can be achieved by adopting more sustainable farming practices and improving energy efficiency throughout the production stages.

The dyeing and finishing processes also represent critical areas in climate impact assessments of textile production. These processes are particularly notorious for their high energy and water consumption, in addition to the environmental harm caused by chemical discharges. Traditional dyeing techniques often require substantial amounts of water and generate significant waste, both in terms of water and harmful effluents. Waterless dyeing technologies, along with innovations such as low-energy dyeing methods, have shown promising results in reducing the environmental impact of these stages. Furthermore, eco-friendly alternatives to conventional chemicals used in dyeing and finishing can further mitigate the negative climate effects of textile production.

Life cycle assessment (LCA) has become a valuable tool for evaluating the overall environmental impacts of textile products. LCA examines every phase of the production process, from raw material extraction to end-of-life disposal, and quantifies its contribution to global warming potential, resource depletion, and pollution. By applying LCA, companies can identify the most carbon-intensive steps in the production process and implement targeted interventions to reduce emissions Svensson, G. (2007). This comprehensive approach allows for a holistic understanding of the climate impact of textile products and enables designers and manufacturers to make informed decisions that promote sustainability.

To address the climate impacts of textile production, various strategies are being explored, including the adoption of renewable energy sources and circular economy models. Switching to renewable energy in production facilities, such as solar or wind power, can drastically reduce the carbon emissions associated with energy use in textile manufacturing. Additionally, integrating circular economy principles, such as recycling textile waste and designing for product longevity, can minimize the need for virgin materials, lowering the overall environmental impact. These approaches, combined with advancements in eco-design, are key to reducing the textile industry's climate footprint and contributing to global climate change mitigation efforts (Van Bommel, 2011).

The textile industry is a significant contributor to global climate change, driven by its high resource consumption and pollutant emissions across various stages of production. From raw material extraction to fabric manufacturing and dyeing, every step has a measurable environmental footprint, making climate impact assessment critical for fostering sustainability in this sector.

Raw material sourcing, particularly for cotton and synthetic fibers, has a substantial climate impact. Cotton cultivation requires vast amounts of water and agrochemicals, leading to soil degradation and water scarcity in agricultural regions. Synthetic fibers like polyester, derived from petroleum, contribute to greenhouse gas emissions during production and are energy-intensive. These raw materials form the foundation of textile production, and their environmental costs cascade through the supply chain.

Manufacturing processes exacerbate the industry's carbon footprint, particularly through energy-intensive spinning, weaving, and knitting operations. Traditional dyeing and finishing techniques further intensify the climate impact by consuming large quantities of water and chemicals while generating wastewater laden with pollutants. Dyeing alone contributes to around 20% of global wastewater and releases greenhouse gases due to fossil fuel dependency in heating water and powering equipment.

The end-of-life phase of textiles, characterized by significant waste generation, also contributes to climate issues. Globally, millions of tons of textiles are discarded annually, with a considerable portion ending up in landfills or incinerators. Synthetic fibers, which are not biodegradable, release microplastics into the environment, while incineration emits carbon dioxide and other harmful pollutants, exacerbating the climate crisis.

Mitigating these impacts requires integrating sustainable practices across the textile lifecycle. The adoption of renewable energy sources, waterless dyeing technologies, and closed-loop recycling systems can reduce emissions and resource depletion. Additionally, fostering circular economies through recycling and designing textiles for durability and reuse can address the growing waste problem. Consumer awareness and regulatory measures also play a vital role in steering the industry toward a low-carbon future.

In conclusion, the climate impact of textile production is a multifaceted challenge with implications for water, air, and soil health. Addressing these issues necessitates systemic changes within the industry, alongside global collaboration to innovate and implement sustainable solutions. As awareness of the environmental cost of textiles grows, so does the opportunity to transform the industry into a model of climate-conscious production.

IMPLICATIONS

The most effective way through which the problem faced by the world due to the chemical pollution done by the textile industry is through the adoption of the conceptual framework. The benefits of adopting this new framework and the technology within this is that it directly attacks the source of the pollution and tries to resolve it. This framework encourages to use the new enhanced technology such as waterless dyeing and closed-loop system that helps in reducing the pollution done by the textile industry and reduces the environmental impact.

The most important and the basic function of this framework is to emphasise on the new, better and advanced technologies. Such as the waterless dyeing technology like CO_2 dyeing helps in reducing the usage of the harmful chemicals and water usage help in reducing the ecological impact. These solutions address one of the biggest causes of water pollution in the textile industry by removing water from the dyeing process, which significantly reduces the number of chemical effluents released into waterways. Additionally, by encouraging a circular economy—where waste is reduced and resources are efficiently reused—this innovation supports sustainability goals.

Sustainable frameworks in the textile industry extend beyond environmental benefits, impacting social conditions and supporting community development. Many textile production sites are located in regions where water scarcity, pollution, and poor labour conditions are prevalent. By reducing pollution, conserving resources, and ensuring ethical labour practices, textile companies can contribute positively to the well-being of local communities. Cleaner production practices and fair labour conditions foster a healthier working environment and contribute to the socioeconomic development of these regions, promoting social responsibility alongside environmental sustainability.

Additionally, the conceptual framework encourages cooperation among stakeholders to emphasize sustainability by facilitating systemic improvements throughout the supply chain. Through the use of closed-loop technologies, which reduce waste and keep chemical-laden materials out of landfills, textiles are made to be durable and recyclable thanks to this cooperative approach. The industry can guarantee that chemicals used in production are managed ethically and lessen their environmental impact by establishing a strong recycling infrastructure.

Additionally, the framework offers a starting point for policy and regulatory assistance targeted at lowering chemical pollution. By enacting laws that restrict the use of dangerous chemicals in the textile industry and encouraging the best waste management techniques, governments may encourage the adoption of sustainable practices. Policymakers can encourage the industry to use cleaner production techniques by establishing strict regulations and offering financial support for the use of eco-friendly technologies.

In conclusion, the textile industry's adoption of a sustainability framework has important ramifications for addressing the problem of chemical contamination. Innovative technology, systemic adjustments, regulatory assistance, and an emphasis on long-term gains can all help the industry move toward a more environmentally friendly future.

FUTURE SCOPE

Future studies into waterless dyeing technologies may result in more economical and effective techniques, opening these processes to a wider variety of producers. New developments include digital dyeing techniques, inkjet printing, and enhanced supercritical CO2 dyeing systems can further cut down on chemical and water consumption without sacrificing output quality. These technologies have the potential to significantly reduce chemical runoff and the overall impact on the environment as they develop and become more commonly used.

Another crucial area for future growth is the implementation of sophisticated closed-loop recycling systems. One of the main obstacles to establishing a circular economy can be overcome by advances in recycling technology, which can enhance the processing of synthetic and blended textiles. Furthermore, sorting procedures can be improved by integrating AI and machine learning, allowing for more effective material recovery. As these technologies develop, they have the potential to drastically cut down on the waste and chemical pollutants that come with the manufacture and disposal of textiles.

Future studies might also concentrate on creating non-toxic, biodegradable polymers that can take the place of dangerous chemicals currently employed in the textile industry. These materials can be made to have as little of an impact on the environment as possible while still providing the functionality needed for textile applications. As these materials become more popular, they have the potential to drastically lower the pollution that comes with the production and disposal of textiles.

The circular economy model will likely play an even larger role in the textile industry's future, with frameworks emphasizing closed-loop systems to minimize waste and maximize resource efficiency. Innovations in recycling technologies, such as fibre-to-fibre recycling, are expected to become mainstream, allowing used textiles to be fully reprocessed into new garments. Additionally, business models like product-as-a-service (leasing and subscription-based apparel) and resale markets could flourish, shifting the focus from ownership to access and creating a sustainable consumption cycle. This transition will be supported by frameworks that encourage durability, repairability, and recyclability in product design, ultimately reducing the need for virgin resources.

In the future, more resources will be allocated toward research and innovation, funded by both private and public investments. Governments and organizations may offer incentives for companies that invest in sustainable textile innovations, such as energy-efficient machinery, waterless dyeing techniques, or carbon capture technologies. Research into these areas can significantly advance sustainable practices, leading to breakthroughs that make eco-friendly textiles more affordable and scalable. Frameworks that prioritize continuous improvement and knowledge-sharing will support this progress, creating a competitive advantage for companies that prioritize sustainability and contribute to climate change mitigation.

In conclusion, there are many aspects to the future of using a conceptual framework for sustainability in the textile sector, and there is a great deal of promise to address the problems associated with chemical contamination. By means of technological developments, governmental backing, customer involvement, and cooperative endeavours, the textile sector can shift towards more environmentally friendly and sustainable methods. In addition to helping the environment, this all-encompassing strategy will increase the industry's adaptability and resilience in a global market that is becoming more environmentally sensitive.

CONCLUSION

Thcrc is a revolutionary chance to address chemical contamination in the textile sector by implementing a conceptual framework for sustainability. The industry may greatly lessen its environmental impact by incorporating cutting-edge technology like closed-loop systems and waterless dyeing methods. While closed-loop methods encourage recycling and reuse, reducing the release of pollutants, waterless dyeing does away with the necessity for dangerous chemicals and unnecessary water use. When combined, these developments can lessen chemical pollution and dependency on virgin resources, transforming the textile industry into a more environmentally friendly, circular model. However, broad cooperation, the construction of infrastructure, and supportive legislation are necessary for these solutions to succeed. In the end, implementing such a sustainability framework presents the industry as a pioneer in conscientious, forward-looking production methods while also assisting it in addressing environmental issues.

One of the key benefits of a conceptual sustainability framework is the promotion of a circular economy within the textile sector. Circular economy principles encourage resource reuse, recycling, and waste reduction, which are essential in an industry traditionally reliant on linear production models. Through circular practices, textile companies can transform waste into valuable resources, design for longevity and recyclability, and reduce dependency on non-rcncwable inputs. These practices not only lessen environmental harm but also create economic opportunities by cutting down on raw material costs and appealing to environmentally conscious consumers.

Supply chain transparency is another essential aspect of the conceptual framework, as it enables companies to track the sustainability of their operations across complex, often global supply chains. Through transparency initiatives, companies can ensure that all suppliers adhere to ethical and environmental standards, promoting fair labour practices and reducing the risk of environmental violations. Blockchain technology and digital monitoring tools have become instrumental in advancing supply chain transparency, providing textile brands and consumers alike with greater visibility into the lifecycle of products.

In summary, a conceptual framework for sustainability equips the textile industry with a comprehensive, structured approach to climate action, fostering a transformation that benefits the environment, businesses, and consumers alike. By adopting circularity, lifecycle assessments, eco-design, certifications, and supply chain transparency, the textile industry can move toward a model that prioritizes both profitability and environmental stewardship. This shift is essential to address the pressing challenges of climate change and resource scarcity, enabling the industry to contribute positively to a more sustainable future. As more companies embrace these frameworks, the textile sector has the potential to lead by example, showing that economic growth and environmental responsibility can go hand in hand in the quest for global sustainability.

REFERENCES

Ara, H., Leen, J. Y. A., & Hassan, S. H. (2019). GMS for sustainability performance in the apparel manufacturing industry: A conceptual framework. *Vision (Basel), 23*(2), 170–179. DOI: 10.1177/0972262919850931

Bag, S., Rahman, M. S., Rogers, H., Srivastava, G., & Pretorius, J. H. C. (2023). Climate change adaptation and disaster risk reduction in the garment industry supply chain network. *Transportation Research Part E, Logistics and Transportation Review, 171*, 103031. DOI: 10.1016/j.tre.2023.103031

Chica-Urzola, J. A. (2024). Strategies for Sustainable Manufacturing in the Textile Sector and Their Impact on the Sustainable Development Goals: Case Studies in Colombia. Sustainability, Stakeholders and Marketing in the Textile Sector: Conceptual and Practical Cases for the Americas, 181.

Diabat, A., Kannan, D., & Mathiyazhagan, K. (2014). Analysis of enablers for implementation of sustainable supply chain management–A textile case. *Journal of Cleaner Production, 83*, 391–403. DOI: 10.1016/j.jclepro.2014.06.081

Giancotti, M., & Mauro, M. (2020). Building and improving the resilience of enterprises in a time of crisis: From a systematic scoping review to a new conceptual framework. *Economia Aziendale Online, 11*(3), 307–339.

Kazancoglu, I., Kazancoglu, Y., Yarimoglu, E., & Kahraman, A. (2020). A conceptual framework for barriers of circular supply chains for sustainability in the textile industry. *Sustainable Development (Bradford), 28*(5), 1477–1492. DOI: 10.1002/sd.2100

Khan, B. A., Kuofie, M. H., & Suman, S. (Eds.). (2021). *Handbook of Research on Future Opportunities for Technology Management Education*. IGI Global. DOI: 10.4018/978-1-7998-8327-2

McGrew, L. L. (2016). Managing change for environmental sustainability: an international comparison of small and medium enterprises in the fabric and textile industry.

Moreira, N., de Santa-Eulalia, L. A., Aït-Kadi, D., Wood–Harper, T., & Wang, Y. (2015). A conceptual framework to develop green textiles in the aeronautic completion industry: A case study in a large manufacturing company. *Journal of Cleaner Production, 105*, 371–388. DOI: 10.1016/j.jclepro.2014.09.056

Oelze, N. (2017). Sustainable supply chain management implementation–enablers and barriers in the textile industry. *Sustainability (Basel), 9*(8), 1435. DOI: 10.3390/su9081435

Pattnaik, P., & Dangayach, G. S. (2019). Sustainability of wastewater management in textile sectors: A conceptual framework. [EEMJ]. *Environmental Engineering and Management Journal, 18*(9), 1947–1965. DOI: 10.30638/eemj.2019.186

Perry, P., & Towers, N. (2013). Conceptual framework development: CSR implementation in fashion supply chains. *International Journal of Physical Distribution & Logistics Management, 43*(5/6), 478–501. DOI: 10.1108/IJPDLM-03-2012-0107

Saha, K., Dey, P. K., & Papagiannaki, E. (2022). Implementing circular economy in the textile and clothing industry. In *Supply chain sustainability in small and medium sized enterprises* (pp. 239–276). Routledge. DOI: 10.4324/9781003018551-10

Seuring, S., & Müller, M. (2008). From a literature review to a conceptual framework for sustainable supply chain management. *Journal of Cleaner Production, 16*(15), 1699–1710. DOI: 10.1016/j.jclepro.2008.04.020

Svensson, G. (2007). Aspects of sustainable supply chain management (SSCM): Conceptual framework and empirical example. *Supply Chain Management, 12*(4), 262–266. DOI: 10.1108/13598540710759781

Van Bommel, H. W. (2011). A conceptual framework for analyzing sustainability strategies in industrial supply networks from an innovation perspective. *Journal of Cleaner Production, 19*(8), 895–904. DOI: 10.1016/j.jclepro.2010.12.015

Yildirim, S., Demirtas, I., & Yildirim, D. C. (2022). A review of alternative economic approaches to achieve sustainable development: the rising digitalization and degrowth post COVID-19. Handbook of research on sustainable development goals, climate change, and digitalization, 288-307.

Yıldırım, S., Gedikli, A., Erdoğan, S., & Yıldırım, D. Ç. (2020). Natural resources rents-financial development nexus: Evidence from sixteen developing countries. *Resources Policy, 68*, 101705. DOI: 10.1016/j.resourpol.2020.101705

Chapter 19
Mapping the Nexus Rule of Law's Influence on Good Governance: A Bibliometric Exploration

Widia Astuti
Universitas Muhammadiyah Makassar, Indonesia

Asra Yanti
Universitas Muhammadiyah Makassar, Indonesia

Muhammad Younus
https://orcid.org/0000-0001-9654-1546
Universitas Muhammadiyah Yogyakarta, Indonesia

Aqmal Reza Amri
Universitas Muhammadiyah Makassar, Indonesia

Andi Luhur Prianto
Universitas Muhammadiyah Makassar, Indonesia

Ahmad Harakan
Universitas Muhammadiyah Makassar, Indonesia

Nursaleh Hartaman
Universitas Muhammadiyah Makassar, Indonesia

Achmad Nurmandi
Universitas Muhammadiyah Yogyakarta, Indonesia

Wahdania Suardi
Universitas Muhammadiyah Yogyakarta, Indonesia

Sherajum Munira
https://orcid.org/0009-0004-4795-9160
Universitas Muhammadiyah Yogyakarta, Indonesia

ABSTRACT

The purpose of this study is to spread knowledge regarding how Indonesia's good governance will be impacted by the rule of law in 2021. Following that, we have determined the effects of the Rule of Law on Indonesian residents. Citizenry and the

DOI: 10.4018/979-8-3693-9909-5.ch019

alterations to their mental processes toward the rule of law. "Qualitative research" is the methodology employed by the researchers working in this field. The World Bank website, Dimension AI, and Google Scholar provided the research's data, which was then presented descriptively. This paper's research helps us understand how a robust legal system might promote effective governance in a system or society. Moreover, how the government can effectively and transparently carry out the rule of law, this paper will only cover the extent to which each aspect of the change is explained in detail; it will also examine potential solutions to problems raised by the public and how best to put them into practice; and lastly, it will discuss the beneficial effects that the change in the laws has had on good governance.

INTRODUCTION

Administrative law is an important aspect of the legal system and public government because it regulates the complex relationship between the state and its citizens. (Hirsch et al., 2022). Administrative Law consists of rules, principles, and procedural frameworks that shape the dynamics of administrative actions and decision-making. (O. Russo, 2023). Over time, the study of State Administrative Law has evolved to reflect changes in the government landscape, legal challenges, and societal demands. (Muhammad & Husen, 2019). For example, in Malaysia, additional regulations function as an important part of the legal system, complementing the legislative functions of the Parliament and the State Assembly.(Nugroho & Rahardjo, 2023). This helps ease the burden on legislative institutions in dealing with the details of laws and obtaining expert views on specific issues. (Susanti, 2022)

In the context of decision-making assisted by technology, principles of administrative law have been implemented to address the potential replacement of certain jobs by robots or technology (Budiarrianto & Sebayang, 2021). The emergence of Global Administrative Law has also become a topic of interest, as it explores potential sources and suitability of global regulatory frameworks (Hukum et al., 2023). In Singapore, the evolution of administrative law has been shaped by the context and system of unique laws in the country (Vallejo, 2021). European Standardization Organization (ESO) and Harmonized European standards are also subject to the principles of administrative law, considering public nature and its legal impact (Dinata, 2022). In Russia, the subject of law and administrative proceedings is a complex, multifaceted legal phenomenon that includes a variety of relationships and social issues, such as public administration, administrative and legal regulations, and police law (Szadok-Bratuń, 2021).

In Canada, administrative law principles have been applied to fracking hydraulics, with courts adopting administrative and privileged principles and expert institution regulations.(Shaheen et al., 2019). Overall, administrative law plays an important role in shaping the dynamics of administrative action and decision-making in various legal systems and countries.(Zdyb, 2019). Its evolution reflects the changing governance landscape governance, legal challenges, and societal demands, ensuring that the relationship between the state and its citizens remains well-governed and balanced.(Lee, 2021)

Currently, Indonesia is in a chaotic condition, in a state of crisis in various fields, including the legal field.(Mumpuni & Kusumawati, 2021). Laws that are expected to provide justice for society turned out to be the opposite.(Muhammad & Husen, 2019). The effectiveness of law enforcement only applies to small communities committing petty crimes.(Delmana, 2019). Meanwhile, perpetrators of major crimes such as corruption, collusion, and nepotism (KKN), which is commonly called white-collar crime, are hard to touch. In this case, people really need courage, especially law enforcement officers, to make deep breakthroughs to resolve the case.(Nugroho & Rahardjo, 2023).

The laws that apply in society have differences between social classes. A community with upward social stratification clearly receives different treatment than a society that does have downward social stratification.(Triwahyuni, 2020). People whose families have a higher position or position have special treatment or honor, which is different from people who come from ordinary or ordinary family backgrounds and do not have a position in society.(Trofimova, 2020). This means that here is the position. In the applicable law, there is an indication that the treatment for violators of the law injustice occurs to law enforcement officers. The law is sharp down, and the law is blunt and above; this phenomenon occurs in almost all areas of the country in Indonesia.(Karunia et al., 2022).

Laws are basically made to create order and peace in society.(Govindaraju et al., 2023). Therefore, a legal system must operate like a law. The series of social organs must complement each other and have a high awareness of applicable law. The paradigm that views law as a system has dominated the thinking of the majority of legal circles, both theorists and among its practitioners, since the birth of the modern state in the 17th century until today, namely, a paradigm that considers law as an order(Priyo Sembodo et al., 2022). Indonesia, which, in fact, is a rule-of-law country, means everything that relates to violations of law or disobedience to existing legal regulations gets strict sanctions for the perpetrators.(Bërdufi & Nezha, 2023). It is recommended that anyone alive as a citizen of society living on this mother earth create order socially well, namely by upholding applicable laws. The law must be enforced without any favoritism in legal cases.(Sakti Koeswanto et al., 2023).

In Weber's view, law is an order that is coercive because the establishment of the legal order (different from other social orders and norms that are not law) is supported entirely by the coercive power of the state. Weber differentiated various legal systems on the basis of substantive and formal rationality. Weber said that there is substantive rationality when it is substantial. The law does consist of general rules in the abstract, which are ready to be deduced for use in punishing various concrete cases. On the other hand, the law is said to have no substantive rationality if every case is resolved on the basis of a unique political or ethical policy in its order(Rao, 2021). Maybe even resolved emotionally, which cannot refer to general rules at all that objectively exist. On the other hand, the law can be said to have rationality formal (irrational) if the law is only obtained through inspiration or through whispers whispered ideas that are said to be accepted by charismatic leaders so that the truth and its feasibility cannot be tested objectively.

Governance is a broad concept with high complexity in its main pillars. Kaufmann (2010) defines government as a set of traditions and institutions that can be used for exercising the power of authority, where the provisions of good governance lead to improvements in institutions related to political stability, absence of terrorism and violence, formulation and proficiency implementation of government policies, better regulatory mechanisms, reduction of corruption and ensuring the supremacy of law that can be recognized as high-quality governance. The institutional framework consists of legal rules and norms that constrain the behavior of policymakers. (Muhammad & Husen, 2019). Legal rules and norms, too, must ensure that government actions do not damage but rather support the functioning of growth economics.(Witkowska & Kucheryavaya, 2021).

LITERATURE REVIEW

Governance

According to Gal, (2020) Governance is a set of rules, processes, and behaviors that influence how power is exercised, how decisions are made, how citizens and stakeholders are involved, and how responsibility is attributed between them. Governance refers to the mechanisms, processes, and institutions through which citizens and groups articulate their interests, exercise their legal rights and obligations, and mediate differences between them. (Maryam, 2016). According to other researchers, governance is the traditions and institutions through which power is exercised.(Siljanovska & Shalevska, 2019). This includes policy and decision-making processes as well as mechanisms for involving citizens and other societal groups.(Valerio et al., 2014)

The growth rate of research publications on good governance has indeed increased rapidly since 2011 until now. Several factors that encourage increased publication of good governance research include:

- Increasingly aware of the fundamental role of good governance in public sector reform and achieving sustainable development goals.
- Support and funding for governance research from international donor institutions such as the IMF, World Bank, and UN is increasing.
- Globalization and accelerated knowledge transfer facilitate the diffusion of good governance concepts and policies.
- Increased availability of governance indicator data since 1996 through Worldwide Governance Indicators.
- The development of sophisticated quantitative methods makes it easier for empirical research to measure the impact of good governance policies.
- The development of multi- and inter-disciplinary research to understand the complexity of governance.

According to Edralin & Pastrana, (2020), the factor driving the increase in good governance research publications is the strengthening consensus that good governance is a prerequisite for administrative efficiency, economic growth, and poverty alleviation. Good governance is becoming an increasingly stringent requirement for international donors and financial institutions to provide aid and loans.(Khodakivskyi, 2021). There are also those who argue that the driving factor for increasing good governance research publications is increasing awareness about corruption and encouraging public sector reform through the application of good governance principles.(Edralin & Pastrana, 2020)

Good Governance

According to Setyowati (2015), in her journal entitled "Implementation of Good Governance Principles in Public Services," good government must fulfill the following principles:

- Information Disclosure. The government must be transparent and provide the widest possible information to the public regarding policies and the use of public resources.
- Public Accountability. The government must be able to account for its performance and decisions to the community.

- Community empowerment. The government needs to encourage community participation and aspirations in decision-making so that community needs can be accommodated.

Winarna (2018), in her journal entitled "Application of Good Governance Principles in Bureaucratic Reform," also mentions the characteristics of good government, namely:

- Responsive. The government must be responsive and quickly handle complaints, problems, and community needs.
- Fair. The government serves all citizens without discrimination and fulfills their civil rights.
- Efficient and Effective. The government uses the minimum possible resources to achieve the best results.
- Accountable. The government is responsible for its decisions and performance.
- Transparent. The government provides information that is open to public scrutiny.

Rule of Law

Legal rules are a set of regulations, norms, and restrictions made by an authorized body. They are binding and coercive, with the aim of regulating human behavior in social and state life.(Tima et al., 2021). According to Permata (2020), in his journal "Legal Politics of Values-Based Regional Regulations Pancasila," legal rules are binding provisions made by authorized institutions to regulate interactions between members of society. Meanwhile, Sugianto (2021), in his journal "The Influence of Legal Rules on Adolescent Behavior," defines legal rules as a set of written regulations that are generally binding and enforced by relevant institutions to minimize social deviation.

The aim of legal rules is to regulate human behavior in society, create order, justice, and legal certainty, protect public interests, and solve social problems in society.(Firmantoro, 2022). There are two forms of legal rules, namely, written and unwritten.(Serohin, 2022). Written forms such as laws, government regulations, and others. Unwritten forms, such as the customs of a particular society(Fredericks & Bradfield, 2023).

Rule of Law in Indonesia

The concept of the "rule of law" relates to the importance of justice and the law. The primary foundation for public life and governance in a state of law is the law. (Tursin & Ahmad, 2000). Everybody must be treated equally and fairly when the law is applied, without exception. Indonesia is a state of law that upholds democratic values. This is reaffirmed in Article 1, paragraph (3) of the 1945 Constitution of the Republic of Indonesia (UUD NRI, 1945), which declares Indonesia to be a state of law.(Busthami, 2017). In recent times, there have been advancements in the application of the rule of law in Indonesia. This has been characterized by strengthening democratic institutions, raising public understanding of the law, and enhancing the caliber of law enforcement personnel.(Akmal et al., 2021).

However, there are still a number of obstacles to the adoption of the rule of law in Indonesia, including corruption. One of the main issues preventing Indonesia from implementing the rule of law is still corruption. Corruption may be detrimental to society and impede progress.(Saputra & Firmansyah, 2023)—regulations' inefficiency. The governance space needs to be effectively regulated by the laws in place. Sometimes, extant regulations conflict, are imprecise, or fail to address demands(Baral & Heinen, 2020)—the caliber of police enforcement personnel. Law enforcement personnel still need to be put up to par. There are instances when law enforcement officials act in an unprofessional manner and lack the bravery to confront corrupt officials—insufficient involvement of the community. The public's involvement in oversight of government operations needs to be higher. Sometimes, people need to be more active in their oversight of government operations and recognize the value of the rule of law(Hukum et al., 2023).

These obstacles must be addressed if Indonesia is to strengthen the application of the rule of law there. Among these initiatives is the bolstering of the fight against corruption.(Abreu et al., 2022). The fight against corruption needs to be taken seriously and persistently. Implement regulatory change. It is necessary to make improvements to the current regulations in order to make them more efficient and relevant. They are raising the standard of law enforcement personnel. Law enforcement personnel need to be properly qualified and equipped. Boost involvement in the community.(Ayu et al., 2019). Encouragement is required for the general public to take part in monitoring government operations.

Table 1. Taxonomy of research data

Papers	Summarized Abstract	Results
Intellectual Dynamics of Good Governance Studies: A Bibliometric Analysis	- Bibliometric analysis of good governance research publications from 1984 to 2020. - Increase in the growth rate of good governance literature since 2011.	- Increase in the growth rate of good governance literature since 2011. - The United States, the United Kingdom, and Indonesia are top contributors.
Rule of law approach to the alleviation of poverty: An analysis on human rights dimension of governance	- The rule of law is essential for poverty alleviation and equitable opportunities. - Emphasis on the human rights dimension of governance and legal empowerment.	- The paper argues that poverty is formed by society itself. - The paper emphasizes the importance of the rule of law and human rights for poverty alleviation.
The Importance of the Rule of Law in Governance	- Paper analyzes governance and the role of law in a constructive manner. - Explores how the rule of law affects governance, especially in developing countries.	- The paper provides a constructive understanding of the rule of law and governance. - The paper examines the impact of the rule of law on governance, particularly in the developing world.
Good Governance, Rights, and State: Quo Vadis Administrative Law?	- Good governance emphasizes openness, transparency, and accountability in state activities. - The concept of governance ensures inclusion and respect for fundamental rights.	- The paper discusses the concept of good governance and its importance in assessing the lawfulness of human rights interference with the activities of the state. - The paper highlights the principles of good governance, such as participation, transparency, accountability, and the rule of law.
Good Governance and The Rule Of Law	- Good governance is essential for both public and private sectors. - The rule of law is crucial for ensuring good governance.	
Good Governance: An Analytical Study	- Importance of good governance and methods/tools to assess it - Impacts of governance on development outcomes in developing countries	- The paper highlights the importance of good governance. - The paper discusses the impacts of governance work on development outcomes.
Good governance and rule of law	- Good governance is essential for both public and private sectors. - The rule of law is crucial for ensuring good governance.	- The rule of law is essential for good governance. - Abuse of power and corruption undermines the rule of law and good governance.
The rule of law, governance, and development	- The rule of law is linked to better development outcomes - Different approaches to the rule of law promotion explored in case studies	- Positive relationship between the rule of law and development outcomes - Mixed approaches and results in the rule of law promotion

continued on following page

Table 1. Continued

Papers	Summarized Abstract	Results
Impact of the Rule of Law as a Fundamental Public Governance Principle on Administrative Law Interpretation in the Czech Republic	- The paper analyzes the use of the rule of law in administrative law interpretation in the Czech Republic. - The courts in the Czech Republic have moved from formalistic interpretation to interpretation through values, including the rule of law.	- The paper analyzes the use of the rule of law principle in interpreting administrative law provisions in the Czech Republic. - The courts in the Czech Republic have moved from purely linguistic interpretation to interpretation through values, including the rule of law.
Political Modernisation: The Rule of Law Perspective on Good Governance	- The paper explores the meaning and reference of "good governance" and its relation to the rule of law. - It presents an inquiry into the rule of law I and the rule of law II and validates these concepts using data from the World Bank Governance project.	- The paper presents an inquiry into the rule of law I and the rule of law II. - Data from the World Bank Governance project is used to validate these concepts.

RESEARCH METHOD

Since the goal of this article is to comprehend how the rule of law affects good governance, the qualitative research method is suitable for this kind of research topic. With the help of this method, we can gain a deep understanding of the intricate social phenomenon and conduct an in-depth investigation of the theme that incorporates unbiased viewpoints and experiences.

Research Locations

Our article's study was conducted using scholarly journals and the World Bank website as sources. However, according to our analysis, academic publications have emerged as a top source of knowledge and data, including in-depth evaluations of expert research. Data relating to Academic Journals were obtained from many online database sources, including Scopus, Google Scholar, Dimension AI, and others. The data underwent peer review to ensure accuracy and dependability.

Data Collection Techniques

As researchers, to collect data, we conducted a literature review, which enabled us to compile the body of current knowledge. We were effective in understanding the data because of the information that was gathered. We identified the holes in the research and the crucial information for our brainstorming. We started by defining the parameters of our study to conduct a literature review; specifically, this involved identifying the keywords we would use to look for information pertaining to the article's issue, such as "good governance," "rule of law," etc. We also restricted the scope of our search to papers released in the previous five years. We first checked to make sure the collected publications were related to our article; then, we quickly perused their abstracts

Figure 1. Filters and keywords used for research

Source. Dimension AI

Data Analysis Techniques

VOSViewer is a great desktop program for bibliometric analysis and visualization of research data. To begin the analysis, we first imported the research data into the VOSViewer program in RIS format. Once the data was imported successfully, we conducted a bibliometric analysis to determine which themes were most influential. The ability to create maps with VOSViewer allowed us to determine which themes were most influential.

Figure 2. Density visualization of research

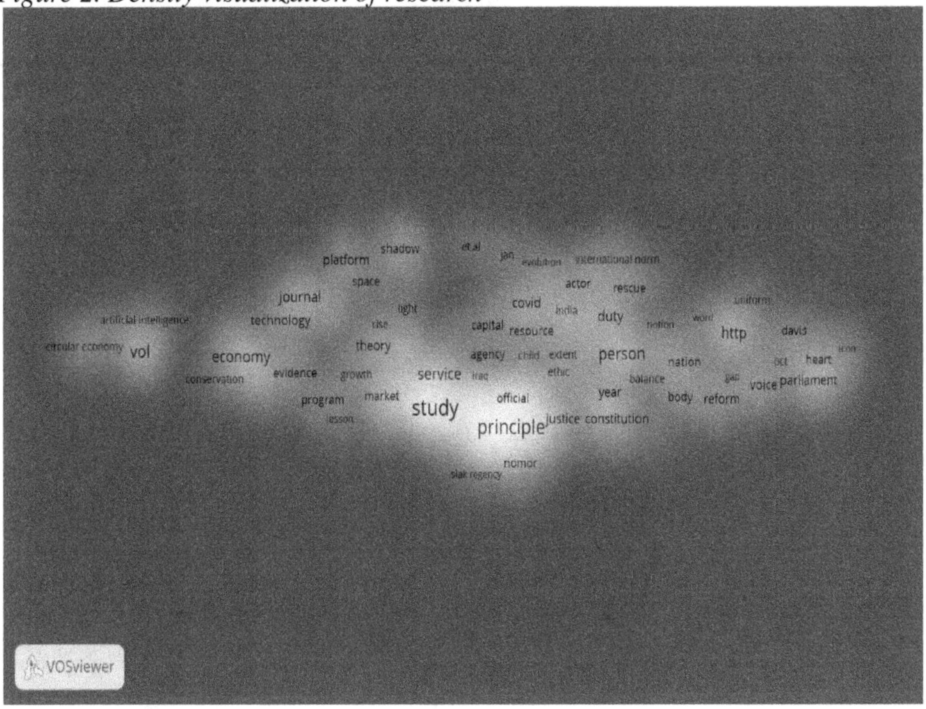

Source. Created By Author

Figure 2 shows the research themes in addressing the rule of law in good governance. There are various related research themes, but only a few are relatively recent and consistently discussed, namely Principle, Study, and Service. Principles are something that researchers prioritize the most. This is because principles in governance such as transparency, accountability, and participation are able to create good governance that can support better governance and also play an important role in directing the organization or entity in a good and sustainable direction. It is expected to improve the quality of public services that are poor in the eyes of the public by applying basic principles such as legal certainty, accountability, transparency, fairness, professionalism, and democracy(Maryam, 2016). Meanwhile, the theme of research related to the study is that by studying the government, it can also find out the rule of law from best practices that exist in other countries in creating governance by comparing governance in other countries(Busthami, 2017). Studies in this field aim to smooth national development to achieve national goals(Akmal et al., 2021). The final theme is service, where service is very important to the rule of law in good governance, as well as with good services useful for meeting the needs or interests of the community(Saputra & Firmansyah, 2023).

RESULT AND DISCUSSION

Figure 3 presents a timeline illustrating the publication of findings on the impact of the rule of law on governance from 2018 to 2023. In 2018, no documents were published, and in 2019, there was a surge in publications, which amounted to 60 documents from 0 documents. In 2020 and 2021, the number of published documents increased by only ten each year, namely 70 and 80 documents.

Figure 3. Publication of documents by years

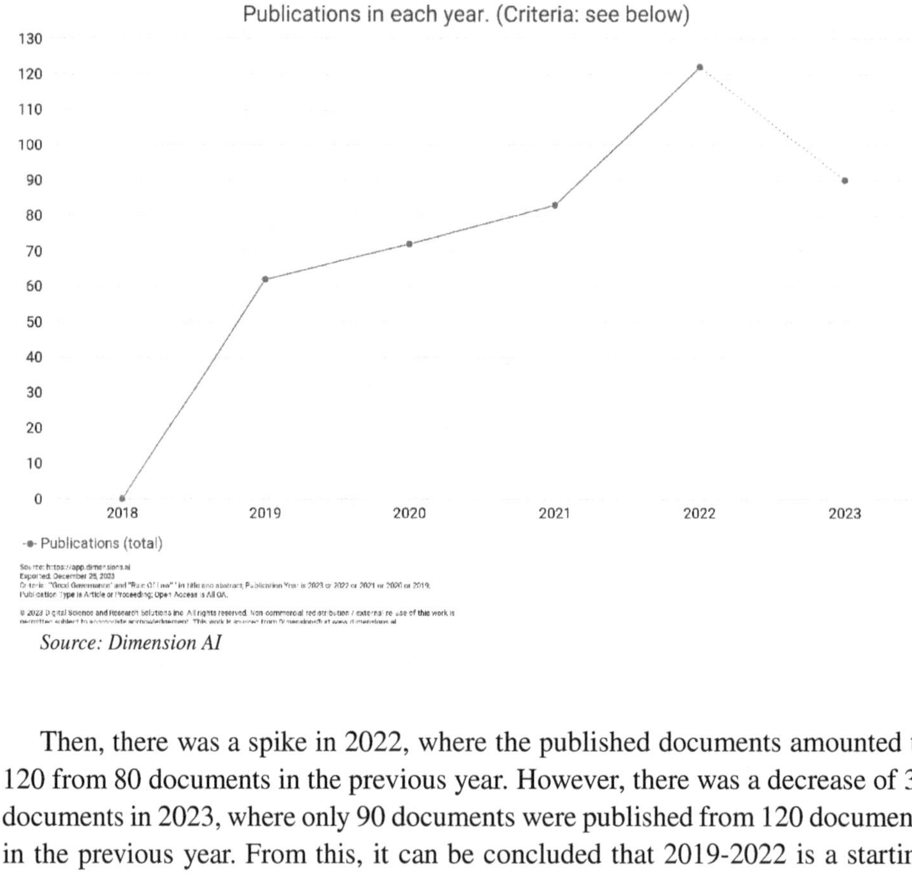

Source: Dimension AI

Then, there was a spike in 2022, where the published documents amounted to 120 from 80 documents in the previous year. However, there was a decrease of 30 documents in 2023, where only 90 documents were published from 120 documents in the previous year. From this, it can be concluded that 2019-2022 is a starting point year to achieve and maintain progress.

Document Publication by Subject Area

The graph from 2019, 2020, 2021, 2022, and 2023 is shown in Figure 4. Out of all the organizations that contribute to publications on the subject of good governance and the rule of law, Human Society is the most influential, with 210 articles. With

187 publications published, Law and Legal Studies holds second place in terms of quantity of publications. Next, with 57 publications, commerce and management tourism are among the top three most often published categories. Economics holds the fourth spot, with 22 publications available. With 17 documents, it holds sixth place in information and computing science.

Figure 4. Publication by subject area

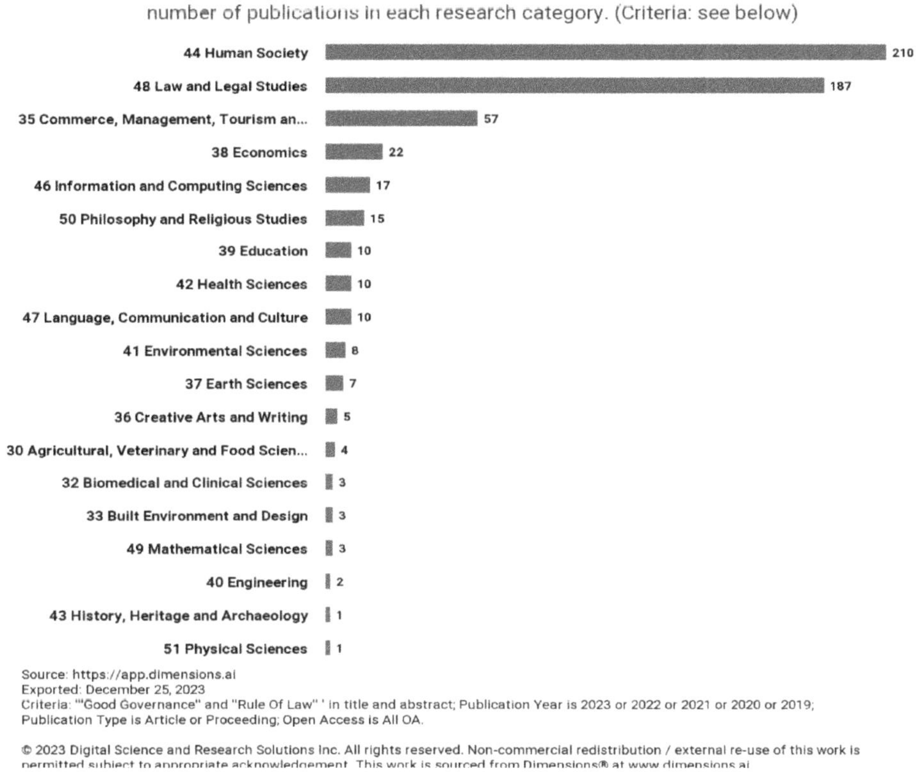

Source. Dimension AI

Furthermore, philosophy and religious studies, in the sixth position, produced 15 documents, continued in the seventh position, namely education, healthy science, and language communication and culture; the publications produced by each of the three fields were the same, totaling ten publications, then in the aspects of environmental science and earth science had one difference, totaling 8 and 7 publications, as well as aspects between creative arts and writing and agricultural veterinary and food science which had one difference, namely five and four publications. In addition, aspects of biomedical and clinical science, built environment and design, and mathematical science produced the same publications with three publications.

She was followed by aspects of engineering with two publications. Finally, in terms of aspects of history, heritage, archaeology, and physical science, only 1 document has been published related to this subject.

Documents Publication by Author

Figure 5. Publication by author

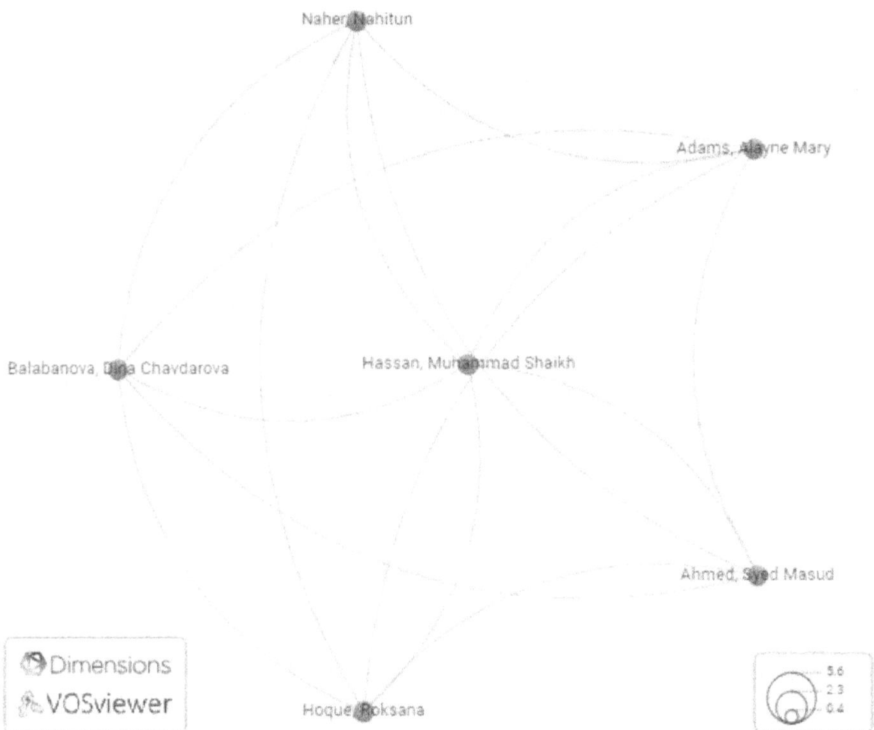

Source Dimension AI

Figure 5 shows the authors' relationship. Hassan, Muhammad Shaikh, of the BRAC University for Comparative Public Law and International Law, is identified by one publication, five total quotations, and five quotation links. With five quote connections, five total quotations, and one publication, Naher, Nahitun of the BRAC University of Bangladesh is indicated with a red cluster. One publication, five total quotations, and five quotation links can be found for Adam, Alayne Mary of McGill University in Canada. Five quotation links, five total quotations, and one publication are provided by Hoque, Roksana of BRAC University. Then, Balabanova, Dina Chavdarova, of the London School of Hygiene & Tropical Medicine, with one publication, five quotation links, and five quotes overall, all highlighted in red. Based on the information provided, it can be deduced that the image above depicts a single cluster, six researchers, fifteen quote linkages, fifteen total quotations, and one group.

Discussion

Using VOSViewer for network visualization, overlay, and density, the network between topics in 324 documents obtained from the Dimensions AI database was bibliometrically performed. A bibliometric network consists of nodes and edges. Nodes can be publication circles, journals, or keywords, while edges show the relationship between nodes, indicating the relationship between two nodes. In addition, distance represents the strength of the relationship; the correlation between nodes becomes more significant the closer the distance between nodes.

When talking about how the rule of law affects good governance, it is important to take legal compliance, accountability, and openness into account. Good governance greatly depends on legal compliance. It has a direct bearing on the management, supervision, and governance of an organization and is related to compliance with relevant rules and laws. On the other side, an essential component of good governance is stakeholder involvement and participation. Involving stakeholders in the decision-making process makes it more inclusive and guarantees that a range of interests and points of view are taken into account. The influence of the rule of law in good governance is depicted in the following VOS viewer analysis findings, which provide a visual representation of the keyword network.

Figure 6. Network visualization or research

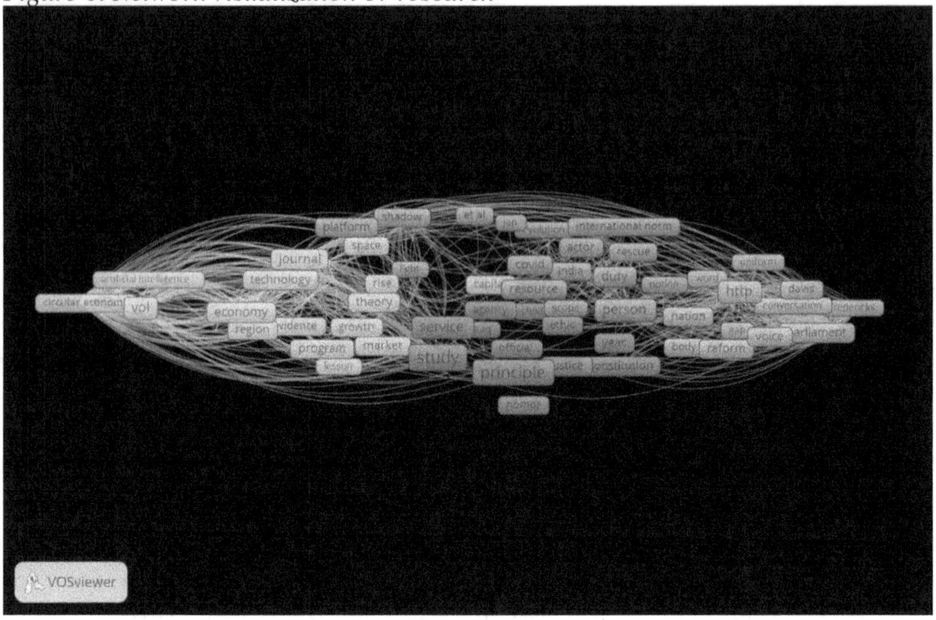

Source, Created by author

Image labels indicate frequently occurring terms or keywords, and clustering is used to understand bibliometric clustering. On the other hand, image mapping is used to gain additional understanding of the bibliometric network. Commonly used terms or keywords are displayed within each circle. The diameter of the circle indicates the number of papers and publications associated with the phrase or keyword in the document title.

Study, Principle, and Service are the largest nodes, as can be seen in the figure above, where each hue is organized into six groups according to the correspondence between the keywords and their illustrations. The first rank is occupied by study, which has 36832 link strength and 455 total occurrences. With 316 occurrences and 9133 link strength, Principle took second place. Service, with 156 occurrences and 15276 overall strength, came in third. Furthermore, in terms of good governance, the nodes with the lowest frequency in order are artificial intelligence (16 occurrences and 6994 link strength), followed by Gap (14 occurrences and 2578 link strength).

The theme mapping on the impact of the rule of law in good governance aims to identify and analyze various topics and research focuses related to the impact of the rule of law in good governance. It is expected that this mapping will reveal various issues and innovations that researchers propose to understand and address challenges in the impact of the rule of law in good governance.

Figure 7: Overlay visualization of research

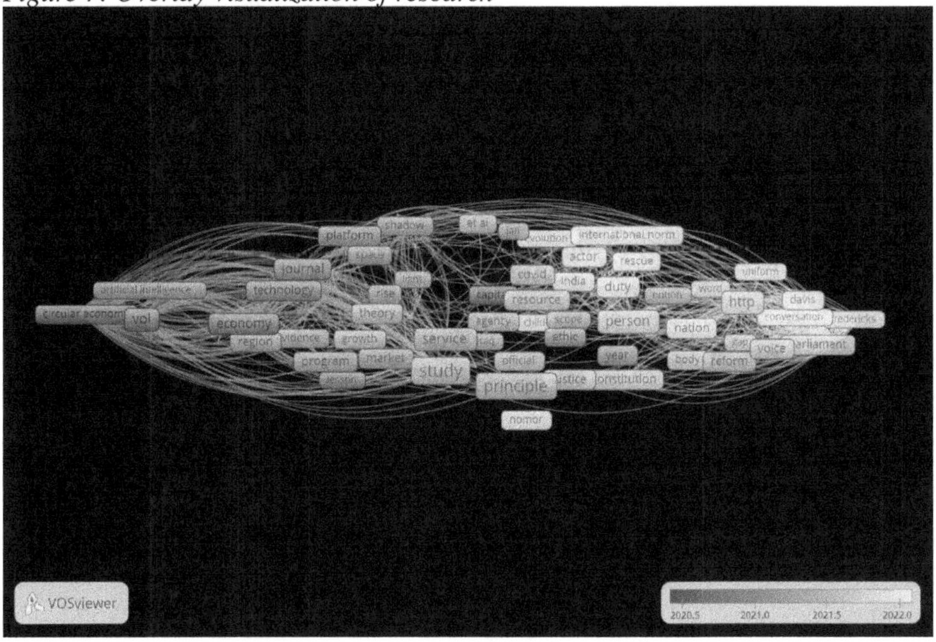

Source. Create By Author

Figure 7 shows the research themes that address the impact of the rule of law on good governance. There are a variety of related research themes, but only a few are relatively recent and consistently discussed, namely Actors, Uniforms, evolution, and international norms. Actors is a longstanding research theme, and the most recently published in 2022 is Actors in the Context of the Rule of Law in Good Governance. Research in this area focuses on Stakeholder Awareness Raising, Transparency and Accountability, and Advocacy for Sustainable Legal Change. Meanwhile, the research theme on Uniforms is becoming increasingly important because the uniform that every employee in governance wears shows order in good governance. The next theme, evolution, highlights the importance of change, which is designed to guide organizational and individual behavior and ensure that good governance is effectively implemented. The final theme, international norms, creates a framework that can influence and shape good governance practices at the global level. The impact of the rule of law on good governance is often closely linked to the engagement and application of international norms.

The discussion of the four research themes, namely Actors, Uniform, Evolution, and international norms, have a significant relationship; actors are the main figures or actors of the rule of law in good governance and can have a significant positive impact. Through the actors in the governance, we can know the duties and functions of each in good governance. In addition, good actors can build public pressure to increase accountability and transparency in business and government practices. With this increased awareness, organizations and governments may feel compelled to strengthen oversight systems, improve policies, and ensure compliance with the rule of law. On the other hand, uniform good governance can reflect good and orderly governance in a country.

International conventions and norms carry broader implications in the context of good governance. Convention implementation can motivate countries and organizations to adopt good governance principles proactively and result in more substantial legal changes. International norms can establish recognized global standards, creating the basis for global adherence to good governance principles. International conventions and norms can lead to harmonization and increased consistency in governance practices at the global level, reducing barriers and supporting effective implementation of the rule of law in the context of good governance.

Finding from World Governance Indicator

Figure 8 displays the Rule of Law indicator score from the World Bank's Worldwide Governance Indicators (WGI). Rule of Law is one of the dimensions of governance measured by the WGI in a percentage range of 0-100.

The graph below shows a comparison of the development of Indonesia and Australia's Rule of Law scores from 1996 to 2020. In 1996, Indonesia's Rule of Law score was 36.18. Then there was a decline in 1998 to 27.50. In 2000, it rose again to 30.85. Then the score fell again in 2002 to 21.39. Then, there was a gradual upward trend until it reached the highest score of 44.71 in 2014, although it decreased slightly to 39.52 in 2016. Then, there was stability in the rule of law score in governance in Indonesia from 2017 until 2022. This indicates that the quality of the rule of law, law enforcement, and equal access to justice in Indonesia continues to improve. However, the achievement of 45.28 still needs to reach the ideal standard of 100. This means there is still much homework for Indonesia to do to strengthen the rule of law and achieve good governance.

Figure 8. Rule of law indicator score

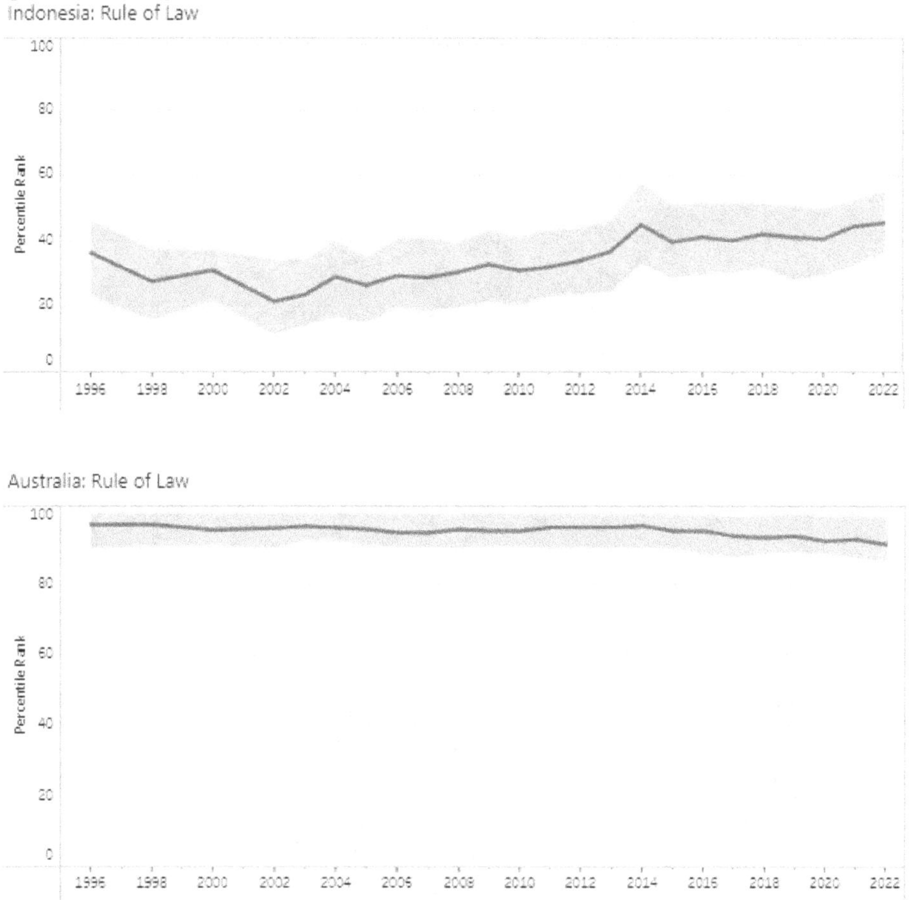

Source By World Bank Website

Meanwhile, from 1996 to 2022, Australia has a stable score, namely the initial score in 1996, which is 96.482, although it has slightly decreased the score in 2022 to 91.038. Australia's score is ideal, which indicates that Australia has implemented the rule of law well in its governance compared to Indonesia, which still needs improvement. However, Australia can be an example of applying the rule of law in governance if it wants to create good governance in a country.

Overall, this figure provides an empirical picture of the development of the rule of law in Indonesia and Australia based on objective data from the World Bank's WGI. This data is useful for tracking Indonesia's progress in realizing credible and accountable governance.

Figure 9. Map visualisation of WGI

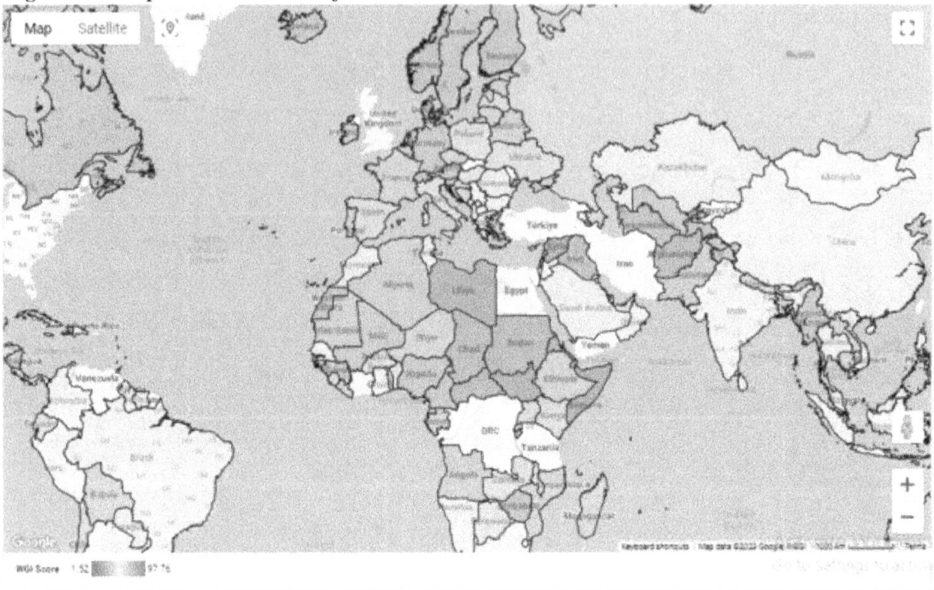

Source By World Bank Website

Figure 9 presents a visualization of the map of Rule of Law rankings and scores around the world based on the World Bank's World Governance Indicators (WGI) 2020 data. The map is categorized into quartiles, which divide countries into seven groups based on their Rule of Law score ranking, which can be classified into different colors, namely bright green, light green, yellow, light orange, bright orange, and red colors. The greener the color indicates, the better the application of the rule of law in the country; conversely, the red color indicates that the application of the rule of law in the country is not going well. It can be seen that there are only a few countries that have good applications of the rule of law, such as Finland, Sweden, and Germany. Followed in the second-best position are Poland, Spain, and Italy, then in the third position with a moderate application of the rule of law are China, Saudi Arabia, and Indonesia, while countries with poor levels of application of the rule of law, such as Somalia, Libya, and Afghanistan. Then, in the last position, the countries without any rules of application of law in governance are Yemen, Iran, and Turkey. Where this is due to the Governments in Yemen, Iran, and Turkey tend to be authoritarian, thus not providing space for public participation in the decision-making process.

It can be seen from the visualization of the figure above that the country Denmark is ranked number one in the world in terms of law and order with a score of 0.89 for compliance with the rule of law is a country with a score of 0.87 on compliance with the rule of law. Having the same score as Denmark, Norway is also the leading country in terms of following law and order and adherence to the law. Finland comes in second place, scoring around 0.87 on compliance with the rule of law. The country emphasizes that the rule of law is not just a common phrase but an essential component of everyday life. These two countries show that the level of peace and security is determined not only by geography but also by commitment and policies.

While Indonesia is still in the middle of implementing the rule of law for good governance in a country like Finland, due to many complex factors, the implementation of the rule of law for good governance in Indonesia has yet to reach an optimal level. First, Indonesia's dynamic and unstable political history has led to difficulties in building a stable and strong legal culture. While providing political freedom, democratic political systems are often characterized by shifting leaders and political instability, which can hinder the success of good governance. In addition, an intractable culture of corruption also contributes to the denial of progress. Indonesia's legal system and law enforcement agencies need improvement to operate effectively and independently. Given that low levels of education and legal understanding can hinder public participation in ensuring transparency and accountability in government, legal education and public legal awareness are also important focus in efforts to improve the application of the rule of law.

CONCLUSION

This study has shed light on good governance in a useful way. We have added to the corpus of information already available on good governance by conducting a thorough analysis of the rule of law. The ramifications of our research encompass pertinent domains or uses of the Rule of Law, emphasizing the advantages and difficulties associated with good governance. However, our research has shown several aspects that are crucial to acknowledge the possible constraints, like diplomatic immunity, postponed justice, and the wealthy and well-connected minority. To expand on our basis and deepen our understanding of good governance, future research should include criminal justice reform, anti-corruption and transparent government, and access to justice. In summary, this study highlights the need for more research and careful discussion in the field of the rule of law to promote developments that would benefit the pertinent parties and communities.

REFERENCES

Abreu, M., Soares, I., & Silva, S. (2022). Governance quality and environmental policy on emergent, resource-rich economies: The case of Brazil. *Energy Reports*, *8*, 70–75. DOI: 10.1016/j.egyr.2022.01.041

Akmal, D. U., Sultan, U., & Tirtayasa, A. (2021). Politik Reformasi Hukum. *Pembentukan Sistem.*, *8*, 21–35.

Ayu, A., Niswaty, R., Darwis, M., & Arhas, S. H. (2019). Applying the principles of good governance in the efforts of guiding out-of-school children at social service offices at Takalar Regency. *Jurnal Office*, *5*(2), 51–58. DOI: 10.26858/jo.v5i2.11821

Baral, N., & Heinen, J. T. (2020). Regulatory Compliance of Community-Based Conservation Organizations: Empirical Evidence from Annapurna Conservation Area, Nepal. *Sustainability (Basel)*, *12*(22), 9420. DOI: 10.3390/su12229420

Bërdufi, N., & Nezha, K. (2023). Understanding Civilian Participation and its Implications: A Case Study on the Inclusion of Syria's Civilian Population in the Context of Violence. *Interdisciplinary Journal of Research and Development*, *10*(2), 68. DOI: 10.56345/ijrdv10n209

Budiarrianto, D., & Sebayang, A. F. (2021). Penerapan Prinsip-Prinsip Good Governance terhadap Pelayanan Publik Desa Rahayu Kecamatan Margaasih Kabupaten Bandung. *Bandung Conference Series: Economics Studies*, *1*(1), 33–40. DOI: 10.29313/bcses.v1i1.49

Busthami, D. (2017). Kekuasaan Kehakiman Dalam Perspektif Negara Hukum. 4

Delmana, L. P. (2019). Pengaruh Penerapan Good Governance Dalam E-Purcashing Terhadap Pencegahan Korupsi. *Jurnal Ilmu Pemerintahan Widya Praja*, *45*(1), 47–62. DOI: 10.33701/jipwp.v45i1.241

Dinata, A. W. (2022). Legal Implications Of Non-Compliance With The Decision Of The State Administrative Court In Terms Of The Implementation Of Regional Autonomy And The Unitary State. *Jurnal Hukum Peratun*, *4*(1), 1–30. DOI: 10.25216/peratun.412021.1-30

Edralin, D., & Pastrana, R. (2020). Nexus between Sustainable Business Practices and the Quest for Peace. *Bedan Research Journal*, *5*(1), 1–58. DOI: 10.58870/berj.v5i1.11

Firmantoro, K. (2022). Negligence Of Customs and Excise in the Implementation of Judicial Decisions on Audit Result Disputes that Can Cause Country Loss and Business Actors (A Study On Court Decision No. 46/Pk/Tun/2011 Which Has Not Been Done By Customs And Excise International Journal of Business. *Economics, and Social Development*, *3*(2), 76–87. DOI: 10.46336/ijbesd.v3i2.248

Fredericks, B., & Bradfield, A. (2023). Many Bodies, One Heart. M/C Journal, 26(1). DOI: 10.5204/mcj.2908

Gal, J. (2020). Corporate governance of insurers in Germany. German National Report. *Zeitschrift Für Die Gesamte Versicherungswissenschaft*, *109*(1), 41–64. DOI: 10.1007/s12297-020-00467-9

Govindaraju, P., Rahman, M. R. A., & Ismail, S. M. (2023). Ekstremisme Ganas Dan Semakan Kehakiman Dalam Kes-Kes Berkaitan Dengannya Dari Perspektif Undang-Undang Tahanan Pencegahan Malaysia : (Violent Extremism and Judicial Review in Related Cases: A Perspective on the Preventive Detention Laws of Malaysia). UUM. *The Journal of Legal Studies*, *14*(1), 297–331. DOI: 10.32890/uumjls2023.14.1.12

Hirsch, B., Schäfer, F.-S., Aristovnik, A., Kovač, P., & Ravšelj, D. (2022). The impact of Digitalized Communication on the effectiveness of Local Administrative Authorities – Findings from Central European Countries in the COVID-19 Crisis. *Journal of Business Economics*, *93*(1–2), 173–192. DOI: 10.1007/s11573-022-01110-y PMID: 38013851

Hukum, S., Kurniawati, R., Lubis, A. F., Indrayana, S., & Kurniawan, B. A. (2023). Rekonstruksi Penelitian Hukum Administrasi Negara melalui Pendekatan Bibliometrik : Sebuah Tinjauan Komprehensif tentang Perkembangan dan Kontribusi Literatur. 2(01), 20–29. https://doi.org/DOI: 10.58812/shh.v2.i01

Karunia, R. L., Solihati, K. D., & Wati, N. K. (2022). Implementation of Good Governance Principles in the Land Transportation Management Center. KnE Social Sciences, 1253–1269–1253–1269. DOI: 10.18502/kss.v7i9.11014

Khodakivskyi, M. (2021). The formation of Ukraine's legal policy at the present stage. Yearly Journal of Scientific Articles. *Pravova Derzhava*, *32*, 496–505. DOI: 10.33663/0869-2491-2021-32-496-505

Lee, A. (2021). In the Shadow of Platforms. M/C Journal, 24(2). DOI: 10.5204/mcj.2750

Maryam, N. S. (2016). Mewujudkan good governance melalui pelayanan publik. Jurnal ilmu politik dan komunikasi, 6(1).

Muhammad, M., & Husen, L. O. (2019). State Civil Apparatus in Indonesia in the Conception of Welfare State: A Study of Legal Material Law Number 5 Year 2014 on State Civil Apparatus. *Asian Social Science*, *15*(3), 64. DOI: 10.5539/ass.v15n3p64

Mumpuni, N. W. R., & Kusumawati, M. P. (2021). Good governance pengelolaan sampah: Komitmen negara terhadap lingkungan hidup. *Jurnal Cakrawala Hukum*, *12*(1), 79–89. DOI: 10.26905/idjch.v12i1.5725

Nugroho, S. S., & Rahardjo, M. (2023). Penerapan Fungsi Pengawasan Dprd Dalam Mewujudkan Pemerintahan Daerah Yang Baik. *Yustisia Merdeka : Jurnal Ilmiah Hukum*, *9*(1), 59–67. DOI: 10.33319/yume.v9i1.218

Priyo Sembodo, G., Nefi, A., & Fakhriah, E. L. (2022). Urgensi Penerapan Prinsip Business Judgment Rule Dalam Peraturan Pemerintah Nomor 63 Tahun 2019 Tentang Investasi Pemerintah. *Jurnal Poros Hukum Padjadjaran*, *3*(2), 185–208. DOI: 10.23920/jphp.v3i2.789

Rao, S. S. (2021). Police: A Public Safety Institution of Governance. *International Journal for Research in Applied Science and Engineering Technology*, *9*(10), 1351–1353. DOI: 10.22214/ijraset.2021.38593

Russo, O., R. (. (2023). Environmental Rights are a Human Right to a Healthy Environment: A Brief Review. *International Journal of Life Science and Agriculture Research*, *03*(12). Advance online publication. DOI: 10.55677/ijlsar/V02I12Y2023-01

Sakti Koeswanto, E., Riswandi, R., & Redi, A. (2023). Implications of Public Trust Due to Weak Law Enforcement Morality. *Edunity : Kajian Ilmu Sosial Dan Pendidikan*, *2*(1), 78–86. DOI: 10.57096/edunity.v1i05.39

Saputra, E. F., & Firmansyah, H. (2023). Politik Hukum dalam Upaya Pemberantasan Tindak Pidana Korupsi melalui Pembaharuan Pengaturan Tindak Pidana Korupsi sebagai Extraordinary Crime dalam KUHP Nasional. 6(2), 4493–4504.

Serohin, V. (2022). Європейське управління як інститут конституційного права Європейського Союзу. Вісник Харківського Національного Університету Імені В. Н. Каразіна. Серія. *Право*, *33*, 28–43. DOI: 10.26565/2075-1834-2022-33-03

Shaheen, Y., Zhiqiang, X., & Zhiqiang, X. (2019). Public Administration in Afghanistan: Challenges and Way Forward. *Open Journal of Social Sciences*, *07*(06), 142–160. DOI: 10.4236/jss.2019.76012

Siljanovska, Z., & Shalevska, E. (2019). Oecd Principles For Good Corporate Governance And Their Impact On The Process Of Standardization. *Knowledge International Journal*, *34*(5), 1491–1497. DOI: 10.35120/kij34051491S

Susanti, A. (2022). Implementation Of The Principles Of Good Governance In Improving Public Services At The Investment Office And One-Stop Integrated Services Of Central Sulawesi Province. *Indonesian Journal of Multidisciplinary Science*, *1*(12), 1598–1611. DOI: 10.55324/ijoms.v1i12.243

Szadok-Bratuń, A. (2021). Fullerowski paradygmat (nie)dobrego prawa i jego aktualność „hic et nunc. *Studia Nad Autorytaryzmem i Totalitaryzmem*, *43*(2), 295–316. DOI: 10.19195/2300-7249.43.2.19

Tima, T. A., Schneider, P., Chanda, S. K., Mozumder, M. M. H., Hossain, M. M., Begum, A., & Shamsuzzaman, M. (2021). Analyses Implementation Realities of Legal Frameworks for Sustainable Management of Tanguar Haor Fisheries Resources in Bangladesh. *Sustainability (Basel)*, *13*(16), 8784. DOI: 10.3390/su13168784

Triwahyuni, A. D. (2020). Kewajiban Dan Tanggung Jawab Notaris Sebagai Pejabat Umum. *Acta Comitas*, *5*(1), 1. DOI: 10.24843/AC.2020.v05.i01.p01

Trofimova, L. (2020). Problematic points of the quality of qualification assessment of judges. *Law Review of Kyiv University of Law*, *2*(2), 198–207. DOI: 10.36695/2219-5521.2.2020.35

Tursin, S. R., & Ahmad, R. (2000). Konsep Supremasi Hukum, Masyarakat Madani Dan Kaitan Keduanya. 66–72.

Valerio, A., Parton, B., & Robb, A. (2014). Entrepreneurship Education and Training Programs around the World: Dimensions for Success. In Entrepreneurship Education and Training Programs around the World: Dimensions for Success. books.google.com. DOI: 10.1596/978-1-4648-0202-7

Vallejo, R. (2021)... *The Private Administrative Law of Technical Standardization.*, *40*(1), 172–229.

Witkowska, M., & Kucheryavaya, E. (2021). The Concept Of Good Governance In The Eu Legislation: Analysis In The Context Of Methodology Of The Civil Law Research. *Methodological Problems Of The Civil Law Researches*, *3*(3), 139–162. DOI: 10.33397/2619-0559-2021-3-3-139-162

Zdyb, M. (2019). Prawo do dobrej administracji. Aksjologiczne aspekty dobrej administracji. *Studia Iuridica Lublinensia*, *28*(2), 107. DOI: 10.17951/sil.2019.28.2.107-133

KEY TERMS AND DEFINITIONS

Governance: Governance refers to the processes, structures, and mechanisms through which power and authority are exercised, decisions are made, and stakeholders are held accountable. It involves the systems that manage resources and ensure order and effectiveness within organizations, governments, or communities. Governance emphasizes transparency, participation, and the rule of law in guiding collective action and policy.

Good Governance: The concept refers to the effective and efficient management of resources and affairs in a manner that is transparent, accountable, participatory, and responsive to the needs of society, ensuring equity and the rule of law.

Rule of Law: A principle under which all members of a society, including the government, are equally subject to publicly disclosed legal codes and processes, ensuring fairness, justice, and the protection of rights.

Better Governance: The continuous improvement of governance practices to achieve higher efficiency, accountability, and effectiveness in service delivery and decision-making, emphasizing citizen-centric and results-oriented approaches.

World Governance Indicator (WGI): A composite measure developed by the World Bank, assessing governance quality in countries through six dimensions: voice and accountability, political stability, government effectiveness, regulatory quality, rule of law, and control of corruption.

Chapter 20
Implementing Sustainability Into Higher Education:
Challenges and Prospects

Chahat Sahani
https://orcid.org/0009-0008-1771-6095
Graphic Era University, India

Navneet Rawat
Graphic Era University, India

ABSTRACT

Higher education has a big influence on sustainability and supports international initiatives to advance sustainable development and the digital revolution. Higher education's digital revolution has a big impact on sustainability since it supports international initiatives to advance sustainable development. Despite the significant regional differences in the SDGs' implementation, it is evident that universities play a critical role in fostering an attitude that makes the concepts of the SDGs easier to spread. This study examines the challenges and benefits of integrating sustainability into higher education. Higher education is crucial to achieving the SDGs, particularly Goal 1 (end poverty in all of its manifestations), Goal 3 (guarantee healthy lifestyles and promote well-being for all at all ages), and Goal 5 (gender equality). Objectives 8 (economic expansion and skilled labor), 12 (conscientious production and consumption), 13 (climate change), and 16 (peace, justice, and robust institutions).

DOI: 10.4018/979-8-3693-9909-5.ch020

INTRODUCTION

The purpose of this study is to review the body of literature on sustainability and sustainable development. In this study, the challenges and benefits of integrating sustainability into higher education are explored. To do this, the study reviews the literature on sustainability and sustainable development. Sustainability and sustainable development are becoming more and more important in the scholarly and policy literature. For more than 20 years, these two ideas have dominated the field of international development policy. Within higher education institutions (HEIs) in India there is vital importance of sustainability in the education sector because these institutions develop sustainable practices to maintain long-term operational effectiveness. (Tiwari, 2024) Sustainability in higher education means integrating environmental, social, and economic concerns into the management of university infrastructure, curriculum and community engagement. It is necessary to have a strong commitment in order to drive the sustainable initiatives and to steer actions and investments towards the sustainability goals. (Hutami et al., 2024) Universities can incorporate sustainability into higher education by integrating sustainable development goals into their governance systems and processes when implementing it. It also stresses the need for universities to become the agents of change, to act as the 'living labs' and the connectors with the civil society. Incorporating sustainability, within education entails blending concerns with social and economic aspects in university functions like infrastructure development and academic programs while actively engaging with the community for a well rounded approach to sustainability efforts is vital, for driving meaningful change and progress. (Hutami et al 2024). The article talks about how universities can promote sustainability, in education by incorporating development goals (SDGs) into their administrative structures and operations. Written by (Supaat et al., 2024) it emphasizes the importance of universities playing a role by acting as spaces and fostering connections, with the community. Introducing sustainability, into education involves six elements. Blending it into the curriculum; fostering research and innovation; managing campus operations efficiently; engaging with the community; establishing leadership and governance; and ensuring assessment and reporting mechanisms are in place. Each of these components is backed by strategies for implementation that draw from a review of existing literature and real world examples of successful sustainability projects. The main goal of this framework is to propel sustainability initiatives within institutions while also encouraging development in the wider community by offering guidance to institutions as well, as policymakers and stakeholders (Alenezi & Alanazi 2024). Implement sustainability, in universities through a strategy that includes integrating sustainability into all aspects of campus life such as policies and practices and incorporating it into programs and research activities to address challenges such as lack of awareness and resistance to change requiring collaborative research efforts and innovative teaching methods for a sustainable future focusing on leadership in driving change in a sustainable manner embracing interdisciplinary collaboration and leveraging research centers and funding to overcome sustainability obstacles, in higher education ensuring sustainability is deeply ingrained within the university environment for long term positive impact adapted from (Aghajani et al., 2024). Bringing sustainability practices into education requires incorporating friendly management strategies and supply chain practices while promoting a culture that values sustainability, within the organization. The research emphasizes that these aspects play a role in enhancing the sustainability efforts of institutions; particularly green innovation acts as a key intermediary factor. By embracing these approaches in their operations and policies private universities in Jakarta can boost their sustainability initiatives. (Sutawaidjaya et al., 2024).

Higher Education Institutions Affecting Sustainability

Higher education institutions are also more likely to embrace sustainability by adopting GHRM practices. This shift is due to the need to respond to environmental issues and to improve economic, environmental, and social performance. The implementation of EMS is enabled by the KAM practices which in turn ensure that the organizational goals are in sync with the stakeholder interests. Hence, these institutions are vital in encouraging sustainable practices that will be beneficial to the organization and the society at large. (Singh & Sihmar, 2024). The focus of Higher Educational Institutions (HEIs) on sustainability initiatives can be enhanced by focusing on environmental sustainability on campus. However, most of these initiatives have been directed at the environmental sustainability pillar while the social and economic sustainability pillars have not been well addressed. The study establishes the challenges that HEIs encounter in embracing sustainability and proposes a framework for an improved role as agents of sustainable development to achieve a more integrated approach to sustainability education and practices. (Shetty, 2024). The set up of Higher Education Institutions (HEIs) in declining resource dependent towns they can enhance the socio-economic sustainability. HEIs create employment, business and infrastructure development. Nevertheless, for long-term sustainability, it is necessary to expand the economic base, promote innovation, and cooperate in the development process. This four pillar approach can be used to transform these towns into vibrant urban centers and reduce the effects of resource depletion. (Chadzamira & Chigara, 2024). The Engineering Sustainability Competency Map was developed to help integrate sustainability into engineering curricula, which is a critical role that higher education institutions play in promoting sustainability. The paper details 29 essential learning outcomes in this area, which the authors argue should be part of all engineering degrees. This map links these outcomes to the Sustainable Development Goals adopted by the United Nations in 2015 and the learning objectives defined by UNESCO in 2017 to ensure a more holistic approach to sustainability education in higher education. (Papell et al., 2024). From concrete actions, sustainability can be affected greatly in Higher education institutions in India through reform of degree programs, curricula and assessment methods to incorporate sustainability principles. They should also encourage energy conservation, minimize material use and engage in remote learning and digitalization. Furthermore, universities can improve their sustainability agenda by developing their own 'carbon brain print' and aligning with international sustainability standards such as the Sustainable Development Goals of the United Nations to be role models in the post-COVID period. (Nagi, 2023). HEIs are key players in encouraging sustainability by raising students' and the community's awareness and concern. They engage in sustainable curricula and activities, developing new generations of problem solvers for environmental issues. However, the study shows that there is a wide gap between the HEIs' awareness and commitment towards sustainability and the actual implementation of sustainable practices that are pro-sustainability. Recommendations include migrating to the cloud, using green IT for website hosting and optimizing website code to close this gap effectively. (Kumbhojkar & Raman, 2024).

Universities and other Higher Education Institutions (HEIs) should lead the way in sustainability, making it a driver of change. They respond to external influencing factors and apply sustainable principles to their functions. In addition, HEIs can further broaden their influence beyond the campus by engaging with communities and other organizations in order to increase the efforts towards sustainability and the SDGs. This dual approach allows the HEIs to act as agents of change organically as well as systematically. (The Bloomsburry Handbook of Sustainability 2023)

The Role of Higher Education in Supporting the Achievement of the SDG'S Implementation

Higher education institutions (HEIs) support the Sustainable Development Goals (SDGs) through strategic initiatives in teaching, research, and engagement. They incorporate the SDGs into the curricula, support SDG-related research, and participate in the SDGs awareness and achievement with internal and external stakeholders. These actions show that the organization is ready to change and develop good action plans for sustainable development, which in turn create a sustainable culture in the educational environment. (Fuchs et al., 2024). Universities are also contributing to the Sustainability by achieving Sustainability through teaching and research by incorporating sustainability measures into them. The systematic review indicates that universities have to work actively towards the SDGs in order to end poverty, address inequality and protect the planet. However, the research shows that there is limited research on the SDG implementation in universities which may be due to the pandemic, low demand, and non-compliance with the international standards which mandate organizations to include SDG integration in their strategies. (Carpio et al., 2024). Research indicates that for development (SD) to progress effectively beyond 2015 and achieve the Development Goals (SDGs) it is crucial, for higher education institutions (HEIs) to actively promote SD initiatives. The key focus areas include technology integration with sustainability practices and holistic education approaches along with business strategies. The survey highlights HEIs and researchers making contributions, in sustainability focused domains. With the possibilities of collaboration, in mind and drawing insights from the study by (Filho 2024) universities can develop strategies to progress in sustainable development research and achieve the SD goals effectively. Higher education institutions play a role in tackling the Sustainable Development Goals (SDGs). They engage with these goals through governance structures and activities such as research and teaching while also acting as a bridge, between knowledge and societal needs. The main focus areas include technology integration and sustainable practices, in education and business methods are topics in this study area; the survey acknowledges leading education institutions (HEIs) and researchers involved in sustainability related studies. The university can develop strategies, for advancing development research and meeting development goals by leveraging the potential of interdisciplinary collaboration (Filho et al., 2024). Higher education institutions (HEIs), in their role as facilitators of the Sustainable Development Goals (SDGs) contribute to development through governance practices and academic engagement that connect with communities and promote research and teaching initiatives in line, with the goals set post 2015 for development. In the realm of areas of study lie technology advancements and sustainable practices intertwined with approaches and business strategies as highlighted in the survey that acknowledges leading higher education institutions (HEIs) and researchers, in sustainability focused fields. Drawing upon the potential of efforts across disciplines paves the way for universities to develop strategic initiatives for advancing research in sustainable development (SD). This in turn plays a role in achieving success, in meeting SD objectives as emphasized by (Filho et al., 2024). Higher education institutions play a role in advancing the Sustainable Development Goals (SDGs). They contribute to development through their governance structures and engagement, in research and teaching activities that connect academia with the community. Research underscores the importance of education institutions in supporting development as a crucial element, for achieving the post 2015 SDGs framework. The university focuses its areas of study, around technology and sustainability assessment as integrative education and business practices while acknowledging the top higher education institutions (HEIs) and researchers in sustainability related fields through a survey report highlighting their achievements. Emphasizing the potential of collaboration allows the university to develop strategies, for advancing sustainable development (SD) research and achieving the SD goals effectively in the future. They serve as catalysts, for driving development objectives both internationally in this context of the report universities are depicted as "living laboratories," partnering, with civil society to exchange their knowledge in sustainable development. Henceforth educational institutions have the power to enact change and encourage practices within communities by incorporating the Sustainable Development Goals (SDGs) into their structures (Supaat et al., 2024). Studies contribute to the discussion, about how educational programs in universities can contribute to achieving the Sustainable Development Goals (SDGs). The research highlights the connection between SDGs 4 (Quality Education) SDGs 17 (Partnerships for the Goals) and the curriculum provided by a university. Additionally the report depicts universities as participants in development efforts, by acting as "living labs," where they engage with civil society to exchange knowledge and expertise. Colleges have the potential to bring about change and encourage practices in society by incorporating the Sustainable Development Goals (SDGs).

The Role of Effective Communication of Sustainability Issues

Sustainability issues should be properly disclosed to the consumers in order to make them buy products that are environmentally friendly. This ensures that the information is verifiable from other organizations and this avoids the misrepresentation known as green washing that damages the consumer's confidence. The study also shows that social information provided by companies on their sustainable actions has a positive effect on consumers' purchase intentions, particularly with regard to pollution and packaging recycling. Therefore, communication is vital in capturing the attention of consumers and making them want to purchase products that are more sustainable. (Bruno, 2024). Therefore, the effective communication of sustainability issues is important as it determines the level of public understanding and participation. In 2024 Lakhan emphasized the importance of sustainability communication, in raising awareness about the impact of activities on the environment and promoting initiatives such as energy conservation and recycling to encourage an optimistic approach, to environmental preservation. This means that effective communication is crucial in the sense of involving and persuading the different interested parties, and thus the creation of a sustainability culture within organizations. The research also reveals that SC is based on theoretical underpinnings such as CSR and ethical consumerism. In this case, companies can be seen to be supporting behavior that is friendly to the sustainability agenda, get better consumer response and in the end contribute to a better sustainability environment. This knowledge helps the marketers to come up with good SC strategies that will be effective in capturing the target audience's attention. (Braga et al., 2024). This means that sustainability communication is important in creating accountability and accountability in organizations. It means that both the good and the bad aspects of sustainability are addressed to build trust. Today's communication strategies are not very effective, especially in the environment of scientific skepticism, which requires specific plans to address various target audiences. Honesty and inconvenient truth are important tools that communicators can use in order to challenge misinformation, influence people's behavior and thus contribute to the sustainable development of the society. (Voci & Karmasin, 2023). This means that communication of sustainability issues is important for establishing credibility and placing the companies within their specific sectors as socially responsible organizations. It encourages the involvement of various stakeholders, and also enhances the organization's accountability and accountability to social responsibilities. In this case, organizations can achieve their goals by supporting social causes and at the same time, gain customers' trust and defend their methods. This approach not only supports the sustainable development and the achievement of long-term and credible effects in the message in a global context. ("Effective Service Management, Organizational Processes and Objectives through Responsible Communication", 2023). This means that sustainability communication is important because it is used to translate existing data into useful information that can be used in decision making. With the help of different communication models, the stakeholders can be able to solve some problems and also have a clear understanding of the sustainability concepts. This guarantees that the right message gets to the right people and, in turn, gets them to act and embrace sustainable practices. At the end of the day, sustainability communication is a very important part of shaping the society's consciousness and action towards the environment and the future. (Cetinic, 2023)

Sustainability Implementation Challenges and Barriers in Higher Education

Some of the issues encountered in the implementation of sustainability in HEIs are; lack of information and awareness of the stakeholders and resistance to change. The study also recommends the need to adopt a holistic approach in order to overcome these challenges as recommended by the study, more so through the use of interdisciplinary research. However, there is still a problem with the integration of sustainability due to the absence of new pedagogical approaches and insufficient leadership for sustainable development. (Aghajani and others, 2024). The survey reveals that the major challenges to sustainability integration in the EU HEIs are financial and awareness deficits. Despite the fact that countries with higher financial constraints have lower knowledge constraints, countries with low scores in both categories are almost equally constrained. This classification helps to understand the specific problems of HEIs in relation to the SDGs and informs the ways to overcome them. It also gives information on the socio-economic characteristics and management structures of universities. (Albanese et al.,2024) The challenges and barriers to sustainability integration in higher education are not well addressed in this study. For the purpose of enhancing the environmental performance, the higher education institutions should adopt sustainable practices such as GHRM as the author notes. Rather, the benefits of GHRM are emphasized instead of the challenges that have been faced. Therefore, although there may be some problems, the abstract of the article does not contain them. (Sihmar and Singh, 2024). The present study also reveals some challenges and barriers to sustainability integration in higher education, including different levels of institutional commitment, lack of integration, and low faculty participation. It also focuses on the international projects like the Impact Ranking as a threat or opportunity and the need for integrated educational systems. The study also concludes that these barriers must be removed for academic institutions across the world to develop sustainable curricula. (Abo-Khalil, 2024) The research also suggests that future work should examine the ability of the assessment tools to bring about real sustainability outcomes, suggesting that current tools may not adequately meet the needs of institutions around the world. The following potential challenges and/or weaknesses are identified in the implementation and/or measurement of sustainability within the context of higher education institutions (HEIs): A major weakness that needs to be addressed in order to integrate sustainability development goals (SDGs) into all departments is the absence of suitable, cross-institutional evaluation tools that are specific to the different sustainability dimensions (Ahmed et al., 2024). The following issues were noted: opposition to change, lack of resources, and apathy of students. These barriers can be overcome through strategic planning and change in the institutional culture. However, there are curricular integration challenges that require institutional leadership and support to achieve. There is a need for top-level leadership and the inclusion of students in decision-making processes to address these challenges and make good sustainability practices a reality. (Park, 2024). As for the challenges and barriers to sustainability integration in the higher education institutions (HEIs), these were identified as: Lack of adequate knowledge on sustainability, absence of integration of social and economic dimensions with environmental management, and resistance to change in institutional structures. Furthermore, the HEIs are likely to have limited resources and lack of support from their stakeholders to implement sustainability strategies effectively. To this end, these challenges must be met to enable the HEIs to carry out their role as drivers of sustainable development. (Shetty, 2024). In its current form, higher education is more inclined towards the technological and capitalist development, which may sometimes ignore the ethical aspects and the common welfare. Some issues include the increasing commercialization of campuses, limited interdisciplinary sustainability education beyond the area of study, and low engagement with the Sustainable Development Goals (SDGs). However, more importantly, there is a need for high-impact practices that can transform learning into action and a collaborative work between different institutions to tackle the multifaceted societal issues effectively.(Husic,2024).Lack of public interest, shortage of funding, absence of management support, and the difficulty of making it sustainable are the main challenges to its implementation in Higher Education Institutions (HEIs). These challenges make sustainability programs and the greater efforts to promote sustainable development in HEIs difficult. For HEIs to continue to contribute to the achievement of their educational goals and to an expanded role in sustainability, these obstacles must be removed. (Alhazmi, 2023). The current models of higher education do not effectively address sustainability issues and challenges. However, it stresses the significance of education and application in developing sustainable communities. Despite the emphasis on the positives of enhancing student's sustainability awareness and participation, it fails to consider the particular problems of higher education institutions. Hence, there is a need for further inquiry to fully understand such challenges. In (2023, Biancardi et al.) The current models of higher education do not effectively address sustainability issues and challenges. However, it stresses the significance of education and practical projects in creating sustainable communities. Despite the emphasis on the positives of increasing students' understanding and engagement with sustainability, it fails to consider the particular problems that higher education institutions face. Hence, there is a need for further inquiry to fully understand such challenges.

THE REVIEW OF THE LITERATURE

Higher Education Sustainability

As the literature review shows, higher education institutions (HEIs) have to engage in sustainable practices in their operations. It outlines how the Sustainable Development Goals (SDGs) and the UN 2030 Agenda act as reference for sustainability actions in HEIs and as a reference to balance development. Also, it examines the impact of properly designed sustainability policies in the area of higher education on community members' sustainable behavior. As it was established during the review of the literature, the number of papers published on sustainable food systems (SFS) education in higher education institutions (HEIs) is growing every year and the policies are suggested to cover different topics such as waste management, environmental sustainability of transportation and sustainable procurement to create a sustainability culture in the university community (Vikram & Tiwari, 2024). As for the universities, they are aware of the challenges that are involved with the food system and the need to address them. The review also focuses on the need to develop SFS education in so far as transdisciplinary and multidisciplinary approaches, stakeholder involvement, and student development for future citizens and professionals are concerned. Moreover, since SFS issues are ethically complex, it emphasizes the importance of both applied and theoretical work on their ethics (Salminen et al., 2024). The paper aims at exploring the following challenges that Nepalese HEIs encounter in the implementation of sustainability strategies: Lack of a changing vision and values that are needed to form sustainable behavior, The socio-economic and cultural differences between the policy makers and the implementers. It identifies the following as the critical pre-requisites for the success of sustainability initiative implementation: Leadership, Learning organization, Sustainability strategies, Human resource management, Organizational structure, Technology, Reward systems and University industry and academic relations (UIA) (Shrestha, 2024). The literature review reveals that sustainability is becoming more vital in education and that institutions are required to address the new and emerging organizational perspective that embraces environmental sustainability alongside economic and social values. This paper aims to contribute to the debate on the implementation of GHRM as a strategic approach to link environmental management with HRM in order to develop an EMS in universities. (Singh & Sihmar, 2024). It identifies the major dimensions of sustainable development of the HEIs: Education, research and campus development and presents good practices of the universities in Europe and the role of the HEIs in educating the future leaders for sustainable development (Popowska, 2024).

Higher Education Institutions Effect on Sustainability

The review of the literature discusses the growing focus on the sustainable performance of the HEIs after the SDGs were adopted in 2017, with an average annual growth rate of publications in this field. Different aspects of sustainable curriculum, teaching, campuses and assessment tools in relation to sustainability in the HEIs have been highlighted by researchers. From the review, there is a gap in the literature in terms of representation of Asian countries; the publications are mainly made in the advanced European countries and the United States. It also suggested that future research should focus on the unexplored areas of sustainability performance, green knowledge sharing and management practices in the HEIs. (Umar et al., 2024). For this study, the systematic literature review and evaluation were conducted on the articles that were published between 2015 and 2021, on the topic of sustainable educational technology in higher education institutions (HEI) and the sustainable factors that affect the challenges in online learning particularly during the recent pandemic. The review was conducted on 44 articles from five journal databases, and the four main themes that emerged from the analysis are technology related, teaching and learning, ethical issues, and instructional support and 12 subthemes, including the proposed conceptual framework and recommendations for future research (Amir et al., 2024). In this review, the concept of sustainability and sustainable leadership of higher education teachers is defined within the context of the reviewed literature and frameworks such as the Triple Bottom Line and the Principles for Responsible Management Education. To this end, it seeks to arrive at a theoretical correlation model between the sustainability of the higher education institutions and the sustainable leadership of the higher education teachers in order to inform the sustainable development of higher education and to provide a practical direction. (Zhang et al., 2024). In this study findings, on information security policies in higher education institutions reveal a lack of measures in place to address the challenges and trends related to policy compliance in settings. The analysis involved a review of 50 research papers published from 2013 to 2023 aimed at identifying the factors that influence compliance behavior. Three significant factors emerged from the review as influencers; 1) Awareness and cultural aspects 2) intentions, 3) The dynamics, between organizational structures and human resource management. Through analyzing and summarizing the research conducted in the decade in a manner this thorough review adds to the existing pool of information, on information security (Ahmed & Abas 2024). The review of literature highlights the increasing significance of sustainability, in education. Underscores the necessity for institutions to adapt to the evolving landscape that embraces social, economic and environmental values. In universities today the implementation of an Environmental Management System (EMS) is enabled through the approach of Green Human Resource Management (GHMR) which establishes a link, between human resource management practices as discussed in "The Handbook of Sustainability" by Bloomsbury, in 2023.The literature review also shows that HEIs play a crucial role in shaping social change and handling environmental issues, and that they should play the role of raising awareness and encouraging sustainable behavior among students. It points out that previous studies have been carried out on particular regions or certain categories of learners thus there is a need for a broad empirical analysis across a number of different HEIs. The analysis highlights a gap, in universities' dedication to sustainability and their level of consciousness well as in their tangible implementation of sustainable strategies. It argues that despite universities advocating for innovation and knowledge enhancement these efforts are not consistently reflected in sustainability actions indicating a need, for further investigation and recommendations. Referenced from (Raman and Kumbhojkars, 2024).

The Role of Higher Education in the Achievement of the SDG's

This review of the literature entailed an examination of the literature and documentation of SD in HEIs with a view to identifying how these institutions can ensure that SD is incorporated into their teaching, research and outreach activities. This review was intended to reveal the best practices and approaches that have been reported in the literature. - The review also helped to reveal the current state of sustainability actions within the QualEnv Consortium HEIs, identifying missing knowledge and areas for improvement, which in turn helped to create action plans for improving the effectiveness of SD activities in these institutions. The systematic literature review was conducted on the implementation of the 17 Sustainable Development Goals (SDGs) in higher education institutions; the articles were published between 2015 and 2023. The review employed a search strategy that incorporated terms such as academic institutions and the SDGs, and the resulting search string was used to gather relevant literature. - The results showed that research on the achievement of SDGs in universities is scarce and the possible explanations for this include the impact of the pandemic, low demand for such research, and lack of international policies that are supposed to enforce the achievement of the SDGs.

METHOD

In the research, secondary sources are used to synthesize current information and to extract important insights from it, and the research design is exploratory and descriptive. A qualitative analysis provides a thorough understanding of the opportunities and challenges in sustaining higher education because it is a dynamic concept. In the course of a systematic review of the literature, journal articles, conference proceedings and institutional reports were examined as part of relevant academic sources. Among the literature, the following were used to select the literature: Databanks from reputable databases such as Google Scholar, Web of Science and Scopus. To ensure their applicability, last ten to fifteen years peer reviewed papers are used. Those that focus on frameworks, integration strategies and sustainability policies in higher education. Evaluations of empirical research on sustainability implementation challenges and success factors. Fifty to sixty research publications were reviewed and the results of those assessments were used to identify important themes. In this study, thematic analysis was employed to systematically analyze the collected data.

The essential actions were;

Coding & Classification: In this case, the persistent themes, such as policy level issues, faculty and student participation, curricular integration, infrastructure, and institutional support were used to categorize relevant studies.

Challenges and Opportunities: Recognized were the opportunities that existed to help a more seamless transition as well as the barriers to the integration of sustainability in higher education.

Comparative Analysis: To identify the best practices and practical insights, the study analyzed several case studies and frameworks.

This study is based on the systematic literature review which provides a comprehensive analysis of the challenges and opportunities of integrating sustainability in higher education. The results of the study are important in that they offer practical implications for educators, policymakers, and institutions who are interested in integrating sustainability into their education systems and curricula, and they contribute to the existing scholarly discussions on sustainable education.

CONCLUSION

Several studies have been carried out to investigate the advantages and disadvantages of integrating sustainability into higher education. Sustainability is crucial to incorporate to ensure that the future generation is well-equipped to deal with world challenges despite all the challenges that come with it. Among these include the connection of educational frameworks with sustainability goals, effective leadership, and a shift in institutional culture. However, there are also many ways to enhance the effective application of sustainable practices to enhance institutional performance and socictal impact. Culture and Vision Related Barriers: This is a problem for many HEIs, the absence of cultural readiness and vision to implement sustainability strategies. This is compounded by socio-economic and cultural differences in the formulation and implementation of policies (Shrestha, 2024). Interdisciplinary Integration: (Abo-Khalil, 2024) stated that although interdisciplinary approaches are necessary and important, they are difficult to implement in practice because they involve a transformation of the traditional educational strategies and the active engagement of the faculty. Resource Allocation: As sustainability entails the incorporation of costly features such as buildings, equipment, and people, many organizations may find it difficult to implement (Shrestha, 2024). Improved Institutional Outcomes: Previous studies have revealed that sustainable marketing and management are beneficial to HEIs as they lead to the creation of value and the support of long-term social and environmental objectives (Mestrovic et al., 2024). Techniques and Structures: (Alenezi and Alanazi,2024) have pointed out that the development of comprehensive frameworks such as those that incorporate Environmental, Social, and Governance (ESG) principles offers organizations valuable guidance on how to move their sustainability initiatives forward. International and Local Initiatives: The following case studies from the United Arab Emirates show that sustainability plans can be linked with international initiatives like the United Nations Sustainable Development Goals to increase the effectiveness of sustainability education (Abo-Khalil, 2024). Although there are many difficulties in integrating sustainability into higher education, there are also many possibilities for enhancing institutional and societal development. Through the implementation of strategic frameworks and a commitment to ongoing development, HEIs may overcome these challenges and help students to develop a sustainability mindset that will enable them to contribute to the solution of global challenges. Universities and other higher education institutions have to contribute to the achievement of the ambitious SDG targets and to raising sustainability awareness among the next generation. Another very important factor that determines the position and reputation of a university in the world is sustainability. Higher education is one of the "changing agents" for sustainability development as it forms the perceptions of adults. However, the geography of attitude formation is not uniform and not everyone has an equal opportunity to get higher education. However, it can be seen that there are several factors that are associated with education, including, on the one hand, low mortality rates for children under five, low pollution, and low employment vulnerability, and on the other hand, high access to clean energy, women's participation in higher education, and external commerce.

SCOPE OF FURTHER STUDY

The last studies have shown that there are several strengths and weaknesses of implementing sustainability in the higher education system. This means that HEIs have to embrace sustainability to contribute to the attainment of sustainable development and address global environmental challenges. However, there are several concerns that have been raised regarding this integration. These challenges can be described as: ignorance and opposition to change. For these barriers to be eliminated and sustainability to be effectively integrated into HEIs, the following recommendations for future research are proposed: developing transdisciplinary studies, innovative teaching strategies, and holistic approaches.

REFERENCES

Abo-Khalil, A. G. (2024). Integrating sustainability into higher education: Challenges and opportunities for universities worldwide. *Heliyon, 10*(9), e29946. Advance online publication. DOI: 10.1016/j.heliyon.2024.e29946 PMID: 38707336

Aghajani, M., Memari, A., Tumpa, R. J., & Ruge, G. (2024). Systematic exploration of sustainability in higher education: A tertiary perspective. *International Journal of Sustainability in Higher Education*. Advance online publication. DOI: 10.1108/IJSHE-02-2024-0095

Agrawal, S., Ray, A., & Rana, N. P. (2024). Effectiveness of sustainability communication on stakeholder engagement: A multi-method qualitative assessment of sustainable development. *Corporate Social Responsibility and Environmental Management*. Advance online publication. DOI: 10.1002/csr.2992

Ahmed, A. A., & Abas, H. (2024). Factors influencing information security policy compliance behavior in higher education institutions: Systematic literature review. *Advances in Social Sciences Research Journal, 11*(7), 260–273. DOI: 10.14738/assrj.117.17308

Ahmed, V., Bahroun, Z., & Anane, C. (2024). Exploring sustainability assessment practices in higher education: A comprehensive review through content and bibliometric analyses. *Sustainability (Basel), 16*(13), 5799. DOI: 10.3390/su16135799

Albanese, M., Busato, F., Cisco, G., Di Giorno, S., & Grasso, C. (2024). Overcoming obstacles: Higher education institutions and the SDGs in the EU – A cluster analysis. *Educational Policy and Innovation Journal, 1*(1), 49–70. DOI: 10.35603/epi24.1.05

Alenezi, M., & Alanazi, F. (2024). Integrating environmental, social, and governance values into higher education curriculum. *International Journal of Evaluation and Research in Education, 13*(5), 3493. DOI: 10.11591/ijere.v13i5.29440

Alhazemi, A. (2024). *Transformative approaches to sustainable education: Technology, leadership and SDGs in higher education institutions*. International Journal of Learning, Teaching and Educational Research., DOI: 10.26803/ijlter.23.5.3

Alhazmi, A. K., Zain, A., Alsakkaf, N., & Othman, Y. (2023). Identification of sustainability barriers in higher education institutions (HEIs) and the role of technology in improving sustainability in HEIs. *T p Chí Khoa H c và Công Ngh - Đ i H c Nguy n T t Thành*.https://doi.org/DOI: 10.20428/jst.v28i1.2126

Amir, A. F., Syed Mohamed, A. T. F., & Juhary, J. (2024). Higher education sustainable online education for institutions: A systematic literature review. *Asia Pacific Journal of Educators and Education, 39*(1), 45–91. DOI: 10.21315/apjee2024.39.1.3

Armas, K. L., & Jugo, R. R. (2024). Education for a sustainable future: Analysis of universities' vision and mission statements in line with global sustainable development goals. *Journal of Lifestyle and SDGs Review, 4*(1), e01790. DOI: 10.47172/2965-730X.SDGsReview.v4.n00.pe01790

Bada, G. M., Adedayo, O., & Olufemi, O. I. (2024). Fostering educational equity through leadership and policy innovations for sustainable development in higher education. *International Journal of Frontline Research in Multidisciplinary Studies, 4*(1), 35–44. DOI: 10.56355/ijfrms.2024.4.1.0031

Biancardi, A., Colasante, A., D'Adamo, I., Daraio, C., Gastaldi, M., & Uricchio, A. F. (2023). Strategies for developing sustainable communities in higher education institutions. *Scientific Reports, 13*(1), 20596. Advance online publication. DOI: 10.1038/s41598-023-48021-8 PMID: 37996505

Bonilla Jurado, D. M., Zumba, E., Lucio-Quintana, A., Yerbabuena-Torres, C., Ramírez-Casco, A., & Guevara, C. (2024). Advancing university education: Exploring the benefits of education for sustainable development. *Sustainability (Basel), 16*(17), 7847. DOI: 10.3390/su16177847

Braga, L. D., Tardin, M., Perin, M. G., & Boaventura, P. (2024). Sustainability communication in marketing: A literature review. *RAUSP Management Journal, 59*(3), 293–311. Advance online publication. DOI: 10.1108/RAUSP-10-2023-0205

Bruno, R. X. (2024). The crucial role of transparent communication in sustainability: Challenges and opportunities. *International Seven Multidisciplinary Journal, 3*(5), 1414–1422. Advance online publication. DOI: 10.56238/isevmjv3n5-007

Carpio, J. E. P., López-Gonzales, J. L., Cano, E. T., & Chaparro, J. E. T. (2024). Implementing the SDGs in higher education: A systematic review. *Academic Journal of Interdisciplinary Studies, 13*(5), 412. DOI: 10.36941/ajis-2024-0174

Cembranel, P., Luchi de Paulo Gewehr, L., Dal Moro, L., Fuchs, P. G., Birch, R., & Salgueirinho Osório de Andrade Guerra, J. B. (2024). The pivotal role of higher education institutions in cultivating a sustainable development goals-centric culture. *International Journal of Sustainability in Higher Education, 25*(7), 1385–1411. Advance online publication. DOI: 10.1108/IJSHE-01-2024-0057

Cetinic, E. (2023). Sustainability communication. In *Sustainability communication* (pp. 113–127). Springer., DOI: 10.1007/978-981-19-8237-8_7

Chadzamira, R., & Chigara, B. (2024). Higher education institutions and the socio-economic sustainability of "declining" towns. In Chadzamira, R., & Chigara, B. (Eds.), *Springer International Publishing* (pp. 127–142)., DOI: 10.1007/978-3-031-45568-1_7

Din, N. N. H. O. (2024). Revolutionizing higher education: Integrating global dynamics through SDG 3 and SDG 4 in higher education institutions. *Eduphoria*, *2*(3), 31–34. DOI: 10.59231/eduphoria/230411

Effective service management, organizational processes, and objectives through responsible communication. (2023). *International Journal of Language and Linguistics.*https://doi.org/DOI: 10.11648/j.ijll.20231104.11

Filho, W. L., Pires Eustachio, J. H., Ávila, L. V., Dinis, M. A. P., Hernandez-Diaz, P. M., Batista, K., Borsari, B., & Abubakar, I. R. (2024). Enhancing the contribution of higher education institutions to sustainable development research: A focus on post-2015 SDGs. *Sustainable Development.* Advance online publication. DOI: 10.1002/sd.3184

Fuchs, P., Aver, G., da Rocha, V.Junior, Cembranel, P., Leal Vieira Cubas, A., de Aguiar Dutra, A. R., & Salgueirinho Osório de Andrade Guerra, J. B. (2024). Strategies for promoting sustainable development in higher education institutions of the QualEnv Consortium: A focus on teaching, research, and outreach. *International Journal of Sustainability in Higher Education.* Advance online publication. DOI: 10.1108/IJSHE-11-2023-0531

Herrada-Lores, S., Palazón, M., Iniesta Bonillo, M. Á., & Estrella-Ramón, A. (2024). The communication of sustainability on social media: The role of dialogical communication. *Journal of Research in Interactive Marketing.* Advance online publication. DOI: 10.1108/JRIM-10-2023-0372

Husic, D. W. (2024). *Reframing sustainability initiatives in higher education.* Sustainable Earth., DOI: 10.1186/s42055-024-00076-9

Hutami, A. D., Qayim, I., Putra, H., Slamet, A. S., Mardiana, R., & Dwiyanti, F. G. (2024). Toward a vision of sustainable university: Linkages between commitment and practices. *Journal of Sustainability Perspectives*, *4*(1), 71–84. DOI: 10.14710/jsp.2024.24793

Jarrar, A. G. (2024). Towards building peaceful and inclusive societies for promoting sustainable development goals (SDGs) in higher education in the Arab region. *International Conference on Teaching and Learning Education (ICTLE).*https://doi.org/DOI: 10.33422/ictle.v1i1.612

Kumbhojkar, N., & Raman, R. (2024). A critical and empirical analysis of the relation between awareness, commitment, and pro-sustainability practices amongst higher education institutions. *IEEE Transactions on Quality, Control, and Engineering in Business and Technology*, *25*, 1–7. DOI: 10.1109/TQCEBT59414.2024.10545066

Lakhan, C. (2024). The language of sustainability: Framing in sustainability communication (Chapter 1). *OSF Preprints*.https://doi.org/DOI: 10.31235/osf.io/36pbn

Nagi, K. (2023). How can higher education institutions in India promote sustainability? *Social Science Research Network*.https://doi.org/DOI: 10.2139/ssrn.4616827

Nasution, M. S., & Sazali, H. (2024). *Peran komunikasi dalam membangun keberlanjutan dan ketahanan lingkungan. Religion, Education, and Social Laa Roiba Journal*. RESLAJ., DOI: 10.47467/reslaj.v6i9.2432

Papell, G. T., Carracedo, F. S., & Coral, J. S. (2024). Education for sustainable development in higher education institutions. In Papell, G. T., Carracedo, F. S., & Coral, J. S. (Eds.), *Springer International Publishing* (pp. 45–60)., DOI: 10.1007/978-3-031-49853-4_4

Park, Y. (2024). Sustainable education practices: Voices from higher education institutions. *Research Square*.https://doi.org/DOI: 10.21203/rs.3.rs-3923563/v1

Prior, D. D., Sandeep, M. S., Zhang, Y., & McCormack, M. (2024). Measuring sustainable development goals (SDGs) in higher education through semantic matching. *Studies in Higher Education*, •••, 1–14. DOI: 10.1080/03075079.2024.2386625

Sachdeva, H., & Agarwal, G. D. (2024). Holistic approach to the Indian higher education system in design and sustainability. *Journal of Operations and Strategic Planning*.https://doi.org/DOI: 10.1177/2516600X241284361

Salminen, J., Friman, M., Mikkonen, K., & Mutanen, A. (2024). Higher education of sustainable food systems: A literature review. *International Journal of Sustainability in Higher Education*, *25*(9), 358–377. DOI: 10.1108/IJSHE-02-2024-0079

Shetty, S. H. (2024). Higher education institutions as a catalyst for sustainability development. In Shetty, S. H. (Ed.), *IGI Global* (pp. 1–16)., DOI: 10.4018/978-1-6684-9859-0.ch001

Shrestha, P. (2024). Sustainability initiatives in higher education institutions: The stakeholder perspectives. *Journal of Applied Research in Higher Education*. Advance online publication. DOI: 10.1108/JARHE-03-2024-0141

Singh, A., & Sihmar, P. (2024). Higher education institution sustainability. *Advances in Higher Education and Professional Development Book Series*, 1–14. https://doi.org/DOI: 10.4018/979-8-3693-2956-6.ch001

Supaat, K., Aziz, S., Hasan, S. Z., Basri, N. E. A., & Shafie, S. H. (2024). Universities and sustainable development: Options to help 'walk the talk.'. *Journal of Ecohumanism*, *3*(8). Advance online publication. DOI: 10.62754/joe.v3i8.4751

Sutawaidjaya, A. H., Yusuf, M., & Hamdi, E. (2024). Building sustainability in higher education through green management and innovation: A case study of private universities in Jakarta. *Journal of Ecohumanism*, *3*(6), 1034–1047. DOI: 10.62754/joe.v3i6.4071

The Bloomsbury handbook of sustainability in higher education. (2023). Bloomsbury., DOI: 10.5040/9781350244375

Umar, S. B., Ahmad, J., Bukhori, M. A. B. M., Ali, K. A. M., & Hussain, W. M. H. W. (2024). A decade in review: Bibliometric analysis of sustainable performance trends in higher education institutes. *Frontiers in Education*, *9*, 1433525. Advance online publication. DOI: 10.3389/feduc.2024.1433525

Vikram, S., & Tiwari, M. (2024). Assessing the awareness level, knowledge, and attitudes of the university community for sustainable development in higher educational institutions in India—A review paper. *Journal of Infrastructure Policy*, *8*(12), 9329. DOI: 10.24294/jipd.v8i12.9329

Voci, D., & Karmasin, M. (2023). Sustainability communication: How to communicate an inconvenient truth in the era of scientific mistrust. *Journal of Communication Management (London)*. Advance online publication. DOI: 10.1108/JCOM-05-2022-0060

Zhang, G., Chen, P., & Xu, S. (2024). Sustainability of higher education institutions and sustainable leadership of higher education teachers: A literature review-based exploration. *Journal of Educational and Social Research*, *14*(3), 150. Advance online publication. DOI: 10.36941/jesr-2024-0063

About the Contributors

Durmuş Çağrı Yıldırım was born in 1981. After he graduated from Department of Economics in Kocaeli University in 2005, he started his master's education in Economics in the same year. He got the degree of Masters in Economic Policy in 2007. In 2011, he got his PhD degree in Economics. While he was a doctorate student, he studied one semester in Almeria University-Spain. Between 2006-2012, he worked as a research assistant at University of Kocaeli. He works at Tekirdag Namık Kemal University since April 2015. He studies on sustainable development, carbon emission, health and education issues, employment, innovation.

Seda Yıldırım currently works at Tekirdag Namık Kemal University in Turkey. She has a doctoral degree (PhD) in marketing. She studies on sustainable development, 2030 sustainable development goals, sustainable consumption patterns, consumption patterns, environmental quality and food security. She is married and has two children. She has over a thousand citations by Google scholar. She has many publications in journals indexed by Web of Science (SSCI, SCI-Exp, ESCI and others) and SCOPUS. She published several book chapters in IGI Global publishing. She won Emerald literati award-Outstanding Paper award 2022.

Vinay Kandpal (Post doc and Ph.D. in Finance) possesses over 18 years of experience in Academics and Industry. He has published more than 60 research papers in leading refereed Journals. Published 6 books in the area of Management. He has presented several works at International Conferences on Smart City at The University of Nice Sophia Antipolis, France, and Oxford Brookes University, UK. He has also presented research works in institutions like IIM A, IIM B, IIM I, IIT Delhi, etc. He is a Full-Time Member of ANAHEI Florida, USA. He works as a Professor in Department of Management Studies, Graphic Era Deemed to be University, Dehradun, Uttarakhand, India. He is on the Editorial and Reviewer Board of Heliyon Business and Economics, Qualitative Research in Financial Market, Annals of Public and Cooperative Economics, International Review of Economics and Finance, World Development Perspective, Journal of Retailing and Consumer Services, Journal of Public Affairs, etc. His area of research is Sustainability, Fintech, Smart City, Risk Management, and Financial Inclusion.

***

Aliza is Senior Research Fellow, working at Faculty of Commerce, Banaras Hindu University, India. Her research interest lies in the field of startup fundings, international business practices, alternative investments, etc. she has received grants from Government of India to while pursuing her doctoral research.

Kapil Ahalawat is a pharmaceutical sales professional, academician and researcher. He is currently working as Assistant Professor at Graphic Era (Deemed to be University) Dehradun. He worked as a Business Manager for a well-known pharmaceutical company before entering academia, where he exceled in pharmaceutical marketing. He has successfully negotiated the dynamic and challenging business environment throughout his 11 years of corporate journey. Realizing the value of education, he entered the academics and has spent the past nearly three years mentoring and teaching students. He has become an invaluable asset for igniting young minds and future professionals due to his special blend of real-world business skills and a passion for education.

Vinayak Anil Bhat is a seasoned academician with 27 years of experience in postgraduate management education. As an Associate Professor at CMS Business School, Jain (Deemed-to-be University), he specializes in Organizational Behavior & Human Resource Management. Holding a Ph.D. in Competency Mapping from Christ University, he is actively involved in student mentoring, examination management, and corporate collaborations. He is affiliated with professional bodies like NHRD Bangalore Chapter and L&D Global. His expertise in competency development and leadership training is complemented by TA 101 and DISC certifications, enhancing his contributions to academic excellence and industry engagement.

Meenakshi Bose is a research and Communication professional having 8 years of working experience in the development sector. Contributed towards the body of research on concerns related to gender and labour. Author of the book – "Valuing Women's Unpaid Work: Experiences from Odisha" published by Aakar Books, New Delhi. Also, I have contributed papers in various publications including edited books, and national and international journals. I hold a doctorate in the field of Management. Currently, I am working as the Manager of Research and Communication with

Bandhan-Konnagar (A CSR Wing of Bandhan Group). I have a wide range of experience in working in different areas of the social development sector in affiliation with UNICEF, UNWFP, KIIT School Rural Management (KIIT University), and many others. Also, I have qualified for CBSE NET in Labour Welfare & HRM. I have a Master's in Business Administration and her second Master's in Women's Studies (MA). I was awarded the Swami Vivekananda Single Girl Child Scholarship for Research in Social Sciences granted by UGC, Government of India. My research interests include labour and gender issues in development, women's participation in the labour force, women's work and women's role in economic development.

Akanksha Choudhary is working as an Assistant Professor at the Centre for Distance and Online Education, in Management Department at Manipal University Jaipur. She has 5 years of Academic experience. Her expertise lies in primary research areas concerning Sustainable Tourism, Ethical Consideration, MSME, Tech advancement in socio-economic arena and Developmental Economics. Her magnum opus lies with publication of quality research articles in the field of handicrafts upliftment, SMEs, sustainable development, IoT AI application, etc. The Author also has published edited book and have contributed to various book chapters in the field of sustainability and sustainable business perspectives. She stands as a proud alumnus from Banasthali University and Banaras Hindu University, currently located at MUJ.

Hrishikesh Desai is an Assistant Professor in the Department of Accounting at Arkansas State University (A-State). He is also the Director of the Master of Accountancy with Data Analytics (MAcc) program at A-State. Prior to joining the faculty, Dr. Desai was a Chazen Visiting Scholar at Columbia Business School and has also worked as a Tutor in the Department of Accounting at the University of Melbourne. Dr. Desai's views have been featured in prominent news articles on Bloomberg News and The Economic Times, as well as on popular personal finance websites such as WalletHub and MoneyGeek.com. He has also received research grants from the AR Department of Human Services (DHS), Neil Griffin College of Business (NGCoB), the AR Department of Higher Education (ADHE), the Arkansas Biosciences Institute (ABI), and the Association of Certified Fraud Examiners (ACFE). At A-State, he teaches the Master of Accountancy classes, "Applied Professional Research" and "Strategic Accounting Technologies II," and the Bachelor of Science class "Introduction to Managerial Accounting." Dr. Desai's research examines a wide range of topics in the following areas: accounting regulation, fraud examination, corporate governance, executive compensation, emerging technologies, financial reporting, judgment and decision-making, inter-firm relationships, sustainability reporting, disclosures via social media, drug pricing, personal finance, and taxation. Prior to entering academia, Dr. Desai worked at boutique public accounting firms, Deutsche Bank Group, General Mills, and Bakhresa Group. Dr. Desai holds a Ph.D. in Accounting from the University of Melbourne. He also holds two Master's degrees in Commerce (M. Com.) from the University of Melbourne and the University of Mumbai. He is a Chartered Accountant (CA) and a member of the Institute of Chartered Accountants of India (ICAI). Additionally, he is a Member of the Arkansas Society of Certified Public Accountants (ARCPA), a CFA (USA) Charterholder and a member of the CFA Society Memphis, a Member of the American Accounting Association (AAA), and the Institute of Management Accountants (IMA). He is also an Associate Member of the ACFE. He is also an Enrolled Agent (EA), admitted to practice before the Internal Revenue Service (IRS), U.S. Department of Treasury. He is also a Member of the National Association of Enrolled Agents (NAEA) and the Tennessee Society of Enrolled Agents (TNSEA). Dr. Desai is an Australian and likes to travel, swim, play tennis, and watch soccer in his free time.

Ömer Esen was born in Istanbul, Turkey in 1981. He is a full-time professor in the Department of Public Finance at the Tekirdağ Namık Kemal University. He received his Ph.D. degrees on the topic of the energy deficit issue of Turkey in the context of sustainable growth in 2013 from the Ataturk University in Erzurum, Turkey. His research interests are primarily in the areas of macroeconomics, energy economics, Turkish economy and structural reforms with a particular focus on the economic growth, macroeconomic stability, sustainable growth, pollution emissions, climate change and public policy. He has published widely on the energy economics, public economics, and applied macroeconomics.

Deepak Gautam is an Assistant Professor at the School of Business, UPES Dehradun, specializing in Operations and Supply Chain Management. His research focuses on sustainability, reverse logistics, and supply chain optimization, particularly within circular economy frameworks for electric vehicle batteries. He has published in high-impact journals such as Business Strategy and the Environment and Journal of Cleaner Production and has presented his work at international conferences, including EURO 22 in Finland. Previously, he was an Assistant Professor at IILM University, where he led industry collaborations and research initiatives, including a NICDC-funded project on Industrial Corridor optimization and a Mitsui India study on sustainable mobility solutions. His expertise spans Operations Research, Mathematical Modeling, Sustainable Supply Chains, and Decision Support Systems. Deepak emphasizes experiential learning, integrating case studies and industry interactions into his teaching. He has taught courses such as Operations Management, Supply Chain Management, and Quality Management while also designing certification programs.

Sachin Ghai is an Academician, Trainer, Startup mentor, and a Business and Life coach with over 2 decades of experience and a proven track-record in teaching and training at various levels. He has a Doctorate in Economics in the area of Urban Development Planning and Management. He holds a Master's degree in Economics and Bachelor's degree in Management & Information Technology. He is presently working in the capacity of Head of the Department - Management Studies and Director - Technology Business Incubator at Graphic Era Deemed to be University, Dehradun.

Priyanka Gupta is an Assistant professor at Graphic Era Deemed to be University. For the last 4 years, she has been contributing in teaching and research. She has developed skills in qualitative and quantitative research methods. Her current research areas are Sustainable Human Resource Management, Artificial Intelligence, Digital Stress, Internet of Things (IoT). Email id: priyankagupta.mgt@geu.ac.in ORCID: SCOPUS AUTHOR ID: 58499631200

Chithambar Gupta V, is currently working as a Professor at School of Management, Presidency University, Bengaluru. He holds a Masters Degree in 'Management' from the "Indian Institute of Science" (IISc), Bengaluru. He has 24 years of rich work experience with 7+ years of Industry and 17+ years of academic experience in the areas of Marketing & General Management. He has won various awards and accolades for his work. Recipient of 'Best Teacher Award' (2015), 'Special Mention Award' (2004) and 'Team Award' (1999) for his excellent contribution towards work. He has published several papers, book chapters in reputed Management Journals. His research interest lies in the areas of Digital Marketing, Services Marketing & Retailing. Two PhD scholars have successfully completed PhD under his guidance.

Vansh Jain is a dedicated BBA student specializing in finance, known for academic excellence and a deep commitment to mastering the principles of financial management. With a strong foundation in analytical skills and a passion for problem-solving, Vansh Jain consistently achieves high grades, showcasing a thorough understanding of key financial concepts. Beyond academics, they actively participate in finance-related events and discussions, demonstrating both leadership and a collaborative spirit.

Sudhanshu Joshi currently working in Operations & Supply Chain Management Area, Doon University, INDIA. His research interest anchored within Digital-Twin,

Cyber Supply Chain Management with special focus on green supply chain network design, sustainable supply chain design, and coordination in humanitarian supply chain network, application of big data analytics in sustainable and humanitarian supply chains, emerging technologies (Including Industry 4.0), Circular Economy, Agriculture Supply Chain and Soft-computing Applications in Supply Chain Management. He is member with leading Societies including CSI, IEEE, POMS, INFORMS. He is a series editor of research note series CRC press, Taylor & Francis

Vidhi Kapruwan is an Assistant Professor at Graphic Era (Deemed to be) University and has 2.4 years of teaching experience. Along with Ms. Vidhi is currently pursuing a Doctor of Philosophy in Organizational Behavior at Graphic Era (Deemed to be University). As an academician and doctoral student, she started to do research in pursuit of knowledge and more understanding of the field. Her research started with exploring various factors related to Organizational Behavior and Sustainability. Recent work done by Ms. Vidhi covers several SDGs. The books to which she has contributed are "Green Futures: Navigating the Path to Environmental Resilience", "Rethinking Resources: Strategies, Policies and Partnership for a Circular Future", "Smart Systems and Innovative Technologies for Smart Cities" and "Effects of Digitalization and Circular Economy on Sustainable Policy and Climate Change Prevention". Ms. Vidhi has communicated her research paper to the journal International Journal of Work Organisation and Emotion. Along with several FDPs, Ms. Vidhi has also done a short-term course on Environmental Issues & Disaster Management organised by UGC. Ms. Vidhi has two patents granted.

Rejaul Karim is an Associate Professor of Finance and Banking in the Department of Business Administration at Varendra University, Rajshahi, Bangladesh and a Research Fellow at the INTI Internataional University, Malaysia. He received his PhD in finance from the Institute of Bangladesh Studies, University of Rajshahi, Bangladesh. He completed the Bachelor of Business Administration and Master of Business Administration in Finance and Banking from the Pabna University of Science and Technology, Pabna, Bangladesh, with outstanding academic performance. Dr. Karim has teaching and research experience for over 10 years in undergraduate and master-level tertiary education. He has a great passion for teaching and researching finance and economics. His research focused on financial performance, financial structure, sustainable finance, the circular economy, and other contemporary issues in finance and economics. Dr. Karim continuously contributes through his field research and publishing many papers in reputed national and international peer-reviewed journals, including book chapters and conference proceedings. Dr. Karim founded and headed BusinessXcel Lab, a center for quantitative business research. This voluntary organization aims to facilitate collaboration among domestic and international undergraduate, graduate, and PhD students and professional researchers. Dr. Karim can be contacted at rkarimreja@gmail.com

Krishna Koppa is an Associate Professor, Faculty of Management Studies, JAIN (Deemed-to-be University). A results-driven academic leader with 27 years of experience in higher education and industry, Dr. Krishna excels in strategic planning, curriculum innovation, and faculty mentorship. His competencies include applied research design and execution, coupled with a proven track record of impactful publications. Expertise spans national and international accreditations and building dynamic learning ecosystems. His research focuses on the intersection of marketing and emerging technologies, aiming to generate actionable insights. He is dedicated to fostering industry-academia collaborations and contributing to scholarly discourse.

Swarnendu Mazumder is a Development professional with four years of experience in research, communication and CSR project management.Key skills include programme design, implementation and evaluation, analysis of policies and institutional strategies in livelihoods, agriculture & allied activities, institutional development, micro finance and natural resource management. Development specialist and strategic planning, health and rights, policy advocacy and CSR

Archan Mitra is a Grade III Assistant Professor & Head of Research at (NICO) NITTE Institute of Communication. He edited various publications and wrote "Cases for Classroom Media and Entertainment Business," an IIM textbook. "Environmental informatics and communication for sustainability" is his doctorate at Visva-Bharati Santiniketan, West Bengal. He is also a qualified Science Communicator and Journalism from Indian Science Communication Society (ISCOS), Corporate Trainer with Amity Institute of Training and Development, and Social Media Network Analyst. He loves environmental communication. He received the Journalistic Club of Kolkata Medal of Honour and PRSI Kolkata Chapter Certificate of Merit. He was a World Bank research assistant at IIM Kashipur. He launched the World Bank's Green MBA MOOC. He publishes in prestigious emergency management magazines. IIM Kashipur, Alabama State University, Birmingham, USA, is his research partner. He used Developing Communities from Within (DCFW) to turn a West Bengal village called "Noachar" into an eco-village.

Utsav Krishan Murari is a gifted writer, dedicated life coach, motivational speaker, passionate educator, and, above all, an exceptional mentor. He holds a doctorate in communication and media studies from the Central University of South Bihar. Graced with a vivid imagination, he is a genuine catalyst for ideas and is always open to trying new things. His one such initiative is Saarthi, which began as

Saarthi Entertainment. Saarthi is a Delhi-based creative media production start-up that wants to leave a legacy of artistry in the production industry.

Achmad Nurmandi is a Professor at the Department of Government Affairs and Administration, Jusuf Kalla School of Government, Universitas Muhammadiyah Yogyakarta, Indonesia. His research interests are on e-government, urban governance and strategic management in the public sector and published in many international journal and Book Chapter, such as International Journal of Public Sector Management, Transforming Government: People, Policy and Process, Jamba Journal of Disaster Studies, Asian Review of Political Science, Journal of Human Behavior in the Social Environment, Public Policy and Administration, Global Encyclopedia of Public Administration, Public Policy, and Governance, Springer. He is currently the Secretary of the Asia Pacific Society for Public Affairs (APSPA). He is also the Editor-in-Chief of Jurnal Studi Pemerintahan (Journal of Government and Politics) and a Guest Editor of International Journal of Public Sector Performance Management and International Journal of Sustainable Society.

P. Aranganathan is currently a Director in Hallmark Business School. He is a doctorate in management and also holds B.E., (Mechanical and Production Engineering), and MBA degrees. He has 22 years of rich teaching, and research experience in Strategic Operations and HR areas. He has authored Four books (Principles of Management, Total Quality Management, Employee Retention, Logistics and Supply Management, and Production & Operations Management) and holds more than 30 number of publications in various international and national peer evaluated high indexed journals

P. Selvakumar is currently an Associate Professor in the Department of Science and Humanities at Nehru Institute of Technology, Coimbatore, Tamil Nadu, India. Since 2023, he has served as the Department IQAC Coordinator at the institute. Prior to joining Nehru Institute of Technology, he held positions as an Assistant Professor in the Department of Science and Humanities various engineering colleges in Coimbatore, Tamil Nadu, India, spanning from 2011 to 2023. Dr.P. Selvakumar's academic and research interests lie primarily in the field of phytochemistry. He has made significant contributions to his field, with an impressive record that includes 30 journal publications, 12 patents, and 20 conference papers. Additionally, he has organized 5 seminars and conferences, further demonstrating his commitment to academic and scientific exchange. Throughout his career, Dr. P.Selvakumar has been recognized for his dedication and excellence in teaching, having been honored with several awards, including the Best Teacher Awards. He holds a prominent position in the academic community, contributing actively to research.

Sanchita Paul is a dedicated and accomplished content strategist with extensive experience in managing diverse content formats across different digital platforms. With a strong background in content strategy, editing, and UX writing, she has honed her skills in branding, social media management, and technical writing. Holding an MA in Communication and Journalism from Visva-Bharati University, Ms. Paul is proficient in tools like Google Analytics, Canva, and MS Office Suite, making her a versatile professional committed to producing high-quality, impactful content to improve brand visibility.

R.Sivarethinamohan is an Associate Professor at Symbiosis Centre for Management Studies, Bengaluru campus, Symbiosis International Deemed University, Pune, India. Previously, He held dean, director, principal, and vice chancellor positions in prestigious institutions and universities. He is a sustainability Leader with 30+ years of experience in corporate finance, climate finance, Data science, Air pollutants, Digital Twin, Biochar, climate change, Blockchain Technology in Environment Sustainability, IoT enabled Smart Cities, Nanotechnology in agriculture and food production health robots, Accounting Fraud Detection. He has participated in and presented research papers at national and international conferences and holds nearly 75 article publications in leading journals indexed in Scopus (40), and Web of Science (18) to his credit. He is on the Editorial Board of over 7 peer-reviewed journals, including some prestigious ones. He has edited a couple of books and published textbooks with Tata McGraw Hill and

R.V. Suresh, Associate Professor in Marketing, Gnanam School of Business, Tamilnadu, India has comprehensive experience across academia, research and corporate spreading over two decades in various institutes of repute. Apart from Patent Publications and Grants of both national and international, he has published articles in peer reviewed journals, contributed chapters in four books, and authored three books. He has been the editorial board member for national and international Book Publishers. His knowledge has been disseminated as Resource Person in various institutions. He was recipient of "Faculty of Academic Excellence Award" in ICSSR sponsored International Conference in 2022. He had exhibited sound administrative skill by holding various positions Viz. President - IIC, Coordinator - IPR Cell, Advisor - Planning & Evaluation committee, Centre for Research and Member in Board of Studies.

Manju Shree Raman has a strong education professional with Ph. D., M.Phil., MBA., from Faculty of Management, Bharathiar University, Coimbatore. She received the prestigious honor "Teachers Transforming India 2022" on World Knowledge Day, April 14, 2022, as well as the "Best Faculty Award" from

Akshaya Institute of Management Studies, Coimbatore, on September 5, 2014. She is a member of the CMA and AIMA. She is also an internal ISO Auditor with certification. She is a member of the Editorial Board, an editor, a reviewer, a chief guest, an outbound trainer, a panel member, and a lecturer at colleges and business schools in India and Ethiopia. She has served as an international relations officer and the initiator of several Memorandums of Understanding and Agreements (MOUs and MOAs) involving India, Ethiopia, Mozambique, Austria, Kenya, and others. She was working as Professor in Ethiopia. She is presently Associate Professor in CHRIST University.

Navneet Rawat is a distinguished academic and researcher specializing in Marketing and Corporate Laws. With a Ph.D., MBA, and BMIT from Hemwati Nandan Bahuguna Garhwal University, he has made significant contributions to education and social development. Recognized with the prestigious "Uttarakhand Pratibha Samman" award, Dr. Navneet Rawat is celebrated for their work in applying marketing principles to non-profit organizations and social causes, earning a reputation as a dedicated social activist and marketer. As the Project Head of "Project SCOPE," he played a pivotal role in advancing virtual education for schools across Uttarakhand. A certified Six Sigma Green Belt (NITCON), Dr. Navneet Rawat is a sought-after keynote speaker and chair at international conferences, workshops, and seminars, inspiring audiences as a social motivational speaker

Md. Mustaqim Roshid is a Bangladeshi citizen who has served as a research associate at BusinessXcell Lab. Mustaqim holds a Bachelor of Business Administration in Marketing from Varendra University and is currently an MBA student in Management Studies at Rajshahi University. His research focuses on business operations, supply chain management, financial performance, and agribusiness, Industry Revolution, Neuromarketing, H2H Marketing, Energy Economics, ESG, SDGs, Circular Economy. To make queries, contact mustaqimroshid@gmail.com.

Guru Basava Aradhya S is a seasoned academician with 17 years of experience in management education. He currently serves as the Director; Professor at Chandigarh School of Business, Mohali. He holds a Ph.D. in Management and has mentored five Ph.D. scholars. His expertise includes HR, marketing, and business analytics. A prolific researcher, he has published in SCOPUS, ABDC, and UGC CARE journals. He has organized numerous international conferences and contributed to NAAC and NBA accreditations. His industry collaborations include faculty exchange programs and consultancy projects. Passionate about academia, he actively fosters innovation, research, and student development.

S. Manikandan is an Assistant Professor of Systems and Operations in Gnanam School of Business,Tamilnadu, India He received his Post Graduation in Management from Anna University, Chennai, M.Tech from PRIST University, Thanjavur and his Under Graduation in B. Tech from SASTRA University, Thanjavur. He is a certified Lean Six Sigma Green Belt holder from AU TVS center for Quality Management, Anna University, Chennai and also an ISO 9001:2008 certified Internal auditor from TATA Projects Quality Service. He has immense experience of 5 years in the area of academics, and 7.5 years in the area of IT. He is the overall in charge of IT facility in GSB.

Dhruv Sabharwal is an Associate Professor and Head of the department at Sharda School of Media Film and Entertainment - Sharda University Greater Noida. Having 16 years of experience in teaching, research and administration. Areas of research are Media research and advertising research.

Vanshika Saklani is a dedicated student at Graphic Era University, India, pursuing a Bachelor's in Psychology Honours within the Department of Humanities and Social Sciences. She is deeply committed to advancing her research skills and expanding her academic knowledge. Through her academic journey, she aims to deepen her knowledge of theories and methodologies, contributing meaningfully to the field. Her focus on research will strengthen her expertise but also equip her with the critical thinking and analytical skills essential for future professional endeavours.

Asheetu Bhatia Sarin is an Assistant Professor Sr. (Management) at Vivekananda Institute of Professional Studies-Technical Campus (VIPS-TC) affiliated to GGSIPU, Delhi. She has over 11 years of teaching experience in the fields of Commerce and Management. She has done a Ph.D. in Behavioral Finance, did her post-graduation (MBA) from ICFAI Business School in Finance and Marketing, and earned her graduate degree in B.Com (H) from Delhi University. She is a lifetime member of the Indian Commerce Association and has to her credit many national and International publications.

Janmejai Kumar Shah is a 2 times UGC NET Management qualified Assistant Professor and PHD research scholar from graphic era deemed to be university dehradun. His research areas are in field of supply chain, sustainability, ESG, digitalisation and decarbonization. H pursued PGDM specialization in International Business From ITM Navi Mumbai

Manu Sharma is an academician and researcher with a notable h-index of 22 on Scopus and 32 on Google Scholar, reflecting her significant contributions to the field of supply chain management. Her research spans a range of high-impact

Scopus journal publications to a count of 50, casing the emerging areas like digital technologies, supply chains, sustainability, Circular Economy, and waste management. During her past three years, Dr. Sharma has published an impressive count of 50+ research papers in high-impact Scopus-indexed journals. Her work is featured in top-tier publications, including A* and ABDC/A category journals such as Transportation Research Part E: Logistics and Transportation Review, Business Strategy and Environment, Journal of Cleaner Production, Sustainable Production and Consumption, and Journal of Retailing and Consumer Studies, Journal of Business Research, Technological Forecasting and Social Change, Journal of Environmental Management and many others. Her research is widely recognized and cited, with a citation count of 3805 on Google Scholar and 1,791 on Scopus. In recognition of her outstanding research and teaching capabilities, Dr. Sharma has been awarded the AICTE sponsored Faculty Development Program ATAL FDP for three consecutive years which she successfully conducted. Additionally, she has recently received sponsorship from the Indian Council of Social Science Research (ICSSR) for an International Conference which held at Department of Management Studies on 28-29 June 2024, underscoring her leadership and influence in the academic community. She has also Visiting Research Fellowship at University of Technology, Sydney, Australia for the year 2023-2024. Her expertise spans qualitative and quantitative research methods, including Multi-Criteria Decision Making, Fuzzy Theory, Multivariate Analysis, Grey Theory, and Soft Computing applications. Dr. Sharma's research focuses on Industry 4.0 technologies, Circular Economy, Net Carbon Zero strategies, Digital Supply Chains, Waste Management, and Sustainability. Her dedication to advancing knowledge in supply chain management, coupled with her commitment to sustainability, and Circular economy transition positions her as a leading researcher in her field. Her contributions not only enhance academic discourse but also provide valuable insights for industry practitioners aiming to implement sustainable practices in supply chains. Email Id: manu.sharma@geu.ac.in

Naresh Kumar Sharma is the Registrar at Graphic Era (Deemed to be) University, Dehradun, and a former Under Secretary at the University Grants Commission (UGC). With nearly 40 years of experience in academic administration, he holds a Ph.D. in Political Science from the University of Rajasthan, India. Dr. Sharma has been actively involved in various UGC schemes and bureaus, including International Collaborations, Research Fellowships, Career-Oriented Programs, Policy and Monitoring of Deemed-to-be and Private Universities, and the inclusion of institutions under the UGC Act, 1956. He played a key role in drafting UGC regulations, such as those for Private Universities, Institutions Deemed to be Universities and College Autonomy. His research publications cover topics such as the development of deemed-to-be universities, the growth and regulation of private universities in India, Sustainable Development Goals (SDGs), environmental sustainability, and education.

S. Christina Sheela is an Associate Professor of Finance at Gnanam School of Business, Tamilnadu, India, has 16 years of teaching experience in various Engineering colleges in Tamil Nadu and 5 years of Industrial experience. She was appreciated with Best faculty award by Indian Academic Researchers Association, Tiruchirappalli (IARA Award -2016) for her research contribution. She was awarded with gold medal as Best Women Achiever – 2012, Trichy from Tiruchirappalli Mayer Mrs.A.Jaya. She has authored a book titled Logistics and Supply Management. She has also published various research articles in national and international peer evaluated high indexed journals

Muskan Singh received the bachelor's degree in commerce in 2019 from Gorakhpur University, Uttar Pradesh, India. She received master's degree in Management in 2021 from H.N.B Garhwal University, Srinagar, Uttarakhand, India. Currently she is pursuing PhD in Management from Graphic Era Deemed to be University, Dehradun, India. Her research interests are in human resource management and organization behaviour.

Agrim Singhal passed out from Amity International school Sector 1 Vasundhara Ghaziabad in the year 2023 with flying colours achieving good marks in the Commerce Stream. I am currently a student of Second Year pursuing Bachelor's degree in Business administration from Graphic Era deemed to be University Dehradun. He has research interests in Marketing and Human Resource Management and wants to pursue his career in Markting Intelligence and Human Resource Management.

Abdul Waaje earned a Master's in Engineering Management from Michigan Tech, along with a Quality Engineering Graduate Certification, and holds a Master's in International Trade and Business from the University of Dhaka and a Bachelor's in Electronics and Telecommunication Engineering from East West University. His expertise lies in supplier quality integration, optimizing procurement, and ensuring quality assurance in production. His research focuses on IoT-based industrial solutions for supply chain optimization. He interned at the world's largest commercial laundry appliances manufacturer, managing nonconforming materials and standardizing supply chain processes. A former Teaching Assistant at Michigan Tech, he is a Lifetime Member of Beta Gamma Sigma and an active member of ASEM and ASCM, contributing to professional and community initiatives.

Seda Yıldırım currently works at Tekirdag Namık Kemal University in Turkey. She works as full-time profrssor at Tekirdag Namık Kemal University in Turkey. She studies on sustainable development, 2030 sustainable development goals, sustainable consumption patterns, consumption patterns, environmental quality and food security. She has over 1900 citiation by Google scholar. She has many articles published by journals indexed in WOS and SCOPUS. She is married and has two children. In 2022, She won Emerald literati award- Outstanding Paper Award 2022. #LiteratiAwards.

Muhammad Younus graduated in 2019 from the Karachi Institute of Economics and Technology University in Karachi, Pakistan, and got a Bachelor's degree in Computer Science. The four-year undergraduate program in Computer Science formed the basis of my perseverance towards achieving an in-depth knowledge in the field. After graduation, I worked for the past few years as a business IT analyst in the product and research development of a logistics startup in Pakistan called TPL Logistics. We aim to Digitize Logistics Operations in Pakistan from the First Mile to the Last Mile. After this, I graduated in 2024 from the Universitas Muhammadiyah Yogyakarta in Yogyakarta, Indonesia, and got a Master's Degree in Government Science. My research interest is transforming governance and the public sector through technology. This Experience has given me the confidence and also kindled a desire to do research. The interdisciplinary nature has convinced me in taking this decision as it is the ideal choice for an exciting research career. At the same time, I am confident of contributing originally to the ongoing work. By working under the guidance of distinguished Team, I shall certainly be able to exploit my pote.

Index

A

B

C

D

E

F

G

H

I

N

O

P

R

S

T

U

W

Z

www.ingramcontent.com/pod-product-compliance
Lightning Source LLC
LaVergne TN
LVHW080354011026
849161LV00009B/2302